OPTICAL FIBER COMMUNICATIONS

McGraw-Hill Series in Electrical Engineering

Consulting Editor

Stephen W. Director, *Carnegie-Mellon University*

Circuits and Systems
Communications and Signal Processing
Control Theory
Electronics and Electronic Circuits
Power and Energy
Electromagnetics
Computer Engineering
Introductory
Radar and Antennas
VLSI

Previous Consulting Editors

Ronald N. Bracewell, Colin Cherry, James F. Gibons, Willis W. Harman, Hubert Heffner, Edward W. Herold, John G. Linvill, Simon Ramo, Ronald A. Rohrer, Anthony E. Siegman, Charles Susskind, Frederick E. Terman, John G. Truxal, Ernst Weber, and John R. Whinnery

Communications and Signal Processing

Consulting Editor

Stephen W. Director, *Carnegie-Mellon University*

OPTICAL FIBER COMMUNICATIONS

Second Edition

Gerd Keiser

GTE Communications
Systems Division

McGraw-Hill, Inc.

New York   St. Louis   San Francisco   Auckland   Bogotá
Caracas   Hamburg   Lisbon   London   Madrid   Mexico   Milan   Montreal
New Delhi   Paris   San Juan   São Paulo   Singapore   Sydney   Tokyo   Toronto

This book was set in Times Roman by Science Typographers, Inc.
The editors were Roger L. Howell and John M. Morriss;
the production supervisor was Denise L. Puryear.
The cover was designed by David Romanoff.
Project supervision was done by Science Typographers, Inc.
R. R. Donnelley & Sons Company was printer and binder.

OPTICAL FIBER COMMUNICATIONS

2 3 4 5 6 7 8 9 0 DOC DOC 9 0 9 8 7 6 5 4 3 2 1

ISBN 0-07-033617-2

Library of Congress Cataloging-in-Publication Data is available: LC Card #90-22882.

ABOUT THE AUTHOR

Gerd Keiser received his B.A. and M.S. degrees in mathematics and physics from the University of Wisconsin in Milwaukee and a Ph.D. degree in solid-state physics from Northeastern University in Boston in 1973. From 1973 to 1974 he carried out research and development of state-of-the-art infrared photodetectors at the Honeywell Electro-Optics Center in Lexington, Massachusetts.

In 1974 Dr. Keiser joined the GTE Government Systems Corporation in Needham Heights, Massachusetts, where he currently is an engineering manager. His work involves analysis, design, and development of digital switches and fiber optic information-transmission equipment and networks.

Dr. Keiser has also served as an Adjunct Professor of Electrical Engineering at Northeastern University. In addition, he has been a guest editor for the *IEEE Communications Magazine*, has organized the IEEE Lecture Series on Optical Fiber Communications, and has presented numerous invited lectures at conferences, local industries, and local professional meetings.

Dr. Keiser has published over 20 technical journal and conference papers and is the author of the book *Local Area Networks* (McGraw-Hill, 1989).

TO CHING-YUN AND NISHLA

CONTENTS

PREFACE

This book, like its previous edition, provides the basic material for an introductory senior or first-year graduate course in the theory and application of optical fiber communication technology. It will also serve well as a working reference for practicing engineers dealing with modern optical fiber communication system designs. To aid students in learning the material and applying it to practical designs, examples are given throughout the book and a collection of over 200 homework problems is included. These problems are an integral and important part of the book. They will help the readers not only to test their comprehension of the material covered, but also to extend and elucidate the text. As an aid to doing these problems, the appendices include brief overviews of relevant units and mathematical relations. To assist the instructor, a problem solutions manual is available from the publisher.

The background required to study the book is that of typical senior-level engineering students. This includes introductory electromagnetic theory, calculus and elementary differential equations, basic concepts of optics as presented in a freshman physics course, and the basic concepts of electronics. Courses in communication theory and solid state physics are not essential, but would be helpful for gaining a full understanding of the material. To refresh the readers' memories, concise reviews of several background topics, such as optics concepts, electromagnetic theory, and semiconductor theory, are included in the main body of the text.

The book is organized to give a clear and logical sequence of topics. It progresses systematically from descriptions of the individual elements of a communication link to analyses of digital and analog network designs, and ends with discussions of advanced communication techniques for higher-performance fiber optic systems. The introductory chapter gives a brief historical background of communication system developments and shows how the operating region of optical fibers fits into the electromagnetic spectrum. It also includes an overview of the fundamental components of an optical fiber link and the types of networks that can be constructed with these devices. In addition, Chapter 1 notes the advantages of optical fibers relative to other transmission media and points to application areas that show rapidly growing interest in optical fiber technology.

Chapter 2 reviews the basic laws and definitions of optics that are relevant to optical fibers. Following a description of optical fiber structures, the geometrical or ray optics concept of light reflection and refraction and the theory of electromagnetic wave propagation are used to show how a fiber guides light. Chapter 2 ends with descriptions of materials, manufacturing methods, strengths, and cabling of optical fibers. As a final topic on optical fibers, Chapter 3 addresses the degradation of light signals arising from attenuation and distortion mechanisms.

To transmit information through a fiber waveguide, an optical source compatible with the fiber size and transmission properties is required. Thus Chapter 4 discusses the structures, materials, and operating characteristics of semiconductor light sources. The question of how to couple the light emitted from these sources is dealt with in Chapter 5. The effects of optical power loss at fiber-to-fiber joints resulting from mechanical misalignments and from mismatched fiber geometries and structures are also described.

When an optical signal emerges from the end of a fiber, a photodetector converts it to an electrical signal. Chapter 6 discusses photodetection principles and examines detector performance parameters for *pin* and avalanche photodiodes, which are the two major types of detectors used in optical fiber systems. Once the optical signal has been reconverted to an electrical signal, the receiver restores the fidelity of the attenuated and distorted signal. The theory and design of receivers for optical fiber links is addressed in Chapter 7.

The next several chapters deal with the latest developments in optical fiber systems. These include explanations of optical power budgets and digital fiber optic link designs in Chapter 8, analog fiber optic link designs in Chapter 9, and the theory and design of coherent transmission systems in Chapter 10. As a final topic, Chapter 11 addresses modern advanced system techniques, such as wavelength division multiplexing, optical fiber networks, and optical switching.

The original concept of this book is attributable to Professor Tri T. Ha, Naval Postgraduate School, who urged me to write it. His suggestions and comments for the second edition were also most helpful. I am especially indebted to Don Rice of Tufts University for numerous technical discussions and for a critical proofreading of the manuscript. Special thanks go to Nishla Keiser for helping prepare the index. In addition, improvements to the text were made from the suggestions of Professor C. T. Chang of San Diego State University, Dr. Marvin Drake and Jonathon Fridman of the Mitre Corporation, Professor Elias Awad of Wentworth Institute of Technology, and the following reviewers for McGraw-Hill: Ted Batchman, University of Oklahoma; John Buck, Georgia Institute of Technology; Ting-Chung Poon, Virginia Institute of Technology; Michael Ruane, Boston University; Henry Taylor, Texas A & M University; and Xiao-Bang Xu, Clemson University. Particularly encouraging for doing the second edition were the many positive comments on the first edition received from users and adopters at numerous institutions. As a final personal note, I am grateful to my wife, Ching-yun, and my daughter, Nishla, for their patience and encouragement during the time I devoted to writing and revising this book.

Gerd Keiser

OPTICAL FIBER COMMUNICATIONS

CHAPTER
1

OVERVIEW
OF OPTICAL
FIBER
COMMUNICATIONS

Ever since ancient times, one of the principal interests of human beings has been to devise communication systems for sending messages from one distant place to another. The fundamental elements of any such communication system are shown in Fig. 1-1. These elements include at one end an information source which inputs a message to a transmitter. The transmitter couples the message onto a transmission channel in the form of a signal which matches the transfer properties of the channel. The channel is the medium bridging the distance between the transmitter and the receiver. This can be either a guided transmission line such as a wire or waveguide, or it can be an unguided atmospheric or space channel. As the signal traverses the channel, it may be progressively both attenuated and distorted with increasing distance. For example, electric power is lost through heat generation as an electric signal flows along a wire, and optical power is attenuated through scattering and absorption by air molecules in an atmospheric channel. The function of the receiver is to extract the weakened and distorted signal from the channel, amplify it, and restore it to its original form before passing it on to the message destination.

1.1 FORMS OF COMMUNICATION SYSTEMS

Many forms of communication systems have appeared over the years. The principal motivations behind each new one were either to improve the transmission fidelity, to increase the data rate so that more information could be sent,

1

FIGURE 1-1
Fundamental elements of a communication system.

or to increase the transmission distance between relay stations. Prior to the nineteenth century all communication systems were of a very low data rate type and basically involved only optical or acoustical means, such as signal lamps or horns. One of the earliest known optical transmission links,[1] for example, was the use of a fire signal by the Greeks in the eighth century B.C. for sending alarms, calls for help, or announcements of certain events. Only one type of signal was used, its meaning being prearranged between the sender and the receiver. In the fourth century B.C. the transmission distance was extended through the use of relay stations, and by approximately 150 B.C. these optical signals were encoded in relation to the alphabet so that any message could then be sent. Improvements of these systems were not actively pursued because of technology limitations. For example, the speed of the communication link was limited since the human eye was used as a receiver, line-of-sight transmission paths were required, and atmospheric effects such as fog and rain made the transmission path unreliable. Thus it generally turned out to be faster and more efficient to send messages by a courier over the road network.

The invention of the telegraph by Samuel F. B. Morse in 1838 ushered in a new epoch in communications—the era of electrical communications.[2] The first commercial telegraph service using wire cables was implemented in 1844 and further installations increased steadily throughout the world in the following years. The use of wire cables for information transmission expanded with the installation of the first telephone exchange in New Haven, Connecticut, in 1878. Wire cable was the only medium for electrical communication until the discovery of long-wavelength electromagnetic radiation by Heinrich Hertz in 1887. The first implementation of this was the radio demonstration by Guglielmo Marconi in 1895.

In the ensuing years an increasingly larger portion of the electromagnetic spectrum was utilized for conveying information from one place to another. The reason for this is that, in electrical systems, data are usually transferred over the communication channel by superimposing the information signal onto a sinusoidally varying electromagnetic wave which is known as the *carrier*. At the destination the information is removed from the carrier wave and processed as desired. Since the amount of information that can be transmitted is directly related to the frequency range over which the carrier wave operates, increasing the carrier frequency theoretically increases the available transmission bandwidth and, consequently, provides a larger information capacity. Thus the trend in electrical communication system developments was to employ progressively higher frequencies (shorter wavelengths), which offered corresponding increases in bandwidth and, hence, an increased information capacity. This activity led, in turn, to the birth of television, radar, and microwave links.

The portion of the electromagnetic spectrum that is used for electrical communications[3] is shown in Fig. 1-2. The transmission media used in this spectrum include millimeter and microwave waveguides, metallic wires, and radio waves. Among the communication systems using these media are the familiar telephone, AM and FM radio, television, CB (citizen's band radio),

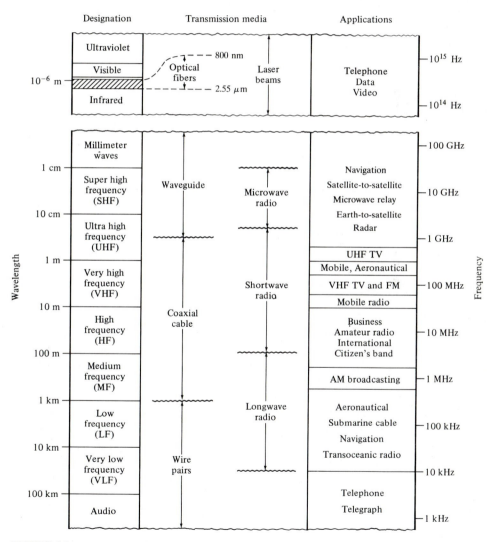

FIGURE 1-2
Examples of communication systems applications in the electromagnetic spectrum. (From Carlson,[3] copyright 1986. Used with permission of McGraw-Hill Book Company.)

radar, and satellite links, all of which have become a part of our everyday lives. The frequencies of these application range from about 300 Hz in the audioband to about 90 GHz for the millimeter band.

Another important portion of the electromagnetic spectrum encompasses the optical region shown in Fig. 1-2. In this region it is customary to specify the band of interest in terms of wavelength, instead of frequency as in the radio region. The optical spectrum ranges from about 50 nm (ultraviolet) to about 100 μm (far infrared), the visible spectrum being the 400- to 700-nm band. Similar to the radio-frequency spectrum, two classes of transmission media can be used: an atmospheric channel or a guided-wave channel.

1.2 THE EVOLUTION OF FIBER OPTIC SYSTEMS

Great interest in communicating at optical frequencies was created in 1960 with the advent of the laser,[4] which made available a coherent optical source. Since optical frequencies are on the order of 5×10^{14} Hz, the laser has a theoretical information capacity exceeding that of microwave systems by a factor of 10^5, which is approximately equal to 10 million TV channels.

With the potential of such wideband transmission capabilities in mind, a number of experiments[5,6] using atmospheric optical channels were carried out in the early 1960s. These experiments showed the feasibility of modulating a coherent optical carrier wave at very high frequencies. However, the high installation expense, the tremendous cost of developing all the necessary components, and the limitations imposed on the atmospheric channel by rain, fog, snow, and dust make such extremely high-speed systems economically unattractive in view of present demand for communication channel capacity. Nevertheless, numerous developments of unguided-channel optical systems operating at baseband frequencies are in progress for short-distance (up to about 1 km) earth-based, long-distance earth-to-space, and satellite-to-satellite links.[7-10]

Concurrent with this work, investigations into optical fibers have been made, since they can provide a much more reliable and versatile optical channel than the atmosphere. Initially, the extremely large losses (more than 1000 dB/km observed in the best optical fibers) made them appear impractical. This changed in 1966, when Kao and Hockman[11] and Werts[12] almost simultaneously speculated that these high losses were a result of impurities in the fiber material, and that the losses could be reduced to the point where optical waveguides would be a viable transmission medium. This was realized in 1970, when Kapron, Keck, and Maurer[13] of the Corning Glass Works fabricated a silica fiber having a 20-dB/km attenuation (a signal-power loss factor of 100 over 1 km). At this attenuation, repeater spacings for optical fiber links become comparable to those of copper systems, thereby making lightwave technology an engineering reality.

In the next two decades researchers worked intensively to reduce the attenuation to 0.16 dB/km (a signal-power loss of 4 percent/km) at a 1550-nm wavelength, which is close to its theoretical value of 0.14 dB/km.

The development and application of optical fiber systems grew from the combination of semiconductor technology, which provided the necessary light sources and photodetectors, and optical waveguide technology. The result was an information-transmission link that had certain inherent advantages over conventional copper systems. Included in these are the following:

1. *Low transmission loss and wide bandwidth.*   Optical fibers have lower transmission losses and wider bandwidths than copper wires. This means that with optical fiber cable systems more data can be sent over longer distances, thereby decreasing the number of wires and reducing the number of repeaters needed for these spans. This reduction in equipment and components decreases the system cost and complexity.

2. *Small size and weight.*   The low weight and the small (hair-sized) dimensions of fibers offer a distinct advantage over heavy, bulky wire cables in crowded underground city ducts or in ceiling-mounted cable trays. This is also of importance in aircraft, satellites, and ships, where small, lightweight cables are advantageous, and in tactical military applications, where large amounts of cable must be unreeled and retrieved rapidly.[14]

3. *Immunity to interference.*   An especially important feature of optical fibers relates to their dielectric nature. This provides optical waveguides with immunity to electromagnetic interference (EMI), such as inductive pickup from signal-carrying wires and lightning. It also ensures freedom from electromagnetic pulse (EMP) effects, which is of particular interest in military applications.

4. *Electrical isolation.*   Since optical fibers are constructed of glass, which is an electrical insulator, there is no need to worry about ground loops, fiber-to-fiber crosstalk is very low, and equipment interface problems are simplified. This also makes the use of fibers attractive in electrically hazardous environments, since the fiber creates no arcing or sparking.

5. *Signal security.*   By using an optical fiber a high degree of data security is afforded, since the optical signal is well confined within the waveguide (with any emanations being absorbed by an opaque jacketing around the fiber). This makes fibers attractive in applications where information security is important, such as banking, computer networks, and military systems.

6. *Abundant raw material.*   Of additional importance is the fact that silica is the principal material of which optical fibers are made. This raw material is abundant and inexpensive, since it is found in ordinary sand. The expense in making the actual fiber arises in the process required to make ultrapure glass from this raw material.

Early applications of fiber optic transmission systems were largely for trunking of telephone lines. These were digital links consisting of time-division-multiplexed (TDM) 64-kb/s voice channels. Figure 1-3 shows the digital transmission hierarchy used in the North American telephone network. The fundamental building block is a 1.544-Mb/s transmission rate known as a T1

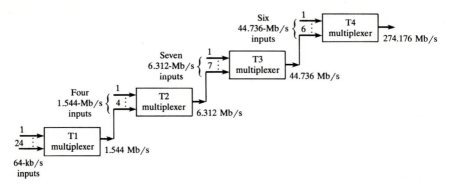

FIGURE 1-3
Digital transmission hierarchy used in the North American telephone network.

rate. It is formed by time-division-multiplexing twenty-four voice channels, each digitized at a 64-kb/s rate. Framing bits are added along with these voice channels to yield the 1.544 Mb/s bit stream. At any level a signal at the designated input rate is multiplexed with other input signals at the same rate.

The system is not restricted to multiplexing voice signals. For example, at the T1 level, any 64-kb/s signal of the appropriate format could be transmitted as one of the 24 input channels shown in Fig. 1-3. As noted in Fig. 1-3 and Table 1-1, the multiplexed rates are designated as T1 (1.544 Mb/s), T2 (6.312 Mb/s), T3 (44.736 Mb/s), and T4 (274.176 Mb/s). Similar hierarchies using different bit-rate levels are employed in Europe and Japan, as Table 1.1 shows.

Slightly different transmission rates are used in a newly developed standard called SONET (Synchronous Optical Network), which defines a synchronous frame structure for sending multiplexed digital traffic over optical fibers.[15] The basic building block and first level of the SONET signal hierarchy is called the *synchronous transport signal—level 1* (STS-1) and has a bit rate of 51.84 Mb/s. Higher-rate SONET signals are obtained by byte-interleaving *N* STS-1 frames which are then scrambled and converted to an *optical carrier—level N* (OC-*N*)

TABLE 1-1
Standard telephone transmission rates in North America, Europe, and Japan

Hierarchy level no.	North America		Europe		Japan	
	Rate Mb/s	No. of voice channels	Rate, Mb/s	No. of voice channels	Rate, Mb/s	No. of voice channels
1	1.544	24	2.048	30	1.544	24
2	6.312	96	8.448	120	6.312	96
3	44.736	672	34.368	480	32.064	480
4	274.176	4032	139.264	1920	97.728	1440
5	—	—	565.148	7680	396.200	5760

TABLE 1-2
Levels of the SONET signal hierarchy

Level	Line rate (Mb / s)
OC-1	51.84
OC-3	155.52
OC-9	466.56
OC-12	622.08
OC-18	933.12
OC-24	1244.16
OC-36	1866.24
OC-48	2488.32

signal. Thus the OC-N signal will have a line rate exactly N times that of an OC-1 signal. Table 1-2 shows the standard SONET OC-N levels.

Beyond the use of fiber optics for telephone trunking lies an enormous world of both digital and analog applications. One is the use of fibers in the subscriber loop (the local network for distribution of telephone and video services).[16] With fibers it is possible to transmit both narrowband and broadband communication services such as telephone, ISDN-telephone, video conferencing, and ultrafast data on a single subscriber line. A key concept is the use of fiber optics for the *integrated services digital network* (ISDN). The ISDN concept encompasses the ability of a digital communication network to handle voice (telephone), facsimile, data communication, videotex, telemetry, and broadcast audio and video services. Transmission rates for these concepts range from 1.7 Gb/s for high-capacity telephone trunking to 10 Gb/s for super-broadband ISDN.

Since the successful breakthrough in reducing optical fiber losses in 1970, more than 10 million kilometers of fibers have been installed worldwide in less than two decades. To achieve this, researchers and engineers have developed several different generations of lightwave transmission systems. Figure 1-4 shows the progress of four of these generations measured in terms of the product of bit-rate times repeaterless transmission distance. The first-generation links operated at 800 nm using GaAs-based optical sources, silicon photodetectors, and multimode fibers. Some of the initial first-generation telephone-system field trials were carried out in 1977 by GTE in Los Angeles[17] and by AT & T in Chicago.[18] Similar links to these were demonstrated in Europe and Japan. Intercity applications in the 1979-to-1983 time period were at 45 and 90 Mb/s in the United States, at 34 and 140 Mb/s in Europe, and at 32 and 100 Mb/s in Japan, with repeater spacings around 10 km.

A shift in the operating wavelength from 800 to 1300 nm allowed a substantial increase in the repeaterless transmission distance for long-haul telephone trunks. For intercity applications this generation first used multimode fibers, but in 1984 it switched exclusively to single-mode fibers, which have a lower loss and a significantly larger bandwidth. Bit rates for long-haul links

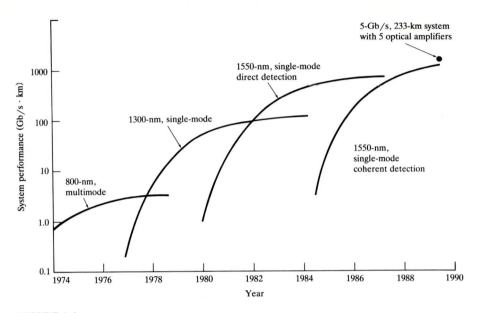

FIGURE 1-4
The evolution of four different generations of lightwave systems. New technology insertion in each generation created higher-capacity systems. The dot shows further developments using optical amplifiers to increase the transmission distance.

typically range between 400 and 600 Mb/s, and in some cases up to 4 Gb/s, over repeater spacings of 40 km. The first undersea fiber optic link, called TAT-8, operates at 296 Mb/s and uses single-mode 1300-nm fibers.[19] This system started operating in 1988. Both multimode and single-mode 1300-nm fibers are used in local area networks, where bit rates range from 10 to 100 Mb/s over distances varying from 500 m to tens of kilometers.[20, 21]

Systems operating at 1550 nm provide the lowest attenuation, but have a much larger signal dispersion than at 1300 nm. Since this dispersion can be overcome using specially fabricated fibers, this generation of fiber optic systems has attracted much attention for high-capacity, long-span terrestrial and undersea transmission links.[22] Both direct-detection and coherent-detection schemes are under investigation for 1550-nm links. Coherent detection[23] offers significant improvements in receiver sensitivity and wavelength selectivity over direct detection, and it allows the use of electronic equalization to compensate for the effects of optical pulse dispersion in the fiber.

The improvement in optical fiber system performance has been extraordinary. In the 1980s the growth rate shown in Figure 1-4, which is measured in terms of the product of bit-rate times repeaterless transmission distance, has been roughly a factor of two per year. Despite this rapid growth and the many successful applications, lightwave technology is not even close to being mature. Beyond these systems there are all-fiber networks, which include all-optical switches, repeaters, and network-access units. Applications include local area networks (LANs), subscriber loops, and TV distribution.[16] In addition there is ongoing research in soliton transmission.[24] A soliton is a nondispersive pulse that makes use of nonlinearity in a fiber to cancel out chromatic dispersion effects. Researchers at AT & T have transmitted solitons at 4 Gb/s over 136 km of conventional optical fiber having approximately 15 ps/(nm · km) of dispersion.[25] This is representative of dispersion-limited fibers which have already been installed. Much longer transmission distances (on the order of thousands of kilometers) are possible using dispersion-shifted fiber.[26]

1.3   ELEMENTS OF AN OPTICAL FIBER TRANSMISSION LINK

An optical fiber transmission link comprises the elements shown in Figure 1-5. The key sections are a transmitter consisting of a light source and its associated drive circuitry, a cable offering mechanical and environmental protection to the optical fibers contained inside, and a receiver consisting of a photodetector plus

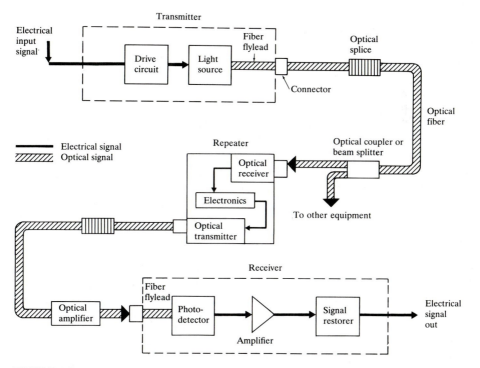

FIGURE 1-5

Major elements of an optical fiber transmission link. The basic components are the transmitter, cable, and receiver. Additional elements include fiber and cable splices, repeaters, beam splitters, and optical amplifiers.

amplification and signal-restoring circuitry. Additional components include optical connectors, splices, couplers or beam splitters, and repeaters. The cabled optical fiber is one of the most important elements in an optical fiber link, as we shall see in Chaps. 2 and 3. In addition to protecting the glass fibers during installation and service, the cable may contain copper wires for powering repeaters which are needed for periodically amplifying and reshaping the signal when the link spans long distances. The cable generally contains several cylindrical hair-thin glass fibers, each of which is an independent communication channel.

Analogous to copper cables, the installation of optical fiber cables can be either aerial, in ducts, undersea, or buried directly in the ground, as Figure 1-6 illustrates. As a result of installation and/or manufacturing limitations, individual cable lengths will range from several hundred meters to several kilometers for long-distance applications. Practical considerations such as reel size and cable weight determine the actual length of a single cable section. The shorter lengths tend to be used when the cables are pulled through ducts. Longer cable lengths are used in aerial or direct-burial applications. The complete long-dis-

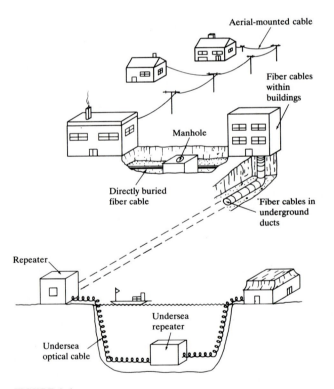

FIGURE 1-6
Optical fiber cables can be installed on poles, in ducts, and undersea, or they can be buried directly in the ground.

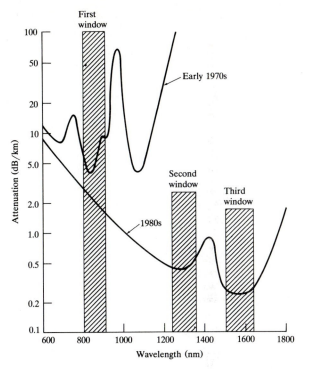

FIGURE 1-7

Optical fiber attenuation as a function of wavelength. Early fiber links exclusively used the 800- to 900-nm range, where there was a local attenuation minimum (the first window). Achievement of lower attenuations created an interest in longer-wavelength second-window (around 1300 nm) and third-window (around 1550 nm) operation.

tance transmission line is formed by splicing or connecting together these individual cable sections.

One of the principal characteristics of an optical fiber is its attenuation as a function of wavelength, as shown in Figure 1-7. Early technology made exclusive use of the 800- to 900-nm wavelength band, since in this region the fibers made at that time exhibited a local minimum in the attenuation curve, and optical sources and photodetectors operating at these wavelengths were available. This region is often referred to as the *first window*. By reducing the concentration of hydroxyl ions and metallic ion impurities in the fiber material, fiber manufacturers were eventually able to fabricate optical waveguides with very low losses in the 1100- to 1600-nm region. This spectral bandwidth is usually referred to as the *long-wavelength region*. Two windows are defined here, these being the *second window* centered around 1300 nm and the *third window* centered around 1550 nm.

Researchers have also looked at new types of fiber materials for use in the 3- to 5-μm wavelength band. In 1975 researchers at the Université de Rennes[27] discovered fluoride glasses having extremely low transmission losses at mid-infrared wavelengths (0.2 to 8 μm, with the lowest loss being around 2.55 μm). The material that researchers have concentrated on is a *heavy-metal fluoride glass* which uses ZrF_4 as the major component and glass network former. Although this glass potentially offers intrinsic minimum losses of 0.01 to 0.001

dB/km, fabricating long lengths of these fibers is difficult. First, ultrapure materials must be used to reach this low loss level. Secondly, fluoride glass is prone to devitrification. Fiber-making techniques have to take this into account to avoid the formation of microcrystallites, which have a drastic effect on scattering losses.[28-33]

Once the cable is installed a light source which is dimensionally compatible with the fiber core is used to launch optical power into the fiber. Semiconductor light-emitting diodes (LEDs) and laser diodes are suitable transmitter sources for this purpose since their light output can be modulated rapidly by simply varying the bias current. The electric input signals to the transmitter can be either of an analog or of a digital form. The transmitter circuitry converts these electric signals to an optical signal by varying the current flow through the light source. An optical source is a *square-law device*, which means that a linear variation in drive current results in a corresponding linear change in the optical output power. In the 800- to 900-nm region the light sources are generally alloys of GaAlAs. At the longer wavelengths (1100 to 1600 nm), an InGaAsP alloy is the principal optical source material. These optical sources are discussed in detail in Chap. 4.

After an optical signal has been launched into the fiber, it will become progressively attenuated and distorted with increasing distance because of scattering, absorption, and dispersion mechanisms in the waveguide. At the receiver the attenuated and distorted modulated optical power emerging from the fiber end will be detected by a photodiode. Analogous to the light source, the photodetector is also a square-law device since it converts the received optical power directly into an electric current output (photocurrent). Semiconductor *pin* and avalanche photodiodes (APDs) are the two principal photodetectors used in a fiber optic link. Both device types exhibit high efficiency and response speed. For applications in which a low-power optical signal is received, an avalanche photodiode is normally used, since it has a greater sensitivity owing to an inherent internal gain mechanism (avalanche effect). Silicon photodetectors are used in the 800- to 900-nm region. A variety of optical detectors are potentially available at the longer wavelengths. The prime material candidate in the 1100- to 1600-nm region is an InGaAs alloy. We address these photodetectors in Chap. 6.

The design of the receiver is inherently more complex than that of the transmitter, since it has to both amplify and reshape the degraded signal received by the photodetector. The principal figure of merit for a receiver is the minimum optical power necessary at the desired data rate to attain either a given error probability for digital systems or a specified signal-to-noise ratio for an analog system. As we shall see in Chap. 7, the ability of a receiver to achieve a certain performance level depends on the photodetector type, the effects of noise in the system, and the characteristics of the successive amplification stages in the receiver.

When an optical signal has traveled a certain distance along a fiber, the signal has become attenuated and distorted to such a degree that a repeater is needed in the transmission line to amplify and reshape the signal. An optical repeater consists of a receiver and a transmitter placed back to back. The receiver section detects the optical signal and converts it to an electric signal, which is amplified, reshaped, and sent to the electric input of the transmitter section. The transmitter section converts this electric signal back to an optical signal and sends it on down the optical fiber waveguide. Chapter 8 presents the conditions that specify how far one can transmit a signal in a digital fiber optic system before a repeater is required. In addition that chapter addresses code formats used for optical signals, optically induced noise effects on system performance, and computer-aided and statistical techniques which can be used for modeling the performance of a digital fiber optic link. Chapter 9 addresses similar considerations for analog systems.

The discussions in Chaps. 8 and 9 deal with direct modulation at the source and direct detection at the receiver. An alternative detection method is to use a coherent system. Here a locally generated optical signal at the receiver is mixed with the incoming information-carrying optical signal. This method, which enhances the signal-to-noise ratio, is outlined in Chap. 10.

To take advantage of the tremendous potential capacity of an optical fiber system, one can employ a number of advanced techniques and networks beyond the simple single-channel point-to-point link. The ones discussed in Chap. 11 include wavelength-division multiplexing, optical-fiber-based local area networks (LANs), the use of all-optical amplifiers, and optical switching.

1.4   USE AND EXTENSION OF THE BOOK

The following chapters are intended as an introduction to the field of optical fiber communcation systems. The sequence of topics systematically takes the reader from a description of the major building blocks to an analysis of a complete optical fiber link. Although the material is introductory, it provides a broad and firm basis with which to analyze and design optical fiber communication systems.

Numerous references are provided at the end of each chapter as a start for delving deeper into any given topic. Since optical fiber communications brings together research and development efforts from many different scientific and engineering disciplines, there are hundreds of articles in the literature relating to the material covered in each chapter. Even though not all of these articles are cited in the references, the selections represent some of the major contributions to the fiber optics field and can be considered as a good introduction to the literature. All references were chosen on the basis of easy accessibility and should be available in a good technical library.

Additional references for up-to-date developments can be found in various conference proceedings. These proceedings are normally obtained from the organizers of the conference or through a major technical library. Some of these conferences are devoted exclusively to fiber optic components and systems.

Examples are given in Refs. 34 through 36. Further supplementary material can be found in the books listed in Refs. 37 through 49.

To help the reader understand and use the material in the book, an overview of units, listings of mathematical formulas, and discussions on decibels and topics from communication theory are given in Apps. A through E.

As is the case for any active scientific and engineering discipline, optical fiber technology is constantly undergoing changes and improvements. New concepts are presently being pursued in optical multiplexing, integrated optics, and device configurations; new materials are being introduced for fibers, sources, and detectors; and measurement techniques are improving. These changes should not have a major impact on the material presented in this book, since it is primarily based on enduring fundamental concepts. The understanding of these concepts will allow a rapid comprehension of any of the new technological developments that will undoubtedly arise.

REFERENCES

1. V. Aschoff, "Optische Nachrichtenübertragung im klassischen Altertum," *Nachrichtentechn. Z. (NTZ)*, vol. 30, pp. 23–28, Jan. 1977.
2. H. Busignies, "Communication channels," *Sci. Amer.*, vol. 227, pp. 99–113, Sept. 1972.
3. A. B. Carlson, *Communication Systems*, McGraw-Hill, New York, 3rd ed., 1986.
4. T. H. Maiman, "Stimulated optical radiation in ruby," *Nature*, vol. 187, pp. 493–494, Aug. 1960.
5. D. W. Berreman, "A lens or lightguide using convectively distorted thermal gradients in gases," *Bell Sys. Tech. J.*, vol. 43, pp. 1469–1475, July 1964.
6. R. Kompfner, "Optical communications," *Science*, vol. 150, pp. 149–155, Oct. 1965.
7. J. H. McElroy, N. McAvoy, E. H. Johnson, J. J. Degnan, F. E. Goodwin, D. M. Henderson, T. A. Nussmeier, L. S. Stokes, B. J. Peyton, and T. Flattau, "CO_2 laser communication systems for near-earth space applications," *Proc. IEEE*, vol. 65, pp. 221–251, Feb. 1977.
8. J. D. Barry, "Design and system requirements imposed by the selection of GaAs/GaAlAs single-mode laser diodes for free space optical communications," *IEEE J. Quantum Electron.*, vol. QE-20, pp. 478–491, May 1984.
9. N. W. Carlson and V. J. Masin, "Evaluation of the diffraction losses for a single-element diode laser transmitter in a free space optical link," *J. Lightwave Tech.*, vol. 6, pp. 413–418, Mar. 1988.
10. V. W. S. Chan, "Space coherent optical communication systems—introduction," *J. Lightwave Tech.*, vol. 5, pp. 633–637, Apr. 1987.
11. K. C. Kao and G. A. Hockman, "Dielectric fiber surface waveguides for optical frequencies," *Proc. IEE*, vol. 133, pp. 1151–1158, July 1966.
12. A. Werts, "Propagation de la lumière cohérente dans les fibres optiques," *L'Onde Électrique*, vol. 46, pp. 967–980, 1966.
13. F. P. Kapron, D. B. Keck, and R. D. Maurer, "Radiation losses in glass optical waveguides," *Appl. Phys. Lett.*, vol. 17, pp. 423–425, Nov. 1980.
14. D. H. Rice and G. E. Keiser, "Applications of fiber optics to tactical communication systems," *IEEE Commun. Mag.*, vol. 23, pp. 46–57, May 1985.
15. R. Ballart and Y.-C. Ching, "SONET: Now it's the standard optical network," *IEEE Commun. Mag.*, vol. 29, pp. 8–15, Mar. 1989.
16. Special Issue on Subscriber Loop Technology, *J. Lightwave Tech.*, vol. 7, Nov. 1989.
17. E. E. Basch, R. A. Beaudette, and H. A. Carnes, "Optical transmission for interoffice trunks," *IEEE Trans. Commun.*, vol. COM-26, pp. 1007–1014, July 1978.
18. M. I. Schwartz, W. A. Reenstra, J. H. Mullins, and J. S. Cook, "Chicago lightwave communications project," *Bell Sys. Tech. J.*, vol. 57, pp. 1881–1888, July–Aug. 1978.

19. P. K. Runge and N. S. Bergano, "Undersea cable transmission systems," in S. E. Miller and I. P. Kaminow, eds., *Optical Fiber Telecommunications—II*, Academic, New York, 1988.
20. G. E. Keiser, *Local Area Networks*, McGraw-Hill, New York, 1989.
21. P. S. Henry, "High-capacity lightwave local area networks," *IEEE Commun. Mag.*, vol. 27, pp. 20–26, Oct. 1989.
22. N. S. Bergano, "Next generation undersea lightwave systems," *Tech. Digest IEEE/OSA Optical Fiber Comm. Conf.*, p. 55, San Francisco, Jan. 1990.
23. T. Okoshi and K. Kikuchi, *Coherent Optical Fiber Communications*, Kluwer Academic, Boston, 1988.
24. A. Hasegawa, *Optical Solitons in Fibers*, Springer-Verlag, New York, 1989.
25. N. A. Olsson, P. A. Andrekson, P. C. Becker, J. R. Simpson, T. Tanbun-Ek, R. A. Logan, H. Presby, and K. Wecht, "4 Gb/s soliton data transmission experiment," *OFC-90 Postdeadline Papers*, PD-4, Jan. 1990.
26. L. F. Mollenauer and K. Smith, "Demonstration of soliton transmission over more than 4000 km in fiber with loss periodically compensated by Raman gain," *Optics Lett.*, vol. 13, pp. 675–677, 1988.
27. M. Poulain, U. Poulain, J. Lucas, and P. Bran, "Verres fluorés au tetrafluorure de zirconium: Propriétés optiques d'un verre dopé au Nd^{3+}," *Materials Res. Bull.*, vol. 10, no. 4, pp. 243–246, 1975.
28. M. G. Drexhage, "Heavy metal fluoride glasses," in M. Tomozawa and R. H. Doremus, eds., *Treatise on Materials Science and Technology, Vol. 26: Glass IV*, Academic, New York, 1985.
29. D. C. Tran, G. H. Sigel, Jr., and B. Bendow, "Heavy metal fluoride glasses and fibers: A review," *J. Lightwave Tech.*, vol. LT-2, pp. 566–586, Oct. 1984.
30. R. C. Folweiler, "Fluoride glasses," *GTE J. Science and Tech.*, vol. 3, no. 1, pp. 25–37, 1989.
31. J. Lucas, "Review: Fluoride glasses," *J. Materials Science*, vol. 24, pp. 1–13, Jan. 1989.
32. T. Katsuyama and H. Matsumura, *Infrared Optical Fibers*, Adam Hilger, Bristol, England, 1989.
33. P. Klocek and G. H. Sigel, Jr., *Infrared Fiber Optics*, SPIE Optical Eng. Press, Bellingham, WA, 1989.
34. *Optical Fiber Communications (OFC) Conf.*, cosponsored annually by the Optical Society of America (OSA), Washington and the Institute of Electrical and Electronic Engineers (IEEE), New York.
35. *Optoelectronic and Fiber Optic Devices and Applications Symp.*, sponsored annually by SPIE–The International Society for Optical Engineering, Bellingham, WA.
36. *European Conference on Optical Fibre Communications (ECOC)*, held annually in Europe; sponsored by various European engineering organizations.
37. D. Gloge, *Optical Fiber Technology*, IEEE Press, New York, 1976. This book contains reprints of fundamental research articles appearing in the literature from 1969 to 1975.
38. C. K. Kao, *Optical Fiber Technology, II*, IEEE Press, New York, 1980. This is a continuation of Ref. 31. It contains reprints of articles appearing in the literature from 1976 to 1979.
39. J. Gowar, *Optical Communication Systems*, Prentice-Hall, Englewood Cliffs, NJ, 1984.
40. S. E. Miller and A. G. Chynoweth, eds., *Optical Fiber Telecommunications*, Academic, New York, 1979.
41. S. E. Miller and I. P Kaminow, eds., *Optical Fiber Telecommunications—II*, Academic, New York, 1988.
42. P. K. Cheo, *Fiber Optics: Devices and Systems*, Prentice-Hall, Englewood Cliffs, NJ, 1985.
43. J. Senior, *Optical Fiber Communications*, Prentice-Hall, Englewood Cliffs, NJ, 1985.
44. C. Baack, *Optical Wideband Transmission Systems*, CRC Press, Boca Raton, FL, 1986.
45. E. E. Basch, ed., *Optical Fiber Transmission*, Sams, Indianapolis, IN, 1987.
46. W. B. Jones, Jr., *Optical Fiber Communication Systems*, Holt, Reinhart and Winston, New York, 1988.
47. G. P. Agrawal, *Nonlinear Fiber Optics*, Academic, New York, 1989.
48. L. B. Jeunhomme, *Single-Mode Fiber Optics*, Dekker, New York, 2nd ed., 1989.
49. C. Yeh, *Handbook of Fiber Optics*, Academic, New York, 1990.

CHAPTER
2

OPTICAL FIBERS: STRUCTURES, WAVEGUIDING, AND FABRICATION

One of the most important components in any optical fiber system is the optical fiber itself, since its transmission characteristics play a major role in determining the performance of the entire system. Some of the questions that arise concerning optical fibers are:

1. What is the structure of an optical fiber?
2. How does light propagate along a fiber?
3. Of what materials are fibers made?
4. How is the fiber fabricated?
5. How are fibers incorporated into cable structures?
6. What is the signal loss or attenuation mechanism in a fiber?
7. Why and to what degree does a signal get distorted as it travels along a fiber?

The purpose of this chapter is to present some of the fundamental answers to the first five questions in order to attain a good understanding of the physical structure and waveguiding properties of optical fibers. Questions 6 and 7 are answered in Chap. 3.

Since fiber optics technology involves the emission, transmission, and detection of light, we begin our discussion by first considering the nature of light and then we shall review a few basic laws and definitions of optics.[1] Following a description of the structure of optical fibers, two methods are used to describe how an optical fiber guides light. The first approach uses the geometrical or ray optics concepts of light reflection and refraction to provide an intuitive picture of the propagation mechanisms. In the second approach light is treated as an electromagnetic wave which propagates along the optical fiber waveguide. This involves solving Maxwell's equations subject to the cylindrical boundary conditions of the fiber.

2.1 THE NATURE OF LIGHT

The concepts concerning the nature of light have undergone several variations during the history of physics. Until the early seventeenth century it was generally believed that light consisted of a stream of minute particles that were emitted by luminous sources. These particles were pictured as traveling in straightlines, and it was assumed that they could penetrate transparent materials but were reflected from opaque ones. This theory adequately described certain large-scale optical effects such as reflection and refraction, but failed to explain finer-scale phenomena such as interference and diffraction.

The correct explanation of diffraction was given by Fresnel in 1815. Fresnel showed that the approximately rectilinear propagation character of light could be interpreted on the assumption that light is a wave motion, and that the diffraction fringes could thus be accounted for in detail. Later the work of Maxwell in 1864 theorized that light waves must be electromagnetic in nature. Furthermore observation of polarization effects indicated that light waves are transverse (that is, the wave motion is perpendicular to the direction in which the wave travels). In this *wave* or *physical optics viewpoint* the electromagnetic waves radiated by a small optical source can be represented by a train of spherical wave fronts with the source at the center as shown in Fig. 2-1. A *wave front* is defined as the locus of all points in the wave train which have the same phase. Generally one draws wave fronts passing either through the maxima or

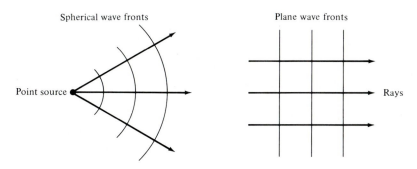

FIGURE 2-1
Representations of spherical and plane wavefronts and their associated rays.

the minima of the wave, such as the peak or trough of a sine wave, for example. Thus the wave fronts (also called *phase fronts*) are separated by one wavelength.

When the wavelength of the light is much smaller than the object (or opening) which it encounters, the wave fronts appear as straight lines to this object or opening. In this case the light wave can be represented as a *plane wave*, and its direction of travel can be indicated by a *light ray* which is drawn perpendicular to the phase front. Thus large-scale optical effects such as reflection and refraction can be analyzed by the simple geometrical process of *ray tracing*. This view of optics is referred to as *ray* or *geometrical optics*. The concept of light rays is very useful because the rays show the direction of energy flow in the light beam.

2.1.1   Linear Polarization

A train of *plane* or *linearly polarized waves* traveling in a direction **k** can be represented in the general form

$$\mathbf{A}(\mathbf{x}, t) = \mathbf{e}_i A_0 \exp[j(\omega t - \mathbf{k} \cdot \mathbf{x})] \tag{2-1}$$

with $\mathbf{x} = x\mathbf{e}_x + y\mathbf{e}_y + z\mathbf{e}_z$ representing a general position vector and $\mathbf{k} = k_x\mathbf{e}_x + k_y\mathbf{e}_y + k_z\mathbf{e}_z$ representing the wave propagation vector.

Here A_0 is the maximum amplitude of the wave, $\omega = 2\pi\nu$, where ν is the frequency of the light, the magnitude of the wavevector **k** is $k = 2\pi/\lambda$, which is known as the *wave propagation constant* with λ being the wavelength of the light, and $\mathbf{e}_i$ is a unit vector lying parallel to an axis designated by i.

Note that the components of the actual (measurable) electromagnetic field represented by Eq. (2-1) are obtained by taking the real part of this equation. For example, if $\mathbf{k} = k\mathbf{e}_z$ and if **A** denotes the electric field **E** with the coordinate axes chosen such that $\mathbf{e}_i = \mathbf{e}_x$, then the real measurable electric field varies harmonically in the x direction and is given by

$$\mathbf{E}_x(z, t) = \mathrm{Re}(\mathbf{E}) = \mathbf{e}_x E_{0x} \cos(\omega t - kz) \tag{2-2}$$

which represents a plane wave traveling in the z direction. The reason for using the exponential form is that it is more easily handled mathematically than equivalent expressions given in terms of sine and cosine. In addition, the rationale for using harmonic functions is that any waveform can be expressed in terms of sinusoidal waves using Fourier techniques.

The electric and magnetic field distributions in a train of plane electromagnetic waves at a given instant in time are shown in Fig. 2-2. The waves are moving in the direction indicated by the vector **k**. Based on Maxwell's equations it is easily shown[2] that **E** and **H** are both perpendicular to the direction of propagation. Such a wave is called a *transverse wave*. Furthermore **E** and **H** are mutually perpendicular, so that **E**, **H**, and **k** form a set of orthogonal vectors.

The plane wave example given by Eq. (2-2) has its electric field vector always pointing in the $\mathbf{e}_x$ direction. Such a wave is *linearly polarized* with

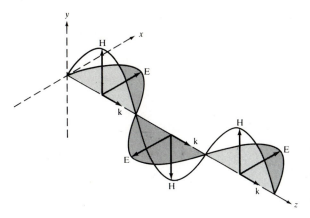

FIGURE 2-2
Electric and magnetic field distributions in a train of plane electromagnetic waves at a given instant in time.

polarization vector $\mathbf{e}_x$. A general state of polarization is described by considering another linearly polarized wave which is independent of the first wave and orthogonal to it. Let this wave be

$$\mathbf{E}_y(z, t) = \mathbf{e}_y E_{0y} \cos(\omega t - kz + \delta) \tag{2-3}$$

where δ is the relative phase difference between the waves. The resultant wave is then simply

$$\mathbf{E}(z, t) = \mathbf{E}_x(z, t) + \mathbf{E}_y(z, t) \tag{2-4}$$

If δ is zero or an integer multiple of 2π, the waves are in phase. Equation (2-4) is then also a linearly polarized wave with a polarization vector making an angle

$$\theta = \arctan \frac{E_{0y}}{E_{0x}} \tag{2-5}$$

with respect to $\mathbf{e}_x$ and having a magnitude

$$E = \left(E_{0x}^2 + E_{0y}^2 \right)^{1/2} \tag{2-6}$$

This case is shown schematically in Fig. 2-3. Conversely, just as any two orthogonal plane waves can be combined into a linearly polarized wave, an arbitrary linearly polarized wave can be resolved into two independent orthogonal plane waves that are in phase.

2.1.2 Elliptical and Circular Polarization

For general values of δ the wave given by Eq. (2-4) is *elliptically polarized*. The resultant field vector $\mathbf{E}$ will both rotate and change its magnitude as a function of the angular frequency ω. From Eqs. (2-2) and (2-3) we can show that (see

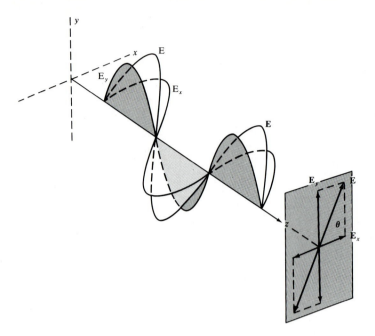

FIGURE 2-3
Addition of two linearly polarized waves having zero relative phase between them.

Prob. 2-5) for a general value of δ

$$\left(\frac{E_x}{E_{0x}}\right)^2 + \left(\frac{E_y}{E_{0y}}\right)^2 - 2\left(\frac{E_x}{E_{0x}}\right)\left(\frac{E_y}{E_{0y}}\right)\cos\delta = \sin^2\delta \qquad (2\text{-}7)$$

which is the general equation of an ellipse. Thus as Fig. 2-4 shows, the endpoint of $\mathbf{E}$ will trace out an ellipse at a given point in space. The axis of the ellipse makes an angle α relative to the x axis given by

$$\tan 2\alpha = \frac{2E_{0x}E_{0y}\cos\delta}{E_{0x}^2 - E_{0y}^2} \qquad (2\text{-}8)$$

To get a better picture of Eq. (2-7), let us align the principal axis of the ellipse with the x axis. Then $\alpha = 0$, or equivalently, $\delta = \pm\pi/2, \pm 3\pi/2, \ldots$, so that Eq. (2-7) becomes

$$\left(\frac{E_x}{E_{0x}}\right)^2 + \left(\frac{E_y}{E_{0y}}\right)^2 = 1 \qquad (2\text{-}9)$$

This is the familiar equation of an ellipse with the origin at the center and semiaxes E_{0x} and E_{0y}.

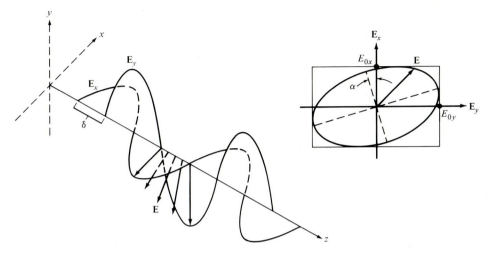

FIGURE 2-4
Elliptically polarized light results from the addition of two linearly polarized waves of unequal amplitude having a nonzero phase difference δ between them.

When $E_{0x} = E_{0y} = E_0$ and the relative phase difference $\delta = \pm\pi/2 + 2m\pi$, where $m = 0, \pm 1, \pm 2, \ldots$, then we have *circularly polarized* light. In this case Eq. (2-9) reduces to

$$E_x^2 + E_y^2 = E_0^2 \tag{2-10}$$

which defines a circle. Choosing the positive sign for δ, Eqs. (2-2) and (2-3) become

$$\mathbf{E}_x(z,t) = \mathbf{e}_x E_0 \cos(\omega t - kz) \tag{2-11}$$

$$\mathbf{E}_y(z,t) = -\mathbf{e}_y E_0 \sin(\omega t - kz) \tag{2-12}$$

In this case the endpoint of **E** will trace out a circle at a given point in space, as Fig. 2-5 illustrates. To see this, consider an observer located at some arbitrary point z_{ref} toward whom the wave is moving. For convenience we will pick this point at $z = \pi/k$ at $t = 0$. Then from Eqs. (2-11) and (2-12) we have

$$\mathbf{E}_x(z,t) = -\mathbf{e}_x E_0 \quad \text{and} \quad \mathbf{E}_y(z,t) = 0$$

so that **E** lies along the negative x axis as Fig. 2-5 shows. At a later time, say $t = \pi/2\omega$, the electric field vector has rotated through 90° and now lies along the positive y axis at z_{ref}. Thus as the wave moves toward the observer with increasing time, the resultant electric field vector **E** rotates *clockwise* at an angular frequency ω. It makes one complete rotation as the wave advances through one wavelength. Such a light wave is *right circularly polarized*.

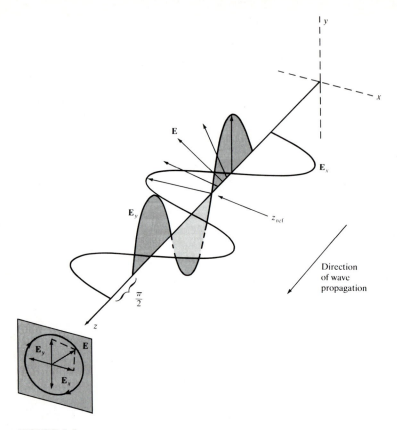

FIGURE 2-5
Addition of two equal-amplitude linearly polarized waves with a relative phase difference $\delta = \pi/2 + 2m\pi$ results in a right circularly polarized wave.

If we choose the negative sign for δ, then the electric field vector is given by

$$\mathbf{E} = E_0\big[\mathbf{e}_x \cos(\omega t - kz) + \mathbf{e}_y \sin(\omega t - kz)\big] \qquad (2\text{-}13)$$

Now $\mathbf{E}$ rotates *counterclockwise* and the wave is *left circularly polarized*.

2.1.3 The Quantum Nature of Light

The wave theory of light adequately accounts for all phenomena involving the transmission of light. However, in dealing with the interaction of light and matter, such as occurs in dispersion and in the emission and absorption of light, neither the particle theory nor the wave theory of light is appropriate. Instead we must turn to quantum theory, which indicates that optical radiation has particle as well as wave properties. The particle nature arises from the observation that light energy is always emitted or absorbed in discrete units called *quanta* or *photons*. In all experiments used to show the existence of photons, the photon energy is found to depend only on the frequency ν. This frequency, in turn, must be measured by observing a wave property of light.

The relationship between the energy E and the frequency ν of a photon is given by

$$E = h\nu \tag{2-14}$$

where $h = 6.625 \times 10^{-34}$ J $\cdot$ s is Planck's constant. When light is incident on an atom, a photon can transfer its energy to an electron within this atom, thereby exciting it to a higher energy level. In this process either all or none of the photon energy is imparted to the electron. The energy absorbed by the electron must be exactly equal to that required to excite the electron to a higher energy level. Conversely, an electron in an excited state can drop to a lower state separated from it by an energy $h\nu$ by emitting a photon of exactly this energy.

2.2 BASIC OPTICAL LAWS AND DEFINITIONS

We shall next review some of the basic optical laws and definitions relevant to optical fibers. A fundamental optical parameter of a material is the *refractive index* (or *index of refraction*). In free space a light wave travels at a speed $c = 3 \times 10^8$ m/s. The speed of light is related to the frequency ν and the wavelength λ by $c = \nu\lambda$. Upon entering a dielectric or nonconducting medium the wave now travels at a speed v, which is characteristic of the material and less than c. The ratio of the speed of light in a vacuum to that in matter is the index of refraction n of the material and is given by

$$n = \frac{c}{v} \tag{2-15}$$

Typical values of n are 1.00 for air, 1.33 for water, 1.50 for glass, and 2.42 for diamond.

The concepts of reflection and refraction can be interpreted most easily by considering the behavior of light rays associated with plane waves traveling in a dielectric material. When a light ray encounters a boundary separating two different media, part of the ray is reflected back into the first medium and the remainder is bent (or refracted) as it enters the second material. This is shown in Fig. 2-6 where $n_2 < n_1$. The bending or refraction of the light ray at the interface is a result of the difference in the speed of light in two materials having different refractive indices. The relationship at the interface is known as Snell's law and is given by

$$n_1 \sin \phi_1 = n_2 \sin \phi_2 \tag{2-16}$$

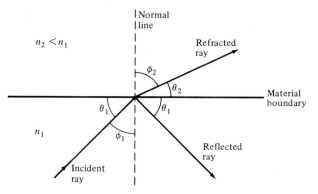

FIGURE 2-6
Refraction and reflection of a light ray at a material boundary.

or equivalently as

$$n_1 \cos \theta_1 = n_2 \cos \theta_2 \qquad (2\text{-}17)$$

where the angles are defined in Fig. 2-6.

According to the law of reflection the angle θ_1 at which the incident ray strikes the interface is exactly equal to the angle the reflected ray makes with the same interface. In addition, the incident ray, the normal to the interface, and the reflected ray all lie in the same plane, which is perpendicular to the interface plane between the two materials. When light traveling in a certain medium is reflected off an optically denser material (one with a higher refractive index), the process is referred to as *external reflection*. Conversely, the reflection of light off of less optically dense material (such as light traveling in glass being reflected at a glass-to-air interface) is called *internal reflection*.

As the angle of incidence θ_1 in an optically denser material (higher refractive index) becomes smaller, the refracted angle θ_2 approaches zero. Beyond this point no refraction is possible and the light rays become *totally internally reflected*. The conditions required for total internal reflection can be determined by using Snell's law [Eq. (2-17)]. Consider Fig. 2-7, which shows a glass surface in air. A light ray gets bent toward the glass surface as it leaves the

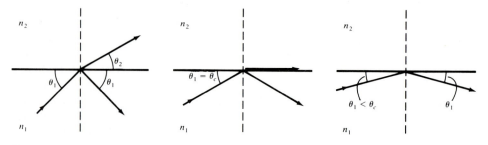

FIGURE 2-7
Representation of the critical angle and total internal reflection at a glass-air interface.

glass in accordance with Snell's law. If the angle of incidence θ_1 is decreased, a point will eventually be reached where the light ray in air is parallel to the glass surface. This point is known as the *critical angle of incidence* θ_c. When the incident angle θ_1 is less than the critical angle, the condition for total internal reflection is satisfied; that is, the light is totally reflected back into the glass with no light escaping from the glass surface. (This is an idealized situation. In practice there is always some tunneling of optical energy through the interface. This can be explained in terms of the electromagnetic wave theory of light which is presented in Sec. 2.4.)

As an example consider the glass-air interface shown in Fig. 2-7. When the light ray in air is parallel to the glass surface then $\theta_2 = 0$ so that $\cos \theta_2 = 1$. The critical angle in the glass is thus

$$\theta_c = \arccos \frac{n_2}{n_1} \tag{2-18}$$

Example 2-1. Using $n_1 = 1.50$ for glass and $n_2 = 1.00$ for air, θ_c is about 48°. Any light in the glass incident on the interface at an angle θ_1 less than 48° is totally reflected back into the glass.

In addition, when light is totally internally reflected, a phase change δ occurs in the reflected wave. This phase change depends on the angle $\theta_1 < \theta_c$ according to the relationships[1]

$$\tan \frac{\delta_N}{2} = \frac{\sqrt{n^2 \cos^2 \theta_1 - 1}}{n \sin \theta_1} \tag{2.19a}$$

$$\tan \frac{\delta_p}{2} = \frac{n\sqrt{n^2 \cos^2 \theta_1 - 1}}{\sin \theta_1} \tag{2.19b}$$

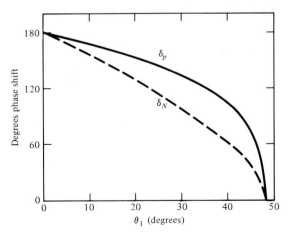

FIGURE 2-8
Phase shifts occurring from the reflection of wave components normal (δ_N) and parallel (δ_p) to the plane of incidence.

Here δ_N and δ_p are the phase shifts of the wave components normal and parallel to the plane of incidence, respectively, and $n = n_1/n_2$. These phase shifts are shown in Fig. 2-8 for a glass-air interface ($n = 1.5$ and $\theta_c = 48°$). The values range from zero immediately at the critical angle to π when $\theta_1 = 0°$.

These basic optical principles will now be used to illustrate how optical power is transmitted along a fiber.

2.3 OPTICAL FIBER MODES AND CONFIGURATIONS

Before going into details on optical fiber characteristics in Sec. 2.3.3, we first present a brief overview of the underlying concepts of optical fiber modes and optical fiber configurations.

2.3.1 Fiber Types

An optical fiber is a dielectric waveguide that operates at optical frequencies. This fiber waveguide is normally cylindrical in form. It confines electromagnetic energy in the form of light to within its surfaces and guides the light in a direction parallel to its axis. The transmission properties of an optical waveguide are dictated by its structural characteristics, which have a major effect in determining how an optical signal is affected as it propagates along the fiber. The structure basically establishes the information-carrying capacity of the fiber and also influences the response of the waveguide to environmental perturbations.

The propagation of light along a waveguide can be described in terms of a set of guided electromagnetic waves called the *modes* of the waveguide. These guided modes are referred to as the *bound* or *trapped* modes of the waveguide. Each guided mode is a pattern of electric and magnetic field lines that is repeated along the fiber at intervals equal to the wavelength. Only a certain discrete number of modes are capable of propagating along the guide. As will be seen in Sec. 2.4, these modes are those electromagnetic waves that satisfy the homogeneous wave equation in the fiber and the boundary condition at the waveguide surfaces.

Although many different configurations of the optical waveguide have been discussed in the literature,[3] the most widely accepted structure is the single solid dielectric cylinder of radius a and index of refraction n_1 shown in Fig. 2-9. This cylinder is known as the *core* of the fiber. The core is surrounded by a solid dielectric *cladding* having a refractive index n_2 that is less than n_1.

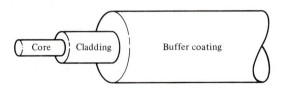

FIGURE 2-9

Schematic of a single-fiber structure. A circular solid core of refractive index n_1 is surrounded by a cladding having a refractive index $n_2 < n_1$. An elastic plastic buffer encapsulates the fiber.

Although, in principle, a cladding is not necessary for light to propagate along the core of the fiber, it serves several purposes. The cladding reduces scattering loss resulting from dielectric discontinuities at the core surface, it adds mechanical strength to the fiber, and it protects the core from absorbing surface contaminants with which it could come in contact.

In low- and medium-loss fibers the core material is generally glass and is surrounded by either a glass or a plastic cladding. Higher-loss plastic-core fibers with plastic claddings are also widely in use. In addition, most fibers are encapsulated in an elastic, abrasion-resistant plastic material. This material adds further strength to the fiber and mechanically isolates or buffers the fibers from small geometrical irregularities, distortions, or roughnesses of adjacent surfaces. These perturbations could otherwise cause scattering losses induced by random microscopic bends that can arise when the fibers are incorporated into cables or supported by other structures.

Variations in the material composition of the core give rise to the two commonly used fiber types shown in Fig. 2-10. In the first case the refractive index of the core is uniform throughout and undergoes an abrupt change (or step) at the cladding boundary. This is called a *step-index fiber*. In the second

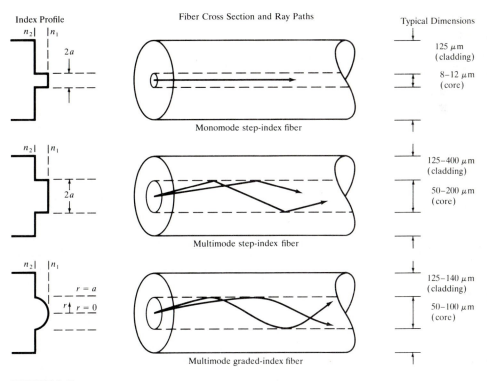

FIGURE 2-10
Comparison of single-mode and multimode step-index and graded-index optical fibers.

case the core refractive index is made to vary as a function of the radial distance from the center of the fiber. This type is a *graded-index fiber*.

Both the step- and the graded-index fibers can be further divided into single-mode and multimode classes. As the name implies, a single-mode fiber sustains only one mode of propagation, whereas multimode fibers contain many hundreds of modes. A few typical sizes of single- and multimode fibers are given in Fig. 2-10 to provide an idea of the dimensional scale. Multimode fibers offer several advantages compared to single-mode fibers. As we shall see in Chap. 5, the larger core radii of multimode fibers make it easier to launch optical power into the fiber and facilitate the connecting together of similar fibers. Another advantage is that light can be launched into a multimode fiber using a light-emitting-diode (LED) source, whereas single-mode fibers must generally be excited with laser diodes. Although LEDs have less optical output power than laser diodes (as we shall discuss in Chap. 4), they are easier to make, are less expensive, require less complex circuitry, and have longer lifetimes than laser diodes, thus making them more desirable in many applications.

A disadvantage of multimode fibers is that they suffer from intermodal dispersion. We shall describe this effect in detail in Chap. 3. Briefly, intermodal dispersion can be described as follows. When an optical pulse is launched into a fiber, the optical power in the pulse is distributed over all (or most) of the modes of the fiber. Each of the modes that can propagate in a multimode fiber travels at a slightly different velocity. This means that the modes in a given optical pulse arrive at the fiber end at slightly different times, thus causing the pulse to spread out in time as it travels along the fiber. This effect, which is known as *intermodal dispersion*, can be reduced by using a graded-index profile in the fiber core. This allows graded-index fibers to have much larger band-widths (data rate transmission capabilities) than step-index fibers. Even higher bandwidths are possible in single-mode fibers, where intermodal dispersion effects are not present.

2.3.2   Rays and Modes

The electromagnetic light field that is guided along an optical fiber can be represented by a superposition of bound or trapped modes. Each of these guided modes consists of a set of simple electromagnetic field configurations which form a standing-wave pattern in the transverse direction (that is, transverse to the waveguide axis). For monochromatic light fields of radian frequency ω, a mode traveling in the positive z direction (that is, along the fiber axis) has a time and z dependence given by

$$e^{j(\omega t - \beta z)}$$

The factor β is the z component of the wave propagation constant $k = 2\pi/\lambda$ and is the main parameter of interest in describing fiber modes. For guided modes β can only assume certain discrete values, which are determined from the requirement that the mode field must satisfy Maxwell's equations and the electric and magnetic field boundary conditions at the core-cladding interface. This is described in detail in Sec. 2.4.

Another method for theoretically studying the propagation characteristics of light in an optical fiber is the geometrical optics or ray-tracing approach. This method provides a good approximation to the light acceptance and guiding properties of optical fibers when the ratio of the fiber radius to the wavelength is large. This is known as the *small-wavelength limit*. Although the ray approach is strictly valid only in the zero-wavelength limit, it is still relatively accurate and extremely valuable for nonzero wavelengths when the number of guided modes is large, that is, for multimode fibers. The advantage of the ray approach is that, compared to the exact electromagnetic wave (modal) analysis, it gives a more direct physical interpretation of the light propagation characteristics in an optical fiber.

Since the concept of a light ray is very different from that of a mode, let us see qualitatively what the relationship is between them. (The mathematical details of this relationship are beyond the scope of this book but can be found in the literature.[4-6]) A guided mode traveling in the z direction (along the fiber axis) can be decomposed into a family of superimposed plane waves that collectively form a standing-wave pattern in the direction transverse to the fiber axis. Since with any plane wave we can associate a light ray that is perpendicular to the phase front of the wave, the family of plane waves corresponding to a particular mode forms a set of rays called a *ray congruence*. Each ray of this particular set travels in the fiber at the same angle relative to the fiber axis. We note here that, since only a certain number M of discrete guided modes exist in a fiber, the possible angles of the ray congruences corresponding to these modes are also limited to the same number M. Although a simple ray picture appears to allow rays at any angle less than the critical angle to propagate in a fiber, the allowable quantized propagation angles result when the phase condition for standing waves is introduced into the ray picture. This is discussed further in Sec. 2.3.5.

Despite the usefulness of the approximate geometrical optics method, a number of limitations and discrepancies exist between it and the exact modal analysis. An important case is the analysis of single-mode or few-mode fibers, which must be dealt with by using electromagnetic theory. Problems involving coherence or interference phenomena must also be solved with an electromagnetic approach. In addition, a modal analysis is necessary when a knowledge of the field distribution of individual modes is required. This arises, for example, when analyzing the excitation of an individual mode or when analyzing the coupling of power between modes at waveguide imperfections (which we shall discuss in Chap. 3).

Another discrepancy between the ray optics approach and the modal analysis occurs when an optical fiber is uniformly bent with a constant radius of curvature. As we shall show in Chap. 3, wave optics correctly predicts that every mode of the curved fiber experiences some radiation loss. Ray optics, on the other hand, erroneously predicts that some ray congruences can undergo total internal reflection at the curve and, consequently, can remain guided without loss.

2.3.3   Step-Index Fiber Structure

We begin our discussion of light propagation in an optical waveguide by considering the step-index fiber. In practical step-index fibers the core of radius a has a refractive index n_1 which is typically equal to 1.48. This is surrounded by a cladding of slightly lower index n_2, where

$$n_2 = n_1(1 - \Delta) \tag{2-20}$$

The parameter Δ is called the *core-cladding index difference* or simply the *index difference*. Values of n_2 are chosen such that Δ is nominally 0.01. Typical values range from 1 to 3 percent for multimode fibers and from 0.2 to 1.0 percent for single-mode fibers. Since the core refractive index is larger than the cladding index, electromagnetic energy at optical frequencies is made to propagate along the fiber waveguide through internal reflection at the core-cladding interface.

2.3.4   Ray Optics Representation

Since the core size of multimode fibers is much larger than the wavelength of the light we are interested in (which is approximately 1 μm), an intuitive picture of the propagation mechanism in an ideal multimode step-index optical waveguide is most easily seen by a simple ray (geometrical) optics representation.[6-11] For simplicity in this analysis we shall consider only a particular ray belonging to a ray congruence which represents a fiber mode. The two types of rays that can propagate in a fiber are meridional rays and skew rays. *Meridional rays* are confined to the meridian planes of the fiber, which are the planes that contain the axis of symmetry of the fiber (the core axis). Since a given meridional ray lies in a single plane, its path is easy to track as it travels along the fiber. Meridional rays can be divided into two general classes: bound rays that are trapped in the core and propagate along the fiber axis according to the laws of geometrical optics, and unbound rays that are refracted out of the fiber core.

Skew rays are not confined to a single plane, but instead tend to follow a helical type path along the fiber as shown in Fig. 2-11. These rays are more difficult to track as they travel along the fiber, since they do not lie in a single plane. Although skew rays constitute a major portion of the total number of guided rays, their analysis is not necessary to obtain a general picture of rays propagating in a fiber. The examination of meridional rays will suffice for this purpose. However, a detailed inclusion of skew rays will change such expressions as the light acceptance ability of the fiber and power losses of light traveling along a waveguide.[6, 9, 10]

A greater power loss arises when skew rays are included in the analyses, since many of the skew rays that geometric optics predicts are trapped in the fiber are actually leaky rays.[5, 12, 13] These leaky rays are only partially confined to

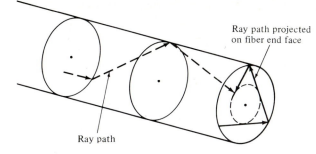

Ray path projected on fiber end face

Ray path

FIGURE 2-11
Ray optics representation of skew rays traveling in a step-index optical fiber core.

the core of the circular optical fiber and attenuate as the light travels along the optical waveguide. This partial reflection of leaky rays cannot be described by pure ray theory alone. Instead, the analysis of radiation loss arising from these types of rays must be described by mode theory. This is explained further in Sec. 2-4.

The meridional ray is shown in Fig. 2-12 for a step-index fiber. The light ray enters the fiber core from a medium of refractive index n at an angle θ_0 with respect to the fiber axis and strikes the core-cladding interface at a normal angle ϕ. If it strikes this interface at such an angle that it is totally internally reflected, the meridional ray follows a zigzag path along the fiber core, passing through the axis of the guide after each reflection.

From Snell's law the minimum angle ϕ_{min} that supports total internal reflection for the meridional ray is given by

$$\sin \phi_{min} = \frac{n_2}{n_1} \tag{2-21}$$

Rays striking the core-cladding interface at angles less than ϕ_{min} will refract out of the core and be lost in the cladding. The condition of Eq. (2-21) can be related to the maximum entrance angle $\theta_{0,max}$ through the relationship

$$n \sin \theta_{0,max} = n_1 \sin \theta_c = \left(n_1^2 - n_2^2\right)^{1/2} \tag{2-22}$$

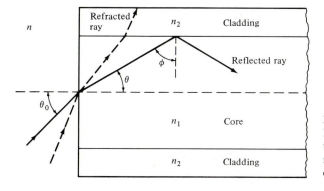

n

Refracted ray

n_2 Cladding

ϕ

Reflected ray

θ

θ_0

n_1 Core

n_2 Cladding

FIGURE 2-12
Meridional ray optics representation of the propagation mechanism in an ideal step-index optical waveguide.

where θ_c is the critical angle. Thus those rays having entrance angles θ_0 less than $\theta_{0,\,\text{max}}$ will be totally internally reflected at the core-cladding interface.

Equation (2-22) also defines the *numerical aperture* NA of a step-index fiber for meridional rays,

$$\text{NA} = n \sin \theta_{0,\,\text{max}} = \left(n_1^2 - n_2^2\right)^{1/2} \simeq n_1 \sqrt{2\Delta} \tag{2-23}$$

The approximation on the right-hand side is valid for the typical case where Δ as defined by Eq. (2-20) is much less than 1. Since the numerical aperture is related to the maximum acceptance angle, it is commonly used to describe the light acceptance or gathering capability of a fiber and to calculate source-to-fiber optical power coupling efficiencies. This is detailed in Chap. 5. The numerical aperture is a dimensionless quantity which is less than unity, with values normally ranging from 0.14 to 0.50.

> **Example 2-2.** Preferred sizes of optical fibers and their corresponding numerical apertures are as follows:
>
> | Core diameter (μm) | Clad diameter (μm) | Numerical aperture |
> | --- | --- | --- |
> | 50 | 125 | 0.19 to 0.25 |
> | 62.5 | 125 | 0.27 to 0.31 |
> | 85 | 125 | 0.25 to 0.30 |
> | 100 | 140 | 0.25 to 0.30 |

2.3.5 Wave Representation

The ray theory appears to allow rays at any angle θ_1 less than the critical angle θ_c to propagate along the fiber. However, when the phase of the plane wave associated with the ray is taken into account, it is seen that only rays at certain discrete angles less than or equal to θ_c are capable of propagating along the fiber.

To see this, consider a light ray in the core incident on the reflective surface at an angle θ as shown in Fig. 2-13. The plane wave associated with this ray is of the form given in Eq. (2-1). As the wave travels it undergoes a phase

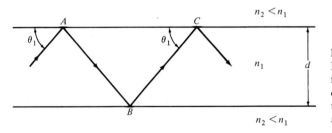

FIGURE 2-13
Lightwave propagating along a fiber waveguide. Phase changes occur both as the wave travels through the fiber medium and at the reflection points.

change δ given by

$$\delta = k_1 s = n_1 k s = \frac{n_1 2 \pi s}{\lambda} \tag{2-24}$$

where k_1 = the propagation constant in the medium of refractive index n_1

  $k = k_1/n_1$ is the free-space propagation constant

  s = the distance traveled along the ray by the wave

The phase of the wave changes not only as the wave travels but also upon reflection from a dielectric interface, as shown in Sec. 2.2.

   In order for the wave associated with a given ray to propagate along the waveguide shown in Fig. 2-13, the phase of the twice reflected wave must be the same as that of the incident wave. That is, the wave must interfere constructively with itself. If this phase condition is not satisfied, the wave will interfere destructively with itself and just die out. Thus the total phase shift that results when the wave traverses the guide twice (from points A to B to C) and gets reflected twice (at points A and B) must be equal to an integer multiple of 2π rad. Using Eqs. (2-24) and (2-19), we let the phase change that occurs over the distance ABC be $\delta_{AC} = n_1 k (2d/\sin \theta_1)$ and the phase changes upon reflection each by (assuming for simplicity that the wave is polarized normal to the plane of incidence)

$$\delta_1 = 2 \arctan \frac{\left(n^2 \cos^2 \theta_1 - 1\right)^{1/2}}{n \sin \theta_1} \tag{2-25}$$

where $n = n_1/n_2$. Then the following condition must be satisfied:

$$\frac{2 n_1 k d}{\sin \theta_1} + 2 \delta_1 = 2 \pi M \tag{2-26}$$

where M is an integer that determines the allowed ray angles for waveguiding.

2.4   MODE THEORY FOR CIRCULAR WAVEGUIDES

To attain a more detailed understanding of the optical power propagation mechanism in a fiber, it is necessary to solve Maxwell's equations subject to the cylindrical boundary conditions of the fiber. This has been carried out in extensive detail in a number of works.[7-10, 14-18] Since a complete treatment is beyond the scope of this book, only a general outline of the analyses will be given here.

   Before we progress with our discussion of mode theory in circular optical fibers, let us first qualitatively examine the appearance of modal fields in the planar dielectric slab waveguide shown in Fig. 2-14. This waveguide is composed of a dielectric slab of refractive index n_1 sandwiched between dielectric material of refractive index $n_2 < n_1$, which we shall call the cladding. This represents the simplest form of an optical waveguide and can serve as a model to gain an

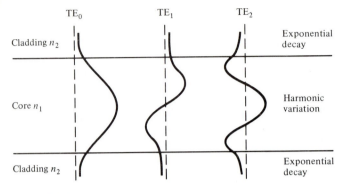

FIGURE 2-14
Electric field distributions for several of the lower-order guided modes in a symmetrical-slab waveguide.

understanding of wave propagation in optical fibers. In fact, a cross-sectional view of the slab waveguide looks the same as the cross-sectional view of an optical fiber cut along its axis. Figure 2-14 shows the field patterns of several of the lower-order modes (which are solutions of Maxwell's equations for the slab waveguide[7-10]). The *order* of a mode is equal to the number of field zeros across the guide. The order of the mode is also related to the angle that the ray congruence corresponding to this mode makes with the plane of the waveguide (or the axis of a fiber); that is, the steeper the angle, the higher the order of the mode. The plots show that the electric field of the guided modes are not completely confined to the central dielectric slab (that is, they do not go to zero at the guide-cladding interface), but, instead, they extend partially into the cladding. The fields vary harmonically in the guiding region of refractive index n_1 and decay exponentially outside of this region. For low-order modes the fields are tightly concentrated near the center of the slab (or the axis of an optical fiber) with little penetration into the cladding region. On the other hand, for higher-order modes the fields are distributed more toward the edges of the guide and penetrate further into the cladding region.

Solving Maxwell's equations shows that, in addition to supporting a finite number of guided modes, the optical fiber waveguide has an infinite continuum of radiation modes that are not trapped in the core and guided by the fiber but are still solutions of the same boundary-value problem. The radiation field basically results from the optical power that is outside the fiber acceptance angle being refracted out of the core. Because of the finite radius of the cladding, some of this radiation gets trapped in the cladding, thereby causing cladding modes to appear. As the core and cladding modes propagate along the fiber, mode coupling occurs between the cladding modes and the higher-order core modes. This coupling occurs because the electric fields of the guided core modes are not completely confined to the core but extend partially into the cladding (see Fig. 2-14) and likewise for the cladding modes. A diffusion of power back and forth between the core and cladding modes thus occurs, which generally results in a loss of power from the core modes. In practice, the cladding modes will be suppressed by a lossy coating which covers the fiber or they will scatter out of the fiber after traveling a certain distance because of roughness on the cladding surface.

In addition to bound and refracted modes, a third category of modes called *leaky modes*[5, 6, 10, 12, 13] is present in optical fibers. These leaky modes are only partially confined to the core region, and attenuate by continuously radiating their power out of the core as they propagate along the fiber. This power radiation out of the waveguide results from a quantum mechanical phenomenon known as the *tunnel effect*. Its analysis is fairly lengthy and beyond the scope of this book. However, it is essentially based on the upper and lower bounds that the boundary conditions for the solutions of Maxwell's equations impose on the propagation constant β. As we shall see in Sec. 2.4.3, a mode remains guided as long as β satisfies the condition

$$n_2 k < \beta < n_1 k$$

where n_1 and n_2 are the refractive indices of the core and cladding, respectively, and $k = 2\pi/\lambda$. The boundary between truly guided modes and leaky modes is defined by the cutoff condition $\beta = n_2 k$. As soon as β becomes smaller than $n_2 k$, power leaks out of the core into the cladding region. Leaky modes can carry significant amounts of optical power in short fibers. Most of these modes disappear after a few centimeters, but a few have sufficiently low losses to persist in fiber lengths of a kilometer.

Although the theory of propagation in optical fibers is well understood, a complete description of the guided and radiation modes is rather complex since it involves six-component hybrid electromagnetic fields having very involved mathematical expressions. A simplification[19-23] of these expressions can be carried out in practice since fibers usually are constructed so that the difference in the core and cladding indices of refraction is very small; that is, $n_1 - n_2 \ll 1$. With this assumption only four field components need to be considered and their expressions become significantly simpler. However, the analysis required for these simplifications is fairly involved, and the interested reader is referred to the literature. Here we shall first follow the standard approach of solving Maxwell's equations for a circular step-index waveguide and then describe the resulting solutions for some of the lower-order modes.

2.4.1 Maxwell's Equations

To analyze the optical waveguide we need to consider Maxwell's equations that give the relationships between the electric and magnetic fields. Assuming a linear, isotropic dielectric material having no currents and free charges, these equations take the form[2]

$$\nabla \times \mathbf{E} = -\frac{\partial \mathbf{B}}{\partial t} \qquad (2\text{-}27a)$$

$$\nabla \times \mathbf{H} = \frac{\partial \mathbf{D}}{\partial t} \qquad (2\text{-}27b)$$

$$\nabla \cdot \mathbf{D} = 0 \qquad (2\text{-}27c)$$

$$\nabla \cdot \mathbf{B} = 0 \qquad (2\text{-}27d)$$

where $\mathbf{D} = \epsilon \mathbf{E}$ and $\mathbf{B} = \mu \mathbf{H}$. The parameter ϵ is the permittivity (or dielectric constant) and μ is permeability of the medium.

A relationship defining the wave phenomena of the electromagnetic fields can be derived from Maxwell's equations. Taking the curl of Eq. (2-27a) and making use of Eq. (2-27b) yields

$$\nabla \times (\nabla \times \mathbf{E}) = -\mu \frac{\partial}{\partial t}(\nabla \times \mathbf{H}) = -\epsilon \mu \frac{\partial^2 \mathbf{E}}{\partial t^2} \qquad (2\text{-}28)$$

Using the vector identity (see App. B)

$$\nabla \times (\nabla \times \mathbf{E}) = \nabla(\nabla \cdot \mathbf{E}) - \nabla^2 \mathbf{E}$$

and using Eq. (2-27c) (that is, $\nabla \cdot \mathbf{E} = 0$), Eq. (2-28) becomes

$$\nabla^2 \mathbf{E} = \epsilon \mu \frac{\partial^2 \mathbf{E}}{\partial t^2} \qquad (2\text{-}29)$$

Similarly, by taking the curl of Eq. (2-27b), it can be shown that

$$\nabla^2 \mathbf{H} = \epsilon \mu \frac{\partial^2 \mathbf{H}}{\partial t^2} \qquad (2\text{-}30)$$

Equations (2-29) and (2-30) are the standard *wave equations*.

2.4.2 Waveguide Equations

Consider electromagnetic waves propagating along a cylindrical fiber shown in Fig. 2-15. For this fiber a cylindrical coordinate system $\{r, \phi, z\}$ is defined with the z axis lying along the axis of the waveguide. If the electromagnetic waves are to propagate along the z axis, they will have a functional dependence of the form

$$\mathbf{E} = \mathbf{E}_0(r, \phi)e^{j(\omega t - \beta z)} \qquad (2\text{-}31)$$

$$\mathbf{H} = \mathbf{H}_0(r, \phi)e^{j(\omega t - \beta z)} \qquad (2\text{-}32)$$

which are harmonic in time t and coordinate z. The parameter β is the z

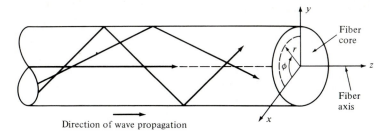

y

Fiber
core

z

Fiber
axis

x

Direction of wave propagation

FIGURE 2-15
Cylindrical coordinate system used for analyzing electromagnetic wave propagation in an optical fiber.

component of the propagation vector and will be determined by the boundary conditions on the electromagnetic fields at the core-cladding interface described in Sec. 2.4.4.

When Eqs. (2-31) and (2-32) are substituted into Maxwell's curl equations, we have, from Eq. (2-27a),

$$\frac{1}{r}\left(\frac{\partial E_z}{\partial \phi} + jr\beta E_\phi\right) = -j\omega\mu H_r, \tag{2-33a}$$

$$j\beta E_r + \frac{\partial E_z}{\partial r} = j\omega\mu H_\phi \tag{2-33b}$$

$$\frac{1}{r}\left[\frac{\partial}{\partial r}(rE_\phi) - \frac{\partial E_r}{\partial \phi}\right] = -j\mu\omega H_z \tag{2-33c}$$

and from Eq. (2-27b)

$$\frac{1}{r}\left(\frac{\partial H_z}{\partial \phi} + jr\beta H_\phi\right) = j\epsilon\omega E_r \tag{2-34a}$$

$$j\beta H_r + \frac{\partial H_z}{\partial r} = -j\epsilon\omega E_\phi \tag{2-34b}$$

$$\frac{1}{r}\left[\frac{\partial}{\partial r}(rH_\phi) - \frac{\partial H_r}{\partial \phi}\right] = j\epsilon\omega E_z \tag{2-34c}$$

By eliminating variables these equations can be rewritten such that, when E_z and H_z are known, the remaining transverse components E_r, E_ϕ, H_r, and H_ϕ can be determined. For example, E_ϕ or H_r can be eliminated from Eqs. (2-33a) and (2-34b) so that the component H_ϕ or E_r, respectively, can be found in terms of E_z or H_z. Doing so yields

$$E_r = -\frac{j}{q^2}\left(\beta\frac{\partial E_z}{\partial r} + \frac{\mu\omega}{r}\frac{\partial H_z}{\partial \phi}\right) \tag{2-35a}$$

$$E_\phi = -\frac{j}{q^2}\left(\frac{\beta}{r}\frac{\partial E_z}{\partial \phi} - \mu\omega\frac{\partial H_z}{\partial r}\right) \tag{2-35b}$$

$$H_r = \frac{-j}{q^2}\left(\beta\frac{\partial H_z}{\partial r} - \frac{\omega\epsilon}{r}\frac{\partial E_z}{\partial \phi}\right) \tag{2-35c}$$

$$H_\phi = \frac{-j}{q^2}\left(\frac{\beta}{r}\frac{\partial H_z}{\partial \phi} + \omega\epsilon\frac{\partial E_z}{\partial r}\right) \tag{2-35d}$$

where $q^2 = \omega^2\epsilon\mu - \beta^2 = k^2 - \beta^2$.

Substitution of Eqs. (2-35c) and (2-35d) into Eq. (2-34c) results in the wave equation in cylindrical coordinates

$$\frac{\partial^2 E_z}{\partial r^2} + \frac{1}{r}\frac{\partial E_z}{\partial r} + \frac{1}{r^2}\frac{\partial^2 E_z}{\partial \phi^2} + q^2 E_z = 0 \tag{2-36}$$

and substitution of Eqs. (2-35a) and (2-35b) into Eq. (2-33c) leads to

$$\frac{\partial^2 H_z}{\partial r^2} + \frac{1}{r}\frac{\partial H_z}{\partial r} + \frac{1}{r^2}\frac{\partial^2 H_z}{\partial \phi^2} + q^2 H_z = 0 \tag{2-37}$$

It is interesting to note that Eqs. (2-36) and (2-37) each contain either only E_z or H_z. This appears to imply that the longitudinal components of **E** and **H** are uncoupled and can be chosen arbitrarily provided that they satisfy Eqs. (2-36) and (2-37). However, in general, coupling of E_z and H_z is required by the boundary conditions of the electromagnetic field components described in Sec. 2-4-4. If the boundary conditions do not lead to coupling between the field components, mode solutions can be obtained in which either $E_z = 0$ or $H_z = 0$. When $E_z = 0$ the modes are called *transverse electric* or TE modes, and when $H_z = 0$ *transverse magnetic* or TM modes result. *Hybrid* modes exist if both E_z and H_z are nonzero. These are designated as HE or EH modes, depending on whether H_z or E_z, respectively, makes a larger contribution to the transverse field. The fact the hybrid modes are present in optical waveguides makes their analysis more complex than in the simpler case of hollow metallic waveguides where only TE and TM modes are found.

2.4.3   Wave Equations for Step-Index Fibers

We now use the above results to find the guided modes in a step-index fiber. A standard mathematical procedure for solving equations such as Eq. (2-36) is to use the separation-of-variables method, which assumes a solution of the form

$$E_z = AF_1(r)F_2(\phi)F_3(z)F_4(t) \tag{2-38}$$

As was already assumed, the time- and z-dependent factors are given by

$$F_3(z)F_4(t) = e^{j(\omega t - \beta z)} \tag{2-39}$$

since the wave is sinusoidal in time and propagates in the z direction. In addition, because of the circular symmetry of the waveguide, each field component must not change when the coordinate ϕ is increased by 2π. We thus assume a periodic function of the form

$$F_2(\phi) = e^{j\nu\phi} \tag{2-40}$$

The constant ν can be positive or negative, but it must be an integer since the fields must be periodic in ϕ with a period of 2π.

Substituting Eq. (2-40) into Eq. (2-38) the wave equation for E_z [Eq. (2-36)] becomes

$$\frac{\partial^2 F_1}{\partial r^2} + \frac{1}{r}\frac{\partial F_1}{\partial r} + \left(q^2 - \frac{\nu^2}{r^2}\right)F_1 = 0 \tag{2-41}$$

which is a well-known differential equation for Bessel functions.[24-26] An exactly identical equation can be derived for H_z.

For the configuration of the step-index fiber we consider a homogeneous core of refractive index n_1 and radius a, which is surrounded by an infinite cladding of index n_2. The reason for assuming an infinitely thick cladding is that the guided modes in the core have exponentially decaying fields outside the core which must have insignificant values at the outer boundary of the cladding. In practice, optical fibers are designed with claddings that are sufficiently thick so that the guided-mode field does not reach the outer boundary of the cladding. To get an idea of the field patterns, the electric field distributions for several of the lower-order guided modes in a symmetrical slab waveguide were shown in Fig. 2.14. The fields vary harmonically in the guiding region of refractive index n_1 and decay exponentially outside of this region.

Equation (2-41) must now be solved for the regions inside and outside the core. For the inside region the solutions for the guided modes must remain finite as $r \to 0$, whereas on the outside the solutions must decay to zero as $r \to \infty$. Thus for $r < a$ the solutions are Bessel functions of the first kind of order ν. For these functions we use the common designation $J_\nu(ur)$. Here $u^2 = k_1^2 - \beta^2$ with $k_1 = 2\pi n_1/\lambda$. The expressions for E_z and H_z inside the core are thus

$$E_z(r < a) = AJ_\nu(ur)e^{j\nu\phi}e^{j(\omega t - \beta z)} \tag{2-42}$$

$$H_z(r < a) = BJ_\nu(ur)e^{j\nu\phi}e^{j(\omega t - \beta z)} \tag{2-43}$$

where A and B are arbitrary constants.

Outside of the core the solutions to Eq. (2-41) are given by modified Bessel functions of the second kind $K_\nu(wr)$, where $w^2 = \beta^2 - k_2^2$ with

$k_2 = 2\pi n_2/\lambda$. The expressions for E_z and H_z outside the core are therefore

$$E_z(r > a) = CK_\nu(wr)e^{j\nu\phi}e^{j(\omega t - \beta z)} \tag{2-44}$$

$$H_z(r > a) = DK_\nu(wr)e^{j\nu\phi}e^{j(\omega t - \beta z)} \tag{2-45}$$

with C and D being arbitrary constants.

The definition of $J_\nu(ur)$ and $K_\nu(wr)$ and various recursion relations are given in App. C. From the definition of the modified Bessel function, it is seen that $K_\nu(wr) \to e^{-wr}$ as $wr \to \infty$. Since $K_\nu(wr)$ must go to zero as $r \to \infty$, it follows that $w > 0$. This, in turn, implies that $\beta \geq k_2$, which represents a cutoff condition. The *cutoff condition* is the point at which a mode is no longer bound to the core region. A second condition on β can be deduced from the behavior of $J_\nu(ur)$. Inside the core the parameter u must be real for F_1 to be real, from which it follows that $k_1 \geq \beta$. The permissible range of β for bound solutions is therefore

$$n_2 k = k_2 \leq \beta \leq k_1 = n_1 k \tag{2-46}$$

where $k = 2\pi/\lambda$ is the free-space propagation constant.

2.4.4   Modal Equation

The solutions for β must be determined from the boundary conditions. The boundary conditions require that the tangential components E_ϕ and E_z of $\mathbf{E}$ inside and outside of the dielectric interface at $r = a$ must be the same and similarly for the tangential components H_ϕ and H_z. Consider first the tangential components of $\mathbf{E}$. For the z component we have, from Eq. (2-42) at the inner core-cladding boundary ($E_z = E_{z1}$) and from Eq. (2-44) at the outside of the boundary ($E_z = E_{z2}$), that

$$E_{z1} - E_{z2} = AJ_\nu(ua) - CK_\nu(wa) = 0 \tag{2-47}$$

The ϕ component is found from Eq. (2-35b). Inside the core the factor q^2 is given by

$$q^2 = u^2 = k_1^2 - \beta^2 \tag{2-48}$$

where $k_1 = 2\pi n_1/\lambda = \omega\sqrt{\epsilon_1\mu}$, while outside the core

$$w^2 = \beta^2 - k_2^2 \tag{2-49}$$

with $k_2 = 2\pi n_2/\lambda = \omega\sqrt{\epsilon_2\mu}$. Substituting Eqs. (2-42) and (2-43) into Eq. (2-35b) to find $E_{\phi1}$, and similarly using Eqs. (2-44) and (2-45) to determine $E_{\phi2}$, yields at $r = a$

$$E_{\phi1} - E_{\phi2} = -\frac{j}{u^2}\left[A\frac{j\nu\beta}{a}J_\nu(ua) - B\omega\mu uJ_\nu'(ua)\right]$$
$$- \frac{j}{w^2}\left[C\frac{j\nu\beta}{a}K_\nu(wa) - D\omega\mu wK_\nu'(wa)\right] = 0 \tag{2-50}$$

where the prime indicates differentiation with respect to the argument.

Similarly, for the tangential components of **H** it is readily shown that at $r = a$

$$H_{z1} - H_{z2} = BJ_\nu(ua) - DK_\nu(wa) = 0 \qquad (2\text{-}51)$$

and

$$H_{\phi 1} - H_{\phi 2} = -\frac{j}{u^2}\left[B\frac{j\nu\beta}{a}J_\nu(ua) + A\omega\epsilon_1 uJ_\nu'(ua)\right]$$

$$-\frac{j}{w^2}\left[D\frac{j\nu\beta}{a}K_\nu(wa) + C\omega\epsilon_2 wK_\nu'(wa)\right] = 0 \qquad (2\text{-}52)$$

Equations (2-47), (2-50), (2-51), and (2-52) are a set of four equations with four unknown coefficients A, B, C, and D. A solution to these equations exists only if the determinant of these coefficients is zero:

$$\begin{vmatrix} J_\nu(ua) & 0 & -K_\nu(wa) & 0 \\ \dfrac{\beta\nu}{au^2}J_\nu(ua) & \dfrac{j\omega\mu}{u}J_\nu'(ua) & \dfrac{\beta\nu}{aw^2}K_\nu(wa) & \dfrac{j\omega\mu}{w}K_\nu'(wa) \\ 0 & J_\nu(ua) & 0 & -K_\nu(wa) \\ -\dfrac{j\omega\epsilon_1}{u}J_\nu'(ua) & \dfrac{\beta\nu}{au^2}J_\nu(ua) & -\dfrac{j\omega\epsilon_2}{w}K_\nu'(wa) & \dfrac{\beta\nu}{aw^2}K_\nu(wa) \end{vmatrix} = 0 \quad (2\text{-}53)$$

Evaluation of this determinant yields the following eigenvalue equation for β:

$$(\mathscr{J}_\nu + \mathscr{K}_\nu)(k_1^2\mathscr{J}_\nu + k_2^2\mathscr{K}_\nu) = \left(\frac{\beta\nu}{a}\right)^2\left(\frac{1}{u^2} + \frac{1}{w^2}\right)^2 \qquad (2\text{-}54)$$

where

$$k_2 \le \beta \le k_1$$

$$\mathscr{J}_\nu = \frac{J_\nu'(ua)}{uJ_\nu(ua)} \quad \text{and} \quad \mathscr{K}_\nu = \frac{K_\nu'(wa)}{wK_\nu(wa)}$$

Upon solving Eq. (2-54) for β it will be found that only discrete values restricted to the range given by Eq. (2-46) will be allowed. Although Eq. (2-54) is a complicated transcendental equation which is generally solved by numerical techniques, its solution for any particular mode will provide all the characteristics of that mode. We shall now consider this equation for some of the lowest-order modes of a step index waveguide.

2.4.5   Modes in Step-Index Fibers

To help describe the modes we shall first examine the behavior of the J-type Bessel functions. These are plotted in Fig. 2-16 for the first three orders. The J-type Bessel functions are similar to harmonic functions since they exhibit oscillatory behavior for real k as is the case for sinusoidal functions. Because of the oscillatory behavior of J_ν there will be m roots of Eq. (2-54) for a given ν value. These roots will be designated by $\beta_{\nu m}$, and the corresponding modes are

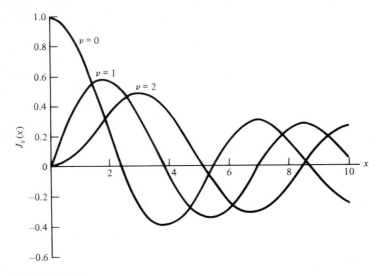

FIGURE 2-16
Variation of the Bessel function $J_\nu(x)$ for the first three orders ($\nu = 0, 1, 2$) plotted as a function of x.

either $\mathrm{TE}_{\nu m}$, $\mathrm{TM}_{\nu m}$, $\mathrm{EH}_{\nu m}$, or $\mathrm{HE}_{\nu m}$. Schematics of the transverse electric field patterns for the four lowest-order modes over the cross-section of a step-index fiber are shown in Fig. 2-17.

For the dielectric fiber waveguide all modes are hybrid modes except those for which $\nu = 0$. When $\nu = 0$ the right-hand side of Eq. (2-54) vanishes and two different eigenvalue equations result. These are

$$\mathscr{J}_0 + \mathscr{K}_0 = 0 \tag{2-55a}$$

or, using the relations for J_ν' and K_ν' for App. C,

$$\frac{J_1(ua)}{uJ_0(ua)} + \frac{K_1(wa)}{wK_0(wa)} = 0 \tag{2-55b}$$

which corresponds to TE_{0m} modes ($E_z = 0$), and

$$k_1^2 \mathscr{J}_0 + k_2^2 \mathscr{K}_0 = 0 \tag{2-56a}$$

or

$$\frac{k_1^2 J_1(ua)}{uJ_0(ua)} + \frac{k_2^2 K_1(wa)}{wK_0(wa)} = 0 \tag{2-56b}$$

which corresponds to TM_{0m} modes ($H_z = 0$). The proof of this is left as an exercise (see Prob. 2-16).

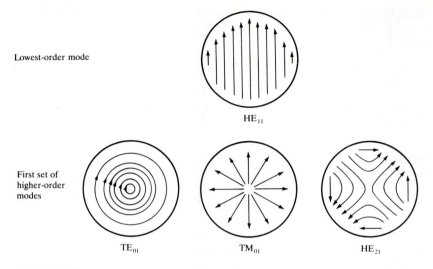

Lowest-order mode

HE_{11}

First set of higher-order modes

TE_{01} TM_{01} HE_{21}

FIGURE 2-17
Cross-sectional views of the transverse electric field vectors for the four lowest-order modes in a step-index fiber.

When $\nu \neq 0$ the situation is more complex and numerical methods are needed to solve Eq. (2-54) exactly. However, simplified and highly accurate approximations based on the principle that the core and cladding refractive indices are nearly the same have been derived by Synder[19] and Gloge.[20, 27] The condition that $n_1 - n_2 \ll 1$ was referred to by Gloge as giving rise to *weakly guided* modes. A treatment of these derivations is given in Sec. 2.4.6.

Let us next examine the cutoff conditions for fiber modes. As was mentioned in relation to Eq. (2-46), a mode is referred to as being cut off when it is no longer bound to the core of the fiber, so that its field no longer decays on the outside of the core. The cutoffs for the various modes are found by solving Eq. (2-54) in the limit $w^2 \rightarrow 0$. This is, in general, fairly complex, so that only the results,[14, 16] which are listed in Table 2-1, will be given here.

TABLE 2-1
Cutoff conditions for some lower-order modes

ν	Mode	Cutoff condition
0	TE_{0m}, TM_{0m}	$J_0(ua) = 0$
1	HE_{1m}, EH_{1m}	$J_1(ua) = 0$
≥ 2	$EH_{\nu m}$	$J_\nu(ua) = 0$
	$HE_{\nu m}$	$\left(\dfrac{n_1^2}{n_2^2} + 1\right)J_{\nu-1}(ua) = \dfrac{ua}{\nu - 1}J_\nu(ua)$

An important parameter connected with the cutoff condition is the *normalized frequency V* (also called the *V-number* or *V-parameter*) defined by

$$V^2 = (u^2 + w^2)a^2 = \left(\frac{2\pi a}{\lambda}\right)^2 (n_1^2 - n_2^2) \tag{2-57}$$

which is a dimensionless number that determines how many modes a fiber can support. The number of modes that can exist in a waveguide as a function of V may be conveniently represented in terms of a *normalized propagation constant* b defined by[20]

$$b = \frac{a^2 w^2}{V^2} = \frac{(\beta/k)^2 - n_2^2}{n_1^2 - n_2^2}$$

A plot of b (in terms of β/k) as a function of V is shown in Fig. 2-18 for a few of the low-order modes. This figure shows that each mode can exist only for values of V that exceed a certain limiting value. The modes are cut off when $\beta/k = n_2$. The HE_{11} mode has no cutoff and ceases to exist only when the core diameter is zero. This is the principle on which the single-mode fiber is based. By appropriately choosing a, n_1, and n_2 so that

$$V = \frac{2\pi a}{\lambda}(n_1^2 - n_2^2)^{1/2} \le 2.405 \tag{2-58}$$

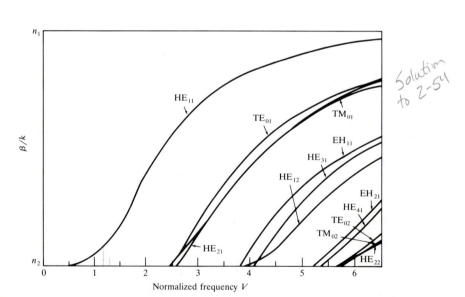

FIGURE 2-18
Plots of the propagation constant (in terms of β/k) as a function of V for a few of the lowest-order modes.

which is the value at which the lowest-order Bessel function J_0 is zero (see Fig. 2-16), all modes except the HE_{11} mode are cut off.

Example 2-3. A step-index fiber has a normalized frequency $V = 26.6$ at a 1300-nm wavelength. If the core radius is 25 μm, let us find the numerical aperture. From Eqs. (2-23) and (2-50) we have

$$V = \frac{2\pi a}{\lambda}\, NA$$

or

$$NA = V\frac{\lambda}{2\pi a} = 26.6\frac{1.3}{2\pi(25)} = 0.22$$

The parameter V can also be related to the number of modes M in a multimode fiber when M is large. An approximate relationship for step-index fibers can be derived from ray theory. A ray congruence incident on the end of a fiber will be accepted by the fiber if it lies within an angle θ defined by the numerical aperture as given in Eq. (2-23):

$$NA = \sin\theta = \left(n_1^2 - n_2^2\right)^{1/2} \tag{2-59}$$

For practical numerical apertures $\sin\theta$ is small so that $\sin\theta \simeq \theta$. The solid acceptance angle for the fiber is therefore

$$\Omega = \pi\theta^2 = \pi\left(n_1^2 - n_2^2\right) \tag{2-60}$$

For electromagnetic radiation of wavelength λ emanating from a laser or a waveguide the number of modes per unit solid angle is given by $2A/\lambda^2$, where A is the area the mode is leaving or entering.[28] The area A in this case is the core cross section πa^2. The factor 2 comes from the fact that the plane wave can have two polarization orientations. The total number of modes M entering the fiber is thus given by

$$M \simeq \frac{2A}{\lambda^2}\Omega = \frac{2\pi^2 a^2}{\lambda^2}\left(n_1^2 - n_2^2\right) = \frac{V^2}{2} \tag{2-61}$$

2.4.6   Linearly Polarized Modes

As may be apparent by now, the exact analysis for the modes of a fiber is mathematically very complex. However, a simpler but highly accurate approximation can be used, based on the principle that in a typical step-index fiber the difference between the indices of refraction of the core and cladding is very small, that is, $\Delta \ll 1$. This is the basis of the *weakly guiding fiber approximation*.[7, 19, 20, 27] In this approximation the electromagnetic field patterns and the propagation constants of the mode pairs $HE_{\nu+1, m}$ and $EH_{\nu-1, m}$ are very similar. This holds likewise for the three modes TE_{0m}, TM_{0m}, and HE_{2m}. This can be seen from Fig. 2-18 with $(\nu, m) = (0, 1)$ and $(2, 1)$ for the mode groupings $\{HE_{11}\}$, $\{TE_{01}, TM_{01}, HE_{21}\}$, $\{HE_{31}, EH_{11}\}$, $\{HE_{12}\}$, $\{HE_{41}, EH_{21}\}$, and $\{TE_{02}, TM_{02}, HE_{22}\}$. The result is that only four field components need to be considered instead of six, and the field description is further simplified by the use of cartesian instead of cylindrical coordinates.

When $\Delta \ll 1$ we have that $k_1^2 \simeq k_2^2 \simeq \beta^2$. Using these approximations, Eq. (2-54) becomes

$$\mathscr{J}_\nu + \mathscr{K}_\nu = \pm \frac{\nu}{a}\left(\frac{1}{u^2} + \frac{1}{w^2}\right) \tag{2-62}$$

Thus Eq. (2-55b) for TE_{0m} modes is the same as Eq. (2-54b) for TM_{0m} modes. Using the recurrence relations for J_ν' and K_ν' given in App. C, we get two sets of equations for Eq. (2-62) for the positive and negative signs. The positive sign yields

$$\frac{J_{\nu+1}(ua)}{uJ_\nu(ua)} + \frac{K_{\nu+1}(wa)}{wK_\nu(wa)} = 0 \tag{2-63}$$

The solution of this equation gives a set of modes called the EH modes. For the negative sign in Eq. (2-62) we get

$$\frac{J_{\nu-1}(ua)}{uJ_\nu(ua)} - \frac{K_{\nu-1}(wa)}{wK_\nu(wa)} = 0 \tag{2-64a}$$

or, alternatively, taking the inverse of Eq. (2-64a) and using the first expressions for $J_\nu(ua)$ and $K_\nu(wa)$ from Sec. C.1.2 and Sec. C.2.2,

$$-\frac{uJ_{\nu-2}(ua)}{J_{\nu-1}(ua)} = \frac{wK_{\nu-2}(wa)}{K_{\nu-1}(wa)} \tag{2-64b}$$

This results in a set of modes called the HE modes.

If we define a new parameter

$$j = \begin{cases} 1 & \text{for TE and TM modes} \\ \nu + 1 & \text{for EH modes} \\ \nu - 1 & \text{for HE modes} \end{cases} \tag{2-65}$$

then Eqs. (2-55b), (2-63), and (2-64b) can be written in the unified form

$$\frac{uJ_{j-1}(ua)}{J_j(ua)} = -\frac{wK_{j-1}(wa)}{K_j(wa)} \tag{2-66}$$

Equations (2-65) and (2-66) show that within the weakly guiding approximation all modes characterized by a common set of j and m satisfy the same characteristic equation. This means that these modes are degenerate. Thus, if an $HE_{\nu+1,m}$ mode is degenerate with an $EH_{\nu-1,m}$ mode (that is, if HE and EH modes of corresponding radial order m and equal circumferential order ν form degenerate pairs), then any combination of an $HE_{\nu+1,m}$ mode with an $EH_{\nu-1,m}$ mode will likewise constitute a guided mode of the fiber.

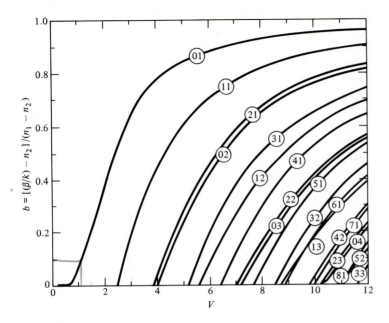

FIGURE 2-19
Plots of the propagation constant b as a function of V for various LP_{jm} modes. (Reproduced with permission from Gloge[20].)

TABLE 2-2
Composition of the lower-order linearly polarized modes

LP-mode designation	Traditional-mode designation and number of modes	Number of degenerate modes
LP_{01}	$HE_{11} \times 2$	2
LP_{11}	$TE_{01}, TM_{01}, HE_{21} \times 2$	4
LP_{21}	$EH_{11} \times 2, HE_{31} \times 2$	4
LP_{02}	$HE_{12} \times 2$	2
LP_{31}	$EH_{21} \times 2, HE_{41} \times 2$	4
LP_{12}	$TE_{02}, TM_{02}, HE_{22} \times 2$	4
LP_{41}	$EH_{31} \times 2, HE_{51} \times 2$	4
LP_{22}	$EH_{12} \times 2, HE_{32} \times 2$	4
LP_{03}	$HE_{13} \times 2$	2
LP_{51}	$EH_{41} \times 2, HE_{61} \times 2$	4

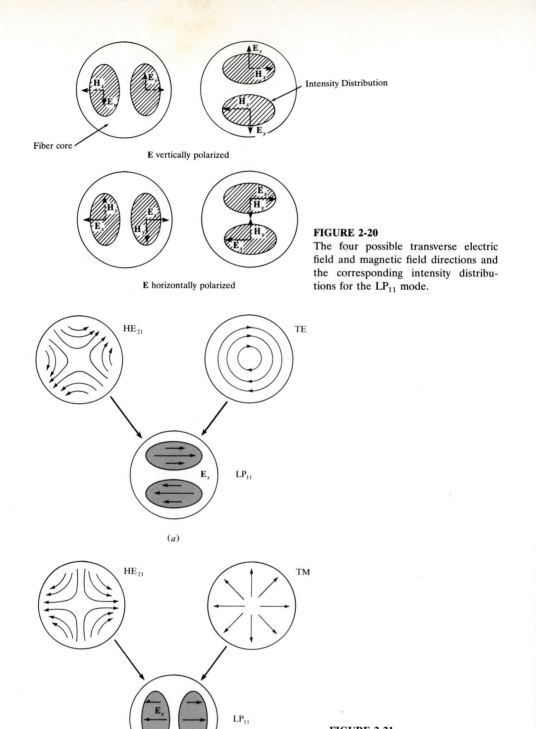

Fiber core

E vertically polarized

Intensity Distribution

E horizontally polarized

FIGURE 2-20
The four possible transverse electric field and magnetic field directions and the corresponding intensity distributions for the LP_{11} mode.

HE_{21} TE

E_x LP_{11}

(a)

HE_{21} TM

E_x LP_{11}

(b)

FIGURE 2-21
Composition of two LP_{11} modes from exact modes and their transverse electric field and intensity distributions.

48

Gloge[20] proposed that such degenerate modes be called *linearly polarized* (LP) modes, and be designated LP_{jm} modes regardless of their TM, TE, EH, or HE field configuration. The normalized propagation constant b as a function of V is given for various LP_{jm} modes in Fig. 2-19. In general we have the following:

1. each LP_{0m} mode is derived from an HE_{1m} mode;

2. each LP_{1m} mode comes from TE_{0m}, TM_{0m}, and HE_{2m} modes;

3. each $LP_{\nu m}$ mode ($\nu \geq 2$) is from an $HE_{\nu+1, m}$ and an $EH_{\nu-1, m}$ mode.

The correspondence between the ten lowest LP modes (that is, those having the lowest cutoff frequencies) and the traditional TM, TE, EH, and HE modes is given in Table 2-2. This table also shows the number of degenerate modes.

A very useful feature of the LP-mode designation is the ability to readily visualize a mode. In a complete set of modes only one electric and one magnetic field component are significant. The electric field vector **E** can be chosen to lie along an arbitrary axis, with the magnetic field vector **H** being perpendicular to it. In addition there are equivalent solutions with the field polarities reversed. Since each of the two possible polarization directions can be coupled with either a $\cos j\phi$ or a $\sin j\phi$ azimuthal dependence, four discrete mode patterns can be obtained from a single LP_{jm} label. As an example, the four possible electric and magnetic field directions and the corresponding intensity distributions for the LP_{11} mode are shown in Fig. 2-20. Figures 2-21*a* and 2-21*b* illustrate how two LP_{11} modes are composed from the exact HE_{21} plus TE_{01} and the exact HE_{21} plus TM_{01} modes, respectively.

2.4.7   Power Flow in Step-Index Fibers

A final quantity of interest for step-index fibers is the fractional power flow in the core and cladding for a given mode. As is illustrated in Fig. 2-14, the electromagnetic field for a given mode does not go to zero at the core-cladding interface, but changes from an oscillating form in the core to an exponential decay in the cladding. Thus the electromagnetic energy of a guided mode is carried partly in the core and partly in the cladding. The further away a mode is from its cutoff frequency the more concentrated its energy is in the core. As cutoff is approached, the field penetrates further into the cladding region and a greater percentage of the energy travels in the cladding. At cutoff the field no longer decays outside the core and the mode now becomes a fully radiating mode.

The relative amounts of power flowing in the core and the cladding can be obtained by integrating the Poynting vector in the axial direction,

$$S_z = \tfrac{1}{2} \, \text{Re}(\mathbf{E} \times \mathbf{H}^*) \cdot \mathbf{e}_z \tag{2-67}$$

over the fiber cross section. Thus the power in the core and cladding, respectively, is given by

$$P_{core} = \frac{1}{2} \int_0^a \int_0^{2\pi} r(E_x H_y^* - E_y H_x^*) \, d\phi \, dr \tag{2-68}$$

$$P_{clad} = \frac{1}{2} \int_a^\infty \int_0^{2\pi} r(E_x H_y^* - E_y H_x^*) \, d\phi \, dr \tag{2-69}$$

where the star denotes the complex conjugate. Gloge[20, 29] has shown that, based on the weakly guided mode approximation, which has an accuracy on the order of the index difference Δ between the core and cladding, the relative core and cladding powers for a particular mode ν are given by

$$\frac{P_{core}}{P} = \left(1 - \frac{u^2}{V^2}\right)\left[1 - \frac{J_\nu^2(ua)}{J_{\nu+1}(ua) J_{\nu-1}(ua)}\right] \tag{2-70}$$

and

$$\frac{P_{clad}}{P} = 1 - \frac{P_{core}}{P} \tag{2-71}$$

where P is the total power in the mode ν. The relationships between P_{core} and P_{clad} are plotted in Fig. 2-22 in terms of the fractional powers P_{core}/P and P_{clad}/P for various LP_{jm} modes. In addition, far from cutoff the average total

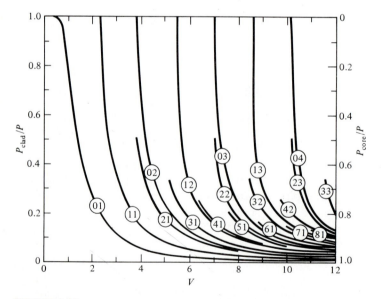

FIGURE 2-22
Fractional power flow in the cladding of a step-index optical fiber as a function of V. When $\nu \neq 1$, the curve numbers νm designate the $HE_{\nu+1, m}$ and $EH_{\nu-1, m}$ modes. For $\nu = 1$, the curve numbers νm give the HE_{2m}, TE_{0m}, and TM_{0m} modes. (Reproduced with permission from Gloge.[20])

power in the cladding has been derived for fibers in which many modes can propagate. Because of this large number of modes, those few modes that are appreciably close to cutoff can be ignored to a reasonable approximation. The derivation assumes an incoherent source, such as a tungsten filament lamp or a light-emitting diode, which, in general, excites every fiber mode with the same amount of power. The total average cladding power is thus approximated by[20]

$$\left(\frac{P_{\text{clad}}}{P}\right)_{\text{total}} = \tfrac{4}{3}M^{-1/2} \tag{2-72}$$

From Fig. 2-22 and Eq. (2-72) it can be seen that, since M is proportional to V^2, the power flow in the cladding decreases as V increases.

> **Example 2-4.** As an example, consider a fiber having a core radius of 25 μm, a core index of 1.48, and $\Delta = 0.01$. At an operating wavelength of 0.84 μm the value of V is 39 and there are 760 modes in the fiber. From Eq. (2-72) approximately 5 percent of the power propagates in the cladding. If Δ is decreased to, say, 0.003 in order to decrease signal dispersion (see Chap. 3), then 242 modes propagate in the fiber and about 9 percent of the power resides in the cladding. For the case of the single-mode fiber, considering the LP_{01} mode (the HE_{11} mode) in Fig. 2-22, it is seen that for $V = 1$ about 70 percent of the power propagates in the cladding, whereas for $V = 2.405$, which is where the LP_{11} mode (the TE_{01} mode) begins, approximately 84 percent of the power is now within the core.

2.5 SINGLE-MODE FIBERS

Single-mode fibers are constructed by letting the dimensions of the core diameter be a few wavelengths (usually 8 to 12) and by having small index differences between the core and the cladding. From Eq. (2-58) with $V = 2.4$, it can be seen that single-mode propagation is possible for fairly large variations in values of the physical core size a and the core-cladding index differences Δ. However, in practical designs of single-mode fibers,[27] the core-cladding index difference varies between 0.2 and 1.0 percent, and the core diameter should be chosen to be just below the cutoff of the first higher-order mode, that is, for V slightly less than 2.4. For example, a typical single-mode fiber may have a core radius of 3 μm and a numerical aperture of 0.1 at a wavelength of 0.8 μm. From Eqs. (2-23) and (2-57) this yields $V = 2.356$.

2.5.1 Mode Field Diameter

For single-mode fibers the geometric distribution of light in the propagating mode, rather than the core diameter and the numerical aperture, is what is important when predicting the performance characteristics of these fibers. Thus a fundamental parameter of a single-mode fiber is the *mode field diameter* (MFD).[30-35] This parameter can be determined from the mode field distribution of the fundamental LP_{01} mode. The mode field diameter is analogous to

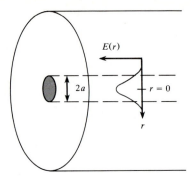

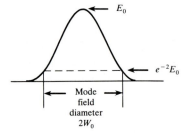

FIGURE 2-23

Distribution of light in a single-mode fiber above its cutoff wavelength. For a gaussian distribution the MFD is given by the $1/e^2$ width of the optical power.

the core diameter in multimode fibers, except that in single-mode fibers not all the light which propagates through the fiber is carried in the core. This is illustrated in Fig. 2-23.

A variety of models for characterizing and measuring the MFD have been proposed.[30-33, 36-40] The main consideration in all these methods is how to approximate the electric field distribution. First let us assume the distribution to be gaussian:[21]

$$E(r) = E_0 \exp\left(-r^2/W_0^2\right) \tag{2-73}$$

where r is the radius, E_0 is the field at zero radius, and W_0 is the width of the electric field distribution. Then one method is take the width $2W_0$ of the MFD to be twice the e^{-1} radius of the optical electric field (which is equivalent to the e^{-2} radius of the optical power) given in Eq. (2-73). The MFD width $2W_0$ of the LP_{01} mode can then be defined as

$$2W_0 = 2\left[\frac{2\int_0^\infty r^3 E^2(r)\,dr}{\int_0^\infty r E^2(r)\,dr}\right]^{1/2} \tag{2-74}$$

where $E(r)$ denotes the field distribution of the LP_{01} mode. This definition is not unique, and several others have been proposed.[31] Also note that in general the mode field varies with the refractive index profile and thus deviates from a gaussian distribution.

2.5.2 Propagation Modes in Single-Mode Fibers

As we saw in Sec. 2.4.6, in any ordinary single-mode fiber there are actually two independent, degenerate propagation modes.[34, 41-43] These modes are very similar, but their polarization planes are orthogonal. These may be chosen arbitrarily as the horizontal (H) and the vertical (V) polarizations as shown in Fig. 2-24. Either one of these two polarization modes constitutes the fundamental HE_{11} mode. In general the electric field of the light propagating along the fiber is a linear superposition of these two polarization modes and depends on the polarization of the light at the launching point into the fiber.

Suppose we arbitrarily choose one of the modes to have its transverse electric field polarized along the x direction and the other independent orthogonal mode to be polarized in the y direction as shown in Fig. 2-24. In ideal fibers with perfect rotational symmetry, the two modes are degenerate with equal propagation constants ($k_x = k_y$) and any polarization state injected into the fiber will propagate unchanged. In actual fibers there are imperfections such as asymmetrical lateral stresses, noncircular cores, and variations in refractive-index profiles. These imperfections break the circular symmetry of the ideal fiber and lift the degeneracy of the two modes. The modes propagate with different phase velocities, and the difference between their effective refractive indices is called the fiber *birefringence*,

$$B_f = n_y - n_x \tag{2-75}$$

Equivalently, we may define the birefringence as

$$\beta = k_0(n_y - n_x) \tag{2-76}$$

where $k_0 = 2\pi/\lambda$ is the free-space propagation constant.

If light is injected into the fiber so that both modes are excited, then one will be delayed in phase relative to the other as they propagate. When this

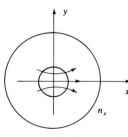

Horizontal mode

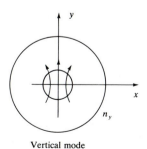

Vertical mode

FIGURE 2-24
Two polarizations of the fundamental HE_{11} mode in a single-mode fiber.

phase difference is an integral multiple of 2π, the two modes will beat at this point and the input polarization state will be reproduced. The length over which this beating occurs is the *fiber beat length*

$$L_p = 2\pi/\beta \tag{2-77}$$

Example 2-5. A single-mode optical fiber has a beat length of 8 cm at 1300 nm. From Eqs. (2-75) to (2-77) we have that the modal birefringence is

$$B_f = n_y - n_x = \frac{\lambda}{L_p} = \frac{1.3 \times 10^{-6} \text{ m}}{8 \times 10^{-2} \text{ m}} = 1.63 \times 10^{-5}$$

or, alternatively,

$$\beta = \frac{2\pi}{L_p} = \frac{2\pi}{0.08 \text{ m}} = 78.5 \text{ m}^{-1}$$

This indicates an intermediate type fiber, since birefringences can vary from $B_f = 1 \times 10^{-3}$ (a typical high-birefringence fiber) to $B_f = 1 \times 10^{-8}$ (a typical low-birefringence fiber).

2.6 GRADED-INDEX FIBER STRUCTURE

In the gradient-index fiber design the core refractive index decreases continuously with increasing radial distance r from the center of the fiber but is generally constant in the cladding. The most commonly used construction for the refractive-index variation in the core is the power law relationship

$$n(r) = \begin{cases} n_1 \left[1 - 2\Delta \left(\dfrac{r}{a} \right)^{\alpha} \right]^{1/2} & \text{for } 0 \leq r \leq a \\ n_1 (1 - 2\Delta)^{1/2} \simeq n_1 (1 - \Delta) = n_2 & \text{for } r \geq a \end{cases} \tag{2-78}$$

Here r is the radial distance from the fiber axis, a is the core radius, n_1 is the refractive index at the core axis, n_2 is the refractive index of the cladding, and the dimensionless parameter α defines the shape of the index profile. The index difference Δ for the graded-index fiber is given by

$$\Delta = \frac{n_1^2 - n_2^2}{2n_1^2} \simeq \frac{n_1 - n_2}{n_1} \tag{2-79}$$

The approximation on the right-hand side of this equation reduces the expression for Δ to that of the step-index fiber given by Eq. (2-20). Thus the same symbol is used in both cases. For $\alpha = \infty$, Eq. (2-78) reduces to the step-index profile $n(r) = n_1$.

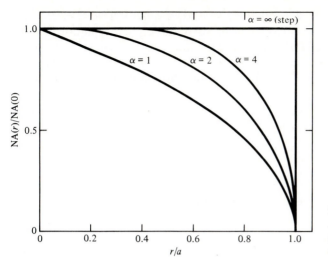

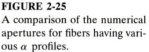

FIGURE 2-25
A comparison of the numerical apertures for fibers having various α profiles.

2.6.1 Graded-Index Numerical Aperture (NA)

The determination of the NA for graded-index fibers is more complex than for step-index fibers. In graded-index fibers the NA is a function of position across the core end face. This is in contrast to the step-index fiber, where the NA is constant across the core. Geometrical optics considerations show that light incident on the fiber core at position r will propagate as a guided mode only if it is within the local numerical aperture $NA(r)$ at that point. The local numerical aperture is defined as[44]

$$NA(r) = \begin{cases} \left[n^2(r) - n_2^2\right]^{1/2} \simeq NA(0)\sqrt{1 - (r/a)^\alpha} & \text{for} \quad r \le a \\ 0 & \text{for} \quad r > a \end{cases} \quad (2\text{-}80)$$

where the axial numerical aperture is defined as

$$NA(0) = \left[n^2(0) - n_2^2\right]^{1/2} = \left(n_1^2 - n_2^2\right)^{1/2} \simeq n_1\sqrt{2\Delta} \quad (2\text{-}81)$$

From Eq. (2-80) it is clear that the NA of a graded-index fiber decreases from $NA(0)$ to zero as r moves from the fiber axis to the core-cladding boundary. A comparison of the numerical apertures for fibers having various α profiles is given in Fig. 2-25.

2.6.2 Modes in Graded-Index Fibers

A modal analysis of an optical fiber based on solving Maxwell's equations can only be carried out rigorously if the core refractive index is uniform, that is, for step-index fibers. In other cases, such as for graded-index fibers, approximation methods are needed. The most widely used analysis of modes in a graded-index fiber is an approximation based on the WKB method[45, 46] (named after Wenzel,

Kramers, and Brillouin), which is commonly used in quantum mechanics. The purpose of the WKB method is to obtain an asymptotic representation for the solution of a differential equation containing a parameter that varies slowly over the desired range of the equation. That parameter in this case is the refractive-index profile $n(r)$, which varies only slightly over distances on the order of an optical wavelength.

Analogous to the step-index fiber, Eq. (2-41) for the radial component of the wave equation must be solved:[10, 11, 29]

$$\frac{d^2F_1}{dr^2} + \frac{1}{r}\frac{dF_1}{dr} + \left[k^2n^2(r) - \beta^2 - \frac{\nu^2}{r^2}\right]F_1 = 0 \qquad (2\text{-}82)$$

where $n(r)$ is given by Eq. (2-78). The general procedure in the WKB method is to let[29]

$$F_1 = Ae^{jkQ(r)} \qquad (2\text{-}83)$$

where the coefficient A is independent of r. Substituting this into Eq. (2-82) gives

$$jkQ'' - (kQ')^2 + \frac{jk}{r}Q' + \left[k^2n^2(r) - \beta^2 - \frac{\nu^2}{r^2}\right] = 0 \qquad (2\text{-}84)$$

where the primes denote differentiation with respect to r. Since $n(r)$ varies slowly over a distance on the order of a wavelength, an expansion of the function $Q(r)$ in powers of λ or, equivalently, in powers of $k^{-1} = \lambda/2\pi$ is expected to converge rapidly. Thus we let

$$Q(r) = Q_0 + \frac{1}{k}Q_1 + \cdots \qquad (2\text{-}85)$$

where $Q_0, Q_1, \ldots$ are certain functions of r. Substituting Eq. (2-85) into Eq. (2-84) and collecting equal powers of k yield

$$\left\{-k^2(Q_0')^2 + \left[k^2n^2(r) - \beta^2 - \frac{\nu^2}{r^2}\right]\right\} + \left(jkQ_0'' - 2kQ_0'Q_1' + \frac{jk}{r}Q_0'\right)$$

$$+ \text{ terms of order } (k^0, k^{-1}, k^{-2}, \ldots) = 0 \qquad (2\text{-}86)$$

A sequence of defining relations for the functions Q_i are obtained by setting to zero the terms in equal powers of k. Thus, for the first two terms of Eq. (2-86),

$$-k^2(Q_0')^2 + \left[k^2n^2(r) - \beta^2 - \frac{\nu^2}{r^2}\right] = 0 \qquad (2\text{-}87)$$

$$jkQ_0'' - 2kQ_0'Q_1' + \frac{jk}{r}Q_0' = 0 \qquad (2\text{-}88)$$

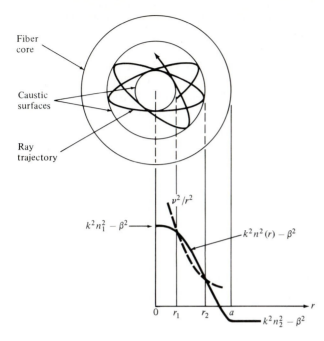

Fiber
core

Caustic
surfaces

Ray
trajectory

ν^2/r^2

$k^2 n_1^2 - \beta^2$

$k^2 n^2(r) - \beta^2$

0 r_1 r_2 a r

$k^2 n_2^2 - \beta^2$

FIGURE 2-26
Cross-sectional projection of a skew ray in a graded-index fiber and the graphical representation of its mode solution from the WKB method. The field is oscillatory between the turning points r_1 and r_2 and is evanescent outside of this region.

Integration of Eq. (2-87) yields

$$kQ_0 = \int_{r_1}^{r_2} \left[k^2 n^2(r) - \beta^2 - \frac{\nu^2}{r^2} \right]^{1/2} dr \qquad (2\text{-}89)$$

A mode is bound in the fiber core only if Q_0 is real. For Q_0 to be real, the radical in the integrand must be greater than zero. In general, for a given mode ν, there are two values r_1 and r_2 for which the radical is zero as is indicated by the limits of integration in Eq. (2-89). Note that these values of r are functions of ν. Guided modes exist for r between these two values. For other values of r the function Q_0 is imaginary, which leads to decaying fields.

To help visualize the solutions to Eq. (2-89), consider the cross-sectional projection of a skew ray in a graded-index fiber shown in Fig. 2-26. The path followed by the ray lies completely within the boundaries of two coaxial cylindrical surfaces, known as the *caustic surfaces*, that have inner and outer radii r_1 and r_2, respectively. The radii r_1 and r_2 are those points at which the radical in the integrand of Eq. (2-89) becomes zero. They are called *turning points*, since the ray turns from increasing to decreasing values of r or vice versa. To evaluate the turning points, consider the functions

$$k^2 n^2(r) - \beta^2$$

and ν^2/r^2 plotted in Fig. 2-26 as solid and dashed curves, respectively. The crossing points of these two curves give the points r_1 and r_2. An oscillating field exists when the solid curve lies above the dashed curve, which indicates bound mode solutions. Evanescent (nonbound decaying mode) fields occur when the solid curve lies below the dashed curve.

To form a bound mode of the graded-index fiber, each wave associated with the ray congruence corresponding to this mode must interfere constructively with itself in such a way as to form a standing-wave pattern in the radial cross-sectional direction. (A full mathematical treatment of this can be found in the literature.[47–48]) This requirement imposes the condition that the phase function Q_0 between r_1 and r_2 must be a multiple of π (that is, an integer number of half-periods), so that

$$m\pi \simeq \int_{r_1}^{r_2} \left[k^2 n^2(r) - \beta^2 - \frac{\nu^2}{r^2} \right]^{1/2} dr \qquad (2\text{-}90)$$

where $m = 0, 1, 2, \ldots$ is the radial mode number that counts the number of half-periods between the turning points. The total number of bound modes $m(\beta)$ can be found by summing Eq. (2-90) over all ν from 0 to ν_{max}, where ν_{max} is the highest-order bound mode for a given value of β. If ν_{max} is a large number, the sum can be replaced by an integral, yielding

$$m(\beta) = \frac{4}{\pi} \int_0^{\nu_{max}} \int_{r_1(\nu)}^{r_2(\nu)} \left[k^2 n^2(r) - \beta^2 - \frac{\nu^2}{r^2} \right]^{1/2} dr \, d\nu \qquad (2\text{-}91)$$

The factor 4 arises from the fact that each combination (m, ν) designates a degenerate group of four modes of different polarization or orientation.[20] If we change the order of integration, the lower limit on r must be $r_1 = 0$ in order to count all the modes, and the upper limit on ν is found from the condition

$$k^2 n^2(r) - \beta^2 - \frac{\nu_{max}^2}{r^2} = 0 \qquad (2\text{-}92)$$

Thus

$$m(\beta) = \frac{4}{\pi} \int_0^{r_2} \int_0^{\nu_{max}} \left[k^2 n^2(r) - \beta^2 - \frac{\nu^2}{r^2} \right]^{1/2} d\nu \, dr \qquad (2\text{-}93)$$

Evaluating the integral over ν with ν_{max} given by Eq. (2-92) yields

$$m(\beta) = \int_0^{r_2} \left[k^2 n^2(r) - \beta^2 \right] r \, dr \qquad (2\text{-}94)$$

To evaluate this further, we choose the index profile $n(r)$ given by Eq. (2-78). The upper limit of integration r_2 is determined from the condition that

$$kn(r) = \beta$$

Combining this condition with Eq. (2-78) gives

$$r_2 = a \left[\frac{1}{2\Delta} \left(1 - \frac{\beta^2}{k^2 n_1^2} \right) \right]^{1/\alpha} \tag{2-95}$$

Using Eqs. (2-78) and (2-95), the number of modes is

$$m(\beta) = a^2 k^2 n_1^2 \Delta \frac{\alpha}{\alpha + 2} \left(\frac{k^2 n_1^2 - \beta^2}{2\Delta k^2 n_1^2} \right)^{(2+\alpha)/\alpha} \tag{2-96}$$

All bound modes in a fiber must have $\beta \geq kn_2$. If this condition does not hold, the mode is no longer perfectly trapped inside the core and loses power by leakage into the cladding. The maximum number of bound modes M is thus found by letting

$$\beta = kn_2 = kn_1(1 - \Delta)$$

where Eq. (2-78) was used for the relationship between n_1 and n_2. Thus

$$M = m(kn_2) = \frac{\alpha}{\alpha + 2} a^2 k^2 n_1^2 \Delta \tag{2-97}$$

gives the total number of bound modes in a graded-index fiber having a refractive-index profile given by Eq. (2-78).

2.7 FIBER MATERIALS

In selecting materials for optical fibers, a number of requirements must be satisfied. For example:

1. It must be possible to make long, thin, flexible fibers from the material.
2. The material must be transparent at a particular optical wavelength in order for the fiber to guide light efficiently.
3. Physically compatible materials having slightly different refractive indices for the core and cladding must be available.

Materials satisfying these requirements are glasses and plastics.
The majority of fibers are made of glass consisting either of silica (SiO_2) or a silicate. The variety of available glass fibers ranges from high-loss glass fibers with large cores used for short-transmission distances to very transparent (low-loss) fibers employed in long-haul applications. Plastic fibers are less widely used because of their substantially higher attenuation than glass fibers. The main use of plastic fibers is in short-distance applications and in abusive environments, where the greater mechanical strength of plastic fibers offers an advantage over the use of glass fibers.

2.7.1   Glass Fibers

Glass is made by fusing mixtures of metal oxides, sulfides, or selenides.[49-54] The resulting material is a randomly connected molecular network rather than a well-defined ordered structure as found in crystalline materials. A consequence of this random order is that glasses do not have well-defined melting points. When glass is heated up from room temperature, it remains a hard solid up to several hundred degrees centigrade. As the temperature increases further, the glass gradually begins to soften until at very high temperatures it becomes a viscous liquid. The expression "melting temperature" is commonly used in glass manufacture. This term refers only to an extended temperature range in which the glass becomes fluid enough to free itself fairly quickly of gas bubbles.

The largest category of optically transparent glasses from which optical fibers are made consists of the oxide glasses. Of these the most common is silica (SiO_2), which has a refractive index of 1.458 at 850 nm. To produce two similar materials having slightly different indices of refraction for the core and cladding, either fluorine or various oxides (referred to as *dopants*) such as B_2O_3, GeO_2, or P_2O_5 are added to the silica. As shown in Fig. 2-27 the addition of GeO_2 or P_2O_5 increases the refractive index whereas doping the silica with fluorine or B_2O_3 decreases it. Since the cladding must have a lower index than the core, examples of fiber compositions are:

1. GeO_2-SiO_2 core; SiO_2 cladding
2. P_2O_5-SiO_2 core; SiO_2 cladding
3. SiO_2 core; B_2O_3-SiO_2 cladding
4. GeO_2-B_2O_3-SiO_2 core; B_2O_3-SiO_2 cladding

Here the notation GeO_2-SiO_2, for example, denotes a GeO_2-doped silica glass.

The principal raw material for silica is sand. Glass composed of pure silica is referred to either as *silica glass, fused silica,* or *vitreous silica.* Some of its desirable properties are a resistance to deformation at temperatures as high as

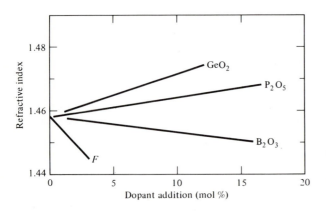

FIGURE 2-27
Variation in refractive index as a function of doping concentration in silica glass.

1000°C, a high resistance to breakage from thermal shock because of its low thermal expansion, good chemical durability, and high transparency in both the visible and infrared regions of interest to fiber optic communication systems. Its high melting temperature is a disadvantage if the glass is prepared from a molten state. However, this problem is partially avoided when using vapor deposition techniques.

2.7.2 Halide Glass Fibers

In 1975 researchers at the Université de Rennes[55] discovered fluoride glasses having extremely low transmission losses at mid-infrared wavelengths (0.2 to 8 μm, with the lowest loss being around 2.55 μm). Fluoride glasses belong to a general family of halide glasses in which the anions are from elements in group VII of the periodic table, namely fluorine, chlorine, bromine, and iodine.

The material that researchers have concentrated on is a *heavy metal fluoride glass*, which uses ZrF_4 as the major component and glass network former. Several other constituents need to be added to make a glass that has moderate resistance to crystallization.[56-61] Table 2-3 lists the constituents and their molecular percentages of a particular fluoride glass referred to as ZBLAN (after its elements ZrF_4, BaF_2, LaF_3, AlF_3, and NaF). This material forms the core of a glass fiber. To make a lower-refractive-index glass, one partially replaces ZrF_4 by HaF_4 to get a ZHBLAN cladding.

Although these glasses potentially offer intrinsic minimum losses of 0.01 to 0.001 dB/km, fabricating long lengths of these fibers is difficult. First, ultrapure materials must be used to reach this low loss level. Secondly, fluoride glass is prone to devitrification. Fiber-making techniques have to take this into account to avoid the formation of microcrystallites, which have a drastic effect on scattering losses.

2.7.3 Active Glass Fibers

Incorporating rare-earth elements (atomic numbers 57 to 71) into a normally passive glass gives the resulting material new optical and magnetic properties. These new properties allow the material to perform amplification, attenuation,

TABLE 2-3
Molecular composition of a ZBLAN fluoride glass

Material	Molecular percentage
ZrF_4	54
BaF_2	20
LaF_3	4.5
AlF_3	3.5
NaF	18

and phase retardation on the light passing through it.[62-64] Doping can be done both for silica and for halide glasses.

Two commonly used materials for fiber lasers are erbium and neodymium. The ionic concentrations of the rare-earth elements are low (on the order of 0.005 to 0.05 mole percent) to avoid clustering effects. By examining the absorption and fluorescence spectra of these materials, one can use an optical source which emits at an absorption wavelength to excite electrons to higher energy levels in the rare-earth dopants. When these excited electrons drop to lower energy levels, they emit light in a narrow optical spectrum at the fluorescence wavelength.

2.7.4   Plastic-Clad Glass Fibers

Optical fibers constructed with glass cores and glass claddings are very important for long-distance applications where the very low losses achievable in these fibers are needed. For short-distance applications (up to several hundred meters), where higher losses are tolerable, the less expensive plastic-clad silica fibers can be used. These fibers are composed of silica cores with the lower-refractive-index cladding being a polymer (plastic) material. These fibers are often referred to as PCS (plastic-clad silica) fibers.

A common material source for the silica core is selected high-purity natural quartz. A common cladding material is a silicone resin having a refractive index of 1.405 at 850 nm. Silicone resin is also frequently used as a protective coating for other types of fibers. Another popular plastic cladding material[65] is perfluoronated ethylene propylene (Teflon FEP). The low refractive index, 1.338, of this material results in fibers with potentially large numerical apertures.

Plastic claddings are only used for step-index fibers. The core diameters are larger (150 to 600 μm) than the standard 50-μm-diameter core of all-glass graded-index fibers, and the larger difference in the core and cladding indices results in a high numerical aperture. This allows low-cost large-area light sources to be used for coupling optical power into these fibers, thereby yielding comparatively inexpensive but lower-quality systems which are quite satisfactory for many applications.

2.7.5   Plastic Fibers

All-plastic multimode step-index fibers are good candidates for fairly short (up to about 100 m) and low-cost links. Although they exhibit considerably greater optical signal attenuations than glass fibers, the toughness and durability of plastic allow these fibers to be handled without special care. The high refractive-index differences that can be achieved between the core and cladding materials yield numerical apertures as high as 0.6 and large acceptance angles of up to 70°. In addition, the mechanical flexibility of plastic allows these fibers to have large cores, with typical diameters ranging from 110 to 1400 μm. These factors permit the use of inexpensive large-area light-emitting diodes which, in conjunction with the less expensive plastic fibers, make an economically attractive system.

Examples of plastic fiber constructions are:

1. A polysterene core (n_1 = 1.60) and a methyl methacrylate cladding (n_2 = 1.49) to give an NA of 0.60
2. A polymethyl methacrylate core (n_1 = 1.49) and a cladding made of its copolymer (n_2 = 1.40) to give an NA of 0.50

2.8 FIBER FABRICATION

Two basic techniques[66-68] are used in the fabrication of all-glass optical waveguides. These are the vapor phase oxidation processes and the direct-melt methods. The direct-melt method follows traditional glass-making procedures in that optical fibers are made directly from the molten state of purified components of silicate glasses. In the vapor phase oxidation process, highly pure vapors of metal halides (e.g., $SiCl_4$ and $GeCl_4$) react with oxygen to form a white powder of SiO_2 particles. The particles are then collected on the surface of a bulk glass by one of four different commonly used processes and are sintered (transformed to a homogeneous glass mass by heating without melting) by one of a variety of techniques to form a clear glass rod or tube (depending on the process). This rod or tube is called a *preform*. It is typically around 10 to 25 mm in diameter and 60 to 120 cm long. Fibers are made from the preform[69-72] by using the equipment shown in Fig. 2-28. The preform is precision-fed into a

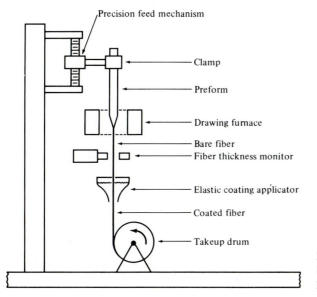

Precision feed mechanism

Clamp

Preform

Drawing furnace

Bare fiber
Fiber thickness monitor

Elastic coating applicator

Coated fiber

Takeup drum

FIGURE 2-28
Schematic of a fiber-drawing apparatus.

circular heater called the *drawing furnace*. Here the preform end is softened to the point where it can be drawn into a very thin filament, which becomes the optical fiber. The turning speed of the takeup drum at the bottom of the draw tower determines how fast the fiber is drawn. This, in turn, will determine the thickness of the fiber, so that a precise rotation rate must be maintained. An optical fiber thickness monitor is used in a feedback loop for this speed regulation. To protect the bare glass fiber from external contaminants such as dust and water vapor, an elastic coating is applied to the fiber immediately after it is drawn.

2.8.1 Outside Vapor Phase Oxidation

The first fiber to have a loss of less than 20 dB/km was made at the Corning Glass Works[73-75] by the *outside vapor phase oxidation* (OVPO) process. This method is illustrated in Fig. 2-29. First, a layer of SiO_2 particles called a *soot* is deposited from a burner onto a rotatying graphite or ceramic mandrel. The glass soot adheres to this bait rod and, layer by layer, a cylindrical, porous glass

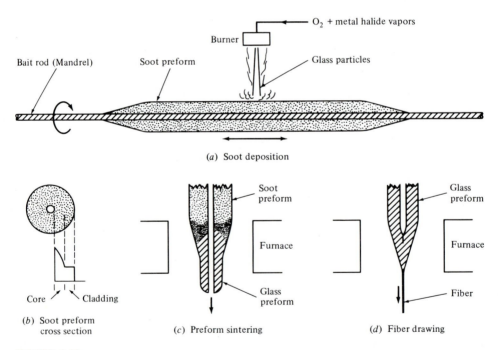

(a) Soot deposition

(b) Soot preform cross section

(c) Preform sintering

(d) Fiber drawing

FIGURE 2-29
Basic steps in preparing a preform by the OVPO process. (*a*) Bait rod rotates and moves back and forth under the burner to produce a uniform deposition of glass soot particles along the rod; (*b*) profiles can be step or graded index; (*c*) following deposition, the soot preform is sintered into a clear glass preform; (*d*) fiber is drawn from the glass preform. (Reproduced with permission from Schultz.[66])

preform is built up. By properly controlling the constituents of the metal halide vapor stream during the deposition process, the glass compositions and dimensions desired for the core and cladding can be incorporated into the preform. Either step- or graded-index preforms can thus be made.

When the deposition process is completed, the mandrel is removed and the porous tube is then vitrified in a dry atmosphere at a high temperature (above 1400°C) to a clear glass preform. This clear preform is subsequently mounted in a fiber-drawing tower and made into a fiber, as shown in Fig. 2-28. The central hole in the tube preform collapses during this drawing process.

2.8.2   Vapor Phase Axial Deposition

The OVPO process described in Sec. 2.8.1 is a lateral deposition method. Another OVPO type process is the *vapor phase axial deposition* method[76, 77] (VAD), illustrated in Fig. 2-30. In this method the SiO_2 particles are formed in the same way as described in the OVPO process. As these particles emerge from the torches, they are deposited onto the end surface of a silica glass rod which acts as a seed. A porous preform is grown in the axial direction by moving

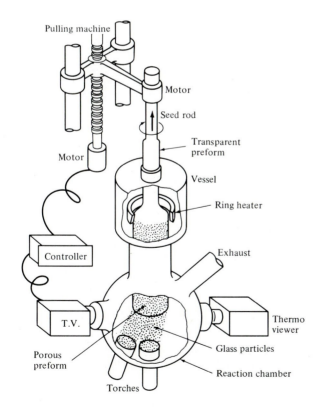

FIGURE 2-30
Apparatus for the VAD (vapor phase axial deposition) process. (Reproduced with permission from Izawa and Inagaki,[76] © 1980, IEEE.)

the rod upward. The rod is also continuously rotated to maintain cylindrical symmetry of the particle deposition. As the porous preform moves upward, it is transformed into a solid, transparent rod preform by zone melting (heating in a narrow localized zone) with the carbon ring heater shown in Fig. 2-30. The resultant preform can then be drawn into a fiber by heating it in another furnace, as shown in Fig. 2-28.

Both step- and graded-index fibers in either multimode or single-mode varieties can be made by the VAD method. The advantages of the VAD method are: (1) the preform has no central hole as occurs in the OVPO process; (2) the preform can be fabricated in continuous lengths which can affect process costs and product yields; and (3) the fact that the deposition chamber and the zone-melting ring heater are tightly connected to each other in the same enclosure allows the achievement of a clean environment.

2.8.3 Modified Chemical Vapor Deposition

The *modified chemical vapor deposition* (MCVD) process shown in Fig. 2-31 was pioneered at Bell Laboratories[53, 78] and widely adopted elsewhere to produce very low-loss graded-index fibers. The glass vapor particles arising from the reaction of the constituent metal halide gases and oxygen flow through the inside of a revolving silica tube. As the SiO_2 particles are deposited, they are sintered to a clear glass layer by an oxyhydrogen torch which travels back and forth along the tube. When the desired thickness of glass has been deposited, the vapor flow is shut off and the tube is heated strongly to cause it to collapse into a solid rod preform. The fiber that is subsequently drawn from this preform rod will have a core that consists of the vapor-deposited material and a cladding consisting of the original silica tube.

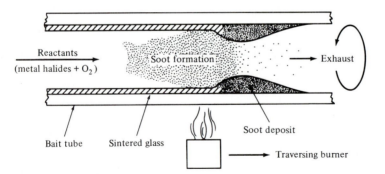

FIGURE 2-31
Schematic of MCVD (modified chemical vapor deposition) process. (Reproduced with permission from Schultz.[66])

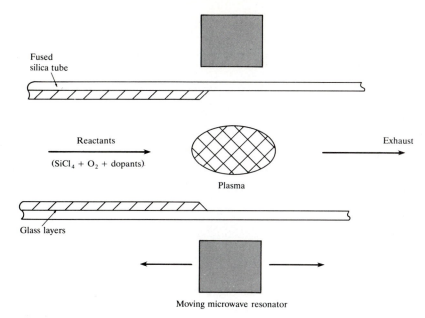

FIGURE 2-32
Schematic of PCVD (plasma-activated chemical vapor deposition) process.

2.8.4 Plasma-Activated Chemical Vapor Deposition

Scientists at Philips Research invented the *plasma-activated chemical vapor deposition process* (PCVD).[79-81] As shown in Fig. 2-32, the PCVD method is similar to the MCVD process in that deposition occurs within a silica tube. However, a nonisothermal microwave plasma operating at low pressure initiates the chemical reaction. With the silica tube held at temperatures in the range of 1000 to 1200°C to reduce mechanical stresses in the growing glass films, a moving microwave resonator operating at 2.45 GHz generates a plasma inside the tube to activate the chemical reaction. This process deposits clear glass material directly on the tube wall; there is no soot formation. Thus no sintering is required. When one has deposited the desired glass thickness, the tube is collapsed into a preform just as in the MCVD case.

2.8.5 Double-Crucible Method

Both silica and halide glass fibers can be made using a direct-melt *double-crucibel* technique.[8,57] In this method, glass rods for the core and cladding materials are first made separately by melting mixtures of purified powders to make the appropriate glass composition. These rods are then used as feedstock for each

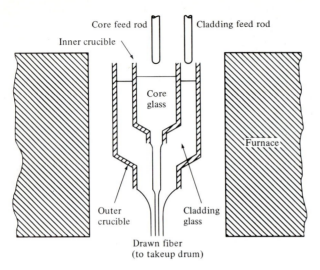

Core feed rod
Cladding feed rod
Inner crucible
Core glass
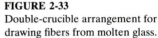
Furnace
Outer crucible
Cladding glass
Drawn fiber
(to takeup drum)

FIGURE 2-33
Double-crucible arrangement for drawing fibers from molten glass.

of two concentric crucibles, as shown in Fig. 2-33. The inner crucible contains molten core glass and the outer one contains the cladding glass. The fibers are drawn from the molten state through orifices in the bottom of the two concentric crucibles in a continuous production process.

Although this method has the advantage of being a continuous process, careful attention must be paid to avoid contaminants during the melting. The main sources of contamination arise from the furnace environment and from the crucible. Silica crucibles are usually used in preparing the glass feed rods, whereas the double concentric crucibles used in the drawing furnace are made from platinum. A detailed description of the crucible design and an analysis of the fiber-drawing process is given by Midwinter.[8]

2.9   MECHANICAL PROPERTIES OF FIBERS

In addition to the transmission properties of optical waveguides, their mechanical characteristics play a very important role when they are used as the transmission medium in optical communication systems.[82–85] Fibers must be able to withstand the stresses and strains that occur during the cabling process and the loads induced during the installation and service of the cable. During cable manufacture and installation the loads applied to the fiber can be either impulsive or gradually varying. Once the cable is in place, the service loads are usually slowly varying ones, which can arise from temperature variations or a general settling of the cable following installation.

Strength and *static fatigue* are the two basic mechanical characteristics of glass optical fibers. Since the sight and sound of shattering glass are quite familiar, one intuitively suspects that glass is not a very strong material. However, the longitudinal breaking stress of pristine glass fibers is comparable to that of metal wires. The cohesive bond strength of the constituent atoms of a

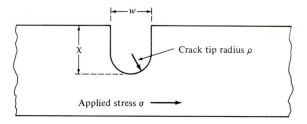

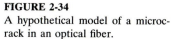

FIGURE 2-34
A hypothetical model of a microcrack in an optical fiber.

glass fiber governs its theoretical intrinsic strength. Maximum tensile strengths of 14 GPa (2×10^6 lb/in.^{2}) have been observed in short-gauge-length glass fibers. This is close to the 20-GPa tensile strength of steel wire. The difference between glass and metal is that, under an applied stress, glass will extend elastically up to its breaking strength, whereas metals can be stretched plastically well beyond their true elastic range. Copper wires, for example, can be elongated plastically by more than 20 percent before they fracture. For glass fibers elongations of only about 1 percent are possible before fracture occurs.

In practice the existence of stress concentrations at surface flaws or microcracks limits the median strength of long glass fibers to the 700- to 3500-MPa (1 to 5×10^5 lb/in.^{2}) range. The fracture strength of a given length of glass fiber is determined by the size and geometry of the severest flaw (the one that produces the largest stress concentration) in the fiber. A hypothetical, physical flaw model is shown in Fig. 2-34. This elliptically shaped crack is generally referred to as a *Griffith microcrack*.[86] It has a width w, a depth χ, and a tip radius ρ. The strength of the crack for silica fibers follows the relation

$$K = Y\chi^{1/2}\sigma \qquad (2\text{-}98)$$

where the stress intensity factor K is given in terms of the stress σ in megapascals applied to the fiber, the crack depth χ in millimeters, and a dimensionless constant Y that depends on flaw geometry. For surface flaws, which are the most critical in glass fibers, $Y = \sqrt{\pi}$. From this equation the maximum crack size allowable for a given applied stress level can be calculated. The maximum values of K depend upon the glass composition but tend to be in the range of 0.6 to 0.9 MN/m$^{3/2}$.

Since an optical fiber generally contains many flaws having a random distribution of size, the fracture strength of a fiber must be viewed statistically. If $F(\sigma, L)$ is defined as the cumulative probability that a fiber of length L will fail below a stress level σ, then, under the assumption that the flaws are independent and randomly distributed in the fiber and that the fracture will occur at the most severe flaw, we have

$$F(\sigma, L) = 1 - e^{-LN(\sigma)} \qquad (2\text{-}99)$$

where $N(\sigma)$ is the cumulative number of flaws per unit length with a strength less than σ. A widely used form for $N(\sigma)$ is the empirical expression proposed by Weibull[87]

$$N(\sigma) = \frac{1}{L_0} \left(\frac{\sigma}{\sigma_0} \right)^m \tag{2-100}$$

where m, σ_0, and L_0 are constants related to the initial inert strength distribution. This leads to the so-called *Weibull expression*

$$F(\sigma, L) = 1 - \exp\left[-\left(\frac{\sigma}{\sigma_0} \right)^m \frac{L}{L_0} \right] \tag{2-101}$$

A plot of a Weibull expression is shown in Fig. 2-35 for measurements performed on long-fiber samples.[85, 88] These data were obtained by testing to destruction a large number of fiber samples. The fact that a single curve can be drawn through the data indicates that the failures arise from a single type of flaw. Earlier works[89] showed a double-curve Weibull distribution with different slopes for short and long fibers. This is indicative of flaws arising from two

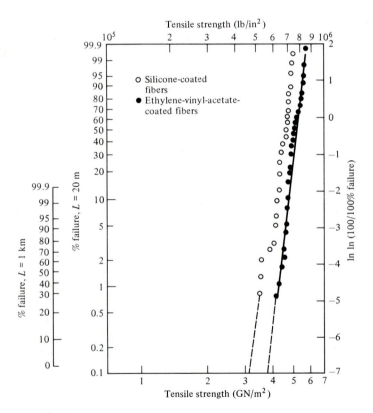

FIGURE 2-35
A Weibull type plot showing the cumulative probability that fibers of 20-m and 1-km lengths will fracture at the indicated applied stress. (Reproduced with permission from Miller, Hart, Vroom, and Bowden.[88])

sources, one from the fiber manufacturing process and the other from fundamental flaws occurring in the glass preform and the fiber. By careful environmental control of the fiber-drawing furnace, numerous 1-km lengths of silica fiber having a single failure distribution and a maximum strength of 3500 MPa have been fabricated.

In contrast to strength, which relates to instantaneous failure under an applied load, *static fatigue* relates to the slow growth of preexisting flaws in the glass fiber under humid conditions and tensile stress.[82, 85] This gradual flaw growth causes the fiber to fail at a lower stress level than that which could be reached under a strength test. A flaw such as the one shown in Fig. 2-34 propagates through the fiber because of chemical erosion of the fiber material at the flaw tip. The primary cause of this erosion is the presence of water in the environment, which reduces the strength of the SiO_2 bonds in the glass. The speed of the growth reaction is increased when the fiber is put under stress. However, based on experimental investigations, it is generally believed (but not yet fully substantiated) that static fatigue does not occur if the stress level is less than approximately 0.20 of the inert strength (in a dry environment, such as a vacuum). Certain fiber materials are more resistant to static fatigue than others, with fused silica being the most resistant of the glasses in water. In general, coatings which are applied to the fiber immediately during the manufacturing process afford a good degree of protection against environmental corrosion.[90]

Another important factor to consider is dynamic fatigue. When an optical cable is being installed in a duct, it experiences repeated stress owing to surging effects. The surging is caused by varying degrees of friction between the optical cable and the duct or guiding tool in a manhole on a curved route. Varying stresses also arise in aerial cables that are set into transverse vibration by the wind. Theoretical and experimental investigations[91] have shown that the time to failure under these conditions is related to the maximum allowable stress by the same lifetime parameters that are found from the cases of static stress and stress that increases at a constant rate.

A high assurance of fiber reliability can be provided by proof testing.[92-94] In this method, an optical fiber is subjected to a tensile load greater than that expected at any time during the cable manufacturing, installation, and service. Any fibers which do not meet the proof test are rejected. Empirical studies of slow crack growth show that the growth rate $d\chi/dt$ is approximately proportional to a power of the stress intensity factor, that is,

$$\frac{d\chi}{dt} = AK^b \tag{2-102}$$

Here A and b are material constants and the stress intensity factor is given by Eq. (2-98). For most glasses b ranges between 15 and 50.

If a proof test stress σ_p is applied for a time t_p, then from Eq. (2-102) we have

$$B\left(\sigma_i^{b-2} - \sigma_p^{b-2}\right) = \sigma_p^b t_p \tag{2-103}$$

where σ_i is the initial inert strength and

$$B = \frac{2}{b-2}\left(\frac{K}{Y}\right)^{2-b}\frac{1}{AY^b} \qquad (2\text{-}104)$$

When this fiber is subjected to a static stress σ_s after proof testing, the time to failure t_s is found from Eq. (2-102) to be

$$B\left(\sigma_p^{b-2} - \sigma_s^{b-2}\right) = \sigma_s^b t_s \qquad (2\text{-}105)$$

Combining Eqs. (2-103) and (2-105) yields

$$B\left(\sigma_i^{b-2} - \sigma_s^{b-2}\right) = \sigma_p^b t_p + \sigma_s^b t_s \qquad (2\text{-}106)$$

To find the failure probability F_s of a fiber after a time t_s after proof testing, we first define $N(t, \sigma)$ to be the number of flaws per unit length which will fail in a time t under an applied stress σ. Assuming that $N(\sigma_i) \gg N(\sigma_s)$, then

$$N(t_s, \sigma_s) \simeq N(\sigma_i) \qquad (2\text{-}107)$$

Solving Eq. (2-106) for σ_i and substituting into Eq. (2-100), we have, from Eq. (2-107),

$$N(t_s, \sigma_s) = \frac{1}{L_0}\left\{\frac{\left[\left(\sigma_p^b t_p + \sigma_s^b t_s\right)/B + \sigma_s^{b-2}\right]^{1/(b-2)}}{\sigma_0}\right\}^m \qquad (2\text{-}108)$$

The failure number $N(t_p, \sigma_p)$ per unit length during proof testing is found from Eq. (2-108) by setting $\sigma_s = \sigma_p$ and letting $t_s = 0$, so that

$$N(t_p, \sigma_p) = \frac{1}{L_0}\left[\frac{\left(\sigma_p^b t_p/B + \sigma_p^{b-2}\right)^{1/(b-2)}}{\sigma_0}\right]^m \qquad (2\text{-}109)$$

Letting $N(t_x, \sigma_x) = N_x$, the failure probability F_s for a fiber after it has been proof-tested is given by

$$F_s = 1 - e^{-L(N_s - N_p)} \qquad (2\text{-}110)$$

Substituting Eqs. (2-108) and (2-109) into Eq. (2-110), we have

$$F_s = 1 - \exp\left(-N_p L\left\{\left[\left(1 + \frac{\sigma_s^b t_s}{\sigma_p^b t_p}\right)\frac{1}{1+C}\right]^{m/(b-2)} - 1\right\}\right) \qquad (2\text{-}111)$$

where $C = B/(\sigma_p^2 t_p)$, and where we have ignored the term

$$\left(\frac{\sigma_s}{\sigma_p}\right)^b \frac{B}{\sigma_s^2 t_p} \ll 1 \qquad (2\text{-}112)$$

This holds because typical values of the parameters in this term are $\sigma_s/\sigma_p \simeq 0.3$ to 0.4, $t_p \simeq 10$ s, $b > 15$, $\sigma_p = 350$ MN/m^2, and $B \simeq 0.05$ to 0.5 (MN/m^2)$^2 \cdot$ s.

The expression for F_s given by Eq. (2-111) is valid only when the proof stress is unloaded immediately, which is not the case in actual proof testing of optical fibers. When the proof stress is released within a finite duration, the C value should be rewritten as

$$C = \gamma \frac{B}{\sigma_p^2 t_p} \qquad (2\text{-}113)$$

where γ is a coefficient of slow-crack-growth effect arising during the unloading period.

2.10 FIBER OPTIC CABLES

In any practical application of optical waveguide technology, the fibers need to be incorporated in some type of cable structure.[95-98] The cable structure will vary greatly, depending on whether the cable is to be pulled into underground or intrabuilding ducts, buried directly in the ground, installed on outdoor poles, or submerged under water. Different cable designs are required for each type of application, but certain fundamental cable design principles will apply in every case. The objectives of cable manufacturers have been that the optical fiber cables should be installable with the same equipment, installation techniques, and precautions as those used in conventional wire cables. This requires special cable designs because of the mechanical properties of glass fibers.

One important mechanical property is the maximum allowable axial load on the cable, since this factor determines the length of cable that can be reliably installed. In copper cables the wires themselves are generally the principal load-bearing members of the cable, and elongations of more than 20 percent are possible without fracture. On the other hand, extremely strong optical fibers tend to break at 4-percent elongation, whereas typical good-quality fibers exhibit long-length breaking elongations of about 0.5 to 1.0 percent. Since static fatigue occurs very quickly at stress levels above 40 percent of the permissible elongation and very slowly below 20 percent, fiber elongations during cable manufacture and installation should be limited to 0.1 to 0.2 percent.

Steel wire which has a Young's modulus of 2×10^4 MPa has been extensively used for reinforcing conventional electric cables and can also be employed for optical fiber cables. For some applications it is desirable to use nonmetallic constructions, either to avoid the effects of electromagnetic induction or to reduce cable weight. In this case plastic strength members and high-tensile-strength organic yarns such as Kevlar® (a product of the DuPont Chemical Corporation) are used. With good fabrication practices the optical fibers are isolated from other cable components, they are kept close to the neutral axis of the cable, and room is provided for the fibers to move when the cable is flexed or stretched.

Another factor to consider is fiber brittleness. Since glass fibers do not deform plastically, they have a low tolerance for absorbing energy from impact

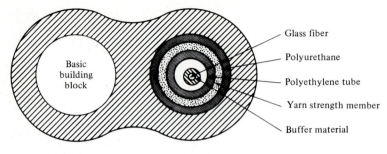

FIGURE 2-36
A hypothetical two-fiber cable design. The basic building block on the left is identical to that shown for the right-hand fiber.

loads. Hence, the outer sheath of an optical cable must be designed to protect the glass fibers inside from impact forces. In addition, the outer sheath should not crush when subjected to side forces, and it should provide protection from corrosive environmental elements. In underground installations, a heavy-gauge-metal outer sleeve may also be required to protect against potential damage from burrowing rodents, such as gophers.

In designing optical fiber cables, several types of fiber arrangements are possible and a large variety of components could be included in the construction. The simplest designs are one- or two-fiber cables intended for indoor use. In a hypothetical two-fiber design shown in Fig. 2-36, a fiber is first coated with a buffer material and placed loosely in a tough, oriented polymer tube, such as polyethylene. For strength purposes this tube is surrounded by strands of aramid yarn which, in turn, is encapsulated in a polyurethane jacket. A final

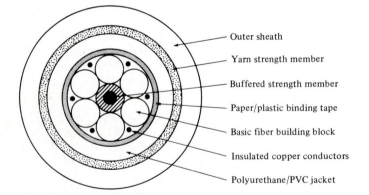

FIGURE 2-37
A typical six-fiber cable created by stranding six basic fiber-building blocks around a central strength member.

outer jacket of polyurethane, polyethylene, or nylon binds the two encapsulated fiber units together.

Larger cables can be created by stranding several basic fiber building blocks (as shown in Fig. 2-36) around a central strength member. This is illustrated in Fig. 2-37 for a six-fiber cable. The fiber units are bound onto the strength member with paper or plastic binding tape, and then surrounded by an outer jacket. If repeaters are required along the route where the cable is to be installed, it may be advantageous to include wires within the cable structure for powering these repeaters. The wires can also be used for fault isolation or as an engineering order wire for voice communications during cable installation.

2.11 SUMMARY

In this chapter we have examined the structure of optical fibers and have considered two mechanisms that show how light propagates along these fibers. In its simplest form an optical fiber is a coaxial cylindrical arrangement of two homogeneous dielectric (glass or plastic) materials. This fiber consists of a central core of refractive index n_1 surrounded by a cladding region of refractive index n_2 that is less than n_1. Ths configuration is referred to as a step-index fiber because the cross-sectional refractive-index profile has a step function at the interface between the core and the cladding.

Graded-index fibers are those in which the refractive-index profile varies as a function of the radial coordinate r in the core but is constant in the cladding. This index profile (r) is often represented as a power law

$$n(r) = \begin{cases} n_1\left[1 - 2\Delta\left(\dfrac{r}{a}\right)^{\alpha}\right]^{1/2} & \text{for} \quad r \le a \\ n_2 & \text{for} \quad r > a \end{cases}$$

where α defines the shape of the index profile, a is the fiber core radius, and Δ is the relative index difference between the maximum value n_1 on the fiber axis and the value n_2 in the cladding:

$$\Delta = \frac{n_1^2 - n_2^2}{2n_1^2} \simeq \frac{n_1 - n_2}{n_1}$$

A commonly used value of the power law exponent is $\alpha = 2$. This special case is referred to as a parabolic graded-index profile. As we shall show in Chap. 3, a graded-index profile reduces signal distortion and thus provides a wider band-width than a step-index fiber.

A general picture of light propagation in an optical fiber can be obtained by considering a ray-tracing (or geometrical optics) model in a slab waveguide. The slab consists of a central region of refractive index n_1 which is sandwiched between two material layers having a lower refractive index n_2. Light rays propagate along the slab waveguide by undergoing total internal reflection in accordance with Snell's law at the interfaces of these two materials.

Although the ray model is adequate for an intuitive picture of how light travels along a fiber, a more comprehensive description of light propagation, signal distortion, and power loss in a cylindrical optical fiber waveguide requires a wave theory approach. In this wave approach, electromagnetic fields (at optical frequencies) traveling in the fiber can be expressed as superpositions of elementary field configurations called the modes of the fiber. A mode of monochromatic light of radian frequency ω traveling in the axial (positive z) direction in a fiber can be described by the factor

$$e^{j(\omega t - \beta z)}$$

where β is the propagation constant of the mode. For guided (bound) modes, β can only assume a finite number of possible solutions. These are found by solving Maxwell's equations for a dielectric medium subject to the boundary conditions of the optical fiber. Here is where the analysis becomes rather complex. The boundary conditions at the core-cladding interface lead to a coupling between the longitudinal components of the **E** and **H** fields. This coupling leads to rather involved hybrid mode solutions. The fact that hybrid modes are present in optical fiber waveguides makes their analysis much more complex than in the case of hollow metallic waveguides, where only transverse electric and transverse magnetic modes are found.

Glasses and plastics are the principal materials of which fibers are made. Of these, silica-based glass is the most widely used material for the following reasons:

1. It is possible to make long, thin, flexible fibers from this material (and also from plastic).
2. Pure silica glass is highly transparent in the 800- to 1600-nm wavelength range. This is necessary in order for the fiber to guide light efficiently. Plastic fibers are less widely used because of their high attenuation compared to highly pure glass fibers.
3. By adding trace amounts of certain elements (known as dopants) to the silcia, physically compatible materials having slightly different refractive indices for the core and cladding are made available.

All-glass optical fibers are generally made by a vapor phase oxidation process. In this process, highly pure vapors of metal halides react with oxygen to form a white powder of SiO_2 particles. These particles are collected on the surface of a bulk glass by one of four processes, and are sintered to form a clear glass rod or tube (depending on the process). This rod or tube is called a preform. Once the preform is made, one end is softened to the point where it can be drawn into a very thin filament which becomes the optical fiber.

The mechanical characteristics of optical fibers must also be carefully considered when fibers are used as the transmission medium in optical communication systems. The optical fibers must be able to withstand the stresses and strains that occur during the cabling process and during the installation and service life of the cable. Strength and static fatigue are the two basic mechanical characteristics of glass optical fibers. Strength relates to the instantaneous failure of a fiber under an applied load, and static fatigue relates to the slow growth of preexisting flaws in the glass fiber under humid conditions and tensile stress.

Since an optical fiber generally contains many flaws having a random distribution of sizes, the fracture strength must be viewed statistically. If $F(\sigma, L)$ is the cumulative probability that a fiber of length L will fail below a stress level σ, then a useful and widely used expression for $F(\sigma, L)$ is the empirical Weibull formula

$$F(\sigma, L) = 1 - \exp\left[-\left(\frac{\sigma}{\sigma_0}\right)^m \frac{L}{L_0}\right]$$

where m, σ_0, and L_0 are constants related to the initial, inert strength distribution.

A high assurance of fiber reliability can be provided by proof testing. In this procedure an optical fiber is subjected for a short time to a tensile load greater than that expected at any time during the cable manufacturing, installation, and service. Any fibers that do not meet the proof test are rejected. Given a knowledge of the initial flaw strength and population, fracture mechanics theory can be used to predict the optical fiber cable failure rate (or equivalently, the cable lifetime) in field conditions where a static stress is expected to be imposed on the cable. If the initial failure distribution has a Weibull form, then the stress corrosion failure distribution in a field environment also has a Weibull form.

Special cable designs are required because of the mechanical properties of glass. These designs can vary greatly depending on where and how the cable is to be used. In general, however, the objectives of cable manufacturers have been to make the optical fiber cables in such a way that they should be installable with the same equipment, installation techniques, and precautions as those used in conventional wire cables.

PROBLEMS

2-1. A typical sheet of paper is 0.003 in. thick. How many wavelengths of 820-nm light (which is widely used in optical fiber systems) will fit into this distance? How does this compare to a 50-μm-diameter optical fiber?

2-2. A wave is specified by $y = 8 \cos 2\pi(2t - 0.8z)$, where y is expressed in micrometers and the propagation constant is given in μm^{-1}. Find (a) the amplitude, (b) the wavelength, (c) the angular frequency, and (d) the displacement at time $t = 0$ and $z = 4$ μm.

2-3. What are the energies in electron volts (eV) of light having wavelengths of 820 nm and 1.3 μm? What are the values of the propagation constant k of these two wavelengths?

2-4. Consider the following two waves X_1 and X_2 having the same frequency ω but different amplitudes a_i and phases δ_i:

$$X_1 = a_1 \cos(\omega t - \delta_1)$$

$$X_2 = a_2 \cos(\omega t - \delta_2)$$

According to the principle of superposition, the resultant wave X is simply the sum of X_1 and X_2. Show that X can be written in the form

$$X = A \cos(\omega t - \phi)$$

where

$$A^2 = a_1^2 + a_2^2 + 2a_1 a_2 \cos(\delta_1 - \delta_2)$$

and

$$\tan \phi = \frac{a_1 \sin \delta_1 + a_2 \sin \delta_2}{a_1 \cos \delta_1 + a_2 \cos \delta_2}$$

2-5. Elliptically polarized light can be represented by the two orthogonal waves given by Eqs. (2-2) and (2-3). Show that elimination of the $(\omega t - kz)$ dependence between them yields

$$\left(\frac{E_x}{E_{0x}}\right)^2 + \left(\frac{E_y}{E_{0y}}\right)^2 - 2\frac{E_x}{E_{0x}} \frac{E_y}{E_{0y}} \cos \delta = \sin^2 \delta$$

which is the equation of an ellipse making an angle α with the x axis, where α is given by Eq. (2-8).

2-6. Light traveling in air strikes a glass plate at an angle $\theta_1 = 33°$, where θ_1 is measured between the incoming ray and the glass surface. Upon striking the glass, part of the beam is reflected and part is refracted. If the refracted and reflected beams make an angle of $90°$ with each other, what is the refractive index of the glass? What is the critical angle for this glass?

2-7. A point source of light is 12 cm below the surface of a large body of water ($n = 1.33$). What is the radius of the largest circle on the water surface through which the light can emerge?

2-8. A $45°$-$45°$-$90°$ prism is immersed in alcohol ($n = 1.45$). What is the minimum index of refraction the prism must have if a ray incident normally on one of the short faces is to be totally reflected at the long face of the prism?

2-9. Calculate the numerical aperture of a step-index fiber having $n_1 = 1.48$ and $n_2 = 1.46$. What is the maximum entrance angle $\theta_{0, \max}$ for this fiber if the outer medium is air with $n = 1$?

2-10. Derive the approximation on the right-hand side of Eq. (2-23).

2-11. (*a*) Verify the expressions for the various phase changes that are used to derive Eq. (2-26).

(*b*) Show that, for the ray that propagates at the critical angle in Fig. 2-13, the integer M in Eq. (2-26) satisfies the condition

$$M = \frac{2n_1^2 d}{\lambda \left(n_1^2 - n_2^2\right)^{1/2}}$$

2-12. Assume the fields of an electromagnetic wave are of the form

$$\mathbf{E}(\mathbf{x}, t) = \mathbf{e}_1 E_0 \exp[j(\omega t - \mathbf{k} \cdot \mathbf{x})]$$

$$\mathbf{B}(\mathbf{x}, t) = \mathbf{e}_2 B_0 \exp[j(\omega t - \mathbf{k} \cdot \mathbf{x})]$$

where $\mathbf{e}_1$ and $\mathbf{e}_2$ are two constant unit vectors with an unspecified orientation.

(a) Using Maxwell's divergence equations, show that $\mathbf{E}$ and $\mathbf{B}$ are both perpendicular to the direction of propagation $\mathbf{k}$. (Such a wave is called a transverse wave.)

(b) From Maxwell's curl equations show that

$$\mathbf{e}_2 = \frac{\mathbf{k} \times \mathbf{e}_1}{k} = \frac{\mathbf{k} \times \mathbf{e}_1}{\omega B_0/E_0}$$

which shows that $(\mathbf{e}_1, \mathbf{e}_2, \mathbf{k})$ form a set of orthogonal vectors and that $\mathbf{E}$ and $\mathbf{B}$ are in phase.

2-13. Show that Maxwell's equations are satisfied by the solution

$$E_x = A \sin(\omega t + kz) \qquad E_y = 0 \qquad E_z = 0$$

$$H_x = 0 \qquad H_y = -A\sqrt{\frac{\epsilon}{\mu}} \sin(\omega t + kz) \qquad H_z = 0$$

In which plane is the wave polarized and in which direction does it travel?

2-14. Derive Eqs. (2-35a) to (2-35d) from Eqs. (2-33) and (2-34). Show that they lead to Eqs. (2-36) and (2-37).

2-15. Evaluate the determinant given by Eq. (2-53) and show that it yields Eq. (2-54).

2-16. Show that for $\nu = 0$ Eq. (2-55b) corresponds to TE_{0m} modes ($E_z = 0$) and that Eq. (2-56b) corresponds to TM_{0m} modes ($H_z = 0$).

2-17. Verify that $k_2^2 \approx k_1^2 \approx \beta^2$ when $\Delta \ll 1$, where k_1 and k_2 are the core and cladding propagation constants, respectively, as defined in Eq. (2-46).

2-18. (a) Using the recurrence relations for J_ν' and K_ν' given in App. C, show that Eqs. (2-63) and (2-64a) result for the positive and negative signs, respectively, in Eq. (2-62).

(b) Show that Eq. (2-64a) can be rewritten as Eq. (2-64b).

2-19. Replot Fig. 2-16 for $J_0(x)$ and $J_1(x)$. On the resulting graph, indicate the range of values of x for the lower-order LP_{1m} modes and the exact lower-order HE_{1m}, TE_{0m}, and TM_{0m} modes.

2-20. Determine the normalized frequency at 0.82 μm for a step-index fiber having a 25-μm core radius, $n_1 = 1.48$, and $n_2 = 1.46$. How many modes propagate in this fiber at 0.82 μm? How many modes propagate at a wavelength of 1.3 μm? What percentage of the optical power flows in the cladding in each case?

2-21. Find the core radius necessary for single-mode operation at 820 nm of a step-index fiber with $n_1 = 1.480$ and $n_2 = 1.478$. What is the numerical aperture and maximum acceptance angle of this fiber?

2-22. A manufacturer wishes to make a silica-core, step-index fiber with $V = 75$ and a numerical aperture NA $= 0.30$ to be used at 820 nm. If $n_1 = 1.458$, what should the core size and cladding index be?

2-23. Draw a design curve of the fractional refractive index difference Δ versus the core radius a for a silica-core ($n_1 = 1.458$), single-mode fiber to operate at 1300 nm. Suppose the fiber we select from this curve has a 5-μm core radius. Is this fiber still single-mode at 820 nm? Which modes exist in the fiber at 820 nm?

2-24. Using the following approximation for W_0 due to Marcuse[99]

$$W_0 = a(0.65 + 1.619V^{-3/2} + 2.879V^{-6})$$

evaluate and plot $E(r)/E_0$ with r ranging from 0 to 3 for values of $V = 1.0$, 1.4, 1.8, 2.2, 2.6, and 3.0.

2-25. Commonly available single-mode fibers have beat lengths in the range 10 cm < L_p < 2 m. What range of refractive index differences does this correspond to for $\lambda = 1300$ nm?

2-26. Plot the refractive-index profiles from n_1 to n_2 as a function of radial distance $r \leq a$ for graded-index fibers that have α values of 1, 2, 4, 8, and ∞ (step index). Assume the fibers have a 25-μm core radius, $n_1 = 1.48$, and $\Delta = 0.01$.

2-27. Verify the steps leading from Eq. (2-82) to Eq. (2-86).

2-28. Show that Eq. (2-94) results from evaluating the integral over v in Eq. (2-93).

2-29. Calculate the number of modes at 820 nm and 1.3 μm in a graded-index fiber having a parabolic-index profile ($\alpha = 2$), a 25-μm core radius, $n_1 = 1.48$, and $n_2 = 1.46$. How does this compare to a step-index fiber?

2-30. Calculate the numerical apertures of: (*a*) a plastic step-index fiber having a core refractive index of $n_1 = 1.60$ and a cladding index $n_2 = 1.49$; (*b*) a step-index fiber having a silica core ($n_1 = 1.458$) and a silicone resin cladding ($n_2 = 1.405$).

2-31. When a preform is drawn into a fiber, the principle of conservation of mass must be satisfied under steady-state drawing conditions. Show that for a solid rod preform this is represented by the expression

$$s = S\left(\frac{D}{d}\right)^2$$

where D and d are the preform and fiber diameters, and S and s are the preform feed and fiber-draw speeds, respectively. A typical drawing speed is 1.2 m/s for a 125-μm outer-diameter fiber. What is the preform feed rate in cm/min for a 9-mm-diameter preform?

2-32. A silica tube with inside and outside radii of 3 and 4 mm, respectively, is to have a certain thickness of glass deposited on the inner surface. What should the thickness of this glass deposition be if a fiber having a core diameter of 50 μm and an outer cladding diameter of 125 μm is to be drawn from this preform?

2-33. (*a*) The density of fused silica is 2.6 g/cm^3. How many grams are needed for a 1-km-long 50-μm-diameter fiber core?

   (*b*) If the core material is to be deposited inside of a glass tube at a 0.5-g/min deposition rate, how long does it take to make the preform for this fiber?

2-34. During fabrication of optical fibers, dust particles incorporated into the fiber surface are prime examples of surface flaws which can lead to reduced fiber strength. What size dust particles are tolerable if a glass fiber having a 20-N/mm$^{3/2}$ stress intensity factor is to withstand a 700-MN/m^2 stress?

2-35. Static fatigue in a glass fiber refers to the condition where a fiber is stressed to a level σ_a, which is much less than the fracture stress associated with the weakest flaw. Initially the fiber will not fail but, with time, cracks in the fiber will grow as a result of chemical erosion at the crack tip. One model for the growth rate of a crack of depth χ assumes a relation of the form given in Eq. (2-102).

(a) Using this equation, show that the time required for a crack of initial depth χ_i to grow to its failure size χ_f is given by

$$t = \frac{2}{(b-2)A(Y\sigma)^b}\left(\chi_i^{(2-b)/2} - \chi_f^{(2-b)/2}\right)$$

(b) For long, static fatigue times (on the order of 20 yr), $K_i^{2-b} \ll K_f^{2-b}$ for large values of b. Show that under this condition the failure time is

$$t = \frac{2K_i^{2-b}}{(b-2)A\sigma^2Y^2}$$

2-36. Derive Eq. (2-106) by starting with Eq. (2-102).

2-37. Derive Eq. (2-111) by using the expressions given in Eqs. (2-108) and (2-109) for the number of flaws per unit length failing in a time t. Verify the relationship given in Eq. (2-112).

2-38. Consider two similar fiber samples of lengths L_1 and L_2 subjected to stress levels of σ_1 and σ_2, respectively. If σ_{1c} and σ_{2c} are the corresponding fast-fracture stress levels for equal failure probability, show that

$$\frac{\sigma_{1c}}{\sigma_{2c}} = \left(\frac{L_2}{L_1}\right)^{1/m}$$

From Fig. 2-35 estimate the value of m for a 10-percent failure probability of these particular ethylene-vinyl-acetate-coated fibers.

REFERENCES

1. See any general physics book or introductory optics book; for example:
   (a) M. V. Klein and T. E. Furtak, *Optics*, Wiley, New York, 2nd ed., 1986.
   (b) E. Hecht and A. Zajac, *Optics*, Addison-Wesley, Reading, Mass., 2nd ed., 1987.
   (c) F. A. Jenkins and H. E. White, *Fundamentals of Optics*, McGraw-Hill, New York, 4th ed., 1976.
2. See any introductory electromagnetics book; for example:
   (a) W. H. Hayt, Jr., *Engineering Electromagnetics*, McGraw-Hill, New York, 5th ed., 1989.
   (b) J. D. Kraus, *Electromagnetics*, McGraw-Hill, New York, 3rd ed., 1984.
   (c) C. R. Paul and S. R. Nasar, *Introduction to Electromagnetic Fields*, McGraw-Hill, New York, 2nd ed., 1987.
3. E. A. J. Marcatili, "Objectives of early fibers: Evolution of fiber types," in S. E. Miller and A. G. Chynoweth, eds., *Optical Fiber Telecommunications*, Academic, New York, 1979.
4. (a) S. J. Maurer and L. B. Felsen, "Ray-optical techniques for guided waves," *Proc. IEEE*, vol. 55, pp. 1718–1729, Oct. 1967.
   (b) L. B. Felsen, "Rays and modes in optical fibers," *Electron. Lett.*, vol. 10, pp. 95–96, Apr. 1974.
5. A. W. Snyder and D. J. Mitchell, "Leaky rays on circular optical fibers," *J. Opt. Soc. Amer.*, vol. 64, pp. 599–607, May 1974.

6. A. W. Snyder and J. D. Love, *Optical Waveguide Theory*, Routledge Chapman and Hall, New York, 1984.

7. D. Marcuse, *Theory of Dielectric Optical Waveguides*, Academic, New York, 1974.

8. J. Midwinter, *Optical Fibers for Transmission*, Wiley, New York, 1979.

9. R. L. Gallawa, *A User's Manual for Optical Waveguide Communications*, publication no. OTR76-83, U.S. Dept. of Commerce, Washington, DC, 1976.

10. H. G. Unger, *Planar Optical Waveguides and Fibers*, Clarendon, Oxford, 1977.

11. D. Keck, "Optical fiber waveguides," in M. K. Barnoski, ed., *Fundamentals of Optical Fiber Communications*, Academic, New York, 2nd ed., 1982.

12. R. Olshansky, "Leaky modes in graded index optical fibers," *Appl. Opt.*, vol. 15, pp. 2773–2777, Nov. 1976.

13. A. Tomita and L. G. Cohen, "Leaky-mode loss of the second propagation mode in single-mode fibers with index well profiles," *Appl. Opt.*, vol. 24, pp. 1704–1707, 1985.

14. E. Snitzer, "Cylindrical dielectric waveguide modes," *J. Opt. Soc. Amer.*, vol. 51, pp. 491–498, May 1961.

15. N. S. Kapany and J. J. Burke, *Optical Waveguides*, Academic, New York, 1972.

16. D. Marcuse, *Light Transmission Optics*, Van Nostrand-Reinhold, New York, 2nd ed., 1982.

17. R. Olshansky, "Propagation in glass optical waveguides," *Rev. Mod. Phys.*, vol. 51, pp. 341–367, Apr. 1979.

18. D. Gloge, "The optical fiber as a transmission medium," *Rep. Progr. Phys.*, vol. 42, pp. 1777–1824, Nov. 1979.

19. A. W. Snyder, "Asymptotic expressions for eigenfunctions and eigenvalues of a dielectric or optical waveguide," *IEEE Trans. Microwave Theory Tech.*, vol. MTT-17, pp. 1130–1138, Dec. 1969.

20. D. Gloge, "Weakly guiding fibers," *Appl. Opt.*, vol. 10, pp. 2252–2258, Oct. 1971.

21. D. Marcuse, "Gaussian approximation of the fundamental modes of graded index fibers," *J. Opt. Soc. Amer.*, vol. 68, pp. 103–109, Jan. 1978.

22. H. M. DeRuiter, "Integral equation approach to the computation of modes in an optical waveguide," *J. Opt. Soc. Amer.*, vol. 70, pp. 1519–1524, Dec. 1980.

23. A. W. Snyder, "Understanding monomode optical fibers," *Proc. IEEE*, vol. 69, pp. 6–13, Jan. 1981.

24. M. Abramowitz and I. A. Stegun, *Handbook of Mathematical Functions*, Dover, New York, 1965.

25. J. Mathews and R. L. Walker, *Mathematical Methods of Physics*, Benjamin, New York, 2nd ed., 1970.

26. W. H. Beyer, *Standard Mathematical Tables*, CRC Press, Boca Raton, FL, 28th ed., 1987.

27. D. Marcuse, D. Gloge, and E. A. J. Marcatili, "Guiding properties of fibers," in S. E. Miller and A. G. Chynoweth, eds., *Optical Fiber Telecommunications*, Academic, New York, 1979.

28. R. M. Gagliardi and S. Karp, *Optical Communications*, Wiley, New York, 1976, chap. 3.

29. D. Gloge, "Propagation effects in optical fibers," *IEEE Trans. Microwave Theory Tech.*, vol. MTT-23, pp. 106–120, Jan. 1975.

30. M. Artiglia, G. Coppa, P. DiVita, M. Potenza, and A. Sharma, "Mode field diameter measurements in single-mode optical fibers," *J. Lightwave Tech.*, vol. 7, pp. 1139–1152, Aug. 1989.

31. T. J. Drapela, D. L. Franzen, A. H. Cherin, and R. J. Smith, "A comparison of far-field methods for determining mode field diameter of single-mode fibers using both gaussian and Petermann definitions," *J. Lightwave Tech.*, vol. 7, pp. 1153–1157, Aug. 1989.

32. K. Petermann, "Constraints for fundamental mode spot size for broadband dispersion-compensated single-mode fibers," *Electron. Lett.*, vol. 19, pp. 712–714, Sept. 1983.

33. M. Ohashi, K.-I. Kitayama, and S. Seikai, "Mode field diameter measurement conditions for fibers by transmitted field pattern methods," *J. Lightwave Tech.*, vol. LT-4, pp. 109–115, Feb. 1986.

34. L. B. Jeunhomme, *Single-Mode Fiber Optics*, Dekker, New York, 2nd ed., 1989.

35. E. G. Neumann, *Single-Mode Fibers I—Fundamentals*, Springer-Verlag, New York, 1988.

36. V. Shah and L. Curtis, "Mode coupling effects of the cutoff wavelength characteristics of dispersion-shifted and dispersion-unshifted single-mode fibers," *J. Lightwave Tech.*, vol. 7, pp. 1181–1186, Aug. 1989.

37. A. R. Tynes, R. M. Derosier, and W. G. French, "Low *V*-number optical fibers: Secondary maxima in the far-field radiation pattern," *J. Opt. Soc. Amer.*, vol. 69, pp. 1587–1596, Nov. 1979.

38. D. L. Franzen and R. Srivastava, "Determining the mode-field diameter of single-mode optical fiber: An interlaboratory comparison," *J. Lightwave Tech.*, vol. LT-3, pp. 1073–1077, Oct. 1985.

39. W. T. Anderson, V. Shah, L. Curtis, A. J. Johnson, and J. P. Kilmer, "Mode-field diameter measurements for single-mode fibers with non-gaussian field profiles," *J. Lightwave Tech.*, vol. LT-5, pp. 211–217, Feb. 1987.

40. W. T. Anderson and D. L. Philen, "Spot-size measurements for single-mode fibers—A comparison of four techniques," *J. Lightwave Tech.*, vol. LT-1, pp. 20–26, Mar. 1983.

41. (*a*) I. P. Kaminow, "Polarization in optical fibers," *IEEE J. Quantum Electron.*, vol. QE-17, pp. 15–22, Jan. 1981.
    (*b*) I. P. Kaminow, "Polarization maintaining fibers," *Appl. Scientific Research*, vol. 41, pp. 257–270, 1984.

42. S. C. Rashleigh, "Origins and control of polarization effects in single-mode fibers," *J. Lightwave Tech.*, vol. LT-1, pp. 312–331, June 1983.

43. (*a*) X.-H. Zheng, W. M. Henry, and A. W. Snyder, "Polarization characteristics of the fundamental mode of optical fibers," *J. Lightwave Tech.*, vol. LT-6, pp. 1300–1305, Aug. 1988.
    (*b*) S.-Y. Huang, J. N. Blake, and B. Y. Kim, "Perturbation effects on mode propagation in highly elliptical core two-mode fibers," *J. Lightwave Tech.*, vol. 8, pp. 23–33, Jan. 1990.
    (*c*) S. J. Garth and C. Pask, "Polarization rotation in nonlinear bimodal optical fibers," *J. Lightwave Tech.*, vol. 8, pp. 129–137, Feb. 1990.

44. D. Gloge and E. Marcatili, "Multimode theory of graded core fibers," *Bell Sys. Tech. J.*, vol. 52, pp. 1563–1578, Nov. 1973.

45. E. Merzbacher, *Quantum Mechanics*, Wiley, New York, 3rd ed., 1988.

46. R. Srivastava, C. K. Kao, and R. V. Ramaswamy, "WKB analysis of planar surface waveguides with truncated index profiles," *J. Lightwave Tech.*, vol. LT-5, pp. 1605–1609, Nov. 1987.

47. J. A. Arnaud, *Beam and Fiber Optics*, Academic, New York, 1976.

48. M. Born and E. Wolf, *Principles of Optics*, Pergamon, Oxford, 6th ed., 1980.

49. N. P. Bansal and R. H. Doremus, *Handbook of Glass Properties*, Academic, New York, 1986.

50. I. Fanderlik, *Optical Properties of Glass*, Elsevier, New York, 1983.

51. J. C. Phillips, "The physics of glass," *Physics Today*, vol. 35, pp. 27–33, Feb. 1982.

52. B. C. Bagley, C. R. Kurkjian, J. W. Mitchell, G. E. Peterson, and A. R. Tynes, "Materials, properties, and choices," in S. E. Miller and A. G. Chynoweth, eds., *Optical Fiber Telecommunications*, Academic, New York, 1979.

53. S. R. Nagel, "Fiber materials and fabrication methods," in S. E. Miller and I. P. Kaminow, eds., *Optical Fiber Telecommunications—II*, Academic, New York, 1988.

54. H. Rawson, *Properties and Applications of Glass*, Elsevier, New York, 1980.

55. M. Poulain, U. Poulain, J. Lucas, and P. Bran, "Verres fluores au tetrafluorure de zirconium: Propriétés optiques d'un verre dopé au Nd^{3+}," *Materials Research Bulletin*, vol. 10, no. 4, pp. 243–246, 1975.

56. M. G. Drexhage, "Heavy metal fluoride glasses," in M. Tomozawa and R. H. Doremus, eds., *Treatise on Materials Science and Technology, Vol. 26: Glass IV*, Academic, New York, 1985.

57. D. C. Tran, G. H. Sigel, Jr., and B. Bendow, "Heavy metal fluoride glasses and fibers: A review," *J. Lightwave Tech.*, vol. LT-2, pp. 566–586, Oct. 1984.

58. R. C. Folweiler, "Fluoride glasses," *GTE J. Science and Tech.*, vol. 3, no. 1, pp. 25–37, 1989.

59. J. Lucas, "Review: Fluoride glasses," *J. Materials Science*, vol. 24, pp. 1–13, Jan. 1989.

60. (*a*) T. Katsuyama and H. Matsumura, *Infrared Optical Fibers*, Adam Hilger, Bristol, England, 1989.
    (*b*) P. Klocek and G. H. Sigel, Jr., *Infrared Fiber Optics*, SPIE Optical Eng. Press, Bellingham, WA, 1989.

61. P. W. France, *Fluoride Glass Optical Fibers*, CRC Press, Boca Raton, FL, 1990.
62. B. J. Ainslie, S. P. Craig, and S. T. Davey, "The absorption and fluorescence spectra of rare-earth ions in silica-based monomode fibers," *J. Lightwave Tech.*, vol. 6, pp. 287–293, Feb. 1988.
63. K. Nagi, A. Oyobe, T. Morikawa, Y. Sasaki, and K. Nakamura, "Gain characteristics of Er^{3+} doped fiber with quasiconfined structure," *J. Lightwave Tech.*, vol. 8, pp. 1319–1322, Sept. 1990.
64. S. P. Craig-Ryan, B. J. Ainslie, and C. A. Millar, "Fabrication of long lengths of low excess loss erbium-doped optical fibre," *Electron. Lett.*, vol. 26, pp. 185–186, Feb. 1990.
65. P. Kaiser, A. C. Hart, Jr., and L. L. Blyler, Jr., "Low-loss FEP-clad silica fibers," *Appl. Opt.*, vol. 14, pp. 156–162, Jan. 1975.
66. P. C. Schultz, "Progress in optical waveguide process and materials," *Appl. Opt.*, vol. 18, pp. 3684–3693, Nov. 1979.
67. W. G. French, R. E. Jaeger, J. B. MacChesney, S. R. Nagel, K. Nassau, and A. D. Pearson, "Fiber preform preparation," in S. E. Miller and A. G. Chynoweth, eds., *Optical Fiber Telecommunications*, Academic, New York, 1979.
68. R. Dorn, A. Baumgärtner, A. Gutu-Nelle, J. Koppenborg, W. Rehm, R. Schneider, and S. Schneider, "Mechanical shaping of preforms for low loss at low cost," *J. Opt. Commun.*, vol. 10, pp. 2–5, Mar. 1989.
69. R. E. Jaeger, A. D. Pearson, J. C. Williams, and H. M. Presby, "Fiber drawing and control," in S. E. Miller and A. G. Chynoweth, eds., *Optical Fiber Telecommunications*, Academic, New York, 1979.
70. U. C. Paek, "High-speed high-strength fiber drawing," *J. Lightwave Tech.*, vol. LT-4, pp. 1048–1060, Aug. 1986.
71. C. Brehm, P. Dupont, G. Lavanant, P. Ledoux, C. LeSergent, C. Reinaudo, J. M. Saugrain, M. Carratt, and R. Jocteur, "Improved drawing conditions for very low loss 1.55-μm dispersion-shifted fiber," *Fiber and Integrated Opt.*, vol. 7, no. 4, pp. 333–341, 1988.
72. P. L. Chu, T. Whitbread, and P. M. Allen, "An on-line fiber drawing tension and diameter measurement device," *J. Lightwave Tech.*, vol. 7, pp. 255–261, Feb. 1989.
73. F. P. Kapron, D. B. Keck, and R. D. Maurer, "Radiation losses in glass optical waveguides," *Appl. Phys. Lett.*, vol. 17, pp. 423–425, Nov. 1970.
74. P. C. Schultz, "Fabrication of optical waveguides by the outside vapor deposition process," *Proc. IEEE*, vol. 68, pp. 1187–1190, Oct. 1980.
75. R. V. VanDewoestine and A. J. Morrow, "Developments in optical waveguide fabrication by the outside vapor deposition process," *J. Lightwave Tech.*, vol. LT-4, pp. 1020–1025, Aug. 1986.
76. T. Izawa and N. Inagaki, "Materials and processes for fiber preform fabrication: Vapor-phase axial deposition," *Proc. IEEE*, vol. 68, pp. 1184–1187, Oct. 1980.
77. H. Murata, "Recent developments in vapor phase axial deposition," *J. Lightwave Tech.*, vol. LT-4, pp. 1026–1033, Aug. 1986.
78. S. R. Nagel, J. B. MacChesney, and K. L. Walker, "Modified chemical vapor deposition," in T. Li, ed., *Optical Fiber Communications, Vol. 1, Fiber Fabrication*, Academic, New York, 1985.
79. (a) P. Geittner, D. Küppers, and H. Lydtin, "Low loss optical fibers prepared by plasma-activated chemical vapor deposition (PCVD)," *Appl. Phys. Lett.*, vol. 28, pp. 645–646, June 1976.
    (b) P. Geittner and H. Lydtin, "Manufacturing optical fibers by the PCVD process," *Philips Tech. Rev. (Netherlands)*, vol. 44, pp. 241–249, May, 1989.
80. T. Hünlich, H. Bauch, R. T. Kersten, V. Paquet, and G. F. Weidmann, "Fiber-preform fabrication using plasma technology: A review," *J. Opt. Commun.*, vol. 8, pp. 122–129, Dec. 1987.
81. H. Lydtin, "PCVD: A technique suitable for large-scale fabrication of optical fibers," *J. Lightwave Tech.*, vol. LT-4, pp. 1034–1038, Aug. 1986.
82. R. D. Maurer, "Behavior of flaws in fused silcia fibers," in C. R. Kurkjian, ed., *Strength of Organic Glasses*, Plenum, New York, 1986.

83. D. Kalish, D. L. Key, C. R. Kurkjian, B. K. Tariyal, and T. T. Wang, "Fiber characterization—mechanical," in S. E. Miller and A. G. Chynoweth, eds., *Optical Fiber Telecommunications*, Academic, New York, 1979.

84. H. H. Yuce, "Chemical effects on fiber reliability," *Technical Digest: IEEE/OSA Optical Fiber Commun. Conf.*, p. 171, San Francisco, CA, Jan. 1990.

85. C. R. Kurkjian, J. T. Krause, and M. J. Matthewson, "Strength and fatigue of silica optical fibers," *J. Lightwave Tech.*, vol. 7, pp. 1360–1370, Sept. 1989.

86. A. A. Griffith, "The phenomena of rupture and flow in solids," *Philos. Trans. Roy. Soc. (London)*, vol. 221A, pp. 163–198, Oct. 1920.

87. W. Weibull, "A statistical theory of the strength of materials," *Ing. Vetenskaps. Akad. Handl. (Proc. Roy. Swed. Inst. Eng. Res.)*, no. 151, 1939; "The phenomenon of rupture in solids," *ibid.*, no. 153, 1939.

88. T. J. Miller, A. C. Hart, W. I. Vroom, and M. J. Bowden, "Silicone and ethylene-vinyl-acetate-coated laser-drawn silica fibers with tensile strengths > 3.5 GN/m^2 (500 kpsi) in > 3 km lengths," *Electron. Lett.*, vol. 14, pp. 603–605, Aug. 1978.

89. H. Schonhorn, C. R. Kurkjian, R. E. Jaeger, H. N. Vazirani, R. V. Albarino, and F. V. DiMarcello, "Epoxy-acrylate coated fused silcia fibers with tensile strengths greater than 500 kpsi (3.5 GN/m^2) in 1-km gauge lengths," *Appl. Phys. Lett.*, vol. 29, pp. 712–714, Dec. 1976.

90. (*a*) K. E. Lu, G. S. Glaesemann, R. V. VanDewoestine, and G. Kar, "Recent developments in hermetically coated optical fiber," *J. Lightwave Tech.*, vol. 6, pp. 240–244, Feb. 1988.
    (*b*) K. E. Lu, "Hermetic coatings," *Technical Digest: IEEE/OSA Optical Fiber Commun. Conf.*, p. 174, San Francisco, CA, Jan. 1990.

91. Y. Katsuyama, Y. Mitsunaga, H. Kobayashi, and Y. Ishida, "Dynamic fatigue of optical fiber under repeated stress," *J. Appl. Phys.*, vol. 53, pp. 318–321, Jan. 1982.

92. F. V. DiMarcello, D. L. Brownlow, R. G. Huff, and A. C. Hart, Jr., "Multikilometer lengths of 3.5-GPa proof-tested fiber," *J. Lightwave Tech.*, vol. 3, pp. 946–949, Oct. 1985.

93. Y. Mitsunaga, Y. Katsuyama, and Y. Ishida, "Reliability assurance for long-length optical fiber based on proof testing," *Electron. Lett.*, vol. 17, pp. 567–568, Aug. 1981.

94. F. A. Donaghy and T. P. Dabbs, "Subthreshold flaws and their failure prediction in long-distance optical fiber cables," *J. Lightwave Tech.*, vol. 6, pp. 226–232, Feb. 1988.

95. B. Wiltshire and M. H. Reeve, "A review of the environmental factors affecting optical cable design," *J. Lightwave Tech.*, vol. 6, pp. 179–185, Feb. 1988.

96. S. Tanaka and M. Honjo, "Long-term reliability of transmission loss in optical fiber cables," *J. Lightwave Tech.*, vol. 6, pp. 210–217, Feb. 1988.

97. P. Bark and D. Lawrence, "Fiber-optic cables: Design, performance characteristics, and field experience," in E. E. Basch, ed., *Optical Fiber Transmission*, Howard W. Sams, Indianapolis, IN, 1987.

98. C. H. Gartside, III, P. D. Patel, and M. R. Santana, "Optical fiber cables," in S. E. Miller and I. P. Kaminow, eds., *Optical Fiber Telecommunications—II*, Academic, New York, 1988.

99. D. Marcuse, "Loss analysis of single-mode fiber splices," *Bell Sys. Tech. J.*, vol. 56, pp. 703–718, May–June 1977.

CHAPTER
3

SIGNAL DEGRADATION IN OPTICAL FIBERS

In Chap. 2 we showed the structure of optical fibers and examined the concepts of how light propagates along a cylindrical dielectric optical waveguide. Here we shall continue the discussion of optical fibers by answering two very important questions:

1. What are the loss or signal attenuation mechanisms in a fiber?
2. Why and to what degree do optical signals get distorted as they propagate along a fiber?

Signal attenuation (also known as *fiber loss* or *signal loss*) is one of the most important properties of an optical fiber, because it largely determines the maximum repeaterless separation between a transmitter and a receiver. Since repeaters are expensive to fabricate, install, and maintain, the degree of attenuation in a fiber has a large influence on system cost. Of equal importance is signal distortion. The distortion mechanisms in a fiber cause optical signal pulses to broaden as they travel along a fiber. If these pulses travel sufficiently far, they will eventually overlap with neighboring pulses, thereby creating errors in the receiver output. The signal distortion mechanisms thus limit the information-carrying capacity of a fiber.

3.1 ATTENUATION

Attenuation of a light signal as it propagates along a fiber is an important consideration in the design of an optical communication system, since it plays a major role in determining the maximum transmission distance between a transmitter and a receiver. The basic attenuation mechanisms[1-4] in a fiber are absorption, scattering, and radiative losses of the optical energy. Absorption is related to the fiber material, whereas scattering is associated both with the fiber material and with structural imperfections in the optical waveguide. Attenuation owing to radiative effects originates from perturbations (both microscopic and macroscopic) of the fiber geometry.

In this section we shall first discuss the units in which fiber losses are measured and then present the physical phenomena giving rise to attenuation.

3.1.1 Attenuation Units

Signal attenuation (or *fiber loss*) is defined as the ratio of the optical output power P_{out} from a fiber of length L to the optical input power P_{in}. This power ratio is a function of wavelength, as is shown by the general attenuation curve in Fig. 3-1. The symbol α is commonly used to express attenuation in decibels per kilometer (see App. D for a discussion of decibels):

$$\alpha = \frac{10}{L} \log\left(\frac{P_{in}}{P_{out}}\right) \tag{3-1}$$

Example 3-1. An ideal fiber would have no loss so that $P_{out} = P_{in}$. This corresponds to a 0-dB attenuation, which is practice is impossible. An actual low-loss fiber may have a 3-dB/km average loss, for example. This means that the optical signal power would decrease by 50 percent over a 1-km length and would decrease by 75 percent (a 6-dB loss) over a 2-km length, since loss contributions expressed in decibels are additive.

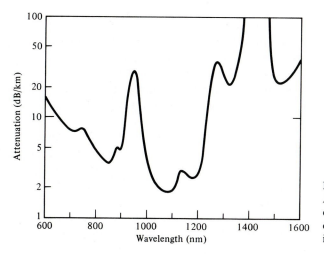

FIGURE 3-1
Attenuation-versus-wavelength curve of a typical early-technology fiber having a high water impurity content.

3.1.2 Absorption

Absorption is caused by three different mechanisms:

1. Absorption by atomic defects in the glass composition
2. Extrinsic absorption by impurity atoms in the glass material
3. Intrinsic absorption by the basic constituent atoms of the fiber material

Atomic defects are imperfections of the atomic structure of the fiber material such as missing molecules, high-density clusters of atom groups, or oxygen defects in the glass structure. Usually absorption losses arising from these defects are negligible compared to intrinsic and impurity absorption effects. However, they can be significant if the fiber is exposed to intense nuclear radiation levels,[5-8] as might occur inside a nuclear reactor, during a nuclear explosion, or in the earth's Van Allen belts.

The dominant absorption factor in fibers prepared by the direct-melt method is the presence of impurities in the fiber material. Impurity absorption results predominantly from transition metal ions such as iron, chromium, cobalt, and copper, and from OH (water) ions. The transition metal impurities which are present in the starting materials used for direct-melt fibers range between 1 and 10 parts per billion (ppb), causing losses from 1 to 10 dB/km. The impurity levels in vapor phase deposition processes are usually one to two orders of magnitude lower. Impurity absorption losses occur either because of electronic transitions between the energy levels associated with the incompletely filled inner subshell of these ions or because of charge transitions from one ion to another. The absorption peaks of the various transition metal impurities tend to be broad, and several peaks may overlap, which further broadens the absorption region.

The presence of OH (water) ion impurities in fiber performs results mainly from the oxyhydrogen flame used for the hydrolysis reaction of the $SiCl_4$, $GeCl_4$, and $POCl_3$ starting materials. Water impurity concentrations of less than a few parts per billion are required if the attenuation is to be less than 20 dB/km. Early optical fibers had high levels of OH ions which resulted in large absorption peaks occurring at 1400, 950, and 725 nm. These are the first, second, and third overtones, respectively, of the fundamental absorption peak of water near 2.7 μm, as shown in Fig. 3-1. Between these absorption peaks there are regions of low attenuation.

The peaks and valleys in the attenuation curve resulted in the assignment of various "transmission windows" to early optical fibers. Significant progress has been made in reducing the residual OH content of fibers to less than 1 ppb. For example, the loss curve of a silica fiber prepared by the VAD method[9] with an OH content of less than 0.8 ppb is shown in Fig. 3-2.

Intrinsic absorption is associated with the basic fiber material (for example, pure SiO_2) and is the principal physical factor that defines the transparency

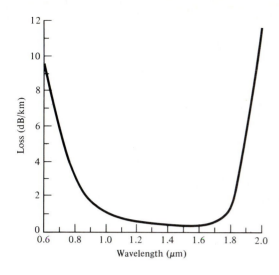

FIGURE 3-2
Attenuation-versus-wavelength curve of a
VAD silica fiber with very low OH content.
(Reproduced with permission from
Moriyama et al.[9])

window of a material over a specified spectral region. It occurs when the material is in a perfect state with no density variations, impurities, material inhomogeneities, etc. Intrinsic absorption thus sets the fundamental lower limit on absorption for any particular material.

Intrinsic absorption results from electronic absorption bands in the ultraviolet region and from atomic vibration bands in the near infrared region. The electronic absorption bands are associated with the band gaps of the amorphous glass materials. Absorption occurs when a photon interacts with an electron in the valence band and excites it to a higher energy level, as is described in Sec. 2.1. The ultraviolet edge of the electron absorption bands of both amorphous and crystalline materials follow the empirical relationship[1,3]

$$\alpha_{uv} = C e^{E/E_0} \tag{3-2a}$$

which is known as Urbach's rule. Here C and E_0 are empirical constants and E is the photon energy. The magnitude and characteristic exponential decay of the ultraviolet absorption are shown in Fig. 3-3. Since E is inversely proportional to the wavelength λ, ultraviolet absorption decays exponentially with increasing wavelength. In particular, the ultraviolet loss contribution in dB/km at any wavelength can be expressed empirically as a function of the mole fraction x of GeO_2 as[10,11]

$$\alpha_{uv} = \frac{154.2x}{46.6x + 60} \times 10^{-2} \exp\left(\frac{4.63}{\lambda}\right) \tag{3-2b}$$

As shown in Fig. 3-3, the ultraviolet loss is small compared to scattering loss in the near infrared region.

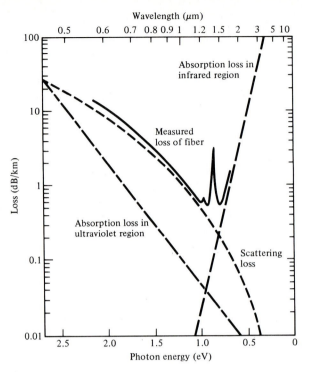

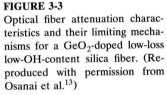

FIGURE 3-3
Optical fiber attenuation characteristics and their limiting mechanisms for a GeO_2-doped low-loss low-OH-content silica fiber. (Reproduced with permission from Osanai et al.[13])

In the near infrared region above 1.2 μm, the optical waveguide loss is predominantly determined by the presence of OH ions and the inherent infrared absorption of the constituent material. The inherent infrared absorption is associated with the characteristic vibration frequency of the particular chemical bond between the atoms of which the fiber is composed. An interaction between the vibrating bond and the electromagnetic field of the optical signal results in a transfer of energy from the field to the bond, thereby giving rise to absorption. This absorption is quite strong because of the many bonds present in the fiber. An empirical expression for the infrared absorption in dB/km for GeO_2-SiO_2 glass is[10, 11]

$$\alpha_{IR} = 7.81 \times 10^{11} \times \exp\left(\frac{-48.48}{\lambda}\right) \qquad (3\text{-}3)$$

These mechanisms result in a wedge-shaped spectral-loss characteristic. Within this wedge losses as low as 0.154 dB/km at 1.55 μm in a single-mode fiber have been measured.[12] A comparison[13] of the infrared absorption induced by various doping materials in low-water-content fibers is shown in Fig. 3-4. This indicates that for operation at longer wavelengths a GeO_2-doped fiber material is the most desirable. Note that the absorption curve shown in Fig. 3-3 is for a GeO_2-doped fiber.

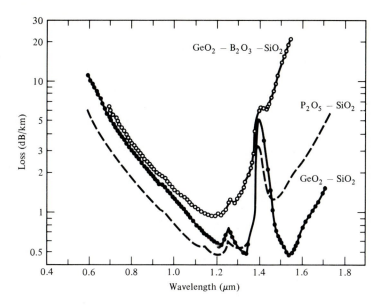

FIGURE 3-4
A comparison of the infrared absorption induced by various doping materials in low-loss silica fibers. (Reproduced with permission from Osanai et al.[13])

3.1.3 Scattering Losses

Scattering losses in glass arise from microscopic variations in the material density, from compositional fluctuations, and from structural inhomogeneities or defects occurring during fiber manufacture. As we saw in Sec. 2.7, glass is composed of a randomly connected network of molecules. Such a structure naturally contains regions in which the molecular density is either higher or lower than the average density in the glass. In addition, since glass is made up of several oxides, such as SiO_2, GeO_2, and P_2O_5, compositional fluctuations can occur. These two effects give rise to refractive-index variations which occur within the glass over distances that are small compared to the wavelength. These index variations cause a Rayleigh-type scattering of the light. Rayleigh scattering in glass is the same phenomenon that scatters light from the sun in the atmosphere, thereby giving rise to a blue sky.

The expressions for scattering-induced attenuation are fairly complex owing to the random molecular nature and the various oxide constituents of glass. For single-component glass the scattering loss at a wavelength λ resulting from density fluctuations can be approximated by[3, 14] (in base e units)

$$\alpha_{\text{scat}} = \frac{8\pi^3}{3\lambda^4}(n^2 - 1)^2 k_B T_f \beta_T \qquad (3\text{-}4a)$$

Here n is the refractive index, k_B is Boltzmann's constant, β_T is the isothermal compressibility of the material, and the fictive temperature T_f is the temperature at which the density fluctuations are frozen into the glass as it solidifies (after having been drawn into a fiber). Alternatively the relation[3, 15] (in base e units)

$$\alpha_{\text{scat}} = \frac{8\pi^3}{3\lambda^4} n^8 p^2 k_B T_f \beta_T \qquad (3\text{-}4b)$$

has been derived, where p is the photoelastic coefficient. A comparison of Eqs. (3-4a) and (3-4b) is given in Prob. 3-5. Note that Eqs. (3-4a) and (3-4b) are given in units of *nepers* (that is, base e units). To change this to decibels for optical power attenuation calculations, multiply these equations by $10 \log e = 4.343$.

For multicomponent glasses the scattering is given by[3]

$$\alpha = \frac{8\pi^3}{3\lambda^4} (\delta n^2)^2 \delta V \qquad (3\text{-}5)$$

where the square of the mean square refractive-index fluctuation $(\delta n^2)^2$ over a volume of δV is

$$(\delta n^2)^2 = \left(\frac{\partial n}{\partial \rho}\right)^2 (\delta \rho)^2 + \sum_{i=1}^{m} \left(\frac{\partial n^2}{\partial C_i}\right)^2 (\delta C_i)^2 \qquad (3\text{-}6)$$

Here $\delta \rho$ is the density fluctuation and δC_i is the concentration fluctuation of the ith glass component. The magnitudes of the composition and density fluctuations are generally not known and must be determined from experimental scattering data. Once they are known the scattering loss can be calculated.

Structural inhomogeneities and defects created during fiber fabrication can also cause scattering of light out of the fiber. These defects may be in the form of trapped gas bubbles, unreacted starting materials, and crystallized regions in the glass. In general the preform manufacturing methods that have evolved have minimized these extrinsic effects to the point where scattering results from them is negligible compared to the intrinsic Rayleigh scattering.

Since Rayleigh scattering follows a characteristic λ^{-4} dependence, it decreases dramatically with increasing wavelength, as is shown in Fig. 3-3. For wavelengths below about 1 μm it is the dominant loss mechanism in a fiber and gives the attenuation-versus-wavelength plots their characteristic downward trend with increasing wavelength. At wavelengths longer than 1 μm, infrared absorption effects tend to dominate optical signal attenuation.

Combining the infrared, ultraviolet, and scattering losses, we get the results shown in Fig. 3-5 for multimode fibers and Fig. 3-6 for single-mode fibers.[16] Both of these figures are for typical commercial-grade silica fibers. The losses of multimode fibers are generally higher than those of single-mode fibers. This is a result of higher dopant concentrations and the accompanying larger scattering loss due to greater compositional fluctuation in multimode fibers. In

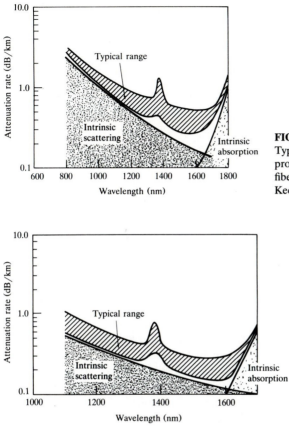

FIGURE 3-5

Typical spectral attenuation range for production-run graded-index multimode fibers. (Reproduced with permission from Keck,[16] © 1985, IEEE.)

FIGURE 3-6

Typical spectral attenuation range for production-run single-mode fibers. (Reproduced with permission from Keck,[16] ©1985, IEEE.)

addition, multimode fibers are subject to higher-order-mode losses owing to perturbations at the core-to-cladding interface.

3.1.4 Bending Losses

Radiative losses occur whenever an optical fiber undergoes a bend of finite radius of curvature.[17-25] Fibers can be subject to two types of bends: (*a*) macroscopic bends having radii that are large compared to the fiber diameter, for example, such as occur when a fiber cable turns a corner, and (*b*) random microscopic bends of the fiber axis that can arise when the fibers are incorporated into cables.

Let us first examine large-curvature radiation losses, which are known as *macrobending losses*, or simply *bending losses*. For slight bends the excess loss is extremely small and is essentially unobservable. As the radius of curvature decreases, the loss increases exponentially until at a certain critical radius the curvature loss becomes observable. If the bend radius is made a bit smaller once

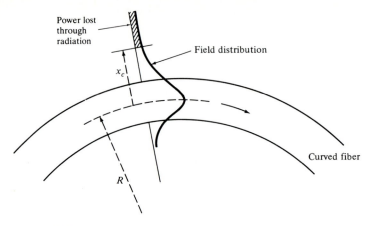

FIGURE 3-7
Sketch of the fundamental mode field in a curved optical waveguide. (Reproduced with permission from E. A. J. Marcatili and S. E. Miller, *Bell Sys. Tech. J.*, vol. 48, p. 2161, Sept. 1969, © 1969, AT & T.)

this threshold point has been reached, the losses suddenly become extremely large.

Qualitatively these curvature loss effects can be explained by examining the modal electric field distributions shown in Fig. 2-14. Recall that this figure shows that any bound core mode has an evanescent field tail in the cladding which decays exponentially as a function of distance from the core. Since this field tail moves along with the field in the core, part of the energy of a propagating mode travels in the fiber cladding. When a fiber is bent, the field tail on the far side of the center of curvature must move faster to keep up with the field in the core, as is shown in Fig. 3-7 for the lowest-order fiber mode. At a certain critical distance x_c from the center of the fiber the field tail would have to move faster than the speed of light to keep up with the core field. Since this is not possible the optical energy in the field tail beyond x_c radiates away.

The amount of optical radiation from a bent fiber depends on the field strength at x_c and on the radius of curvature R. Since higher-order modes are bound less tightly to the fiber core than lower-order modes, the higher-order modes will radiate out of the fiber first. Thus the total number of modes that can be supported by a curved fiber is less than in a straight fiber. Gloge[18] has derived the following expression for the effective number of modes N_{eff} that are guided by a curved multimode fiber of radius a:

$$N_{\text{eff}} = N_\infty \left\{ 1 - \frac{\alpha + 2}{2\alpha\Delta} \left[\frac{2a}{R} + \left(\frac{3}{2n_2 kR} \right)^{2/3} \right] \right\} \tag{3-7}$$

where α defines the graded-index profile, Δ is the core-cladding index difference, n_2 is the cladding refractive index, $k = 2\pi/\lambda$ is the wave propagation constant, and

$$N_\infty = \frac{\alpha}{\alpha + 2} (n_1 ka)^2 \Delta \tag{3-8}$$

is the total number of modes in a straight fiber [see Eq. (2-97)].

> **Example 3-2.** As an example, let us find the radius of curvature R at which the number of modes decreases by 50 percent in a graded-index fiber. For this fiber let $\alpha = 2$, $n_2 = 1.5$, $\Delta = 0.01$, $a = 25~\mu$m, and let the wavelength of the guided light be 1.3 μm. Solving Eq. (3-7) yields $R = 1.0$ cm.

Another form of radiation loss in optical waveguides results from mode coupling caused by random microbends of the optical fiber.[25-30] *Microbends* are repetitive small-scale fluctuations in the radius of curvature of the fiber axis, as is illustrated in Fig. 3-8. They are caused either by nonuniformities in the manufacturing of the fiber or by nonuniform lateral pressures created during the cabling of the fiber. The latter effect is often referred to as *cabling* or *packaging losses*. An increase in attenuation results from microbending because the fiber curvature causes repetitive coupling of energy between the guided modes and the leaky or nonguided modes in the fiber.

One method of minimizing microbending losses is by extruding a compressible jacket over the fiber. When external forces are applied to this configuration, the jacket will be deformed but the fiber will tend to stay relatively straight, as shown in Fig. 3-9. For a multimode graded-index fiber having a core

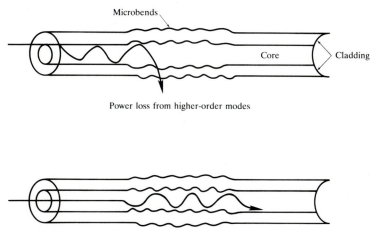

FIGURE 3-8
Small-scale fluctuations in the radius of curvature of the fiber axis leads to microbending losses. Microbends can shed higher-order modes and can cause power from low-order modes to couple to higher-order modes.

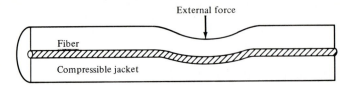

FIGURE 3-9
A compressible jacket extruded over a fiber reduces microbending resulting from external forces.

radius a, outer radius b (excluding the jacket), and index difference Δ, the microbending loss α_M of a jacketed fiber is reduced from that of an unjacketed fiber by a factor[31]

$$F(\alpha_M) = \left[1 + \pi\Delta^2\left(\frac{b}{a}\right)^4 \frac{E_f}{E_j}\right]^{-2} \qquad (3\text{-}9)$$

Here E_j and E_f are the Young's moduli of the jacket and fiber, respectively. The Young's modulus of common jacket materials ranges from 20 to 500 MPa. The Young's modulus of fused silica glass is about 65 GPa.

3.1.5   Core and Cladding Losses

Upon measuring the propagation losses in an actual fiber, all the dissipative and scattering losses will be manifested simultaneously. Since the core and cladding have different indices of refraction and therefore differ in composition, the core and cladding generally have different attenuation coefficients denoted α_1 and α_2, respectively. If the influence of modal coupling is ignored,[32] the loss for a mode of order (ν, m) for a step-index waveguide is

$$\alpha_{\nu m} = \alpha_1 \frac{P_{core}}{P} + \alpha_2 \frac{P_{clad}}{P} \qquad (3\text{-}10)$$

where the fractional powers P_{core}/P and P_{clad}/P are shown in Fig. 2-22 for several low-order modes. Using Eq. (2-71), this can be written as

$$\alpha_{\nu m} = \alpha_1 + (\alpha_2 - \alpha_1)\frac{P_{clad}}{P} \qquad (3\text{-}11)$$

The total loss of the waveguide can be found by summing over all modes weighted by the fractional power in that mode.

For the case of a graded-index fiber the situation is much more complicated. In this case, both the attenuation coefficients and the modal power tend to be functions of the radial coordinate. At a distance r from the core axis the loss is[32]

$$\alpha(r) = \alpha_1 + (\alpha_2 - \alpha_1)\frac{n^2(0) - n^2(r)}{n^2(0) - n_2^2} \qquad (3\text{-}12)$$

where α_1 and α_2 are the axial and cladding attenuation coefficients, respectively, and the n's are defined by Eq. (2-78). The loss encountered by a given mode is then

$$\alpha_{gi} = \frac{\int_0^\infty \alpha(r)p(r)r\,dr}{\int_0^\infty p(r)r\,dr} \qquad (3\text{-}13)$$

where $p(r)$ is the power density of that mode at r. The complexity of the multimode waveguide has prevented an experimental correlation with a model. However, it has generally been observed that the loss increases with increasing mode number.[25,33]

3.2 SIGNAL DISTORTION IN OPTICAL WAVEGUIDES

An optical signal becomes increasingly distorted as it travels along a fiber. This distortion is a consequence of intramodal dispersion and intermodal delay effects. These distortion effects can be explained by examining the behavior of the group velocities of the guided modes, where the *group velocity* is the speed at which energy in a particular mode travels along the fiber.

Intramodal dispersion is pulse spreading that occurs within a single mode. It is a result of the group velocity being a function of the wavelength λ. Since intramodal dispersion depends on the wavelength, its effect on signal distortion increases with the spectral width of the optical source. This spectral width is the band of wavelengths over which the source emits light. It is normally characterized by the root-mean-square (rms) spectral width σ_λ (see Fig. 4-12). For light-emitting diodes (LEDs) the rms spectral width is approximately 5 percent of a central wavelength. For example, if the peak emission wavelength of an LED source is 850 nm, a typical source spectral width would be 40 nm; that is, the source emits most of its optical power in the 830- to 870-nm wavelength band. Laser diode optical sources have much narrower spectral widths, typical values being 1 to 2 nm.

The two main causes of intramodal dispersion are:

1. *Material dispersion*, which arises from the variation of the refractive index of the core material as a function of wavelength. (Material dispersion is sometimes referred to as *chromatic dispersion* or *spectral dispersion*, since this is the same effect by which a prism spreads out a spectrum.) This causes a wavelength dependence of the group velocity of any given mode; that is, pulse spreading occurs even when different wavelengths follow the same path.

2. *Waveguide dispersion*, which occurs because a single-mode fiber only confines about 80 percent of the optical power to the core. Dispersion thus arises,

since the 20 percent of the light propagating in the cladding travels faster than the light confined to the core. The amount of waveguide dispersion depends on the fiber design, since the modal propagation constant β is a function of a/λ (the optical fiber dimension relative to the wavelength λ; here a is the core radius.)

The other factor giving rise to pulse spreading is *intermodal delay*, which is a result of each mode having a different value of the group velocity at a single frequency.

Of these three, waveguide dispersion usually can be ignored in multimode fibers. However, this effect can be significant in single-mode fibers. The full effects of these three distortion mechanisms are seldom observed in practice, since they tend to be mitigated by other factors, such as nonideal index profiles, optical power-launching conditions (different amounts of optical power launched into the various modes), nonuniform mode attenuation, and mode mixing in the fiber and in splices; and by statistical variations in these effects along the fiber. In this section we shall first discuss the general effects of signal distortion and then examine the various dispersion mechanisms.

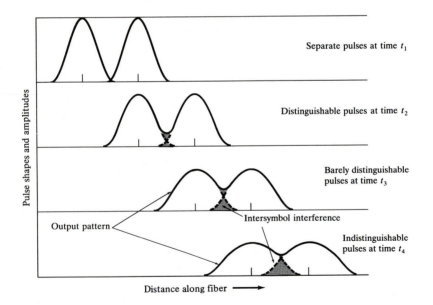

FIGURE 3-10
Broadening and attenuation of two adjacent pulses as they travel along a fiber: (*a*) originally the pulses are separate; (*b*) the pulses overlap slightly and are clearly distinguishable; (*c*) the pulses overlap significantly and are barely distinguishable; (*d*) eventually the pulses strongly overlap and are indistinguishable.

3.2.1 Information Capacity Determination

A result of the dispersion-induced signal distortion is that a light pulse will broaden as it travels along the fiber. As shown in Fig. 3-10 this pulse broadening will eventually cause a pulse to overlap with neighboring pulses. After a certain amount of overlap has occurred, adjacent pulses can no longer be individually distinguished at the receiver and errors will occur. Thus the dispersive properties determine the limit of the information capacity of the fiber.

A measure of the information capacity of an optical waveguide is usually specified by the *bandwidth-distance product* in MHz · km. For a step-index fiber the various distortion effects tend to limit the bandwidth-distance product to about 20 MHz · km. In graded-index fibers the radial refractive-index profile can be carefully selected so that pulse broadening is minimized at a specific operating wavelength. This had led to bandwidth-distance products as high as 2.5 GHz · km. Single-mode fibers can have capacities well in excess of this. A comparison of the information capacities of various optical fibers with the capacities of typical coaxial cables used for UHF and VHF transmission is shown in Fig. 3-11. The curves are shown in terms of signal attenuation versus data rate. The flatness of the attenuation curves for the fibers extend up to the microwave spectrum.

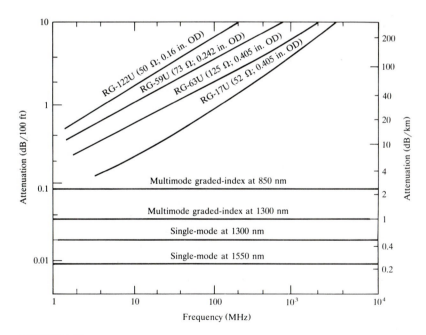

FIGURE 3-11

A comparison of the attenuation as a function of frequency or data rate of various coaxial cables and several types of high-bandwidth optical fibers.

The information-carrying capacity can be determined by examining the deformation of short light pulses propagating along the fiber. The following discussion on signal distortion is thus carried out primarily from the standpoint of pulse broadening, which is representative of digital transmission.

3.2.2   Group Delay

Let us examine a signal that modulates an optical source. We shall assume that the modulated optical signal excites all modes equally at the input end of the fiber. Each mode thus carries an equal amount of energy through the fiber. Furthermore, each mode contains all of the spectral components in the wavelength band over which the source emits. The signal may be considered as modulating each of these spectral components in the same way. As the signal propagates along the fiber, each spectral component can be assumed to travel independently, and to undergo a time delay or *group delay* per unit length in the direction of propagation given by[34]

$$\frac{\tau_g}{L} = \frac{1}{V_g} = \frac{1}{c}\frac{d\beta}{dk} = -\frac{\lambda^2}{2\pi c}\frac{d\beta}{d\lambda} \tag{3-14}$$

Here L is the distance traveled by the pulse, β is the propagation constant along the fiber axis, $k = 2\pi/\lambda$, and the *group velocity*

$$V_g = c\left(\frac{d\beta}{dk}\right)^{-1}$$

is the velocity at which the energy in a pulse travels along a fiber.

Since the group delay depends on the wavelength, each spectral component of any particular mode takes a different amount of time to travel a certain distance. As a result of this difference in time delays, the optical signal pulse spreads out with time as it is transmitted over the fiber. The quantity we are thus interested in is the amount of pulse spreading that arises from the group delay variation.

If the spectral width of the optical source is not too wide, the delay difference per unit wavelength along the propagation path is approximately $d\tau_g/d\lambda$. For spectral components which are $\delta\lambda$ apart and which lie $\delta\lambda/2$ above and below a central wavelength λ_0, the total delay difference $\delta\tau$ over a distance L is

$$\delta\tau = \frac{d\tau_g}{d\lambda}\delta\lambda \tag{3-15}$$

If the spectral width $\delta\lambda$ of an optical source is characterized by its root-mean-square (rms) value σ_λ (see Fig. 4-12), then the pulse spreading can be approximated by the rms pulse width

$$\sigma_g = \frac{d\tau_g}{d\lambda}\sigma_\lambda = -\frac{L\sigma_\lambda}{2\pi c}\left(2\lambda\frac{d\beta}{d\lambda} + \lambda^2\frac{d^2\beta}{d\lambda^2}\right) \tag{3-16}$$

The factor

$$D = \frac{1}{L}\frac{d\tau_g}{d\lambda}$$ (3-17)

is designated as the *dispersion*. It defines the pulse spread as a function of wavelength and is measured in picoseconds per kilometer per nanometer. It is a result of material and waveguide dispersion. In many theoretical treatments of intramodal dispersion it is assumed for simplicity that material dispersion and waveguide dispersion can be calculated separately and then added to give the total dispersion of the mode. In reality these two mechanisms are intricately related, since the dispersive properties of the refractive index (which gives rise to material dispersion) also effects the waveguide dispersion. However, an examination[35] of the interdependence of material and waveguide dispersion has shown that, unless a very precise value is desired, a good estimate of the total intramodal dispersion can be obtained by calculating the effect of signal distortion arising from one type of dispersion in the absence of the other, and then adding the results. Material dispersion and waveguide dispersion are therefore considered separately in the next two sections.

3.2.3 Material Dispersion

Material dispersion occurs because the index of refraction varies as a function of the optical wavelength. This is exemplified in Fig. 3-12 for silica.[36] As a consequence, since the group velocity V_g of a mode is a function of the index of refraction, the various spectral components of a given mode will travel at different speeds, depending on the wavelength.[37] Material dispersion is, therefore, an intramodal dispersion effect, and is of particular importance for single-mode waveguides and for LED systems (since an LED has a broader output spectrum than a laser diode).

To calculate material-induced dispersion, we consider a plane wave propagating in an infinitely extended dielectric medium that has a refractive index $n(\lambda)$ equal to that of the fiber core. The propagation constant β is thus given by

$$\beta = \frac{2\pi n(\lambda)}{\lambda}$$ (3-18)

Substituting this expression for β into Eq. (3-14) with $k = 2\pi/\lambda$ yields the group delay τ_{mat} resulting from material dispersion:

$$\tau_{mat} = \frac{L}{c}\left(n - \lambda\frac{dn}{d\lambda}\right)$$ (3-19)

Using Eq. (3-16), the pulse spread σ_{mat} for a source of spectral width σ_λ is found by differentiating this group delay with respect to wavelength and

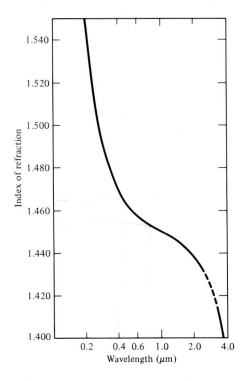

FIGURE 3-12
Variations in the index of refraction as a function of the optical wavelength for silica. (Reproduced with permission from I. H. Malitson, *J. Opt. Soc. Amer.*, vol. 55, pp. 1205–1209, Oct. 1965.)

multiplying by σ_λ to yield

$$\sigma_{\text{mat}} = \frac{d\tau_{\text{mat}}}{d\lambda}\sigma_\lambda = -\frac{L}{c}\lambda\frac{d^2n}{d\lambda^2}\sigma_\lambda = D_{\text{mat}}(\lambda)L\sigma_\lambda \qquad (3\text{-}20)$$

where $D_{\text{mat}}(\lambda)$ is the *material dispersion*.

A plot of Eq. (3-20) for unit length L and unit optical source spectral width σ_λ is given in Fig. 3-13 for the silica material shown in Fig. 3-12. From Eq. (3-20) and Fig. 3-13 it can be seen that material dispersion can be reduced either by choosing sources with narrower spectral output widths (reducing σ_λ) or by operating at longer wavelengths.[38]

> **Example 3-3.** As an example, consider a typical GaAlAs LED having a spectral width of 40 nm at an 800-nm peak output so that $\sigma_\lambda/\lambda = 5$ percent. As can be seen from Fig. 3-13 and Eq. (3-20), this produces a pulse spread of 4.4 ns/km. Note that material dispersion goes to zero at 1.27 μm for pure silica.

3.2.4 Waveguide Dispersion

The effect of waveguide dispersion on pulse spreading can be approximated by assuming that the refractive index of the material is independent of wavelength. Let us first consider the group delay, that is, the time required for a mode to

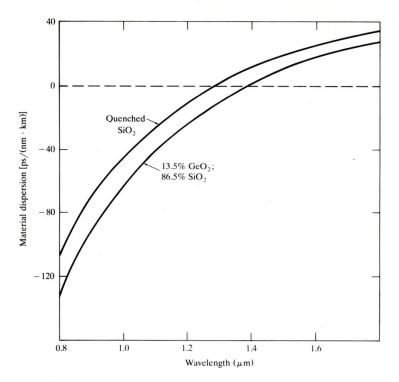

FIGURE 3-13
Material dispersion as a function of optical wavelength for pure silica and 13.5% GeO_2/86.5% SiO_2. (Reproduced with permission from Fleming.[38])

travel along a fiber of length L. To make the results independent of fiber configuration,[37] we shall express the group delay in terms of the normalized propagation constant b defined by

$$b = 1 - \left(\frac{ua}{V}\right)^2 = \frac{\beta^2/k^2 - n_2^2}{n_1^2 - n_2^2} \tag{3-21}$$

For small values of the index difference $\Delta = (n_1 - n_2)/n_1$, Eq. (3-21) can be approximated by

$$b = \frac{\beta/k - n_2}{n_1 - n_2} \tag{3-22}$$

Solving Eq. (3-22) for β, we have

$$\beta = n_2 k(b\Delta + 1) \tag{3-23}$$

With this expression for β and using the assumption that n_2 is not a function of wavelength, we find that the group delay τ_{wg} arising from waveguide dispersion is

$$\tau_{\text{wg}} = \frac{L}{c}\frac{d\beta}{dk} = \frac{L}{c}\left[n_2 + n_2\Delta\frac{d(kb)}{dk}\right] \tag{3-24}$$

The modal propagation constant β is obtained from the eigenvalue equation expressed by Eq. (2-54), and is generally given in terms of the normalized frequency V defined by Eq. (2-57). We shall therefore use the approximation

$$V = ka\left(n_1^2 - n_2^2\right)^{1/2} \simeq kan_2\sqrt{2\Delta}$$

which is valid for small values of Δ, to write the group delay in Eq. (3-24) in terms of V instead of k, yielding

$$\tau_{\text{wg}} = \frac{L}{c}\left[n_2 + n_2\Delta\frac{d(Vb)}{dV}\right] \tag{3-25}$$

The first term in Eq. (3-25) is a constant and the second term represents the group delay arising from waveguide dispersion. The factor $d(Vb)/dV$ can be expressed as[37]

$$\frac{d(Vb)}{dV} = b\left[1 - \frac{2J_\nu^2(ua)}{J_{\nu+1}(ua)J_{\nu-1}(ua)}\right]$$

where u is defined by Eq. (2-48) and a is the fiber radius. This factor is plotted in Fig. 3-14 as a function of V for various LP modes. The plots show that, for a fixed value of V, the group delay is different for every guided mode. When a light pulse is launched into a fiber, it is distributed among many guided modes. These various modes arrive at the fiber end at different times depending on their group delay, so that a pulse spreading results. For multimode fibers the waveguide dispersion is generally very small compared to material dispersion and can therefore be neglected.

3.2.5   Signal Distortion in Single-Mode Fibers

For single-mode fibers waveguide dispersion is of importance and can be of the same order of magnitude as material dispersion. To see this, let us compare the two dispersion factors. The pulse spread σ_{wg} occurring over a distribution of wavelengths σ_λ is obtained from the derivative of the group delay with respect to wavelength,[37]

$$\sigma_{\text{wg}} = \sigma_\lambda\frac{d\tau_{\text{wg}}}{d\lambda} = \sigma_\lambda LD_{\text{wg}}(\lambda)$$

$$= -\frac{V}{\lambda}\sigma_\lambda\frac{d\tau_{\text{wg}}}{dV} = -\frac{n_2L\Delta\sigma_\lambda}{c\lambda}V\frac{d^2(Vb)}{dV^2} \tag{3-26}$$

where $D_{\text{wg}}(\lambda)$ is the *waveguide dispersion*. The factor $Vd^2(Vb)/dV^2$ is plotted as a function of V in Fig. 3-15 for the fundamental LP_{01} mode shown in Fig. 3-14.

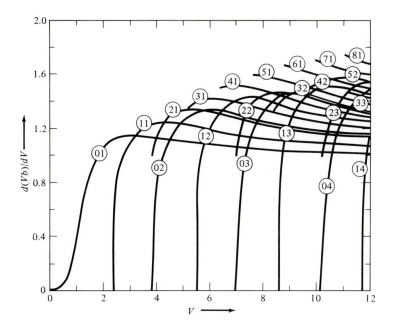

FIGURE 3-14
The group delay arising from waveguide dispersion as a function of the V number for a step-index optical fiber. The curve numbers jm designate the LP_{jm} modes. (Reproduced with permission from Gloge.[37])

This factor reaches a maximum at $V = 1.2$, but runs between 0.2 and 0.1 for a practical single-mode operating range of $V = 2.0$ to 2.4. Thus for values of $\Delta = 0.01$ and $n_2 = 1.5$,

$$\frac{\sigma_{wg}}{L} \simeq -\frac{0.003\sigma_\lambda}{c\lambda} \tag{3-27}$$

Comparing this with the material-dispersion-induced pulse spreading from Eq. (3-20) for $\lambda = 900$ nm, where

$$\frac{\sigma_{mat}}{L} \simeq -\frac{0.02\sigma_\lambda}{c\lambda} \tag{3-28}$$

it is clear that material dispersion dominates at lower wavelengths. However, at longer wavelengths such as at 1.3 μm, which is the spectral region of extremely low material dispersion in silica, waveguide dispersion can become the dominating pulse-distorting mechanism. Examples of the magnitudes of material and waveguide dispersions are given in Fig. 3-16 for a fused-silica-core single-mode fiber having $V = 2.4$. In this figure the approximation that material and waveguide dispersions are additive was used.

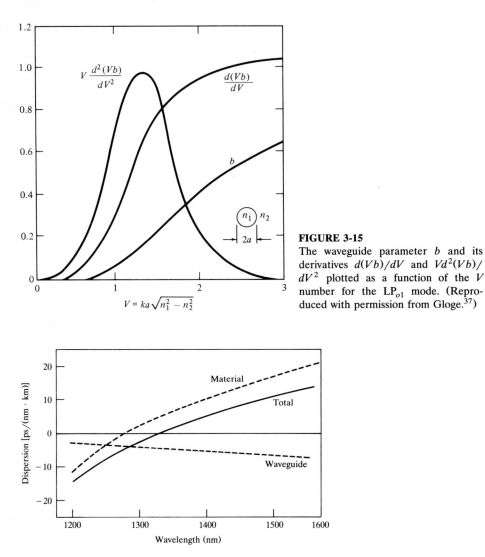

FIGURE 3-15
The waveguide parameter b and its derivatives $d(Vb)/dV$ and $Vd^2(Vb)/dV^2$ plotted as a function of the V number for the LP_{01} mode. (Reproduced with permission from Gloge.[37])

FIGURE 3-16
Examples of the magnitudes of material and waveguide dispersion as a function of optical wavelength for a single-mode fused-silica-core fiber. (Reproduced with permission from Keck,[16] © 1985, IEEE.)

3.2.6   Intermodal Distortion

The final factor giving rise to signal degradation is intermodal distortion, which is a result of different values of the group delay for each individual mode at a single frequency. To see this pictorially, consider the meridional ray picture given for the step-index fiber in Fig. 2-12. The steeper the angle of propagation of the ray congruence, the higher is the mode number and, consequently, the slower the axial group velocity. This variation in the group velocities of the different modes results in a group delay spread or intermodal distortion. This distortion mechanism is eliminated by single-mode operation, but is important in multimode fibers. The pulse broadening arising from intermodal distortion is the difference between the travel time T_{max} of the longest ray congruence paths (the highest-order mode) and the travel time T_{min} of the shortest ray congruence paths (the fundamental mode). This is simply obtained from ray tracing and is given by

$$\sigma_{mod} = T_{max} - T_{min} = \frac{n_1 \Delta L}{c} \tag{3-29}$$

Example 3-4. For values of $n_1 = 1.5$ and $\Delta = 0.01$ the modal spread is $\sigma_{mod} = 0.015L/c$. Comparing this with Eqs. (3-27) and (3-28) with a relative spectral width of $\sigma_\lambda/\lambda = 4$ percent for a light-emitting diode, we have $\sigma_{wg} = 1.2 \times 10^{-4}L/c$ and $\sigma_{mat} = 8 \times 10^{-4}L/c$, which shows that σ_{mod} dominates the pulse spreading by about an order of magnitude in step-index fibers. The relative dominance of σ_{mod} is even greater if a laser diode light source, which has a narrower spectral output width than an LED, is used.

3.3 PULSE BROADENING IN GRADED-INDEX WAVEGUIDES

The analysis of pulse broadening in graded-index waveguides is more involved owing to the radial variation in core refractive index. The feature of this grading of the refractive-index profile is that it offers multimode propagation in a relatively large core together with the possibility of very low intermodal delay distortion. This combination allows the transmission of high data rates over long distances while still maintaining a reasonable degree of light launching and coupling ease. The reason for this low intermodal distortion can be seen by examining the light ray congruence propagation paths shows in Fig. 2-10. Since the index of refraction is lower at the outer edges of the core, light rays will travel faster in this region than in the center of the core where the refractive index is higher. This can be seen from the fundamental relationship $v = c/n$, where v is the speed of light in a medium of refractive index n. Thus the ray congruence characterizing the higher-order mode will tend to travel further than the fundamental ray congruence, but at a faster rate. The higher-order mode will thereby tend to keep up with the lower-order mode, which, in turn, reduces the spread in the modal delay.

The root-mean-square (rms) pulse broadening σ in a graded-index fiber can be obtained from the sum[39]

$$\sigma = \left(\sigma_{intermodal}^2 + \sigma_{intramodal}^2 \right)^{1/2} \tag{3-30}$$

where $\sigma_{intermodal}$ is the rms pulse width resulting from intermodal delay distortion and $\sigma_{intramodal}$ is the rms pulse width resulting from pulse broadening within each mode. To find the intermodal delay distortion, we use the relationship connecting intermodal delay to pulse broadening derived by Personick,[39]

$$\sigma_{\text{intermodal}} = \left(\langle \tau_g^2 \rangle - \langle \tau_g \rangle^2 \right)^{1/2} \tag{3-31}$$

where the group delay τ_g of a mode is given by Eq. (3-14) and the quantity $\langle A \rangle$ is defined as the average of the variable A_{vm} over the mode distribution, that is, it is given by

$$\langle A \rangle = \sum_{v,m} \frac{P_{vm} A_{vm}}{M} \tag{3-32}$$

where P_{vm} is the power contained in the mode of order (v, m).

The group delay

$$\tau_g = \frac{L}{c} \frac{\partial \beta}{\partial k} \tag{3-33}$$

is the time it takes energy in a mode having a propagation constant β to travel a distance L. To evaluate τ_g we solve Eq. (2-96) for β, which yields

$$\beta = \left[k^2 n_1^2 - 2\left(\frac{\alpha + 2}{\alpha} \frac{m}{a^2} \right)^{\alpha/(\alpha+2)} (n_1^2 k^2 \Delta)^{2/(\alpha+2)} \right]^{1/2} \tag{3-34}$$

or, equivalently,

$$\beta = kn_1 \left[1 - 2\Delta \left(\frac{m}{M} \right)^{\alpha/(\alpha+2)} \right]^{1/2}$$

where m is the number of guided modes having propagation constants between $n_1 k$ and β, and M is the total number of possible guided modes given by Eq. (2-97). Substituting Eq. (3-34) into Eq. (3-33), keeping in mind that n_1 and Δ also depend on k, we obtain

$$\tau = \frac{L}{c} \frac{kn_1}{\beta} \left[N_1 - \frac{4\Delta}{\alpha + 2} \left(\frac{m}{M} \right)^{\alpha/(\alpha+2)} \left(N_1 + \frac{n_1 k}{2\Delta} \frac{\partial \Delta}{\partial k} \right) \right]$$

$$= \frac{LN_1}{c} \frac{kn_1}{\beta} \left[1 - \frac{\Delta}{\alpha + 2} \left(\frac{m}{M} \right)^{\alpha/(\alpha+2)} (4 + \epsilon) \right] \tag{3-35}$$

where we have used Eq. (2-97) for M and have defined the quantities

$$N_1 = n_1 + k \frac{\partial n_1}{\partial k} \tag{3-36a}$$

$$\epsilon = \frac{2n_1 k}{N_1 \Delta} \frac{\partial \Delta}{\partial k} \tag{3-36b}$$

As we noted in Eq. (2-46), guided modes only exist for values of β lying between

kn_2 and kn_1. Since n_1 differs very little from n_2, that is,

$$n_2 = n_1(1 - \Delta)$$

where $\Delta \ll 1$ is the core-cladding index difference, it follows that $\beta \simeq n_1 k$. Thus we can use the relationship

$$y = \Delta \left(\frac{m}{M} \right)^{\alpha/(\alpha+2)} \ll 1 \qquad (3\text{-}37)$$

in order to expand Eq. (3-35) in a power series in y. Using the approximation

$$\frac{kn_1}{\beta} = (1 - 2y)^{-1/2} \simeq 1 + y + \frac{3y^2}{2} \qquad (3\text{-}38)$$

we have that

$$\tau = \frac{N_1 L}{c} \left[1 + \frac{\alpha - 2 - \epsilon}{\alpha + 2} \Delta \left(\frac{m}{M} \right)^{\alpha/(\alpha+2)} \right.$$

$$+ \left. \frac{3\alpha - 2 - 2\epsilon}{2(\alpha + 2)} \Delta^2 \left(\frac{m}{M} \right)^{2\alpha/(\alpha+2)} + O(\Delta^3) \right] \qquad (3\text{-}39)$$

Equation (3-39) shows that to first order in Δ, the group delay difference between modes is zero if

$$\alpha = 2 + \epsilon \qquad (3\text{-}40)$$

Since ϵ is generally small, this indicates that minimum intermodal distortion will result from core refractive-index profiles which are nearly parabolic, that is, $\alpha \simeq 2$.

If we assume that all modes are equally excited, that is, $P_{\nu m} = P$ for all modes, and if the number of fiber modes is assumed to be large, then the summation in Eq. (3-32) can be replaced by an integral. Using these assumptions, Eq. (3-39) can be substituted into Eq. (3-31) to yield[40]

$$\sigma_{\text{intermodal}} = \frac{LN_1\Delta}{2c} \frac{\alpha}{\alpha + 1} \left(\frac{\alpha + 2}{3\alpha + 2} \right)^{1/2}$$

$$\times \left[c_1^2 + \frac{4c_1 c_2(\alpha + 1)\Delta}{2\alpha + 1} + \frac{16\Delta^2 c_2^2(\alpha + 1)^2}{(5\alpha + 2)(3\alpha + 2)} \right]^{1/2} \qquad (3\text{-}41)$$

where we have used the abbreviations

$$c_1 = \frac{\alpha - 2 - \epsilon}{\alpha + 2}$$

$$c_2 = \frac{3\alpha - 2 - 2\epsilon}{2(\alpha + 2)} \qquad (3\text{-}42)$$

To find the intramodal pulse broadening, we use the definition[40]

$$\sigma_{\text{intramodal}}^2 = \left(\frac{\sigma_\lambda}{\lambda}\right)^2 \left\langle \left(\lambda \frac{d\tau_g}{d\lambda}\right)^2 \right\rangle \tag{3-43}$$

where σ_λ is the rms spectral width of the optical source. Equation (3-39) can be used to evaluate $\lambda \, d\tau_g/d\lambda$. If we neglect all terms of second and higher order in Δ, we obtain

$$\lambda \frac{d\tau_g}{d\lambda} = -\frac{L}{c}\lambda^2 \frac{d^2 n_1}{d\lambda^2} + \frac{N_1 L \Delta}{c} \frac{\alpha - 2 - \epsilon}{\alpha + 2} \frac{2\alpha}{\alpha + 2} \left(\frac{m}{M}\right)^{\alpha/(\alpha+2)} \tag{3-44}$$

Here we have kept only the largest terms; that is, terms involving factors such as $d\Delta/d\lambda$ and $\Delta \, dn_1/d\lambda$ are negligibly small. Both terms in Eq. (3-44) contribute to $\lambda \, d\tau_g/d\lambda$ for large values of α, since $\lambda^2 \, d^2 n_1/d\lambda^2$ and Δ are the same order of magnitude. However, the second term in Eq. (3-44) is small compared to the first term when α is close to 2.

To evaluate $\sigma_{\text{intramodal}}$ we again assume that all modes are equally excited and that the summation in Eq. (3-32) can be replaced by an integral. Thus substituting Eq. (3-44) into Eq. (3-43) we have[40, 41]

$$\sigma_{\text{intramodal}} = \frac{L}{c} \frac{\sigma_\lambda}{\lambda} \left[\left(-\lambda^2 \frac{d^2 n_1}{d\lambda^2}\right)^2 \right.$$
$$\left. - N_1 c_1 \Delta \left(2\lambda^2 \frac{d^2 n_1}{d\lambda^2} \frac{\alpha}{\alpha + 1} - N_1 c_1 \Delta \frac{4\alpha^2}{(\alpha + 2)(3\alpha + 2)}\right) \right]^{1/2}$$

$$\tag{3-45}$$

Olshansky and Keck[40] have evaluated σ as a function of α at $\lambda = 900$ nm for a titania-doped silica fiber having a numerical aperture of 0.16. This is shown in Fig. 3-17. Here the uncorrected curve assumes $\epsilon = 0$ and includes only intermodal dispersion (no material dispersion). The inclusion of the effect of ϵ shifts the curve to higher values of α. The effect of the spectral width of the optical source on the rms pulse width is clearly demonstrated in Fig. 3-17. The light sources shown are an LED, an injection laser diode, and a distributed-feedback laser having rms spectral widths of 15, 1, and 0.2 nm, respectively. The data transmission capacities of these sources are approximately 0.13, 2, and 10 (Gb · km)/s, respectively.

The value of α which minimizes pulse distortion depends strongly on wavelength. To see this, let us examine the structure of a graded-index fiber. A simple model of this structure is to consider the core to be composed of concentric cylindrical layers of glass, each of which has a different material composition. For each layer the refractive index has a different variation with wavelength λ since the glass composition is different in each layer. Consequently, a fiber with a given index profile α will exhibit different pulse spreading according to the source wavelength used. This is generally called *profile dispersion*. An example of this is given in Fig. 3-18 for a GeO_2-SiO_2 fiber.[42] This

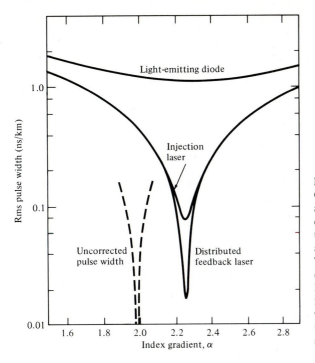

FIGURE 3-17
Calculated rms pulse spreading in a graded-index fiber versus the index parameter α at 900 nm. The uncorrected pulse curve is for $\epsilon = 0$ and assumes mode dispersion only. The other curves include material dispersion for an LED, an injection laser diode, and a distributed-feedback laser having spectral widths of 15, 1, and 0.2 nm, respectively. (Reproduced with permission from Olshansky and Keck.[40])

shows that the optimum value of α decreases with increasing wavelength. Suppose one wishes to transmit at 900 nm. A fiber having an optimum profile α_{opt} at 900 nm should exhibit a sharp bandwidth peak at that wavelength. Fibers with undercompensated profiles, characterized by $\alpha > \alpha_{opt}(900 \text{ nm})$, tend to have a peak bandwidth at a shorter wavelength. On the other hand, overcompensated fibers which have an index profile $\alpha < \alpha_{opt}(900 \text{ nm})$ become optimal at a longer wavelength.

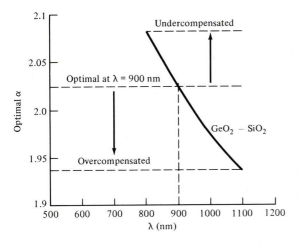

FIGURE 3-18
Profile dispersion effect on the optimum value of α as a function of wavelength for a GeO_2-SiO_2 graded-index fiber. (Reproduced with permission from Cohen, Kaminow, Astle, and Stulz,[42] © 1978, IEEE.)

If the effect of material dispersion is ignored (that is, for $dn_1/d\lambda = 0$), an expression for the optimum index profile can be found from the minimum of Eq. (3-41) as a function of α. This occurs at[40]

$$\alpha_{opt} = 2 + \epsilon - \Delta\frac{(4 + \epsilon)(3 + \epsilon)}{5 + 2\epsilon} \qquad (3\text{-}46)$$

If we take $\epsilon = 0$ and $dn_1/d\lambda = 0$, then Eq. (3-41) reduces to

$$\sigma_{opt} = \frac{n_1\Delta^2 L}{20\sqrt{3}\,c} \qquad (3\text{-}47)$$

This can be compared with the dispersion in a step-index fiber by setting $\alpha = \infty$ and $\epsilon = 0$ in Eq. (3-41), yielding

$$\sigma_{step} = \frac{n_1\Delta L}{c}\frac{1}{2\sqrt{3}}\left(1 + 3\Delta + \frac{12\Delta^2}{5}\right)^{1/2} \simeq \frac{n_1\Delta L}{2\sqrt{3}\,c} \qquad (3\text{-}48)$$

Thus, under the assumptions made in Eqs. (3-47) and (3-48),

$$\frac{\sigma_{step}}{\sigma_{opt}} = \frac{10}{\Delta} \qquad (3\text{-}49)$$

Hence, since typical values of Δ are 0.01, Eq. (3-49) indicates that the capacity of a graded-index fiber is about three orders of magnitude larger than that of a step-index fiber. For $\Delta = 1$ percent the rms pulse spreading in a step-index fiber is about 14 ns/km, whereas that for a graded-index fiber is calculated to be 0.014 ns/km. In practice these values are greater because of manufacturing difficulties. For example, it has been shown[43] that, although theory predicts a bandwidth of about 8 GHz · km, in practice very slight deviations of the refractive-index profile from its optimum shape, owing to unavoidable manufacturing tolerances, can decrease the fiber bandwidth dramatically. This is illustrated in Fig. 3-19 for fibers with $\Delta = 1\%$, 1.3%, and 2%. A change in α of a few percent can decrease the bandwidth by an order of magnitude.

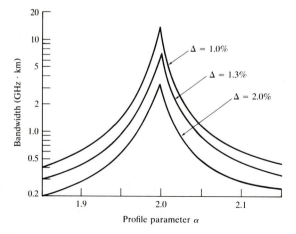

FIGURE 3-19
Variations in bandwidth resulting from slight deviations in the refractive-index profile for a graded-index fiber with $\Delta = 1$, 1.3, and 2 percent. (Reproduced with permission from Marcuse and Presby.[43])

3.4 MODE COUPLING

In real systems pulse distortion will increase less rapidly after a certain initial length of fiber because of mode coupling and differential mode loss.[44-46] In this initial length of fiber, coupling of energy from one mode to another arises because of structural imperfections, fiber diameter and refractive-index variations, and cabling-induced microbends. The mode coupling tends to average out the propagation delays associated with the modes, thereby reducing intermodal dispersion. Associated with this coupling is an additional loss, which is designated by h and which has units of dB/km. The result of this phenomenon is that, after a certain coupling length L_c, the pulse distortion will change from an L dependence to a $(L_c L)^{1/2}$ dependence.

The improvement in pulse spreading caused by mode coupling over the distance $Z < L_c$ is related to the excess loss hZ incurred over this distance by the equation

$$hZ \left(\frac{\sigma_c}{\sigma_0} \right)^2 = C \qquad (3\text{-}50)$$

Here C is a constant, σ_0 is the pulse width increase in the absence of mode coupling, σ_c is the pulse broadening in the presence of strong mode coupling, and hZ is the excess attenuation resulting from mode coupling. The constant C in Eq. (3-50) is independent of all dimensional quantities and refractive indices. It depends only on the fiber profile shape, the mode-coupling strength, and the modal attenuations.

The effect of mode coupling on pulse distortion can be significant for long fibers, as is shown in Fig. 3-20 for various coupling losses in a graded-index fiber. The parameters of this fiber are $\Delta = 1$ percent, $\alpha = 4$, and $C = 1.1$. The coupling loss h must be determined experimentally, since a calculation would require a detailed knowledge of the mode coupling introduced by the various

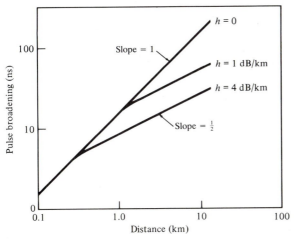

FIGURE 3-20
Mode-coupling effects on pulse distortion in long fibers for various coupling losses.

waveguide perturbations. Measurements of bandwidth as a function of distance have produced values of L_c ranging from about 100 to 550 m.

An important point to note is that extensive mode coupling and power distribution can occur at connectors, splices, and other passive components in an optical link, which can have a significant effect on the overall system bandwidth.[47-49]

3.5 DESIGN OPTIMIZATION OF SINGLE-MODE FIBERS

Since telecommunication companies use single-mode fibers as the principal optical transmission medium in their networks, and because of the importance of single-mode fibers in microwave-speed localized applications,[50] this section addresses their basic design and operational properties. Some of the attributes of single-mode fibers include a long expected installation lifetime, very low attenuation, high-quality signal transfer because of the absence of modal noise, and the largest available bandwidth-distance product. Here we shall examine design-optimization characteristics, cutoff wavelength, dispersion, mode-field diameter, and bending loss.

3.5.1 Refractive-Index Profiles

In the design of single-mode fibers, dispersion behavior is a major distinguishing feature, since this is what limits long-distance and very high-speed transmission. Comparing Figs. 3-3 and 3-16, we see that whereas the dispersion of a single-mode silica fiber is lowest at 1300 nm, its attenuation is a minimum at 1550 nm, where the dispersion is higher. Ideally, for achieving a maximum transmission distance of a high-capacity link, the dispersion null should be at the wavelength of minimum attenuation. To achieve this, one can adjust the basic fiber parameters to shift the zero dispersion minimum to longer wavelengths.

The basic material dispersion is hard to alter significantly, but it is possible to modify the waveguide dispersion by changing from a simple step-index core profile design to more complicated index profiles.[13, 51-57] Researchers have thus examined a variety of core and cladding refractive-index configurations for altering the behavior of single-mode fibers. Figure 3-21 shows representative refractive index profiles of the three main categories, these being 1300-nm-optimized fibers, dispersion-shifted fibers, and dispersion-flattened fibers. To get a better feeling of their geometry, Fig. 3-22 shows the three-dimensional index profiles for several different types of single-mode fibers.

The most popular single-mode fibers used in telecommunication networks are near-step-index fibers, which are dispersion-optimized for operation at 1300 nm. These *1300-nm-optimized single-mode fibers* are either of the *matched-cladding*[13, 51, 52] or the *depressed-cladding*[53, 54] design, as shown in Figs. 3-21*a*, 3-22*a* and 3-22*b*. Matched-cladding fibers have a uniform refractive index throughout the cladding. Typical mode-field diameters are 9.5 μm and core-to-cladding

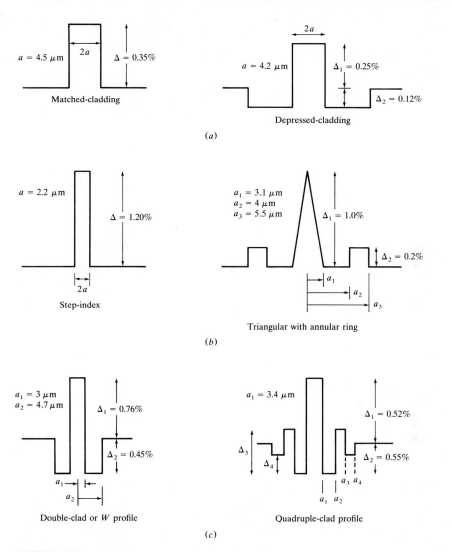

FIGURE 3-21
Representative index profiles for (*a*) 1300-nm-optimized, (*b*) dispersion-shifted, and (*c*) dispersion-flattened single-mode fibers.

index differences are around 0.37 percent. In depressed-cladding fibers the cladding portion next to the core has a lower index than the outer cladding region. Mode-field diameters are around 9 μm, and typical positive and negative index differences are 0.25 and 0.12 percent, respectively.

As we saw from Eqs. (3-20) and (3-26), whereas material dispersion depends only on the composition of the material, waveguide dispersion is a function of the core radius, the refractive-index difference, and the shape of the

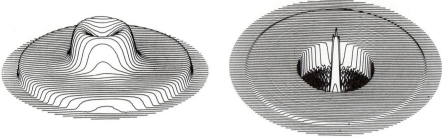

(a)   (b)

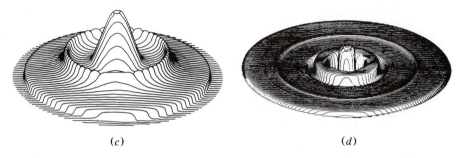

(c)   (d)

FIGURE 3-22
Three-dimensional refractive index profiles for (a) matched-cladding 1300-nm-optimized, (b) depressed-cladding 1300-nm-optimized, (c) triangular dispersion-shifted, and (d) quadruple-clad dispersion-flattened single-mode fibers. [(a) and (c) courtesy of Corning, Inc.; (b) courtesy of York Technology; (d) reproduced with permission from H. Lydtin, *J. Lightwave Tech.*, vol. LT-4, pp. 1034–1038, Aug. 1986, © 1986, IEEE.]

refractive-index profile. Thus the waveguide dispersion can vary dramatically with the fiber design parameters. By shifting the waveguide dispersion to longer wavelengths and assuming a constant value for the material dispersion, addition of waveguide and material dispersion then can produce zero total dispersion at 1550 nm. The resulting optical waveguides are known as *dispersion-shifted fibers*.[50, 52, 55–58] Examples of low-dispersion refractive-index profiles for single-mode fibers at 1550 nm are shown in Figs. 3-21b and 3-22c. Figure 3-23 gives the resultant total-dispersion curve.

An alternative is to reduce fiber dispersion by spreading the dispersion minimum out over a broader range. This approach is known as *dispersion flattening*. Dispersion-flattened fibers[59–61] are more complex to design than dispersion-shifted fibers, because dispersion must be considered over a range of wavelengths. However, they offer desirable characteristics over a much broader range of wavelengths, and thus could be used for optical wavelength-division multiplexing (see Chap. 11). Figures 3-21c and 3-22d show typical cross-sectional and three-dimensional refractive-index profiles, respectively. Figure 3-23 gives the resultant dispersion characteristic.

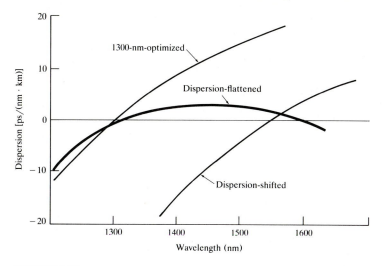

FIGURE 3-23
Typical dispersion characteristics for (*a*) 1300-nm-optimized, (*b*) dispersion-shifted, and (*c*) dispersion-flattened single-mode fibers.

3.5.2 Cutoff Wavelength

The cutoff wavelength of the first higher-order mode (LP_{11}) is an important transmission parameter for single-mode fibers, since it separates the single-mode from the multimode regions.[62-66] As we saw from Eq. (2-58), single-mode operation occurs above the theoretical cutoff wavelength given by

$$\lambda_{c;th} = \frac{2\pi a}{V}\left(n_1^2 - n_2^2\right)^{1/2} \tag{3-51}$$

with $V = 2.405$ for step-index fibers. At this wavelength only the LP_{01} mode (that is, the HE_{11} mode) should propagate in the fiber.

Since in the cutoff region the field of the LP_{11} mode is widely spread across the fiber cross-section (that is, it is not tightly bound to the core), its attenuation is strongly affected by fiber bends, length, and cabling. Recommendation G.652 of the CCITT[66] and the EIA-455-80 Standard[67] define an effective cutoff wavelength λ_c for a 2-m length of fiber containing a single 14-cm-radius loop to be the wavelength at which the difference in loss of the higher-order LP_{11} mode and the fundamental LP_{01} mode is about 20 dB. Recommended values of λ_c range from 1100 to 1280 nm to avoid modal noise and dispersion problems in the 1300-nm region. The choice of length is somewhat flexible. Many manufacturers choose a 2-m length, whereas AT & T uses a 5-m length for the depressed-cladding design.[57]

As λ_c is stabilizing in short fiber lengths, it decreases as

$$d\lambda_c = -m \log \frac{L}{2} \tag{3-52}$$

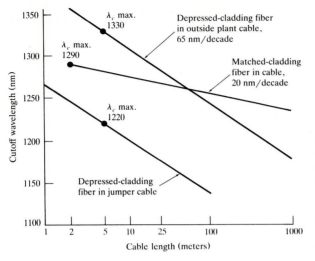

FIGURE 3-24
Cutoff wavelength as a function of fiber length for depressed-cladding (defined at a 5-m length) and matched-cladding (defined at a 2-m length) single-mode fibers. (Reproduced with permission from Kalish and Cohen,[57] © 1987, AT & T.)

where m depends on the fiber type and L is the fiber length in meters. Values of m range from 20 to 60 nm, with matched-cladding fibers tending to be lower than depressed-cladding fibers. Figure 3-24 shows some typical cutoff wavelengths for matched-cladding and depressed-cladding single-mode fibers.

3.5.3 Dispersion

As noted in Sec. 3.5.1, the total dispersion in single-mode fibers consists mainly of material and waveguide dispersions. The dispersion D is represented by[68-69]

$$D(\lambda) = \frac{1}{L}\frac{d\tau}{d\lambda} \qquad (3\text{-}53)$$

which is expressed in ps/(nm · km). The total broadening σ of an optical pulse over a length of fiber L is given by

$$\sigma = D(\lambda)L\sigma_\lambda \qquad (3\text{-}54)$$

where σ_λ is the wavelength spread of the source. To measure the dispersion, one examines the pulse delay over a wide wavelength range. At the zero-dispersion point the pulse delay will go through a minimum. To calculate the dispersion near 1300 nm, the EIA has recommended fitting a three-term Sellmeier equation of the form[67]

$$\tau = A + B\lambda^2 + C\lambda^{-2} \qquad (3\text{-}55)$$

to the pulse-delay data. For zero dispersion near 1550 nm, a five-term Sellmeier equation of the form

$$\tau = A + B\lambda^4 + C\lambda^2 + D\lambda^{-2} + E\lambda^{-4} \qquad (3\text{-}56)$$

is recommended. Considering only the three-term equation, we take the derivative of the fitted curve

$$\frac{d\tau}{d\lambda} = 2B\lambda - 2C\lambda^{-3} \tag{3-57}$$

to get the zero-dispersion wavelength λ_0. Letting S_0 be the slope of $D(\lambda)$ at λ_0, we then have

$$D(\lambda) = \frac{\lambda S_0}{4}\left[1 - \left(\frac{\lambda_0}{\lambda}\right)^4\right] \tag{3-58}$$

where S_0 is given in ps/(nm$^2 \cdot$ km).

When measuring a set of fibers, one will get values of λ_0 ranging from $\lambda_{0,min}$ to $\lambda_{0,max}$. Figure 3-25 shows the range of expected dispersion values for such a set of fibers. Typical values of S_0 are 0.092 ps/(nm$^2 \cdot$ km) for standard single-mode fibers, and between 0.06 and 0.08 ps/(nm$^2 \cdot$ km) for dispersion-shifted fibers. Alternatively, the CCITT Recommendation G.652 has specified this as a maximum dispersion of 3.5 ps/(nm $\cdot$ km) in the 1285- to 1330-nm region.

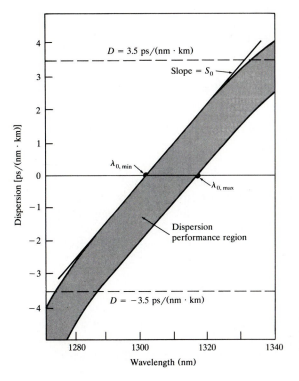

FIGURE 3-25
Example of a dispersion performance curve for a set of single-mode fibers. The two slightly curved lines are found by solving Eq. (3-58). S_0 is the slope of $D(\lambda)$ at the zero-dispersion wavelength λ_0.

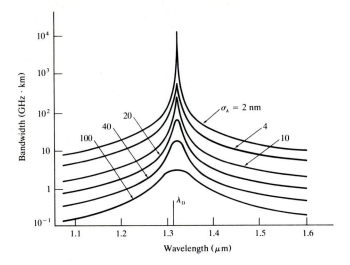

FIGURE 3-26
Examples of bandwidth versus wavelength for different source spectral widths σ_λ in a single-mode fiber having a dispersion minimum at 1300 nm. (Reproduced with permission from Reed, Cohen, and Shang,[56] © 1987, AT & T.)

Figure 3-26 illustrates the importance of controlling dispersion in single-mode fibers. As optical pulses travel down a fiber, temporal broadening occurs because material and waveguide dispersion cause different wavelengths in the optical pulse to propagate with different velocities. Thus, as Eq. (3-54) implies, the broader the spectral width σ_λ of the source, the greater the pulse dispersion will be. The effect of source width on fiber bandwidth is clearly seen in Fig. 3-26.

3.5.4 Mode-Field Diameter

Section 2.5.1 gives the definition of the mode-field diameter in single-mode fibers. One uses the mode-field diameter in describing the functional properties of a single-mode fiber, since it takes into account the wavelength-dependent field penetration into the cladding. This is shown in Fig. 3-27 for 1300-nm-optimized, dispersion-shifted, and dispersion-flattened single-mode fibers.

3.5.5 Bending Loss

Macrobending and microbending losses are important in the design of single-mode fibers.[19-25] These losses are principally evident in the 1550-nm region, and show up as a rapid increase in attenuation when the fiber is bent smaller

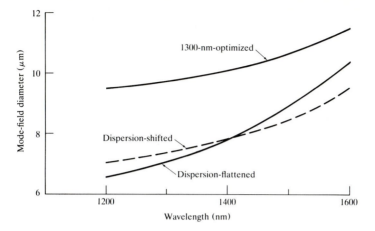

FIGURE 3-27
Typical mode-field diameter variations with wavelength for (*a*) 1300-nm-optimized, (*b*) dispersion-shifted, and (*c*) dispersion-flattened single-mode fibers.

than a certain bend radius. Single-mode fibers are more susceptible to bending losses the lower the cutoff wavelength relative to the operating wavelength. For example, in a fiber which is optimized for operation at 1300 nm, both the microbending and macrobending losses are greater at 1550 nm than at 1300 nm by a factor of 3 to 5, as Fig. 3-28 illustrates. A fiber thus might be transmitting at 1300 nm but have a significant loss at 1550 nm.

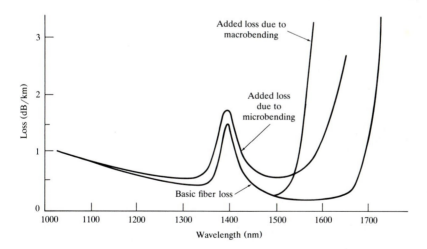

FIGURE 3-28
Representative increases in single-mode fiber attenuation owing to microbending and macrobending effects. (Reproduced with permission from Kalish and Cohen,[57] © 1987, AT & T.)

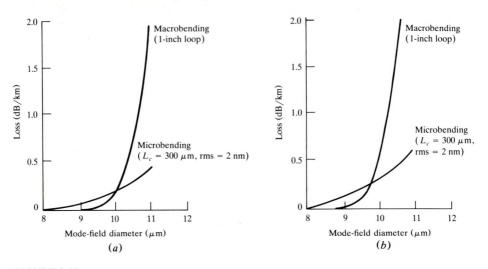

FIGURE 3-29
Calculated increase in attenuation at 1310 nm from microbending and macrobending effects as a function of mode-field diameter for (a) depressed-cladding single-mode fiber (V = 2.514) and (b) matched-cladding single-mode fiber (V = 2.373). The microbending calculations assume a correlation length L_c (microbending repetition rate) of 300 nm and a 2-nm deformation amplitude. (Reproduced with permission from Kalish and Cohen,[57] © 1987, AT & T.)

The bending losses are primarily a function of the mode-field diameter. Generally, the smaller the mode-field diameter (that is, the tighter the confinement of the mode to the core), the smaller the bending loss. This is true for both matched-clad and depressed-clad fibers, as Fig. 3-29 shows.

Figure 3-30 gives examples of calculated and measured macrobending losses.[70] By specifying bend-radius limitations, one can largely avoid high

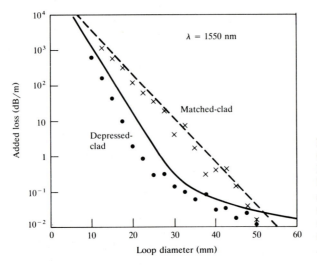

FIGURE 3-30
Calculated and measured bending losses at 1550 nm as a function of loop diameter for matched-cladding (dashed line and crosses) and depressed-cladding (solid line and dots) single-mode fibers. (Reproduced with permission from Andreason.[70])

macrobending losses. Manufacturers usually recommend a minimum fiber or cable bend diameter of 40 to 50 mm (1.6 to 2.0 in.). This is consistent with typical bend diameters of 50 to 75 mm found in fiber-splice enclosures, in equipment bays, or on optoelectronic packages. Since single-mode fibers are designed to have little or no additional attenuation at 1550 nm from bend diameters greater than 50 mm, bending loss should not be a limiting performance factor in correctly installed cables.

3.6 SUMMARY

Attenuation of a light signal as it propagates along a fiber is an important consideration in the design of an optical communication system, since it plays a major role in determining the maximum transmission distance between a transmitter and a receiver. The basic attenuation mechanisms are absorption, scattering, and radiative losses of optical energy. The major causes of absorption are extrinsic absorption by impurity atoms and intrinsic absorption by the basic constituent atoms of the fiber material. Intrinsic absorption, which sets the fundamental lower limit on attenuation for any particular material, results from electronic absorption bands in the ultraviolet region and from atomic vibration bands in the near-infrared region. Scattering losses arise from microscopic variations in the material density, from compositional fluctuations, and from structural inhomogeneities or defects occurring during fiber manufacture. Scattering follows a Rayleigh λ^{-4} dependence, which gives attenuation-versus-wavelength plots their characteristic downward trend with increasing wavelength.

Radiative losses occur whenever an optical fiber undergoes a bend. They can arise from macroscopic bends, such as when a fiber cable turns a corner, or from microscopic bends (microbends) of the fiber axis occurring in fiber manufacturing, during cabling, or from temperature-induced shrinking of the fiber. Of these, microbends are the most troublesome, so that special care must be taken during manufacturing, cabling, and installation to minimize them.

In addition to being attenuated, an optical signal becomes increasingly distorted as it travels along a fiber. This distortion is a consequence of intramodal and intermodal dispersion effects. Intramodal dispersion is pulse spreading that occurs within an individual mode and thus is of importance in single-mode fibers. Its two main causes are:

1. Material dispersion, which arises from the variation of the refractive index of the core material as a function of wavelength.
2. Waveguide dispersion, which depends on the fiber design, since the modal propagation constant β is a function of a/λ (the optical fiber dimension relative to the wavelength λ, where a is the core radius)

In multimode fibers, pulse distortions also occur, since each mode travels at a different group velocity (which is known as intermodal delay distortion). Mode delay is the dominant pulse-distorting mechanism in step-index fibers. On the other hand, intermodal delay distortion can be made very small by careful tailoring of the core refractive-index profile. Material dispersion thus tends to be the dominant pulse-distorting effect in graded-index fibers.

PROBLEMS

3-1. A certain optical fiber has an attenuation of 1.5 dB/km at 1300 nm. If 0.5 mW of optical power is initially launched into the fiber, what is the power level in microwatts after 8 km?

3-2. An optical signal has lost 55 percent of its power after traversing 3.5 km of fiber. What is the loss in dB/km of this fiber?

3-3. A continuous 12-km-long optical fiber link has a loss of 1.5 dB/km. (*a*) What is the minimum optical power level that must be launched into the fiber to maintain an optical power level of 0.3 μW at the receiving end? (*b*) What is the required input power if the fiber has a loss of 2.5 dB/km?

3-4. Consider a step-index fiber with a SiO_2-GeO_2 core having a mole fraction 0.08 of GeO_2. Plot Eqs. (3-2*b*) and (3-3) from 500 nm to 5 μm, and compare the results with the curves in Fig. 3-5.

3-5. The optical power loss resulting from Rayleigh scattering in a fiber can be calculated from either Eq. (3-4*a*) or Eq. (3-4*b*). Compare these two equations for silica (n = 1.460 at 630 nm), given that the fictive temperature T_f is 1400 K, the isothermal compressibility β_T is 6.8×10^{-12} cm²/dyn, and the photoelastic coefficient is 0.286. How does this agree with measured values ranging from 3.9 to 4.8 dB/km at 633 nm?

3-6. Consider graded-index fibers having index profiles α = 2.0, cladding refractive indices n_2 = 1.50, and index differences Δ = 0.01. Using Eq. (3-7), plot the ratio N_{eff}/N_∞ for bend radii less than 10 cm at λ = 1 μm for fibers having core radii of 4, 25, and 100 μm.

3-7. Three common fiber jacket materials are Elvax® 265 (E_j = 21 MPa) and Hytrel® 4056 (E_j = 58 MPa), both made by DuPont, and Versalon® 1164 (E_j = 104 MPa) made by General Mills. If the Young's modulus of a glass fiber is 64 GPa, plot the reduction in microbending loss as a function of the index difference Δ when fibers are coated with these materials. Make these plots for Δ values ranging from 0.1 to 1.0 percent and for a fiber cladding-to-core ratio of b/a = 2.

3-8. Assume that a step-index fiber has a V number of 6.0.
(*a*) Using Fig. 2-22, estimate the fractional power P_{clad}/P traveling in the cladding for the six lowest-order LP modes.
(*b*) If the fiber in (*a*) is a glass-core, glass-clad fiber having core and cladding attenuations of 3.0 and 4.0 dB/km, respectively, find the attenuations for each of the six lowest-order modes.
(*c*) Suppose the fiber in (*a*) is a glass-core, polymer-clad fiber having core and cladding attenuations of 5 and 1000 dB/km, respectively. Find the attenuations for each of the six lowest-order modes.

3-9. Assume a given mode in a graded-index fiber has a power density $p(r)$ = $P_0 \exp(-Kr^2)$, where the factor K depends on the modal power distribution.
(*a*) Letting $n(r)$ in Eq. (3-12) be given by Eq. (2-78) with α = 2, show that the this mode is

$$\alpha_{gi} = \alpha_1 + \frac{\alpha_2 - \alpha_1}{Ka^2}$$

Since $p(r)$ is a rapidly decaying function of r and since $\Delta \ll 1$, for ease of calculation assume that the top relation in Eq. (2-78) holds for all values of r.

(b) Choose K such that $p(a) = 0.1 P_0$, that is, 10 percent of the power flows in the cladding. Find α_{gi} in terms of α_1 and α_2.

3-10. For wavelengths less than 1.0 μm the refractive index n satisfies a Sellmeier relation of the form[70]

$$n^2 = 1 + \frac{E_0 E_d}{E_0^2 - E^2}$$

where $E = hc/\lambda$ is the photon energy and E_0 and E_d are, respectively, material oscillator energy and dispersion energy parameters. In SiO_2 glass $E_0 = 13.4$ eV and $E_d = 14.7$ eV. Show that, for wavelengths between 0.20 and 1.0 μm, the values of n found from the Sellmeier relation are in good agreement with those shown in Fig. 3-12.

3-11. Derive the group delay given by Eq. (3-19) from the propagation constant β given in Eq. (3-18).

3-12. (a) An LED operating at 850 nm has a spectral width of 45 nm. What is the pulse spreading in ns/km due to material dispersion? What is the pulse spreading when a laser diode having a 2-nm spectral width is used?

(b) Find the material-dispersion-induced pulse spreading at 1550 nm for an LED with a 75-nm spectral width.

3-13. (a) Using Eqs. (2-48), (2-49), and (2-57) show that the normalized propagation constant defined by Eq. (3-21) can be written in the form

$$b = \frac{\beta^2/k^2 - n_2^2}{n_1^2 - n_2^2}$$

(b) For small core-cladding refractive-index differences show that the expression for b derived in (a) reduces to

$$b = \frac{\beta/k - n_2}{n_1 - n_2}$$

from which it follows that

$$\beta = n_2 k(b\Delta + 1)$$

3-14. Using the approximation $V \simeq kan_2 \sqrt{2\Delta}$, show that Eq. (3-25) results from Eq. (3-24).

3-15. Derive Eq. (3-29) by using a ray-tracing method.

3-16. Verify that Eq. (3-41) reduces to Eq. (3-48) for the case $\alpha = \infty$ and $\epsilon = 0$.

3-17. Show that, when the effect of material dispersion is ignored and for $\epsilon = 0$, Eq. (3-41) reduces to Eq. (3-47).

3-18. Make a plot on log-log paper of the rms pulse broadening in a parabolic graded-index fiber ($\alpha = 2$) as a function of the optical source spectral width σ_λ in the range 0.10 to 100 nm for peak operating wavelengths of 850 nm and 1300 nm. Let $\Delta = 0.01$, $N_1 = 1.46$, and $\epsilon = 0$ at both wavelengths. Assume the factor $\lambda^2 d^2 n/d\lambda^2$ is 0.025 at 850 nm and 0.004 at 1300 nm.

3-19. Repeat Prob. 3-18 for a graded-index single-mode fiber with $\Delta = 0.001$.

3-20. Derive Eq. (3-35) by substituting Eq. (3-34) into Eq. (3-33).

3-21. Using the approximation given by Eq. (3-38), show that Eq. (3-35) can be rewritten as Eq. (3-39).

3-22. Derive Eq. (3-44) from Eq. (3-39).

3-23. Verify the expression given in Eq. (3-45).

3-24. Starting with Eq. (3-55), derive the dispersion expression given in Eq. (3-58).

REFERENCES

1. B. C. Bagley, C. R. Kurkjian, J. W. Mitchell, G. E. Peterson, and A. R. Tynes, "Materials, properties, and choices," in S. E. Miller and A. G. Chynoweth, eds., *Optical Fiber Telecommunications*, Academic, New York, 1979.
2. P. Kaiser and D. B. Keck, "Fiber types and their status," in S. E. Miller and I. P. Kaminow, eds., *Optical Fiber Telecommunications—II*, Academic, New York, 1988.
3. R. Olshansky, "Propagation in glass optical waveguides," *Rev. Mod. Phys.*, vol. 51, pp. 341–367, Apr. 1979.
4. D. Gloge, "The optical fibre as a transmission medium," *Rpts. Prog. Phys.*, vol. 42, pp. 1777–1824, Nov. 1979.
5. A. Iino and J. Tamura, "Radiation resistivity in silica optical fibers," *J. Lightwave Tech.*, vol. 6, pp. 145–149, Feb. 1988.
6. J. R. Haber, E. Mies, J. R. Simpson, and S. Wong, "Assessment of radiation-induced loss for AT & T fiber-optic transmission systems in the terrestrial environment," *J. Lightwave Tech.*, vol. 6, pp. 150–154, Feb. 1988.
7. R. H. West, "A local view of radiation effects in fiber optics," *J. Lightwave Tech.*, vol. 6, pp. 155–164, Feb. 1988.
8. (*a*) E. J. Friebele, E. W. Taylor, G. Turguet de Beauregard, J. A. Wall, and C. E. Barnes, "Interlaboratory comparison of radiation-induced attenuation in optical fibers," *J. Lightwave Tech.*, vol. 6, pp. 165–171, Feb. 1988.
   (*b*) E. W. Taylor, E. J. Friebele, H. Henschel, R. H. West, J. A. Krinsky, and C. E. Barnes, "Interlaboratory comparison of radiation-induced attenuation in optical fibers. Part II: Steady state," *J. Lightwave Tech.*, vol. 8, pp. 967–976, June 1990.
   (*c*) E. J. Friebele, P. B. Lyons, J. Blackburn, H. Henschel, A. Johan, J. A. Krinsky, A. Robinson, W. Schneider, D. Smith, E. W. Taylor, G. Turguet de Beauregard, R. H. West, and P. Zagarino, "Interlaboratory comparison of radiation-induced attenuation in optical fibers. Part III: Transient exposures," *J. Lightwave Tech.*, vol. 8, pp. 977–989, June 1990.
9. T. Moriyama, O. Fukuda, K. Sanada, K. Inada, T. Edahiro, and K. Chida, "Ultimately low OH content VAD optical fibers," *Electron Lett.*, vol. 16, pp. 699–700, Aug. 1980.
10. V. Miya, Y. Terunuma, T. Hosaka, and T. Miyashita, "Ultra low loss single-mode fibers at 1.55 μm," *Electron. Lett.*, vol. 15, pp. 106–108, 1979.
11. (*a*) S. R. Nagel, J. B. MacChesney, and K. L. Walker, "An overview of the MCVD process and performance," *IEEE J. Quantum Electron.*, vol. QE-18, pp. 459–476, Apr. 1982.
   (*b*) S. R. Nagel, "Fiber materials and fabrication methods," in S. E. Miller and I. P. Kaminow, eds., *Optical Fiber Telecommunications—II*, Academic, New York, 1988.
12. H. Kanamori, H. Yokota, G. Tanaka, M. Watanabe, Y. Ishiguro, I. Yoshida, T. Kakii, S. Itoh, Y. Asano, and S. Tanaka, "Transmission characteristics and reliability of pure-silica-core single-mode fibers," *J. Lightwave Tech.*, vol. LT-4, pp. 1144–1150, Aug. 1986.
13. H. Osanai, T. Shioda, T. Moriyama, S. Araki, M. Horiguchi, T. Izawa, and H. Takata, "Effects of dopants on transmission loss of low OH content optical fibers," *Electron. Lett.*, vol. 12, pp. 549–550, Oct. 1976.
14. R. Maurer, "Glass fibers for optical communications," *Proc. IEEE*, vol. 61, pp. 452–462, Apr. 1973.

15. D. A. Pinnow, T. C. Rich, F. W. Ostermeyer, and M. DiDomenico, Jr., "Fundamental optical attenuation limits in the liquid and gassy state with application to fiber optical waveguide material," *Appl. Phys. Lett.*, vol. 22, pp. 527–529, May 1973.
16. D. B. Keck, "Fundamentals of optical waveguide fibers, *IEEE Commun. Magazine*, vol. 23, pp. 17–22, May 1985.
17. D. Marcuse, "Curvature loss formula for optical fibers," *J. Opt. Soc. Amer.*, vol. 66, pp. 216–220, Mar. 1976.
18. D. Gloge, "Bending loss in multimode fibers with graded and ungraded core index," *Appl. Opt.*, vol. 11, pp. 2506–2512, Nov. 1972.
19. A. J. Harris and P. F. Castle, "Bend loss measurements on high numerical aperture single-mode fibers as a function of wavelength and bend radius," *J. Lightwave Tech.*, vol. LT-4, pp. 34–40, Jan. 1986.
20. S. J. Garth, "Fields in a bent single-mode fibre," *Electron. Lett.*, vol. 23, pp. 373–374, Apr. 9, 1987.
21. G. L. Tangonan, H. P. Hsu, V. Jones, and J. Pikulski, "Bend loss measurements for small mode field diameter fibers," *Electron Lett.*, vol. 25, pp. 142–143, Jan. 19, 1989.
22. A. H. Badar, T. S. M. Maclean, B. K. Gazey, J. F. Miller, and H. Ghafoori-Shiraz, "Radiation from circular bends in multimode and single-mode optical fibres," *IEE Proc.*, vol. 136, pt. J, pp. 147–151, June 1989.
23. N. Kamikawa and C.-T. Chang, "Losses in small-radius bends in single-mode fibres," *Electron. Lett.*, vol. 25, pp. 947–949, July 20, 1989.
24. L. B. Jeunhomme, *Single-Mode Fiber Optics*, Dekker, New York, 2nd ed., 1989.
25. J. D. Love, "Application of low-loss criterion to optical waveguides and devices." *IEE Proc.*, vol. 136, pt. J, pp. 225–228, Aug. 1989.
26. W. B. Gardner, "Microbending loss in optical fibers," *Bell Sys. Tech. J.*, vol. 54, pp. 457–465, Feb. 1975.
27. J. Sakai and T. Kimura, "Practical microbending loss formula for single mode optical fibers," *IEEE J. Quantum Electron.*, vol. QE-15, pp. 497–500, June 1979.
28. K. Furuya and Y. Suematsu, "Random-bend loss in single-mode and parabolic-index multimode optical fiber cables," *Appl. Opt.*, vol. 19, pp. 1493–1500, May 1980.
29. J. H. Povlsen and S. B. Andreasen, "Analysis on splice, microbending, macrobending, and Rayleigh losses in GeO_2-doped dispersion-shifted single-mode fibers," *J. Lightwave Tech.*, vol. LT-4, pp. 706–710, July 1986; *J. Lightwave Tech.*, vol. 6, p. 1447, Sept. 1988.
30. E. Suhir, "Effect of initial curvature on low temperature microbending in optical fibers," *J. Lightwave Tech.*, vol. 6, pp. 1321–1327, Aug. 1988.
31. D. Gloge, "Optical fiber packaging and its influence on fiber straightness and loss," *Bell Syst. Tech. J.*, vol. 54, pp. 245–262, Feb. 1975.
32. D. Gloge, "Propagation effects in optical fibers," *IEEE Trans. Microwave Theory Tech.*, vol. MTT-23, pp. 106–120, Jan. 1975.
33. A. W. Snyder and J. D. Love, *Optical Waveguide Theory*, Routledge Chapman and Hall, New York, 1983.
34. D. Gloge, E. A. J. Marcatili, D. Marcuse, and S. D. Personick, "Dispersion properties of fibers," in S. E. Miller and A. G. Chynoweth, eds., *Optical Fiber Telecommunications*, Academic, New York, 1979.
35. D. Marcuse, "Interdependence of waveguide and material dispersion," *Appl. Opt.*, vol. 18, pp. 2930–2932, Sept. 1979.
36. R. P. Kapron and D. B. Keck, "Pulse transmission through a dielectric optical waveguide," *Appl. Opt.*, vol. 10, pp. 1519–1523, July 1971.
37. D. Gloge, "Weakly guiding fibers," *Appl. Opt.*, vol. 10, pp. 2252–2258, Oct. 1971; "Dispersion in weakly guiding fibers," *Appl. Opt.*, vol. 10, pp. 2442–2445, Nov. 1971.
38. J. W. Fleming, "Material dispersion in lightguide glasses," *Electron. Lett.*, vol. 14, pp. 326–328, May 25, 1978.
39. S. D. Personick, "Receiver design for digital fiber optic communication systems," *Bell Sys. Tech. J.*, vol. 52, pp. 843–874, July-Aug. 1973.

40. R. Olshansky and D. Keck, "Pulse broadening in graded index optical fibers," *Appl. Opt.*, vol. 15, pp. 483–491, Feb. 1976.

41. G. Einarsson, "Pulse broadening in graded index optical fibers: Correction," *Appl. Opt.*, vol. 25, p. 1030, Apr. 1986.

42. L. Cohen, I. Kaminow, H. Astle, and L. Stulz, "Profile dispersion effects on transmission bandwidths in graded index optical fibers," *IEEE J. Quantum Electron.*, vol. QE-14, pp. 37–41, Jan. 1978.

43. D. Marcuse and H. M. Presby, "Effects of profile deformation on fiber bandwidth," *Appl. Opt.*, vol. 18, pp. 3758–3763, Nov. 1979; *Appl. Opt.*, vol. 19, p. 188, Jan. 1980.

44. (*a*) R. Olshansky, "Mode coupling effects in graded index optical fibers," *Appl. Opt.*, vol. 14, pp. 935–945, Apr. 1975.
    (*b*) S. Geckeler, "Pulse broadening in optical fibers with mode mixing," *Appl. Opt.*, vol. 18, pp. 2192–2198, July 1979.

45. M. J. Hackert, "Evolution of power distributions in fiber optic systems: Development of a measurement strategy," *Fiber & Integrated Optics.*, vol. 8, pp. 163–167, 1989.

46. D. Marcuse, *Principles of Optical Fiber Measurements*, Academic, New York, 1981.

47. P. K. Cheo, *Fiber Optics: Devices and Systems*, Prentice-Hall, Englewood Cliffs, NJ, 1985.

48. D. Rice and G. Keiser, "Short-haul fiber-optic link connector loss," *33rd International Wire & Cable Symp.*, pp. 190–192, Reno, Nev., Nov. 1984.

49. (*a*) A. R. Michelson, M. Eriksrud, S. Aamlid, and N. Ryen, "Role of the fusion splice in the concatenation problem," *J. Lightwave Tech.*, vol. LT-2, pp. 126–138, Apr. 1984.
    (*b*) P. J. W. Severin and W. H. Bardoel, "Differential mode loss and mode conversion in passive fiber optic components," *J. Lightwave Tech.*, vol. LT-4, pp. 1640–1646, Nov. 1986.

50. (*a*) Special Issue on Lightwave Systems and Components, *IEEE Commun. Mag.*, vol. 27, Oct. 1989.
    (*b*) Special Issue on Optical Fiber Video Delivery Systems, *IEEE Mag. Lightwave Commun.*, vol. 1, no. 1, Feb. 1990.
    (*c*) Special Issue on Applications of Lightwave Technology to Microwave Devices, Circuits, and Systems, *IEEE Trans. Microwave Theory Tech.*, vol. 38, May 1990.

51. J. C. Lapp, V. A. Bhagavatula, and A. J. Morrow, "Segmented-core single-mode fiber optimized for bending performance," *J. Lightwave Tech.*, vol. 6, pp. 1462–1465, Oct. 1988.

52. B. J. Ainsle and C. R. Day, "A review of single-mode fibers with modified dispersion characteristics," *J. Lightwave Tech.*, vol. LT-4, pp. 967–979, Aug. 1986.

53. D. P. Jablonowski, U. C. Paek, and L. S. Watkins, "Optical fiber manufacturing techniques," *AT & T Tech. J.*, vol. 66, pp. 33–44, Jan./Feb. 1987.

54. P. D. Lazay and D. Pearson, "Developments in single-mode fiber design, materials, and performance at Bell Laboratories," *IEEE J. Quantum Electron.*, vol. QE-18, pp. 504–510, Apr. 1982.

55. M. Y. El-Ibiary, "Parameter optimization in graded-index dispersion-shifted single-mode fibers," *J. Lightwave Tech.*, vol. LT-4, pp. 364–367, Mar. 1986.

56. W. A. Reed, L. G. Cohen, and H. T. Shang, "Tailoring optical characteristics of dispersion-shifted lightguides for applications near 1.55 μm," *AT & T Tech. J.*, vol. 65, pp. 105–122, Sept./Oct. 1986.

57. D. Kalish and L. G. Cohen, "Single-mode fiber: From research and development to manufacturing," *AT & T Tech. J.*, vol. 66, pp. 19–32, Jan./Feb. 1987.

58. T. D. Croft, J. E. Ritter, and V. A. Bhagavatula, "Low-loss dispersion-shifted single-mode fiber manufactured by the OVD process," *J. Lightwave Tech.*, vol. LT-5, pp. 931–934, Oct. 1985.

59. P. K. Bachmann, D. Leers, H. Wehr, D. U. Wiechert, J. A. Van Steenwijk, D. L. A. Tjaden, and E. R. Wehrhahn, "Dispersion-flattened single-mode fibers prepared with PCVD: Performance, limitations, design optimization," *J. Lightwave Tech.*, vol. LT-4, pp. 858–863, July 1986.

60. V. A. Bhagavatula, M. S. Spotz, W. F. Love, and D. B. Keck, "Segmented-core single-mode fibers with low loss and low dispersion," *Electron. Lett.*, vol. 19, pp. 317–318, Apr. 28, 1983.

61. L. G. Cohen, W. L. Mammel, and S. J. Jang, "Low-loss quadruple-clad single-mode lightguides with dispersion below 2 ps/km · nm over the 1.28 μm–1.65 μm wavelength range," *Electron. Lett.*, vol. 18, pp. 1023–1024, 1982.

62. W. T. Anderson and T. A. Lenahan, "Length dependence of the effective cutoff wavelength in single-mode fibers," *J. Lightwave Tech.*, vol. LT-2, pp. 238–242, June 1984.

63. D. L. Franzen, "Determining the effective cutoff wavelength of single-mode fibers: An interlaboratory comparison," *J. Lightwave Tech.*, vol. LT-3, pp. 128–134, Feb. 1985.

64. V. Shah, "Curvature dependence of the effective cutoff wavelength in single-mode fibers," *J. Lightwave Tech.*, vol. LT-5, pp. 35–43, Jan. 1987.

65. V. Shah and L. Curtis, "Mode coupling effects of the cutoff wavelength characteristics of dispersion-shifted and dispersion-unshifted single-mode fibers," *J. Lightwave Tech.*, vol. 7, pp. 1181–1186, Aug. 1989.

66. CCITT (International Telephone and Telegraph Consultative Committee) Recommendation G.652; CCITT COM XV-146-E, ATT, "Relation between fiber and cable cutoff wavelength," 1987.

67. EIA-455-80, "Cutoff wavelength of uncabled single-mode fiber by transmitted power," Electronic Industries Association, Oct. 1988.

68. A. J. Barlow, R. S. Jones, and K. W. Forsyth, "Technique for direct measurement of single-mode fiber chromatic dispersion," *J. Lightwave Tech.*, vol. LT-5, pp. 1207–1213, Sept. 1987.

69. EIA-455-175, "Chromatic dispersion measurement of optical fibers by the differential phase-shift method," Electronic Industries Association, Nov. 1987.

70. S. B. Andreason, "New bending loss formula explain bends on loss curve," *Electron. Lett.*, vol. 23, pp. 1138–1139, Oct. 8, 1987.

71. M. DiDomenico, Jr., "Material dispersion in optical fiber waveguides," *Appl. Opt.*, vol. 11, pp. 652–654, Mar. 1972.

CHAPTER
4

OPTICAL
SOURCES

The principal light sources used for fiber optic communications applications are heterojunction-structured semiconductor *laser diodes* (also referred to as *injection laser diodes* or ILDs) and *light-emitting diodes* (LEDs). A *heterojunction* consists of two adjoining semiconductor materials with different band-gap energies. These devices are suitable for fiber transmission systems because they have adequate output power for a wide range of applications, their optical power output can be directly modulated by varying the input current to the device, they have a high efficiency, and their dimensional characteristics are compatible with those of the optical fiber. Comprehensive treatments of the major aspects of LEDs and laser diodes are presented in the books by Kressel and Butler,[1] Casey and Panish,[2] and Thompson.[3] Shorter reviews covering the operating principles of these devices are also available,[4-8] to which the reader is referred for details.

The intent of this chapter is to give an overview of the pertinent characteristics of fiber-compatible luminescent sources. The first section discusses semiconductor material fundamentals which are relevant to light source operation. The next two sections present the output and operating characteristics of LEDs and laser diodes, respectively. This is followed by sections discussing the temperature responses of optical sources, their linearity characteristics, and their reliability under various operating conditions.

We shall see in this chapter that the light-emitting region of both LEDs and laser diodes consists of a *pn* junction constructed of direct-band-gap III-V semiconductor materials. When this junction is forward biased, electrons and holes are injected into the *p* and *n* regions, respectively. These injected minority carriers can recombine either radiatively, in which case a photon of energy $h\nu$ is emitted, or nonradiatively, whereupon the recombination energy is dissipated in the form of heat. This *pn* junction is thus known as the *active* or *recombination region*.

A major difference between LEDs and laser diodes is that the optical output from an LED is incoherent, whereas that from a laser diode is coherent. In a coherent source the optical energy is produced in an optical resonant cavity. The optical energy released from this cavity has spatial and temporal coherence, which means it is highly monochromatic and the output beam is very directional. In an incoherent LED source no optical cavity exists for wavelength selectivity. The output radiation has a broad spectral width, since the emitted photon energies range over the energy distribution of the recombining electrons and holes, which usually lie between 1 and $2k_BT$ (k_B is Boltzmann's constant and T is the absolute temperature at the *pn* junction). In addition, the incoherent optical energy is emitted into a hemisphere according to a cosine power distribution and thus has a large beam divergence.

In choosing an optical source compatible with the optical waveguide, various characteristics of the fiber such as its geometry, its attenuation as a function of wavelength, its group delay distortion (bandwidth), and its modal characteristics must be taken into account. The interplay of these factors with the optical source power, spectral width, radiation pattern, and modulation capability needs to be considered. The spatially directed coherent optical output from a laser diode can be coupled into either single-mode or multimode fibers. In general, LEDs are used with multimode fibers, since normally the incoherent optical power from an LED can only be coupled into a multimode fiber in sufficient quantities to be useful. However, some applications have used specially fabricated LEDs with single-mode fibers for data transmission at bit rates up to 1.2 Gb/s over several kilometers.[9–14]

4.1 TOPICS FROM SEMICONDUCTOR PHYSICS

Since the material in this chapter assumes a rudimentary knowledge of semiconductor physics, various relevant definitions are given here for semiconductor material properties. This includes the concepts of energy bands, intrinsic and extrinsic materials, *pn* junctions, and direct and indirect bandgaps. Further details can be found in Refs. 15–19.

4.1.1 Energy Bands

Semiconductor materials have conduction properties that lie somewhere between those of metals and insulators. As an example material, we consider silicon (Si), which is located in the fourth column (group IV) of the periodic table of elements. A Si atom has four electrons in its outer shell, by which it makes covalent bonds with its neighboring atoms in a crystal.

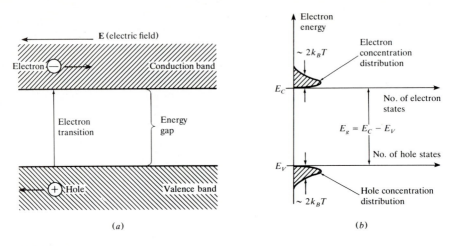

FIGURE 4-1
(*a*) Energy-level diagrams showing the excitation of an electron from the valence band (of energy E_V) to the conduction band (of energy E_C). The resultant free electron and free hole move under the influence of an external electric field *E*. (*b*) Equal electron and hole concentrations in an intrinsic semiconductor created by the thermal excitation of electrons across the band gap.

The conduction properties can be interpreted with the aid of the *energy-band diagrams* shown in Fig. 4-1*a*. In a pure crystal at low temperatures the *conduction band* is completely empty of electrons and the *valence band* is completely full. These two bands are separated by an *energy gap*, or *band gap*, in which no energy levels exist. As the temperature is raised, some electrons are thermally excited across the band gap. For Si this excitation energy must be greater than 1.1 eV, which is the band-gap energy. This gives rise to a concentration *n* of free electrons in the conduction band, which leaves behind an equal concentration *p* of vacancies, or *holes*, in the valence band, as is shown schematically in Fig. 4-1*b*. Both the free electrons and the holes are mobile within the material, so that both can contribute to electrical conductivity; that is, an electron in the valence band can move into a vacant hole. This action makes the hole move in the opposite direction to the electron flow, as is shown in Fig. 4-1*a*.

The concentration of electrons and holes is known as the *intrinsic carrier concentration n_i* and is given by

$$n = p = n_i = K \exp\left(-\frac{E_g}{2k_B T}\right) \tag{4-1}$$

where

$$K = 2\left(2\pi k_B T/h^2\right)^{3/2} \left(m_e m_h\right)^{3/4}$$

is a constant that is characteristic of the material. Here *T* is the absolute temperature, k_B is Boltzmann's constant, h is Planck's constant, and m_e and m_h are the effective masses of the electrons and holes, respectively, which can be smaller by a factor of 10 or more than the free-space electron rest mass of 9.11×10^{-31} kg.

Example 4-1. Given the following parameter values for GaAs at 300 K:

$$\text{electron rest mass } m = 9.11 \times 10^{-31} \text{ kg}$$

$$\text{effective electron mass } m_e = 0.068m = 6.19 \times 10^{-32} \text{ kg}$$

$$\text{effective hole mass } m_h = 0.56m = 5.10 \times 10^{-31} \text{ kg}$$

$$\text{band-gap energy } E_g = 1.42 \text{ eV}$$

then from Eq. (4-1) we find that the intrinsic carrier concentration is

$$n_i = 2.62 \times 10^{12} \text{ m}^{-3} = 2.62 \times 10^6 \text{ cm}^{-3}$$

The conduction can be greatly increased by adding traces of impurities from the group V elements (such as P, As, Sb). This process is called *doping*. These elements have five electrons in the outer shell. When they replace a Si atom, four electrons are used for covalent bonding, and the fifth, loosely bound electron is available for conduction. As shown in Fig. 4-2a, this gives rise to an occupied level just below the conduction band called the *donor level*. The impurities are called *donors* because they can give up an electron to the conduction band. This is reflected by the increase in the free-electron concentration in the conduction band, as shown in Fig. 4-2b. Since in this type of

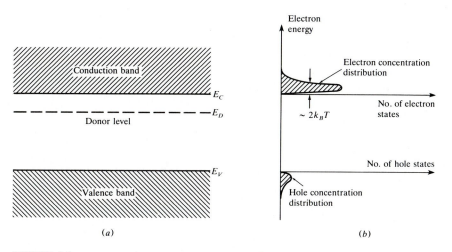

(a)   (b)

FIGURE 4-2
(a) Donor level in an *n*-type material; (b) the ionization of donor impurities creates an increased electron concentration distribution.

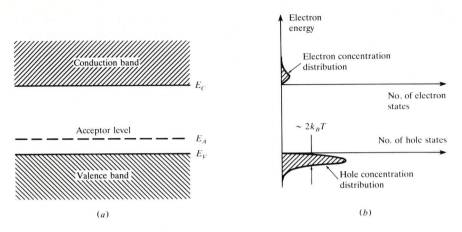

FIGURE 4-3
(*a*) Acceptor level in a *p*-type material; (*b*) the ionization of acceptor impurities creates an increased hole concentration distribution.

material the current is carried by (negative) electrons, it is called *n-type* material.

The conduction can also be increased by adding group III elements, which have three electrons in the outer shell. In this case, three electrons make covalent bonds, and a hole with properties identical to that of the donor electron is created. As shown in Fig. 4-3*a*, this gives rise to an unoccupied level just above the valence band. Conduction occurs when electrons are excited from the valence band to this *acceptor level* (so called because the impurity atoms have accepted electrons from the valence band). Correspondingly, the free-hole concentration increases in the valence band, as shown in Fig. 4-3*b*. This material is called *p-type* because the conduction is a result of (positive) hole flow.

4.1.2   Intrinsic and Extrinsic Material

A material containing no impurities is called *intrinsic* material. Because of thermal vibrations of the crystal atoms, some electrons in the valence band gain enough energy to be excited to the conduction band. This *thermal generation* process produces free electron-hole pairs. In the opposite *recombination* process, a free electron releases its energy and drops into a free hole in the valence band. The generation and recombination rates are equal in equilibrium. If n is the electron concentration and p is the hole concentration, then, for an intrinsic material,

$$pn = p_0 n_0 = n_i^2 \tag{4-2}$$

where p_0 and n_0 refer to the equilibrium hole and electron concentrations, respectively, and n_i is the carrier density of the intrinsic material.

The introduction of small quantities of chemical impurities into a crystal produces an *extrinsic* semiconductor. Since the electrical conductivity is proportional to the carrier concentration, two types of charge carriers are defined for this material:

1. *Majority carriers* refer either to electrons in *n*-type material or to holes in *p*-type material
2. *Minority carriers* refer either to holes in *n*-type material or to electrons in *p*-type material.

The operation of semiconductor devices is essentially based on the *injection* and *extraction* of minority carriers.

> **Example 4-2.** Consider an *n*-type semiconductor which has been doped with a net concentration of N_D donor impurities. Let n_N and p_N be the electron and hole concentrations, where the subscript N is used to denote *n*-type semiconductor characteristics. In this case holes are created exclusively by thermal ionization of intrinsic atoms. This process generates equal concentrations of electrons and holes, so that the hole concentration in an *n*-type semiconductor is
>
> $$p_N = p_i = n_i$$
>
> Since conduction electrons are generated by both impurity and intrinsic atoms, the total conduction-electron concentration n_N is
>
> $$n_N = N_D + n_i = N_D + p_N$$
>
> Substituting Eq. (4-2) for p_N (which states that in equilibrium the product of the electron and hole concentrations equals the square of the intrinsic carrier density, so that $p_N = n_i^2/n_N$), we have
>
> $$n_N = \frac{N_D}{2}\left(\sqrt{1 + \frac{4n_i^2}{N_D^2}} + 1\right)$$
>
> If $n_i \ll N_D$, which is generally the case, then to a good approximation
>
> $$n_N = N_D \quad \text{and} \quad p_N = n_i^2/N_D$$

4.1.3 The *pn* Junctions

Doped *n*- or *p*-type semiconductor material by itself serves only as a conductor. To make devices out of these semiconductors, it is necessary to use both types of materials (in a single, continuous crystal structure). The junction between the two material regions, which is known as the *pn junction*, is responsible for the useful electrical characteristics of a semiconductor device.

When a *pn* junction is created, the majority carriers diffuse across it. This causes electrons to fill holes in the *p* side of the junction and causes holes to appear on the *n* side. As a result an electric field (or *barrier potential*) appears across the junction, as is shown in Fig. 4-4. This field prevents further net

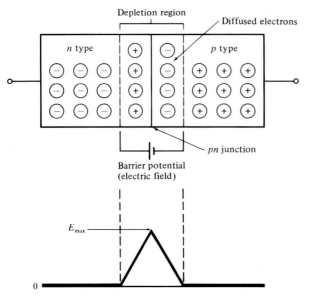

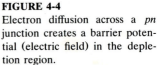

FIGURE 4-4
Electron diffusion across a *pn* junction creates a barrier potential (electric field) in the depletion region.

movements of charges once equilibrium has been established. The junction area now has no mobile carriers, since its electrons and holes are locked into a covalent bond structure. This region is called either the *depletion region* or the *space charge region*.

When an external battery is connected to the *pn* junction with its positive terminal to the *n*-type material and its negative terminal to the *p*-type material, the junction is said to be *reverse-biased*. This is shown in Fig. 4-5. As a result of the reverse bias, the width of the depletion region will increase on both the *n* side and the *p* side. This effectively increases the barrier potential and prevents any majority carriers from flowing across the junction. However, minority carriers can move with the field across the junction. The minority carrier flow is

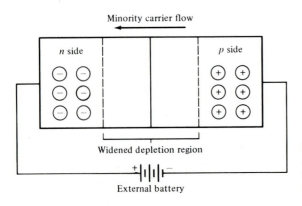

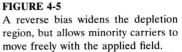

FIGURE 4-5
A reverse bias widens the depletion region, but allows minority carriers to move freely with the applied field.

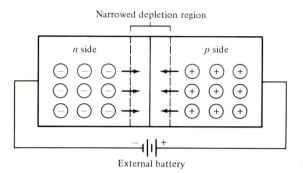

Narrowed depletion region

n side

p side

External battery

FIGURE 4-6
Lowering the barrier potential with a forward bias allows majority carriers to diffuse across the junction.

small at normal temperatures and operating voltages, but it can be significant when excess carriers are created as, for example, in an illuminated photodiode.

When the *pn* junction is *forward-biased*, as shown in Fig. 4-6, the magnitude of the barrier potential is reduced. Conduction band electrons on the *n* side and valence bond holes on the *p* side are, thereby, allowed to diffuse across the junction. Once across, they significantly increase the minority carrier concentrations, and the excess carriers then recombine with the oppositely charged majority carriers. The recombination of excess minority carriers is the mechanism by which optical radiation is generated.

4.1.4 Direct and Indirect Band Gaps

In order for electron transitions to take place to or from the conduction band with the absorption or emission of a photon, respectively, both energy and momentum must be conserved. Although a photon can have considerable energy, its momentum $h\nu/c$ is very small.

Semiconductors are classified either as *direct-band-gap* or *indirect-band-gap* materials depending on the shape of the band gap as a function of the momentum k, as shown in Fig. 4-7. Let us consider recombination of an electron and hole accompanied by the emission of a photon. The simplest and most probable recombination process will be that where the electron and hole have the same momentum value (see Fig. 4-7*a*). This is a direct-band-gap material.

For indirect-band-gap materials, the conduction band minimum and the valence band maximum energy levels occur at different values of momentum, as shown in Fig. 4-7*b*. Here band-to-band recombination must involve a third particle to conserve momentum, since the photon momentum is very small. *Phonons* (i.e., crystal lattice vibrations) serve this purpose.

4.1.5 Semiconductor Device Fabrication

In fabricating semiconductor devices, the crystal structure of the various material regions must be carefully taken into account. In any crystal structure, single

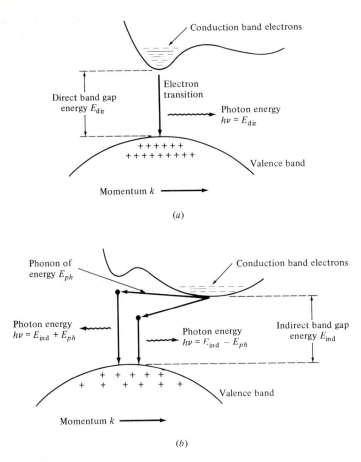

FIGURE 4-7
(a) Electron recombination and the associated photon emission for a direct-band-gap material; (b) electron recombination for indirect-band-gap materials requires a phonon of energy E_{ph} and momentum k_{ph}.

atoms (e.g., Si or Ge) or groups of atoms (e.g., NaCl or GaAs) are arranged in a repeated pattern in space. This periodic arrangement defines a *lattice*, and the spacing between the atoms or groups of atoms is called the *lattice spacing* or the *lattice constant*. Typical lattice spacings are a few angstroms.

Semiconductor devices are generally fabricated by starting with a crystalline substrate which provides mechanical strength for mounting the device and for making electric contacts. A technique of crystal growth by chemical reaction is then used to grow thin layers of semiconductor materials on the substrate. These materials must have lattice structures that are identical to those of the substrate crystal. In particular, the lattice spacings of adjacent materials should be closely matched to avoid temperature-induced stresses and strains at the material interfaces. This type of growth is called *epitaxial*, which is derived from the Greek words *epi* meaning "on" and *taxis* meaning "arrangement," that is, it is an arrangement of atoms from one material on another material. An important characteristic of epitaxial growth is that it is relatively simple to change the impurity concentration of successive material layers, so that a layered semiconductor device can be fabricated in a continuous process. Epitaxial layers can be formed by growth techniques of either vapor phase, liquid phase, or molecular beam.[15-16, 20]

4.2 LIGHT-EMITTING DIODES (LEDs)

For optical communication systems requiring bit rates less than approximately 100 to 200 Mb/s together with multimode fiber-coupled optical power in the tens of microwatts, semiconductor light-emitting diodes (LEDs) are usually the best light source choice. LEDs require less complex drive circuitry than laser diodes since no thermal or optical stabilization circuits are needed (see Sec. 4.3.6), and they can be fabricated less expensively with higher yields.

4.2.1 LED Structures

To be useful in fiber transmission applications an LED must have a high radiance output, a fast emission response time, and a high quantum efficiency. Its *radiance* (or *brightness*) is a measure in watts of the optical power radiated into a unit solid angle per unit area of the emitting surface. High radiances are necessary to couple sufficiently high optical power levels into a fiber, as shown in detail in Chap. 5. The emission response time is the time delay between the application of a current pulse and the onset of optical emission. As we discuss in Secs. 4.2.4 and 4.3.5, this time delay is the factor limiting the bandwidth with which the source can be modulated directly by varying the injected current. The quantum efficiency is related to the fraction of injected electron-hole pairs that recombine radiatively. This is defined and described in detail in Sec. 4.2.3.

To achieve a high radiance and a high quantum efficiency, the LED structure must provide a means of confining the charge carriers and the stimulated optical emission to the active region of the *pn* junction where radiative recombination takes place. Carrier confinement is used to achieve a high level of radiative recombination in the active region of the device, which yields a high quantum efficiency. Optical confinement is of importance for preventing absorption of the emitted radiation by the material surrounding the *pn* junction.

To achieve carrier and optical confinement, LED configurations such as homojunctions and single and double heterojunctions have been widely investigated. The most effective of these predominantly in use at this time is the configuration shown in Fig. 4-8. This is referred to as a *double-heterostructure* (or *heterojunction*) device because of the two different alloy layers on each side of the active region. This configuration evolved from earlier studies on laser diodes. By means of this sandwich structure of differently composed alloy layers,

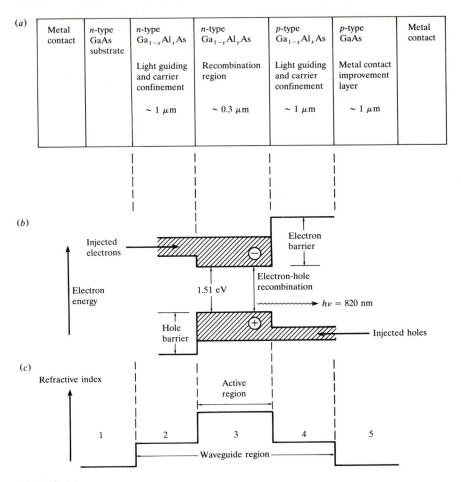

FIGURE 4-8
(*a*) Cross-sectional drawing (not to scale) of a typical GaAlAs double-heterostructure light emitter. In this structure $x > y$ to provide for both carrier confinement and optical guiding. (*b*) Energy-band diagram showing the active region, and the electron and hole barriers which confine the charge carriers to the active layer. (*c*) Variations in the refractive index; the lower index of refraction of the material in regions 1 and 5 creates an optical barrier around the waveguide region because of the higher band-gap energy of this material.

both the carriers and the optical field are confined in the central active layer. The band-gap differences of adjacent layers confine the charge carriers (Fig. 4-8*b*), while the differences in the indices of refraction of adjoining layers confine the optical field to the central active layer (Fig. 4-8*c*). This dual confinement leads to both high efficiency and high radiance. Other parameters influencing the device performance include optical absorption in the active region (self-absorption), carrier recombination at the heterostructure interfaces,

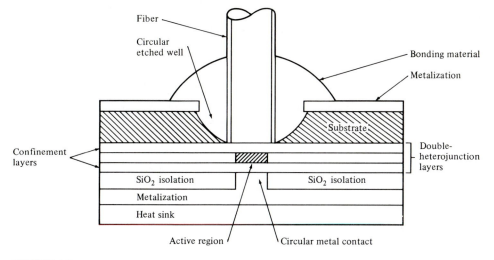

FIGURE 4-9
Schematic (not to scale) of a high-radiance surface-emitting LED. The active region is limited to a circular section having an area compatible with the fiber-core end face.

doping concentration of the active layer, injection carrier density, and active-layer thickness. We shall see the effects of these parameters in the following sections.

The two basic LED configurations being used for fiber optics are *surface emitters* (also called *Burrus* or *front emitters*) and *edge emitters.*[21] In the surface emitter the plane of the active light-emitting region is oriented perpendicularly to the axis of the fiber, as shown in Fig. 4-9. In this configuration a well is etched through the substrate of the device, into which a fiber is then cemented in order to accept the emitted light. The circular active area in practical surface emitters is nominally 50 μm in diameter and up to 2.5 μm thick. The emission pattern is essentially isotropic with a 120° half-power beam width.

This isotropic pattern from a surface emitter is called a *lambertian pattern*. In this pattern the source is equally bright when viewed from any direction, but the power diminishes as $\cos \theta$, where θ is the angle between the viewing direction and the normal to the surface (this is because the projected area one sees decreases as $\cos \theta$). Thus the power is down to 50 percent of its peak when $\theta = 60°$, so that the total half-power beam width is 120°.

The edge emitter depicted in Fig. 4-10 consists of an active junction region, which is the source of the incoherent light, and two guiding layers. The guiding layers both have a refractive index which is lower than that of the active region but higher than the index of the surrounding material. This structure forms a waveguide channel that directs the optical radiation toward the fiber core. To match the typical fiber core diameters (50 to 100 μm), the contact

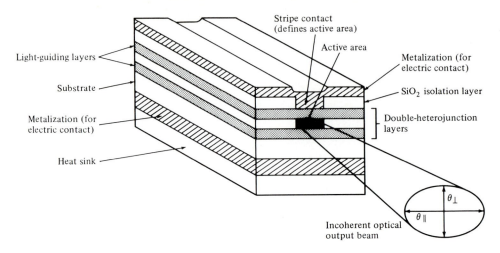

FIGURE 4-10
Schematic (not to scale) of an edge-emitting double-heterojunction LED. The output beam is lambertian in the plane of the *pn* junction ($\theta_{\|} = 120°$) and highly directional perpendicular to the *pn* junction ($\theta_{\perp} \simeq 30°$).

stripes for the edge emitter are 50 to 70 μm wide. Lengths of the active regions usually range from 100 to 150 μm. The emission pattern of the edge emitter is more directional than that of the surface emitter, as is illustrated in Fig. 4-10. In the plane parallel to the junction where there is no waveguide effect, the emitted beam is lambertian (varying as $\cos\theta$) with a half-power width of $\theta_{\|} = 120°$. In the plane perpendicular to the junction the half-power beam width $\theta_{\perp}$ has been made as small as 25 to 35° by a proper choice of the waveguide thickness.[21]

4.2.2 Light Source Materials

The semiconductor material that is used for the active layer of an optical source must have a direct band gap. In a direct-band-gap semiconductor electrons and holes can recombine directly across the band gap without needing a third particle to conserve momentum. Only in direct-band-gap material is the radiative recombination sufficiently high to produce an adequate level of optical emission. Although none of the normal single-element semiconductors are direct-gap materials, many binary compounds are. The most important of these are the so-called III-V materials. These are made from compounds of a group III element (such as Al, Ga, or In) and a group V element (such as P, As, or Sb). Various ternary and quaternary combinations of binary compounds of these elements are also direct-gap materials and are suitable candidates for optical sources.

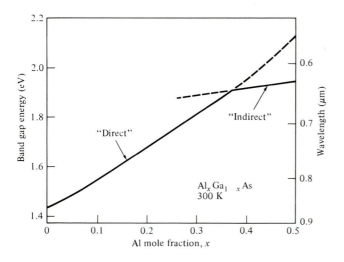

FIGURE 4-11
Band-gap energy and output wavelength as a function of aluminum mole fraction x for $Al_xGa_{1-x}As$ at room temperature. (Reproduced with permission from Miller, Marcatili, and Lee, *Proc. IEEE*, vol. 61, pp. 1703–1751, Dec. 1973, © 1973, IEEE.)

For operation in the 800- to 900-nm spectrum the principal material used is the ternary alloy $Ga_{1-x}Al_xAs$. The ratio x of aluminum arsenide to gallium arsenide determines the band gap of the alloy and, correspondingly, the wavelength of the peak emitted radiation. This is illustrated in Fig. 4-11. The value of x for the active-area material is usually chosen to give an emission wavelength of 800 to 850 nm. An example of the emission spectrum of a $Ga_{1-x}Al_xAs$ LED with $x = 0.08$ is shown in Fig. 4-12. The peak output power occurs at 810 nm and the half-power spectral width σ_λ is 36 nm.

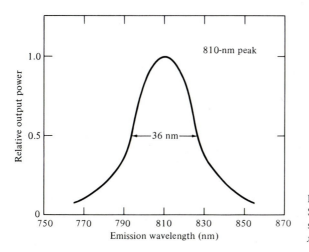

FIGURE 4-12
Spectral emission pattern of a representative $Ga_{1-x}Al_xAs$ LED with $x = 0.08$.

At longer wavelengths the quaternary alloy $In_{1-x}Ga_xAs_yP_{1-y}$ is one of the primary material candidates. By varying the mole fractions x and y in the active area, LEDs with peak output powers at any wavelength between 1.0 and 1.7 μm can be constructed. For simplicity the notations GaAlAs and InGaAsP are generally used unless there is an explicit need to know the values of x and y. Other notations such as AlGaAs, (Al, Ga)As, (GaAl)As, GaInPAs, and $In_xGa_{1-x}As_yP_{1-y}$ are also found in the literature. From the last notation it is obvious that depending on the preference of the particular author, the values of x and $1 - x$ for the same material could be interchanged in different articles in the literature.

The alloys GaAlAs and InGaAsP are chosen to make semiconductor light sources because it is possible to match the lattice parameters of the heterostructure interfaces by using a proper combination of binary, ternary, and quaternary materials. A very close match between the crystal lattice parameters of the two adjoining heterojunctions is required to reduce interfacial defects and to minimize strains in the device as the temperature varies. These factors directly affect the radiative efficiency and lifetime of a light source. Using the fundamental quantum-mechanical relationship between energy E and frequency ν

$$E = h\nu = \frac{hc}{\lambda}$$

the peak emission wavelength λ in micrometers can be expressed as a function of the band-gap energy E_g in electron volts by the equation

$$\lambda \, (\mu m) = \frac{1.240}{E_g \, (eV)} \tag{4-3}$$

The relationships between the band-gap energy E_g and the crystal lattice spacing (or lattice constant) a_0 for various III-V compounds is plotted in Fig. 4-13.

A heterojunction with matching lattice parameters is created by choosing two material compositions having the same lattice constant but different band-gap energies (the band-gap differences are used to confine the charge carriers). In the ternary alloy GaAlAs the band-gap energy E_g and the crystal lattice spacing a_0 are determined by the dashed line in Fig. 4-13, connecting the materials GaAs ($E_g = 1.43$ eV and $a_0 = 5.64$ Å) and AlAs ($E_g = 2.16$ eV and $a_0 = 5.66$ Å). The energy gap in electron volts for values of x between zero and 0.37 (the direct-band-gap region) can be found from the empirical equation[1]

$$E_g = 1.424 + 1.266x + 0.266x^2 \tag{4-4}$$

Given the value of E_g in electron volts, the peak emission wavelength in micrometers is found from Eq. (4-3).

Example 4-3. Consider a $Ga_{1-x}Al_xAs$ laser with $x = 0.07$. From Eq. (4-4) we have $E_g = 1.51$ eV, so that Eq. (4-3) yields $\lambda = 0.82 \, \mu$m.

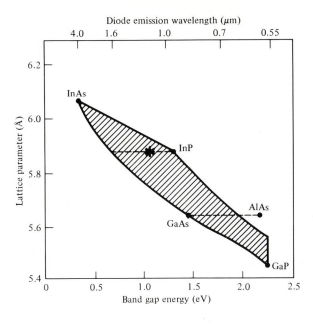

FIGURE 4-13
Relationships between the crystal lattice spacing, energy gap, and diode emission wavelength at room temperature. The shaded area is for the quaternary alloy InGaAsP. The star ∗ is for $In_{0.8}Ga_{0.2}As_{0.35}P_{0.65}$ ($E_g \simeq 1.1$ eV) lattice-matched to InP. (From *Optical Fibre Communications* by Tech. Staff of CSELT, © 1980. Used with the permission of McGraw-Hill Book Co.)

The band-gap energy and lattice-constant range for the quaternary alloy InGaAsP are much larger, as shown by the shaded area in Fig. 4-13. These materials are generally grown on an InP substrate, so that lattice-matched configurations are obtained by selecting a compositional point along the top dashed line in Fig. 4-13, which passes through the InP point. Along this line the compositional parameters x and y follow the relationship $y \simeq 2.20x$ with $0 \le x \le 0.47$. For $In_{1-x}Ga_xAs_yP_{1-y}$ compositions that are lattice-matched to InP, the band gap in eV varies as

$$E_g = 1.35 - 0.72y + 0.12y^2 \qquad (4\text{-}5)$$

Band-gap wavelengths from 0.92 to 1.65 μm are covered by this material system.

Example 4-4. Consider the alloy $In_{0.74}Ga_{0.26}As_{0.57}P_{0.43}$ (that is, $x = 0.26$ and $y = 0.57$). Then from Eq. (4-5) we have $E_g = 0.97$ eV, so that Eq. (4-3) yields $\lambda = 1.27 \, \mu$m.

4.2.3 Internal Quantum Efficiency

An excess of electrons and holes in p- and n-type material, respectively, (referred to as *minority carriers*) is created in a semiconductor light source by carrier injection at the device contacts. The excess densities of electrons Δn and holes Δp are equal, since the injected carriers are formed and recombine in pairs in accordance with the requirement for charge neutrality in the crystal. When carrier injection stops, the carrier density returns to the equilibrium value. In general, the excess carrier density decays exponentially with time according to the relation

$$\Delta n = \Delta n_0 \, e^{-t/\tau} \tag{4-6}$$

where Δn_0 is the initial injected excess electron density and the time constant τ is the carrier lifetime. This lifetime is one of the most important operating parameters of an electro-optic device. Its value can range from milliseconds to fractions of a nanosecond depending on material composition and device defects.

As noted earlier, the excess carriers can recombine either radiatively or nonradiatively. In radiative recombination a photon of energy $h\nu$, which is approximately equal to the band-gap energy, is emitted. When an electron-hole pair recombines nonradiatively, the energy is released in the form of heat (lattice vibrations). The *internal quantum efficiency* in the active region is the fraction of electron-hole pairs that recombine radiatively. If the radiative recombination rate per unit volume is R_r and the nonradiative recombination rate is R_{nr}, the internal quantum efficiency η_0 is the ratio of the radiative recombination rate to the total recombination rate,

$$\eta_0 = \frac{R_r}{R_r + R_{nr}} \tag{4-7}$$

For exponential decay of excess carriers, the radiative recombination lifetime is $\tau_r = \Delta n / R_r$ and the nonradiative recombination lifetime is $\tau_{nr} = \Delta n / R_{nr}$. Thus the internal quantum efficiency is

$$\eta_0 = \frac{1}{1 + (\tau_r/\tau_{nr})} = \frac{\tau}{\tau_r} \tag{4-8}$$

where the *bulk recombination lifetime* τ is

$$\frac{1}{\tau} = \frac{1}{\tau_r} + \frac{1}{\tau_{nr}} \tag{4-9}$$

In a heterojunction structure, nonradiative recombination at the boundaries of the different semiconductor layers resulting from crystal lattice mismatches tends to decrease this lifetime, which, in turn, decreases the internal efficiency. Before we examine this, let us first define some terms. In a semiconductor the flow of electrons or holes gives rise to an electric current i. This is given by

$$i_e = qD_e \frac{\partial(\Delta n)}{\partial x} \quad \text{and} \quad i_h = qD_h \frac{\partial(\Delta p)}{\partial x}$$

for electrons and holes, respectively. This current is a result of nonuniform carrier distributions in the material and flows even in the absence of an applied electric field. The constants D_e and D_h are the electron and hole diffusion coefficients (or constants), respectively, which are expressed in units of centimeters squared per second. As the charge carriers diffuse through the material, some will disappear by recombination. On the average, they move a distance L_e or L_h for electrons and holes, respectively. This distance is known as the *diffusion length*, and is determined from the diffusion coefficient and the lifetime of the material through the relationship

$$L_D = (D\tau)^{1/2}$$

Here we denote a general diffusion length by L_D and its corresponding diffusion coefficient by D.

A heterojunction is commonly characterized by the quantity SL_D/D, which is the ratio of the interfacial recombination velocity S (expressed in centimeters per second) and the bulk diffusion velocity D/L_D. For a completely reflecting boundary $S = 0$, an ohmic contact is characterized by $S = \infty$, and $S = D/L_D$ describes an interface which cannot be distinguished from a continuation of the bulk material. The criterion for high efficiency in an LED is that S should be less than 10^4 cm/s at the heterojunction interfaces. Experimental data show that $S \approx 5 \times 10^3$ cm/s in practical heterostructure devices.[22]

The reduction in the bulk lifetime owing to nonradiative heterointerface recombination can be found from the solution to the one-dimensional steady-state continuity equation[1, 22]

$$D\frac{d^2[\Delta n(x)]}{dx^2} - \frac{\Delta n(x)}{\tau} = 0 \tag{4-10}$$

where $\Delta n(x)$ is the density of the excess electrons per cubic centimeter at the position x in the active layer of thickness d as measured from the *pn* junction (see Fig. 4-14). Assuming the same surface recombination velocity S at both heterointerfaces, the boundary conditions are

$$\frac{d(\Delta n)}{dx}\bigg|_{x=0} = -\frac{J}{qD} + \frac{S}{D}\Delta n(0) \tag{4-11}$$

and

$$\frac{d(\Delta n)}{dx}\bigg|_{x=d} = -\frac{S}{D}\Delta n(d) \tag{4-12}$$

The term $J/(qD)$ gives the number of carriers injected across the *pn* junction at $x = 0$, and $S\Delta n/D$ represents the number of carriers recombining at the

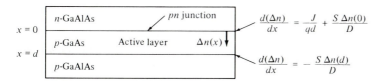

FIGURE 4-14
Boundary conditions at the *pn* junction for a double-heterostructure LED.

interface. The parameter J is the current density. Assuming a solution of the form

$$\Delta n(x) = Ae^{x/L_D} + Be^{-x/L_D}$$

where A and B are constants and $L_D = \sqrt{D\tau}$, Eqs. (4-11) and (4-12) yield

$$\Delta n(x) = \frac{JL_D}{qD} \left\{ \frac{\cosh[(d - x)/L_D] + (L_DS/D)\sinh[(d - x)/L_D]}{[(L_DS/D)^2 + 1]\sinh(d/L_D) + (2L_DS/D)\cosh(d/L_D)} \right\}$$

$$(4\text{-}13)$$

In the active region the average electron density is given by

$$\overline{\Delta n} = \frac{1}{d} \int_0^d \Delta n(x)\, dx = \frac{J}{q} \frac{\tau_{\text{eff}}}{d} \tag{4-14}$$

Here

$$\tau_{\text{eff}} = \tau \left\{ \frac{\sinh(d/L_D) + (L_DS/D)(\cosh d/L_D - 1)}{[(L_DS/D)^2 + 1]\sinh(d/L_D) + (2L_DS/D)\cosh(d/L_D)} \right\} \tag{4-15}$$

is the average effective carrier lifetime when surface recombination is important. When interface recombination is the dominant nonradiative recombination process, the surface recombination velocity S is much smaller than the bulk diffusion velocity D/L_D $(= L_D/\tau)$. Using this condition, that is, $L_DS/D \ll 1$, and the fact that the active-layer thickness d is equal to or smaller than the diffusion length L_D, Eq. (4-15) reduces to

$$\frac{1}{\tau_{\text{eff}}} = \frac{1}{\tau} + \frac{2S}{d} \tag{4-16}$$

Equation (4-16) gives the lifetime reduction caused by interfacial recombination. This lifetime reduction, in turn, decreases the internal quantum efficiency.

A further reduction of the lifetime and the internal efficiency results from self-absorption in the recombination region. This internal absorption of the luminescence is a result of strong energy-dependent absorption near the band gap of a direct-gap semiconductor. The self-absorption becomes important when the interfacial recombination rate is small and the doping in the active region is low. A high efficiency can be achieved in double-heterostructure LEDs because the self-absorption is reduced as a result of the thin active regions that are characteristic of these devices. Self-absorption is more severe in p-type material than in n-type material. Thus its effects can also be minimized by gathering light out of a device from the n side.

If α_λ is the absorption coefficient at a wavelength λ of the active-layer material, then the peak optical power in the pn junction at this wavelength is given by[22]

$$P = \frac{hc}{\lambda \tau_r} \int_0^d \Delta n(x) e^{-\alpha_\lambda x} dx \tag{4-17}$$

Substituting Eq. (4-13) into Eq. (4-17) yields

$$P = \frac{hc}{\lambda q} \eta_i^{dh} J \tag{4-18}$$

where

$$\eta_i^{dh} = \frac{\eta_0}{2} \left\{ \left[\left(\frac{L_D S}{D} \right)^2 + 1 \right] \sinh \frac{d}{L_D} + \frac{2 L_D S}{D} \cosh \frac{d}{L_D} \right\}^{-1}$$

$$\times \left(\frac{1 + L_D S/D}{1 + \alpha_\lambda L_D} \left\{ 1 - \exp \left[\frac{-d(1 + \alpha_\lambda L_D)}{L_D} \right] \right\} e^{d/L_D} \right.$$

$$\left. - \frac{1 - L_D S/D}{1 - \alpha_\lambda L_D} \left\{ 1 - \exp \left[\frac{d(1 - \alpha_\lambda L_D)}{L_D} \right] \right\} e^{-d/L_D} \right) \tag{4-19}$$

is the reduced internal quantum efficiency when interface recombinations and self-absorption are significant. The factor η_0 is given by Eq. (4-8), which holds for $S = 0$ and $\alpha_\lambda = 0$. The superscript dh emphasizes that this expression holds for a double-heterostructure LED with the light being gathered from the n-side passive layer. Similar expressions for homojunction and single-heterostructure devices can be found in the literature.[22, 23]

Similar to the derivation of Eq. (4-15), the total recombination lifetime $\tau_{eff}(\alpha_\lambda)$ when self-absorption is taken into account can be found by considering the average electron density in the active region

$$\overline{\Delta n(\alpha_\lambda)} = \frac{1}{d} \int_0^d \Delta n(x) e^{-\alpha_\lambda x} dx = \frac{J}{q} \frac{\tau_{eff}(\alpha_\lambda)}{d} \tag{4-20}$$

A comparison of Eqs. (4-17), (4-18), and (4-20) shows that

$$\tau_{eff}(\alpha_\lambda) = \eta_i^{dh} \tau_r \tag{4-21}$$

where η_i^{dh} is given by Eq. (4-19).

Examples of the relative internal quantum efficiency reduction η_i^{dh}/η_0 determined from Eq. (4-19) are shown in Fig. 4-15 as a function of the normalized active-layer width d/L_D for two values of the surface recombination velocity and with $D = 80$ cm^2/s in GaAs. The values of α_λ and L_D depend on the active-area doping concentration and were chosen as follows: $\alpha_\lambda = 10^3$ cm^{-1} and $L_D = 10^{-4}$ cm for a 10^{19}-cm^{-3} p-type doping; $\alpha_\lambda = 4 \times 10^3$ cm^{-1} and $L_D = 5 \times 10^{-4}$ cm for a 10^{18}-cm^{-3} doping; and $\alpha_\lambda = 7 \times 10^3$ cm^{-1} and $L_D = 7 \times 10^{-4}$ cm for a 10^{17}-cm^{-3} doping. This results in the $\alpha_\lambda L_D$ products of 0.1, 2, and 5 shown in Fig. 4-15. These curves show that surface recombination severely degrades both the efficiency and the total recombination lifetime.

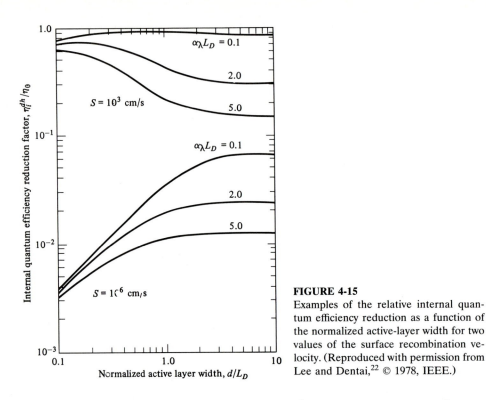

FIGURE 4-15
Examples of the relative internal quantum efficiency reduction as a function of the normalized active-layer width for two values of the surface recombination velocity. (Reproduced with permission from Lee and Dentai,[22] © 1978, IEEE.)

Interfacial recombination dominates for $S \geq 10^5$ cm/s, while for $S \leq 10^3$ cm/s self-absorption becomes significant. An optimum active-region thickness for surface emitters occurs between 2.0 and 2.5 μm for $S = 10^4$ cm/s, whereas thinner values of d are desirable for edge emitters.[5]

4.2.4   Modulation Capability

The frequency response of an LED is limited by its diffusion capacitance because of the storage of injected carriers in the active region of the diode. If the drive current is modulated at a frequency ω, the intensity of the optical output will vary as[24]

$$I(\omega) = I_0\left[1 + (\omega\tau_{\text{eff}})^2\right]^{-1/2} \qquad (4\text{-}22)$$

where I_0 is the intensity emitted at zero modulation frequency and τ_{eff} is the effective carrier lifetime given by Eq. (4-15). Parasitic diode space charge capacitance can cause a delay of the carrier injection into the active junction, and consequently could delay the optical output.[25,26] This delay is negligible if a small, constant forward bias is applied to the diode. Under this condition Eq. (4-22) is valid and the modulation response is limited only by the carrier recombination time.

The modulation bandwidth of an LED is defined in electrical terms as the 3-dB bandwidth of the detected electric power resulting from the modulated portion of the optical signal. Since the detected electric signal power $p(\omega)$ is proportional to $I^2(\omega)$, the modulation bandwidth is defined as the frequency band over which $p(\omega) = p(0)/2$. This is equivalent to setting $I^2(\omega) = I^2(0)/2$. Using Eq. (4-22), the 3-dB modulation bandwidth $\Delta\omega$ is given by

$$\Delta\omega = \frac{1}{\tau_{\mathrm{eff}}} \tag{4-23}$$

Sometimes the modulation bandwidth of an LED is given in terms of the 3-dB bandwidth of the modulated optical power, that is, $I(\omega) = \frac{1}{2}I(0)$. This naturally gives an apparent but erroneous increase in modulation bandwidth by a factor of $3^{1/2}$.

For relatively lightly doped active regions (2×10^{17} cm^{-3}) and active-region thicknesses d less than or equal to a carrier diffusion length L_D, the nonradiative recombination process is dominated by interfacial recombination. In this case the bulk lifetime τ in the active region can be approximated by the radiative lifetime τ_r. Under this condition Eq. (4-16) becomes

$$\frac{1}{\tau_{\mathrm{eff}}} = \frac{1}{\tau_r} + \frac{2S}{d} \tag{4-24}$$

An expression for the radiative lifetime can be found from the following considerations. The radiative carrier lifetime is related to the sum of the initial carrier concentration $n_0 + p_0$ (where n_0 and p_0 are the electron and hole concentrations, respectively, at thermal equilibrium) and the injected electron-hole pair density Δn through the expression

$$\tau_r = \left[B_r(n_0 + p_0 + \Delta n) \right]^{-1} \tag{4-25}$$

The radiative recombination coefficient B_r is a material characteristic which depends on the doping concentration. Its value can range from 0.46×10^{-10} to 7.2×10^{-10} cm^3/s. In the steady-state condition the average injected electron concentration Δn is given by

$$\Delta n = \frac{J\tau_r}{qd} \tag{4-26}$$

where J is the injected current density.

Substituting this into Eq. (4-25) for Δn and solving for τ_r yields

$$\tau_r = \frac{\left[(n_0 + p_0)^2 + 4J/B_r qd \right]^{1/2} - (n_0 + p_0)}{2J/qd} \tag{4-27}$$

When the carrier injection is low compared to the background concentration, that is, $\Delta n \ll n_0 + p_0$, Eq. (4-27) reduces to

$$\tau_r \simeq \left[B_r(n_0 + p_0) \right]^{-1} \tag{4-28}$$

In this case the lifetime is a constant independent of J.

In the other extreme case, at very high carrier injection levels $\Delta n \gg n_0 + p_0$ and the lifetime becomes

$$\tau_r = \left(\frac{qd}{B_r}\right)^{1/2} J^{-1/2} \tag{4-29}$$

At low doping levels (approximately 2×10^{17} cm^{-3}) high carrier injection increases both the efficiency and the modulation speed. For surface emitters the upper current density limit of 7.5 kA/cm^2 at which the LED can safely operate is determined by thermal heating.[22] For a 2-μm-thick active layer the modulation bandwidth is approximately 25 MHz. This increases to about 60 MHz for an 0.3-μm active layer. A higher modulation bandwidth at the same drive level can occur for edge emitters, which operate under the condition where the lifetime is predominantly controlled by the injected carrier density.[5]

4.2.5   Transient Response

A closed-form approximation of the transient response of high-radiance double-heterojunction LEDs has been derived for practical engineering analyses of optical fiber communication systems.[27, 28] The basic assumption of this approximation is that the junction space charge capacitance C_s varies much more slowly with current than the diffusion capacitance C_d and can therefore be considered constant. Typical values of C_s range from 350 to 1000 pF for low to moderately high currents. Under this assumption, the rise time to the half-current point (which is also the half-power point) of the LED is

$$t_{1/2} = \frac{C_s}{\beta I_p} \ln \frac{I_p}{I_s} + \tau \ln 2 \tag{4-30}$$

and the 10- to 90-percent rise time is

$$t_{10-90} = \left(\frac{C_s}{\beta I_p} + \tau\right) \ln 9 \tag{4-31}$$

In these expressions $\beta = q/(2k_B T)$, I_p is the amplitude of the current step function used to drive the LED, I_s is the diode saturation current, and τ is the minority carrier lifetime. These expressions show that the rise time decreases with increasing current. In the high-current limit the rise times depend only on the carrier lifetime, so that $t_{1/2} = \tau \ln 2$ and $t_{10-90} = \tau \ln 9$. The rise time can be reduced considerably by external means such as *current peaking*.[25, 27] If the current rises to a point in excess of the desired level for a very short time and then decays back to the desired level, as is shown in Fig. 4-16, a reduction in rise time to $t_{10-90} = 0.55\tau$ has been achieved.[27] Similarly the 90- to 10-percent fall time can be made considerably shorter than $\tau \ln 9$ by applying a negative bias for a short time after the pulse to sweep out the injected carriers. In addition, as noted in Sec. 4.2.2, the application of a small, constant forward bias on the LED minimizes the delay time of the onset of optical output.[25, 26]

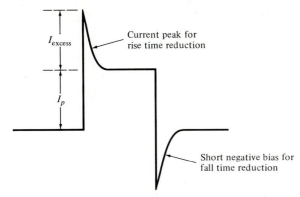

FIGURE 4-16
Current waveform showing a short positive peak for rise-time reduction and a short negative bias for fall-time reduction.

4.2.6 Power-Bandwidth Product

An important parameter to consider for LEDs is the *power-bandwidth product*. This can be found by multiplying both sides of Eq. (4-18) by $\Delta\omega = 1/\tau_{\text{eff}}$:

$$\Delta\omega\, P = \frac{1}{\tau_{\text{eff}}} \frac{hc}{q\lambda} \eta_i^{\text{dh}} J \qquad (4\text{-}32)$$

Using Eq. (4-21), this becomes

$$\Delta\omega\, P = \frac{hc}{q\lambda} \frac{1}{\tau_r} J \qquad (4\text{-}33)$$

which is constant for a given current injection level. For example, suppose the doping in the active layer is increased. This decreases the effective carrier lifetime, which results in an increase in LED bandwidth. However, this bandwidth increase is accompanied by a decrease in power by the same proportionality factor, since η_i^{dh} is equal to τ_{eff}/τ_r. Thus, for a fixed injection level, the net power-bandwidth product remains unchanged. This means that faster LEDs generally emit less power than slow ones.

4.3 LASER DIODES

Lasers come in many forms with dimensions ranging from the size of a grain of salt to one that will occupy an entire room. The lasing medium can be a gas, a liquid, an insulating crystal (solid state) or a semiconductor. For optical fiber systems the laser sources used almost exclusively are semiconductor laser diodes. They are similar to other lasers, such as the conventional solid-state and gas lasers, in that the emitted radiation has spatial and temporal coherence; that is, the output radiation is highly monochromatic and the light beam is very directional.

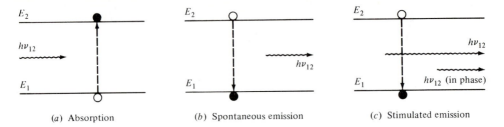

(a) Absorption   (b) Spontaneous emission   (c) Stimulated emission

FIGURE 4-17
The three key transition processes involved in laser action. The open circle represents the initial state of the electron and the heavy dot represents the final state. Incident photons are shown on the left of each diagram and emitted photons are shown on the right.

Despite their differences the basic principle of operation is the same for each type of laser. Laser action is the result of three key processes. These are photon absorption, spontaneous emission, and stimulated emission. These three processes are represented by the simple two-energy-level diagrams in Fig. 4-17, where E_1 is the ground-state energy and E_2 is the excited-state energy. According to Planck's law, a transition between these two states involves the absorption or emission of a photon of energy $h\nu_{12} = E_2 - E_1$. Normally the system is in the ground state. When a photon of energy $h\nu_{12}$ impinges on the system, an electron in state E_1 can absorb the photon energy and be excited to state E_2, as shown in Fig. 4-17a. Since this is an unstable state, the electron will shortly return to the ground state, thereby emitting a photon of energy $h\nu_{12}$. This occurs without any external stimulation and is called *spontaneous emission*. These emissions are isotropic and of random phase, and thus appear as a narrowband gaussian output.

The electron can also be induced to make a downward transition from the excited level to the ground-state level by an external stimulation. As shown in Fig. 4-17c, if a photon of energy $h\nu_{12}$ impinges on the system while the electron is still in its excited state, the electron is immediately stimulated to drop to the ground state and give off a photon of energy $h\nu_{12}$. This emitted photon is in phase with the incident photon, and the resultant emission is known as *stimulated emission*.

In thermal equilibrium the density of excited electrons is very small. Most photons incident on the system will therefore be absorbed, so that stimulated emission is essentially negligible. Stimulated emission will exceed absorption only if the population of the excited states is greater than that of the ground state. This condition is known as *population inversion*. Since this is not an equilibrium condition, population inversion is achieved by various "pumping" techniques. In a semiconductor laser, population inversion is accomplished by injecting electrons into the material at the device contacts to fill the lower energy states of the conduction band.

4.3.1   Laser Diode Modes and Threshold Conditions

For optical fiber communication systems requiring bandwidths greater than approximately 200 MHz, the semiconductor injection laser diode is preferred over the LED. Laser diodes typically have response times less than 1 ns, have optical bandwidths of 2 nm or less, and, in general, are capable of coupling several milliwatts of useful luminescent power into optical fibers with small cores and small numerical apertures. Virtually all laser diodes in use and under investigation at present are multilayered heterojunction devices. As mentioned in Sec. 4.2, the double-heterojunction LED configuration evolved from the successful demonstration of both carrier and optical confinement in heterojunction injection laser diodes. The more rapid evolvement and utilization of LEDs as compared to laser diodes lies in the inherently simpler construction, the smaller temperature dependence of the emitted optical power, and the absence of catastrophic degradation in LEDs (see Sec. 4.5). The construction of laser diodes is more complicated, mainly because of the additional requirement of current confinement in a small lasing cavity.

Stimulated emission in semiconductor lasers arises from optical transitions between distributions of energy states in the valence and conduction bands. This differs from gas and solid-state lasers, in which radiative transitions occur between discrete isolated atomic or molecular levels. The radiation in the laser diode is generated within a Fabry-Perot resonator cavity,[1-3] shown in Fig. 4-18, as in most other types of lasers. However, this cavity is much smaller, being approximately 250 to 500 μm long, 5 to 15 μm wide, and 0.1 to 0.2 μm thick. These dimensions are commonly referred to as the *longitudinal*, *lateral*, and *transverse dimensions* of the cavity, respectively.

In the laser diode Fabry-Perot resonator a pair of flat, partially reflecting mirrors are directed toward each other to enclose the cavity. The mirror facets are constructed by making two parallel cleaves along natural cleavage planes of the semiconductor crystal. The purpose of these mirrors is to provide strong *optical feedback* in the longitudinal direction, thereby converting the device into an oscillator with a gain mechanism that compensates for optical losses in the cavity. The laser cavity can have many resonant frequencies. The device will oscillate (thereby emitting light) at those resonant frequencies for which the gain is sufficient to overcome the losses. The sides of the cavity are simply formed by roughening the edges of the device to reduce unwanted emissions in these directions.

In another laser diode type, commonly referred to as the *distributed-feed-back* (DFB) *laser*,[1,2,7,29-31] the cleaved facets are not required for optical feedback. A typical DFB laser configuration is given in Fig. 4-19. The fabrication of this device is similar to the Fabry-Perot types, except that the lasing action is obtained from Bragg reflectors (gratings) or periodic variations of the refractive index (called *distributed-feedback corrugations*) which are incorporated into the multilayer structure along the length of the diode. This is discussed in more detail in Sec. 4.3.4.

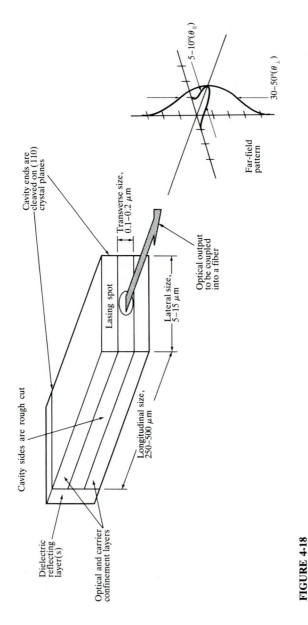

FIGURE 4-18

Fabry-Perot resonator cavity for a laser diode. The cleaved crystal ends function as partially reflecting mirrors. The unused end (the rear facet) can be coated with a dielectric reflector to reduce optical loss in the cavity. Note that the light beam emerging from the laser forms a vertical ellipse, even though the lasing spot at the active-area facet is a horizontal ellipse.

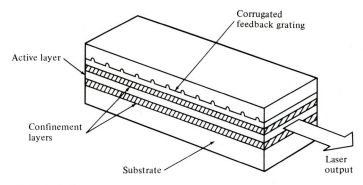

FIGURE 4-19
Structure of a distributed-feedback (DFB) laser diode.

In general, the full optical output is only needed from the front facet of the laser, that is, the one to be aligned with an optical fiber. In this case a dielectric reflector can be deposited on the rear laser facet to reduce the optical loss in the cavity, to reduce the threshold current density (the point at which lasing starts), and to increase the external quantum efficiency. Reflectivities greater than 98 percent have been achieved with a six-layer reflector.[32-34]

The optical radiation within the resonance cavity of a laser diode sets up a pattern of electric and magnetic field lines called the *modes of the cavity* (see Secs. 2.3 and 2.4 for details on modes). These can conveniently be separated into two independent sets of transverse electric (TE) and transverse magnetic (TM) modes. Each set of modes can be described in terms of the longitudinal, lateral, and transverse half-sinusoidal variations of the electromagnetic fields along the major axes of the cavity. The *longitudinal modes* are related to the length L of the cavity and determine the principal structure of the frequency spectrum of the emitted optical radiation. Since L is much larger than the lasing wavelength of approximately 1 μm, many longitudinal modes can exist. *Lateral modes* lie in the plane of the *pn* junction. These modes depend on the side wall preparation and the width of the cavity, and determine the shape of the lateral profile of the laser beam. *Transverse modes* are associated with the electromagnetic field and beam profile in the direction perpendicular to the plane of the *pn* junction. These modes are of great importance, since they largely determine such laser characteristics as the radiation pattern (the angular distribution of the optical output power) and the threshold current density.

To determine the lasing conditions and the resonant frequencies, we express the electromagnetic wave propagating in the longitudinal direction (along the axis normal to the mirrors) in terms of the electric field phasor

$$E(z,t) = I(z)e^{j(\omega t - \beta z)} \tag{4-34}$$

where $I(z)$ is the optical field intensity, ω is the optical radian frequency, and β is the propagation constant (see Sec. 2.3.2).

Lasing is the condition at which light amplification becomes possible in the laser diode. The requirement for lasing is that a population inversion be achieved. This condition can be understood by considering the fundamental relationship between the optical field intensity I, the absorption coefficient α_λ, and the gain coefficient g in the Fabry-Perot cavity. The stimulated emission rate into a given mode is proportional to the intensity of the radiation in that mode. The radiation intensity at a photon energy $h\nu$ varies exponentially with the distance z that it traverses along the lasing cavity according to the relationship

$$I(z) = I(0)\exp\{[\Gamma g(h\nu) - \bar{\alpha}(h\nu)]z\} \tag{4-35}$$

where $\bar{\alpha}$ is the effective absorption coefficient of the material in the optical path and Γ is the *optical confinement factor* (the fraction of optical power in the active layer).

Optical amplification of selected modes is provided by the feedback mechanism of the optical cavity. In the repeated passes between the two partially reflecting parallel mirrors, a portion of the radiation associated with those modes having the highest optical gain coefficient is retained and further amplified during each trip through the cavity.

Lasing occurs when the gain of one or several guided modes is sufficient to exceed the optical loss during one roundtrip through the cavity, that is, for $z = 2L$. During this roundtrip only the fractions R_1 and R_2 of the optical radiation are reflected from the two laser ends 1 and 2, respectively, where R_1 and R_2 are the mirror reflectivities. Thus Eq. (4-35) becomes

$$I(2L) = I(0)R_1 R_2 \exp\{2L[\Gamma g(h\nu) - \bar{\alpha}(h\nu)]\} \tag{4-36}$$

At the lasing threshold a steady-state oscillation takes place, and the magnitude and phase of the returned wave must be equal to those of the original wave. This gives the conditions

$$I(2L) = I(0) \tag{4-37}$$

for the amplitude and

$$e^{-j2\beta L} = 1 \tag{4-38}$$

for the phase. Equation (4-38) gives information concerning the resonant frequencies of the Fabry-Perot cavity. This is discussed further in Sec. 4.3.2. From Eq. (4-37) we can find which modes have sufficient gain for sustained oscillation and we can find the amplitudes of these modes. Thus from Eq. (4-37), the condition for reaching the lasing-threshold optical gain g_{th} is the point at which the gain g is greater than or equal to the total loss α_t in the cavity:

$$\Gamma g_{th} \geq \alpha_t = \bar{\alpha} + \frac{1}{2L} \ln\left(\frac{1}{R_1 R_2}\right) \tag{4-39}$$

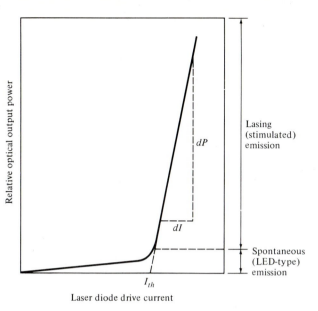

FIGURE 4-20
Relationship between optical output power and laser diode drive current. Below the lasing threshold the optical output is a spontaneous LED type emission.

Example 4-5. For GaAs $R_1 = R_2 = 0.32$ for uncoated facets (that is, 32 percent of the radiation is reflected at a facet) and $\bar{\alpha} \simeq 10$ cm^{-1}. This yields $\Gamma g_{th} = 33$ cm^{-1} for a laser diode of length $L = 500$ μm.

The mode that satisfies Eq. (4-39) reaches threshold first. Theoretically, at the onset of this condition, all additional energy introduced into the laser should augment the growth of this particular mode. In practice, various phenomena lead to the excitation of more than one mode.[1] Studies on the conditions of longitudinal single-mode operation have shown that important factors are thin active regions and a high degree of temperature stability.[34]

The relationship between optical output power and diode drive current is presented in Fig. 4-20. At low diode currents only spontaneous radiation is emitted. Both the spectral range and the lateral beam width of this emission are broad like that of an LED. A dramatic and sharply defined increase in the power output occurs at the lasing threshold. As this transition point is approached, the spectral range and the beam width both narrow with increasing drive current. The final spectral width of approximately 1 nm and the fully narrowed lateral beam width of nominally 5 to 10° are reached just past the threshold point. The *threshold current* I_{th} is conventionally defined by extrapolation of the lasing region of the power-versus-current curve, as shown in Fig. 4-20. At high power outputs the slope of the curve decreases because of junction heating.

The *external differential quantum efficiency* η_{ext} is defined as the number of photons emitted per radiative electron-hole pair recombination above threshold. Under the assumption that above threshold the gain coefficient remains fixed at g_{th}, η_{ext} is given by[1]

$$\eta_{ext} = \frac{\eta_i(g_{th} - \bar{\alpha})}{g_{th}} \tag{4-40}$$

Here η_i is the internal quantum efficiency. This is not a well-defined quantity in laser diodes, but most measurements show that $\eta_i \simeq 0.6$ to 0.7 at room temperature. Experimentally η_{ext} is calculated from the straight-line portion of the curve for the emitted optical power P versus drive current I, which gives

$$\eta_{ext} = \frac{q}{E_g} \frac{dP}{dI} = 0.8065\lambda \ (\mu m) \frac{dP \ (mW)}{dI \ (mA)} \tag{4-41}$$

where E_g is the band-gap energy in electron volts, dP is the incremental change in the emitted optical power in milliwatts for an incremental change dI in the drive current (in milliamperes), and λ is the emission wavelength in micrometers. For standard semiconductor lasers, external differential quantum efficiencies of 15 to 20 percent per facet are typical. High-quality devices have differential quantum efficiencies of 30 to 40 percent.

4.3.2   Resonant Frequencies

Now let us return to Eq. (4-38) to examine the resonant frequencies of the laser. The condition in Eq. (4-38) holds when

$$2\beta L = 2\pi m \tag{4-42}$$

where m is an integer. Using $\beta = 2\pi n/\lambda$ for the propagation constant from Eq. (2-46), we have

$$m = \frac{L}{\lambda/2n} = \frac{2Ln}{c}f \tag{4-43}$$

where $c = f\lambda$. This states that the cavity resonates (that is, a standing-wave pattern exists within it) when an integer number m of half-wavelengths spans the region between the mirrors.

Since in all lasers the gain is a function of frequency (or wavelength, since $c = f\lambda$), there will be a range of frequencies (or wavelengths) for which Eq. (4-43) holds. Each of these frequencies corresponds to a mode of oscillation of the laser. Depending on the laser structure, any number of frequencies can satisfy Eqs. (4-37) and (4-38). Thus some lasers are single-mode and some are multimode. The relationship between gain and frequency can be assumed to have the gaussian form

$$g(\lambda) = g(0)\exp\left[-\frac{(\lambda - \lambda_0)^2}{2\sigma^2}\right] \tag{4-44}$$

where λ_0 is the wavelength at the center of the spectrum, σ is the spectral width of the gain, and the maximum gain $g(0)$ is proportional to the population inversion.

Let us now look at the frequency, or wavelength, spacing between the modes of a multimode laser. Here we only consider the longitudinal modes. Note, however, that for each longitudinal mode there may be several transverse modes that arise from one or more reflections of the propagating wave at the sides of the resonator cavity.[1,3] To find the frequency spacing, consider two successive modes of frequencies f_{m-1} and f_m represented by the integers $m - 1$ and m. From Eq. (4-43) we have

$$m - 1 = \frac{2Ln}{c} f_{m-1}$$

and

$$m = \frac{2Ln}{c} f_m$$

Subtracting these two equations yields

$$1 = \frac{2Ln}{c} (f_m - f_{m-1}) = \frac{2Ln}{c} \Delta f \qquad (4\text{-}45)$$

from which we have the frequency spacing

$$\Delta f = \frac{c}{2Ln} \qquad (4\text{-}46)$$

This can be related to the wavelength spacing $\Delta\lambda$ through the relationship $\Delta f/f = \Delta\lambda/\lambda$, yielding

$$\Delta\lambda = \frac{\lambda^2}{2Ln} \qquad (4\text{-}47)$$

Thus given Eqs. (4-44) and (4-47), the output spectrum of a multimode laser follows the typical gain-versus-frequency plot given in Fig. 4-21, where the exact

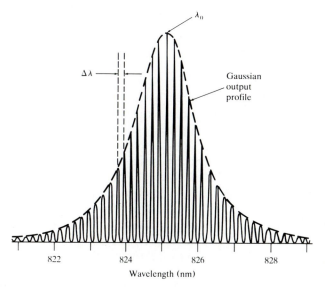

FIGURE 4-21
Typical spectrum from a gain-guided GaAlAs/GaAs laser diode. (Reproduced with permission from K. Petermann and G. Arnold, *IEEE J. Quantum Electron.*, vol. 18, pp. 543–555, Apr. 1982, © IEEE, 1982.)

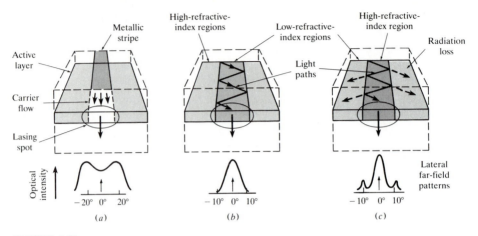

FIGURE 4-22
Three fundamental structures for confining optical waves in the lateral direction: (*a*) in the gain-induced guide, electrons injected via a metallic stripe contact alter the index of refraction of the active layer; (*b*) the positive-index waveguide has a higher refractive index in the central portion of the active region; (*c*) the negative-index waveguide has a lower refractive index in the central portion of the active region. (Reproduced with permission from Botez,[4a] © IEEE, 1985.)

number of modes, their heights, and their spacings depend on the laser construction.

> **Example 4-6.** A GaAs laser operating at 850 nm has a 500-μm length and a refractive index $n = 3.7$. What are the frequency and wavelength spacings? If at the half-power point $\lambda - \lambda_0 = 2$ nm, what is the spectral width σ of the gain?
>
> From Eq. (4-46) we have $\Delta f = 81$ GHz, and from Eq. (4-47) we find that $\Delta\lambda = 0.2$ nm. Using Eq. (4-44) with $g(\lambda) = 0.5g(0)$ yields $\sigma = 1.70$ nm.

4.3.3 Laser Diode Structures and Radiation Patterns

A basic requirement for efficient operation of laser diodes is that, in addition to transverse optical and carrier confinement between heterojunction layers, the current flow must be restricted laterally to a narrow stripe along the length of the laser. Numerous novel methods of achieving this with varying degrees of success have been proposed, but all strive for the same goals of limiting the number of lateral modes so that lasing is confined to a single filament, stabilizing the lateral gain, and ensuring a relatively low threshold current.

Figure 4-22 shows the three basic *optical-confinement methods* used for bounding laser light in the lateral direction.[4] In the first structure a narrow electrode stripe (less than 8 μm wide) runs along the length of the diode. The injection of electrons and holes into the device alters the refractive index of the active layer directly below the stripe. The profile of these injected carriers creates a weak, complex waveguide that confines the light laterally. This type of device is commonly referred to as a *gain-guided laser*. Although these lasers can emit optical powers exceeding 100 mW, they have strong instabilities and can have highly astigmatic, two-peaked beams as shown in Fig. 4-22*a*.

More stable structures use the configurations shown in Fig. 4-22*b* and *c*. Here dielectric waveguide structures are fabricated in the lateral direction. The variations in the real refractive index of the various materials in these structures control the lateral modes in the laser. Thus these devices are called *index-guided lasers*. If a particular index-guided laser supports only the fundamental transverse mode and the fundamental longitudinal mode, it is known as a *single-mode laser*. Such a device emits a single, well-collimated beam of light having an intensity profile which is a bell-shaped gaussian curve.

Index-guided lasers can have either positive-index or negative-index wave-confining structures. In a *positive-index waveguide*, the central region has a higher refractive index than the outer regions. Thus all of the guided light is reflected at the dielectric boundary, just as it is at the core-cladding interface in an optical fiber. By proper choice of the change in refractive index and the width of the higher-index region, one can make a device which supports only the fundamental lateral mode.

In a *negative-index waveguide* the central region of the active layer has a lower refractive index than the outer regions. At the dielectric boundaries, part of the light is reflected and the rest is refracted into the surrounding material and is thus lost. This radiation loss appears in the far-field radiation pattern as narrow side lobes to the main beam as shown in Fig. 4-22*c*. Since the fundamental mode in this device has less radiation loss than any other mode, it is the first to lase. The positive-index laser is the more popular of these two structures.

Index-guided lasers can be made using any one of four fundamental structures. These are the buried heterostructure, a selectively diffused construction, a varying-thickness structure, and a bent-layer configuration. To make the *buried heterostructure* (BH) laser shown in Fig. 4-23, one etches a narrow mesa stripe (1 to 2 μm wide) in double-heterostructure material. The mesa is then embedded in high-resistivity, lattice-matched *n*-type material with an appropriate band gap and low refractive index. This material is GaAlAs in 800- to 900-nm lasers with a GaAs active layer, and is InP for 1300- to 1600-nm lasers with an InGaAsP active layer. This configuration thus strongly traps generated light in a lateral waveguide. A number of variations of this fundamental structure have been used to fabricate high-performing laser diodes.[4,8]

The *selectively diffused construction* is shown in Fig. 4-24*a*. Here a chemical dopant, such as zinc for GaAlAs lasers and cadmium for InGaAsP lasers, is diffused into the active layer immediately below the metallic contact stripe. The dopant changes the refractive index of the active layer to form a lateral waveguide channel. In the *varying-thickness structure* shown in Fig. 4-24*b*, a channel (or other topological configuration such as a mesa or terrace) is etched into the substrate. Layers of crystal are then regrown into the channel using liquid phase epitaxy. This process fills in the depressions and partially dissolves

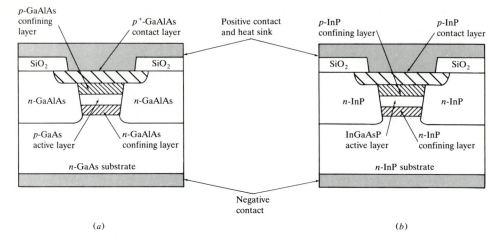

FIGURE 4-23
(*a*) Short-wavelength (800–900 nm) GaAlAs and (*b*) long-wavelength (1300–1600 nm) InGaAsP buried heterostructure laser diodes.

the protrusions, thereby creating variations in the thicknesses of the active and confining layers. When an optical wave encounters a local increase in the thickness as shown in Fig. 4-24*b*, the thicker area acts as a positive-index waveguide of higher-index material. In the *bent-layer structure* a mesa is etched into the substrate as shown in Fig. 4-24*c*. Semiconductor material layers are grown onto this structure using vapor phase epitaxy to exactly replicate the mesa configuration. The active layer has a constant thickness with lateral bends. As an optical wave travels along the flat top of the mesa in the active area, the lower-index material outside of the bends confines the light along this lateral channel.

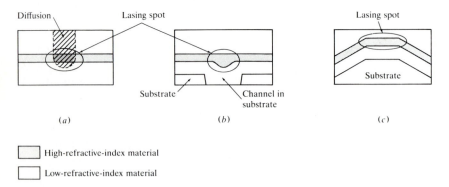

FIGURE 4-24
Positive-index optical-wave-confining structure of the (*a*) selectively diffused, (*b*) varying-thickness, and (*c*) bent-layer types. (Adapted with permission from Botez,[4a] © IEEE, 1985.)

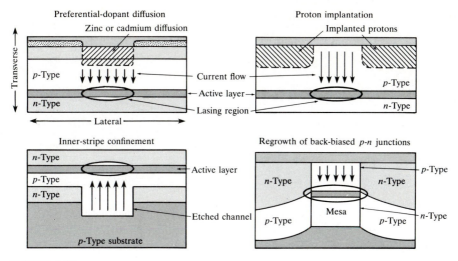

FIGURE 4-25
Four basic methods for achieving current confinement in laser diodes: (*a*) preferential-dopant diffusion, (*b*) proton implantation, (*c*) inner-stripe confinement, and (*d*) regrowth of back-biased *pn* junctions. (Adapted with permission from Botez,[4a] © IEEE, 1985.)

In addition to confining the optical wave to a narrow lateral stripe to achieve continuous high optical output power, one also needs to restrict the drive current tightly to the active layer so more than 60 percent of the current contributes to lasing. Figure 4-25 shows the four basic *current-confinement methods*. In each method the device architecture blocks current on both sides of the lasing region. This is achieved either by high-resistivity regions or by reverse-biased *pn* junctions, which prevent the current from flowing while the device is forward-biased under normal conditions. For structures with a continuous active layer, the current can be confined either above or below the lasing region. The diodes are forward-biased so that current travels from the *p*-type to the *n*-type regions. In the *preferential-dopant diffusion* method, partially diffusing a *p*-type dopant (Zn or Cd) through an *n*-type capping layer establishes a narrow path for the current, since back-biased *pn* junctions block the current outside the diffused region. The *proton implantation* method creates regions of high resistivity, thus restricting the current to a narrow path between these regions. The *inner-stripe confinement* technique grows the lasing structure above a channel etched into planar material. Back-biased *pn* junctions restrict the current on both sides of the channel. When the active layer is discontinuous, as in a buried heterostructure, current can be blocked on both sides of the mesa by growing *pn* junctions that are reverse-biased when the device is operating. A laser diode can use more than one current-confining technique.

In a double-heterojunction laser the highest-order transverse mode that can be excited depends on the waveguide thickness and on the refractive-index differentials at the waveguide boundaries.[1] If the refractive-index differentials

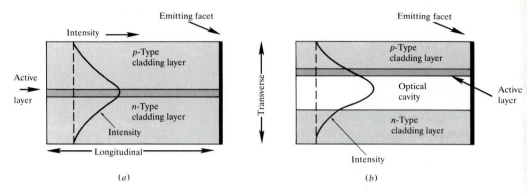

FIGURE 4-26
Two methods of increasing the lasing spot size transverse to the *pn* junction: (*a*) the thin-active-layer (TAL) structure, and (*b*) the large-optical-cavity configuration. (Adapted with permission from Botez,[4a] © IEEE, 1985.)

are kept at approximately 0.08, then only the fundamental transverse mode will propagate if the active area is thinner than 1 μm.

When designing the width and thickness of the optical cavity, a tradeoff must be made between current density and output beam width. As either the width or the thickness of the active region is increased, a narrowing occurs of the lateral or transverse beam widths, respectively, but at the expense of an increase in the threshold current density. Most positive-index waveguide devices have a lasing spot 3 μm wide by 0.6 μm high. This is significantly greater than the active-layer thickness, since about half the light travels in the confining layers. Such lasers can operate reliably only up to continuous-wave (CW) output powers of 3 to 5 mW. Here the transverse and lateral half-power beam widths shown in Fig. 4-18 are about $\theta_\perp \simeq 30$ to 50° and $\theta_\parallel \simeq 5$ to 10°, respectively.

To achieve higher output powers, one must increase the lasing spot size transverse to the *pn* junction to prevent the mirror facets from being degraded at high continuous power levels. To increase the transverse spot size, one can make the active layer very thin as shown in Fig. 4-26*a*, so that most of the optical energy spreads out into the confining (cladding) layers. Decreasing the active-layer thickness from a standard value of about 0.15 μm to 0.05 μm nearly doubles the transverse size of the lasing spot from 0.6 to 1 μm. These *thin-active-layer* (TAL) lasers can operate reliably at 20 to 25 mW.

An alternative method is to fabricate a laser with a *large optical cavity* (LOC), or guide layer, just below the active layer,[4,35] as illustrated in Fig. 4-26*b*. The cavity material has a refractive index between that of the active layer above and that of the *n*-type cladding layer below. Most of the light propagates in the large optical cavity while obtaining gain from the active layer above. This architecture increases the transverse spot size to almost 1.5 μm. For the same beam width at half intensity, the lasing spot in LOC structures is 50 to 70 percent larger than the spot in TAL structures. At a 50-percent duty cycle, these LOC devices can operate reliably up to 40 mW.

4.3.4 Single-Mode Lasers

For high-speed, long-distance communications one needs single-mode lasers, which must contain only a single longitudinal mode and a single transverse mode. Consequently the spectral width of the optical emission is very narrow.

One way of restricting a laser to have only one longitudinal mode is to reduce the length L of the lasing cavity to the point where the frequency separation Δf of the adjacent modes given in Eq. (4-46) is larger than the laser transition line width, that is, only a single longitudinal mode falls within the gain bandwidth of the device. For example, for a Fabry-Perot cavity all longitudinal modes have nearly equal losses and are spaced by about 1 nm in a 250-μm-long cavity at 1300-nm. By reducing L from 250 to 25 μm, the mode spacing increases from 1 nm to 10 nm. However, these lengths make the device hard to handle, and they are limited to optical output powers of only a few milliwatts.[36]

Alternative devices were thus developed. Among these are the surface-emitting laser[37] and structures having a built-in frequency-selective resonator. In the *surface-emitting laser* (SEL), shown in Fig. 4-27, the active area is less than 10 μm thick (that is, less than 10 μm long) and thus acts as a short vertical cavity. Two variations on the basic vertical-cavity surface-emitting laser have been considered to improve its performance. These geometries are configured in the same way as in edge-emitting lasers, but the optical radiation is directed

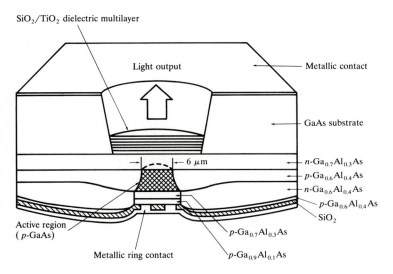

FIGURE 4-27
Structure of GaAlAs surface-emitting laser. (Adapted with permission from Kinoshita and Iga,[37] © IEEE, 1987.)

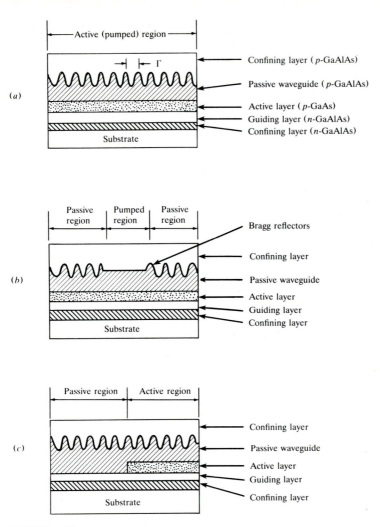

FIGURE 4-28
Three types of laser structures using built-in frequency-selective resonator gratings: (*a*) distributed feedback (DFB) laser, (*b*) distributed-Bragg-reflector (DBR) laser, and (*c*) distributed-reflector (DR) laser.

towards the surface either by 45° mirrors[38] or by second-order distributed Bragg reflectors.[39]

Three types of laser configurations using a built-in *frequency-selective reflector* are shown in Fig. 4-28. In each case the frequency-selective reflector is a corrugated grating which is a passive waveguide layer adjacent to the active region. The optical wave propagates parallel to this grating. The operation of these types of lasers is based on the distributed Bragg phase-grating reflector.[40, 41] A phase grating is essentially a region of periodically varying refractive index

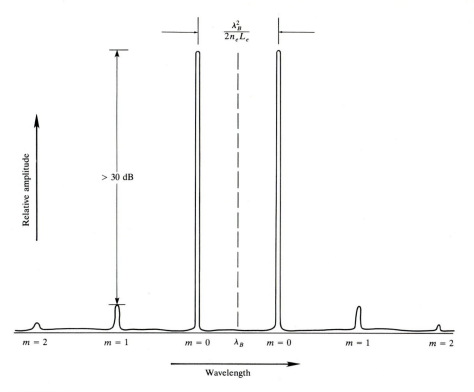

FIGURE 4-29
Output spectrum symmetrically distributed around λ_B in an idealized distributed-feedback (DFB) laser diode.

that causes two counterpropagating traveling waves to couple. The coupling is a maximum for wavelengths close to the Bragg wavelength λ_B, which is related to the period Λ of the corrugations by

$$\lambda_B = \frac{2n_e\Lambda}{k} \qquad (4\text{-}48a)$$

where n_e is the effective refractive index of the mode and k is the order of the grating. First-order gratings ($k = 1$) provide the strongest coupling, but sometimes second-order gratings are used, since their larger corrugation period make fabrication easier. Lasers based on this architecture exhibit good single-mode longitudinal operation with low sensitivity to drive-current and temperature variations.

In the *distributed-feedback* (DFB) laser[29-31, 40] the grating for the wavelength selector is formed over the entire active region. As shown in Fig. 4-29, in an ideal DFB laser, the longitudinal modes are spaced symmetrically around λ_B

at wavelengths given by

$$\lambda = \lambda_B \pm \frac{\lambda_B^2}{2 n_e L_e}\left(m + \tfrac{1}{2}\right) \tag{4-48b}$$

where $m = 0, 1, 2, \ldots$ is the mode order and L_e is the effective grating length. The amplitudes of successively higher-order lasing modes are greatly reduced from the zero-order amplitude; e.g., the first-order mode ($m = 1$) is usually more than 30 dB down from the zero-order amplitude ($m = 0$).

Theoretically, in a DFB laser having both ends antireflection-coated, the two zero-order modes on either side of the Bragg wavelength should experience the same lowest threshold gain and would lase simultaneously in an idealized symmetrical structure. However, in practice the randomness of the cleaving process lifts the degeneracy in the modal gain and results in single-mode operation. This facet asymmetry can be further increased by putting a high-reflection coating on one end and a low-reflection coating on the other, for example, around 2 percent on the front facet and 30 percent on the rear facet. Variations on the DFB design have been the introduction of a $\pi/2$ optical phase shift (that is, a quarter wavelength) in the corrugation at the center of the optical cavity to make the laser oscillate near the Bragg wavelength, since reflections occur most effectively at this wavelength.[42, 43]

For the *distributed-Bragg-reflector* (DBR) laser[31, 44] the gratings are located at the ends of the normal active layer of the laser to replace the cleaved end mirrors used in the Fabry-Perot optical resonator (Fig. 4-28b). The *distributed-reflector* laser[45] consists of active and passive distributed reflectors (Fig. 4-28c). This structure improves the lasing properties of conventional DFB and DBR lasers, and has a high efficiency and high output capability.

4.3.5 Modulation of Laser Diodes

The two principal methods used to vary the optical output from laser diodes are pulse modulation used in digital systems and amplitude modulation used for analog data transmission. The basic limitation on the modulation rate of laser diodes depends on the carrier and photon lifetime parameters associated with the operation of the laser. These are the spontaneous (radiative) and stimulated carrier lifetimes and the photon lifetime. The *spontaneous lifetime* τ_{sp} is discussed in Sec. 4.2 and is a function of the semiconductor band structure and the carrier concentration. At room temperature the radiative lifetime τ_r is about 1 ns in GaAs-based materials for dopant concentrations on the order of 10^{19} cm^{-3}. The *stimulated carrier lifetime* τ_{st} depends on the optical density in the lasing cavity and is on the order of 10 ps. The *photon lifetime* τ_{ph} is the average time that the photon resides in the lasing cavity before being lost either by absorption or by emission through the facets. In a Fabry-Perot cavity the photon lifetime is[1]

$$\tau_{\text{ph}}^{-1} = \frac{c}{n}\left(\bar{\alpha} + \frac{1}{2L}\ln\frac{1}{R_1 R_2}\right) = \frac{c}{n}g_{\text{th}} \qquad (4\text{-}49)$$

For a typical value of $g_{\text{th}} = 50$ cm^{-1} and a refractive index in the lasing material of $n = 3.5$, the photon lifetime is approximately $\tau_{\text{ph}} = 2$ ps. This value sets the upper limit to the modulation capability of the laser diode.

A laser diode can readily be pulse-modulated, since the photon lifetime is much smaller than the carrier lifetime. If the laser is completely turned off after each pulse, the spontaneous carrier lifetime will limit the modulation rate. This is because at the onset of a current pulse of amplitude I_p, a period of time t_d given by (see Prob. 4-23)

$$t_d = \tau \ln \frac{I_p}{I_p + (I_B - I_{\text{th}})} \qquad (4\text{-}50)$$

is needed to achieve the population inversion necessary to produce a gain that is sufficient to overcome the optical losses in the lasing cavity. In Eq. (4-50) the parameter I_B is the bias current and τ is the average lifetime of the carriers in the recombination region when the total current $I = I_p + I_B$ is close to I_{th}. From Eq. (4-50) it is clear that the delay time can be eliminated by dc-biasing the diode at the lasing threshold current. Pulse modulation is then carried out by modulating the laser only in the operating region above threshold. In this region the carrier lifetime is now shortened to the stimulated emission lifetime, so that high modulation rates are possible.

When using a laser diode for high-speed transmission systems, the modulation frequency can be no larger than the frequency of the relaxation oscillations of the laser field. The relaxation oscillation depends on both the spontaneous lifetime and the photon lifetime. Theoretically, assuming a linear dependence of the optical gain on carrier density, the relaxation oscillation occurs approximately at[1]

$$f = \frac{1}{2\pi}\frac{1}{(\tau_{\text{sp}}\tau_{\text{ph}})^{1/2}}\left(\frac{I}{I_{\text{th}}} - 1\right)^{1/2} \qquad (4\text{-}51)$$

Since τ_{sp} is about 1 ns and τ_{ph} is on the order of 2 ps for a 300-μm-long laser, then when the injection current is about twice the threshold current, the maximum modulation frequency is a few gigahertz. An example of a laser having a relaxation-oscillation peak at 3 GHz is shown in Fig. 4-30. A more exact analysis of the relaxation oscillation requires considering a nonlinear dependence of optical gain on electron density due to spectral hole burning or dynamic carrier heating.[8, 46]

Analog modulation of laser diodes is carried out by making the drive current above threshold proportional to the baseband information signal. A

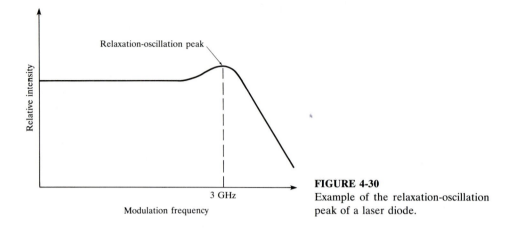

FIGURE 4-30
Example of the relaxation-oscillation peak of a laser diode.

requirement for this modulation scheme is that a linear relation exist between the light output and the current input. However, signal degradation resulting from nonlinearities that are a consequence of the transient response characteristics of laser diodes makes the implementation of analog intensity modulation susceptible to both intermodulation and cross-modulation effects. This can be alleviated by using pulse code modulation or through special distortion compensation techniques, as is discussed in more detail in Sec. 4.4.

4.3.6 Temperature Effects

An important factor to consider in the application of laser diodes is the temperature dependence of the threshold current $I_{th}(T)$. This parameter increases with temperature in all types of semiconductor lasers because of various complex temperature-dependent factors.[47] The complexity of these factors prevents the formulation of a single equation holding for all devices and temperature ranges. However, the temperature variation of I_{th} can be approximated by the empirical expression[8, 48, 49]

$$I_{th}(T) = I_z e^{T/T_0} \tag{4-52}$$

where T_0 is a measure of the relative temperature insensitivity and I_z is a constant. For a conventional stripe geometry GaAlAs laser diode T_0 is typically 120 to 165°C in the vicinity of room temperature. An example of a laser diode with $T_0 = 135$°C and $I_z = 52$ mA is shown in Fig. 4-31. The variation in I_{th} with temperature is 0.8 percent/°C, as is shown in Fig. 4-32. Smaller dependences of I_{th} on temperature have been demonstrated for GaAlAs quantum-well heterostructure lasers. For these lasers T_0 can be as high as 437°C. The temperature dependence of I_{th} for this device is also shown in Fig. 4-32. The threshold

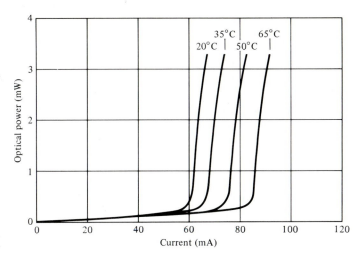

FIGURE 4-31
Temperature-dependent behavior of the optical output power as a function of the bias current for a
particular laser diode.

variation for this particular laser type is 0.23 percent/°C. Experimental values[8]
of T_0 for 1300-nm InGaAsP lasers are typically 60 to 80 K (333 to 353°C).

For the laser diode shown in Fig. 4-31 the threshold current increases by a
factor of about 1.4 between 20 and 60°C. In addition, the lasing threshold can
change as the laser ages. Consequently, if a constant optical output power level
is to be maintained as the temperature of the laser changes or as the laser ages,
it is necessary to adjust the dc bias current level. Possible methods for achieving
this automatically are optical feedback schemes,[50–53] temperature-matching
transistors,[54] and threshold-sensing circuits.[55]

Optical feedback can be carried out by using a photodetector to either
sense the variation in optical power emitted from the rear facet of the laser or
tap off and monitor a small portion of the fiber-coupled power emitted from the
front facet. The photodetector compares the optical power output with a
reference level and adjusts the dc bias current level automatically to maintain a
constant peak light output relative to the reference. The photodetector used
must have a stable long-term responsivity which remains constant over a wide
temperature range. For operation in the 800- to 900-nm region, a silicon *pin*
photodiode generally exhibits these characteristics (see Chap. 6).

An example of a feedback-stabilizing circuit[50] that can be used for a
digital transmitter is shown in Fig. 4-33. In this scheme the light emerging from
the rear facet of the laser is monitored by a *pin* photodiode. With this circuit
the electric input signal pattern is compared to the optical output level of the
laser diode. This effectively prevents the feedback circuit from erroneously
raising the bias current level during long sequences of digital zeros or during a

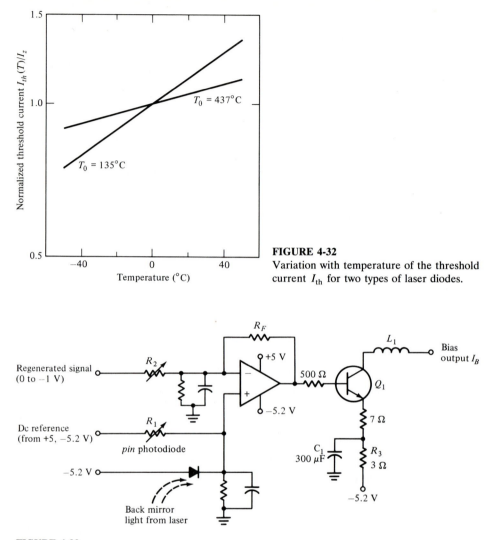

FIGURE 4-32
Variation with temperature of the threshold current I_{th} for two types of laser diodes.

FIGURE 4-33
Example of a bias circuit that provides feedback stabilization of laser output power. (Reproduced with permission from Shumate, Chen, and Dorman,[50] © 1978, The American Telephone and Telegraph Company.)

period in which there is no input signal on the channel. In this circuit the dc reference through resistor R_1 sets the bias current at the proper operating point during long sequences of zeros. When this bias current is added to the laser drive current, the desired peak output power from the laser is obtained. The resistor R_2 balances the signal reference current against the *pin* photocurrent for a 50-percent duty ratio at 25°C. As the lasing threshold changes because of

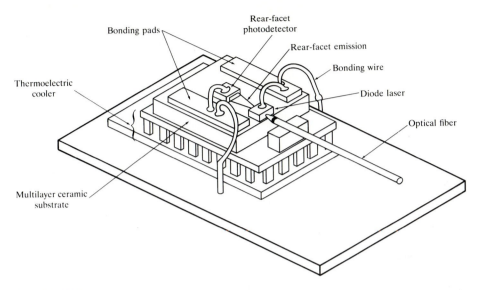

FIGURE 4-34
Construction of a laser-diode transmitter using a thermoelectric cooler for temperature stabilization.

aging or temperature variations, the bias current I_B is automatically adjusted to maintain a balance between the data reference and the *pin* photocurrent. A more sophisticated version of this circuit[51] simultaneously and independently controls both the bias current and the modulation current.

Another standard method of stabilizing the optical output of a laser diode is to use a miniature thermoelectric cooler. This device maintains the laser at a constant temperature and thus stabilizes the output level. Normally a thermo-electric cooler is used in conjunction with a rear-facet detector feedback loop, as is shown in Fig. 4-34.

4.4 LIGHT SOURCE LINEARITY

High-radiance LEDs and laser diodes are well-suited optical sources for wide-band analog applications provided a method is implemented to compensate for any nonlinearities of these devices. In an analog system the time-varying electric analog signal $s(t)$ is used to modulate directly an optical source about a bias current point I_B, as shown in Fig. 4-35. With no signal input the optical power output is P_t. When the signal $s(t)$ is applied, the optical output power $P(t)$ is

$$P(t) = P_t[1 + ms(t)] \qquad (4\text{-}53)$$

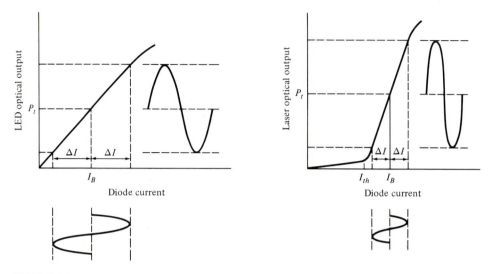

FIGURE 4-35
Bias point and amplitude modulation range for analog applications of LEDs and laser diodes.

Here m is the *modulation index* (or *modulation depth*) defined by

$$m = \frac{\Delta I}{I_B'} \tag{4-54}$$

where $I_B' = I_B$ for LEDs and $I_B' = I_B - I_{th}$ for laser diodes. The parameter ΔI is the variation in current about the bias point. To prevent distortions in the output signal, the modulation must be confined to the linear region of the curve for optical output versus drive current. Furthermore if ΔI is greater than I_B' (that is, m is greater than 100 percent), the lower portion of the signal gets cut off and severe distortion will result. Typical m values for analog applications range from 0.25 to 0.50.

In analog applications any device nonlinearities will create frequency components in the output signal that were not present in the input signal.[56] Two important nonlinear effects are harmonic and intermodulation distortions. If the signal input to a nonlinear device is a simple cosine wave $x(t) = A \cos \omega t$, the output will be

$$y(t) = A_0 + A_1 \cos \omega t + A_2 \cos 2\omega t + A_3 \cos 3\omega t + \cdots \tag{4-55}$$

That is, the output signal will consist of a component at the input frequency ω plus spurious components at zero frequency, at the second harmonic frequency 2ω, at the third harmonic frequency 3ω, and so on. This effect is known as *harmonic distortion*. The amount of nth-order distortion in decibels is given by

$$n\text{th-order harmonic distortion} = 20 \log \frac{A_n}{A_1} \tag{4-56}$$

To determine *intermodulation distortion*, the modulating signal of a non-linear device is taken to be the sum of two cosine waves $x(t) = A_1 \cos \omega_1 t + A_2 \cos \omega_2 t$. The output signal will then be of the form

$$y(t) = \sum_{m,n} B_{mn} \cos(m\omega_1 + n\omega_2) \tag{4-57}$$

where m and $n = 0, \pm 1, \pm 2, \pm 3, \ldots$. This signal includes all the harmonics of ω_1 and ω_2 plus cross-product terms such as $\omega_2 - \omega_1$, $\omega_2 + \omega_1$, $\omega_2 - 2\omega_1$, $\omega_2 + 2\omega_1$, etc. The sum and difference frequencies give rise to the intermodulation distortion. The sum of the absolute values of the coefficients m and n determines the order of the intermodulation distortion. For example, the second-order intermodulation products are at $\omega_1 \pm \omega_2$ with amplitude B_{11}; the third-order intermodulation products are at $\omega_1 \pm 2\omega_2$ and $2\omega_1 \pm \omega_2$ with amplitudes B_{12} and B_{21}; and so on. (Harmonic distortions are also present wherever either $m \neq 0$ and $n = 0$ or when $m = 0$ and $n \neq 0$. The corresponding amplitudes are B_{m0} and B_{0n}, respectively.) In general, the odd-order intermodulation products having $m = n \pm 1$ (such as $2\omega_1 - \omega_2$, $2\omega_2 - \omega_1$, $3\omega_1 - 2\omega_2$, etc.) are the most troublesome, since they may fall within the bandwidth of the channel. Of these only the third-order terms are usually important, since the amplitudes of higher-order terms tend to be significantly smaller. If the operating frequency band is less than an octave, all other intermodulation products will fall outside the passband and can be eliminated with appropriate filters in the receiver.

Nonlinear distortions in LEDs are due to effects depending on the carrier injection level, radiative recombination, and other subsidiary mechanisms, as is described in detail by Asatani and Kimura.[57] In certain laser diodes, such as gain-guided devices, there can be nonlinearities in the curve for optical power output versus diode current, as is illustrated in Fig. 4-36. These nonlinearities

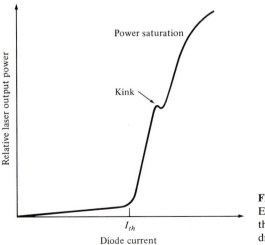

FIGURE 4-36
Example of a kink and power saturation in the curve for optical output power versus drive current of a laser diode.

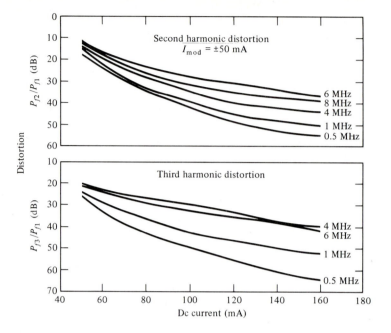

FIGURE 4-37
Second-order and third-order harmonic distortions as a function of bias current in a GaAlAs LED for several modulation frequencies. The distortion is given in terms of the power P_{fn} as the nth harmonic relative to the power P_{f1} at the modulation frequency f_1. (Reproduced with permission from Dawson.[58])

are a result of inhomogeneities in the active region of the device and also arise from power switching between the dominant lateral modes in the laser. They are generally referred to as "kinks." These kinks are generally not seen in modern laser diodes using the structures described in Secs. 4.3.3 and 4.3.4. Power saturation (as indicated by a downward curving of the output-versus-current curve) can occur at high output levels because of active-layer heating.

Total harmonic distortions[58-61] in GaAlAs LEDs and laser diodes tend to be in the range of 30 to 40 dB below the output at the fundamental modulation frequency for modulation depths around 0.5. The second- and third-order harmonic distortions as a function of bias current for several modulation frequencies are shown in Fig. 4-37 for a GaAlAs double-heterojunction LED.[58] The harmonic distortions decrease with increasing bias current but become large at higher modulation frequencies. The intermodulation distortion curves (not shown) follow the same characteristics as those in Fig. 4-37, but are 5 to 8 dB worse.

A number of compensation techniques for linearization of optical sources in analog communication systems have been investigated. These methods include circuit techniques such as complementary distortion,[57,61] negative

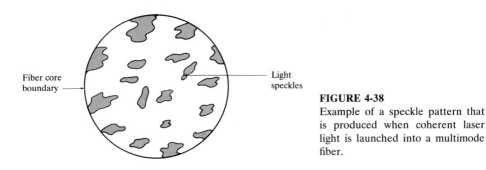

FIGURE 4-38
Example of a speckle pattern that is produced when coherent laser light is launched into a multimode fiber.

feedback,[62] selective harmonic compensation,[63] and quasi-feedforward compensation,[64] and the use of pulse position modulation (PPM) schemes.[65] One of the most successful circuit design techniques is the quasi-feedforward method, with which a 30- to 40-dB reduction in total harmonic distortion has been achieved.

4.5 MODAL, PARTITION, AND REFLECTION NOISE

Three significant factors associated with the operating characteristics of laser diodes arise in high-speed digital and analog applications. These phenomena are called modal or speckle noise,[66-73] mode-partition noise,[73-77] and reflection noise.[73, 78-81] These factors are important because they can introduce receiver output noise, which can be particularly serious for analog systems (see Chaps. 8 and 9).

When light from a coherent laser is launched into a multimode fiber, a number of propagating modes of the fiber are normally excited. As long as these modes retain their relative phase coherence, the radiation pattern seen at the end of the fiber (or at any point along the fiber) takes on the form of a speckle pattern. This is the result of constructive and destructive interference between propagating modes at any given plane. An example of this is shown in Fig. 4-38. The number of speckles in the pattern approximates the number of propagating modes. As the light travels along the fiber, a combination of mode-dependent losses, changes in phase between modes, and fluctuations in the distribution of energy among the various fiber modes will change the modal interference and result in a different speckle pattern. *Modal* or *speckle noise* occurs when any losses that are speckle-pattern-dependent are present in a link. Examples of such losses are splices, connectors, microbends, and photodetectors with nonuniform responsivity across the photosensitive area. Noise is generated when the speckle pattern *changes in time* so as to vary the optical power transmitted through the particular loss element. The continually changing speckle pattern that falls on the photodetector thus produces a time-varying noise in the received signal, which degrades receiver performance. Narrowband, high-coherence sources, such as single-mode lasers, result in more modal noise

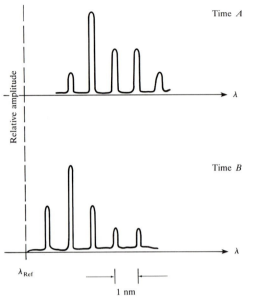

Relative amplitude

Time *A*

λ

Time *B*

λ

λ_Ref

1 nm

FIGURE 4-39
Time-resolved dynamic spectra of a laser diode. Different modes or groups of modes dominate the optical output at different times. The modes are approximately 1 nm apart.

than broadband sources. Incoherent sources such as LEDs do not produce modal noise. The use of single-mode fibers eliminates this problem.

Mode-partition noise is associated with intensity fluctuations in the longitudinal modes of a laser diode. This is the dominant noise in single-mode fibers. The output from a laser diode generally can come from more than one longitudinal mode, as shown in Fig. 4-39. The optical output may arise from all of the modes simultaneously, or it may switch from one mode (or group of modes) to another randomly in time. Intensity fluctuations can occur among the various modes in a multimode laser even when the total optical output does not vary as exhibited in Fig. 4-39. Since the output pattern of a laser diode is highly directional, the light from these fluctuating modes can be coupled into a fiber with high coupling efficiency. Each of the longitudinal modes that is coupled into the fiber has a different attenuation and time delay, because each is associated with a slightly different wavelength (see Sec. 3.3). Since the power fluctuations among the dominant modes can be quite large, in systems with high fiber dispersion significant variations in signal levels can occur at the receiver.

Reflection noise is associated with laser diode output linearity distortion caused by some of the light output being reflected back into the laser cavity from fiber joints. This reflected power couples with the lasing modes, thereby causing their phases to vary. This produces a periodically modulated noise spectrum that is peaked on the low-frequency side of the intrinsic noise profile. The fundamental frequency of this noise is determined by the roundtrip delay of the light from the laser to the reflecting point and back again. Depending on the roundtrip time, these reflections can create noise peaks in the frequency region where optical fiber data transmission systems operate, even though the lasers themselves are very noise-free at these frequencies. Reflection noise problems can be greatly reduced by using optical isolators between the laser diode and the optical fiber transmission line or by using index-matching fluid in the gaps at fiber-to-fiber joints to eliminate reflections at the fiber-to-air interfaces.

4.6 RELIABILITY CONSIDERATIONS

The reliability of double-heterojunction LEDs and laser diodes is of importance in that these devices are of greatest interest in optical communications systems. The lifetimes of these sources are affected by both operating conditions and fabrication techniques. Thus it is important to understand the relationships between light source operation characteristics, degradation mechanisms, and system reliability requirements. A comprehensive review of the reliability of GaAlAs laser diodes has been presented by Ettenberg and Kressel,[82] to which the reader is referred for further details and an extensive list of references. The extension of these results to LEDs is straightforward.[83-84] Reliabilities of InGaAsP LEDs and lasers are given in Refs. 85–88.

Lifetime tests of optical sources are done either at room temperature or at elevated temperatures to accelerate the degradation process. A commonly used elevated temperature is 70°C. At present there is no standard method for determining the lifetime of an optical source. The two most popular techniques either maintain the light output constant by increasing the bias current automatically or keep the current constant and monitor the optical output level. In the first case, the end of life of the device is assumed to be reached when the source can no longer put out a specified power at the maximum current value for CW (continuous-wave) operation. In the second case, the lifetime is determined by the time taken for the optical output power to decrease by 3 dB.

Degradation of light sources can be divided into three basic categories: internal damage and ohmic contact degradation, which hold for both lasers and LEDs, and damage to the facets of laser diodes.

The limiting factor on LED and laser diode lifetime is internal degradation. This effect arises from the migration of crystal defects into the active region of the light source. These defects decrease the internal quantum efficiency and increase the optical absorption. Fabrication steps which can be taken to minimize internal degradation include the use of substrates with low surface dislocation densities (less than 2×10^3 dislocations/cm^2), keeping work-damaged edges out of the diode current path, and minimizing stresses in the active region (to less than 10^8 dyn/cm^2).

For high-quality sources having lifetimes which follow a slow internal-degradation mode, the optical power P decreases with time according to the exponential relationship

$$P = P_0 e^{-t/\tau_m} \qquad (4\text{-}58)$$

Here P_0 is the initial optical power at time $t = 0$, and τ_m is a time constant for the degradation process, which is approximately twice the -3-dB mean time to failure. Since the operating lifetime depends on both the current density J and the junction temperature T, internal degradation can be accelerated by increasing either one of these parameters.

The operating lifetime τ_s has been found experimentally to depend on the current density J through the relation

$$\tau_s \propto J^{-n} \tag{4-59}$$

where $1.5 \le n \le 2.0$. For example, by doubling the current density, the lifetime decreases by a factor of 3 to 4. Since the degradation rate of optical sources increases with temperature, an Arrhenius relationship of the form

$$\tau_s = K e^{E_A / k_B T} \tag{4-60}$$

has been sought. Here E_A is an activation energy characterizing the lifetime τ_s, k_B is Boltzmann's constant, T is the absolute temperature at which τ_s was evaluated, and K is a constant. The problem in establishing such an expression is that several competing factors are likely to contribute to the degradation, thereby making it difficult to estimate the activation energy E_A. Activation energies for laser degradation reported in the literature have ranged from 0.3 to 1.0 eV. For practical calculations, a value of 0.7 eV is generally used. However, this value is subject to change as more long-term statistical data at various temperatures are obtained.

Equations (4-59) and (4-60) indicate that, to increase the light source lifetime, it is advantageous to operate these devices at as low a current and temperature as is practicable. Examples[85] of the luminescent output of InGaAsP LEDs as a function of time for different temperatures are shown in Fig. 4-40. At temperatures below 120°C the output power remains almost constant over the entire measured 15,000-h (1.7-yr) operating time. At higher temperatures the power output drops as a function of time. For example, at 230°C the optical power has dropped to one-half its initial value (a 3-dB decrease) after approximately 3000 h (4.1 months) of operation. The activation energy of these lasers is about 1.0 eV.

A second fabrication-related degradation mechanism is ohmic contact deterioration. In LEDs and laser diodes the thermal resistance of the contact between the light source chip and the device heat sink occasionally increases with time. This effect is a function of the solder used to bond the chip to the heat sink, the current density through the contact, and the contact temperature. An increase in the thermal resistance results in a rise in the junction temperature for a fixed operating current. This, in turn, leads to a decrease in the optical output power. However, careful designs and implementations of high-quality bonding procedures have minimized effects resulting from contact degradation.

Facet damage is a degradation problem that exists for laser diodes. This degradation reduces the laser mirror reflectivity and increases the nonradiative carrier recombination at the laser facets. The two types of facet damage that

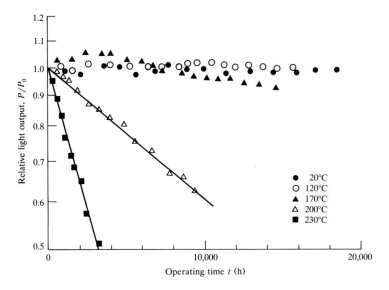

FIGURE 4-40

Normalized output power as a function of operating time for five ambient temperatures. P_0 is the initial optical output power. (Reproduced with permission from Yamakoshi et al.[83])

can occur are generally referred to as *catastrophic facet degradation* and *facet erosion*. Catastrophic facet degradation is mechanical damage of the facets which may arise after short operating times of laser diodes at high optical power densities. This damage tends to reduce greatly the facet reflectivity, thereby increasing the threshold current and decreasing the external quantum efficiency. The fundamental cause of catastrophic facet degradation has not yet been determined, but it has been observed to be a function of the optical power density and the pulse length.

Facet erosion is a gradual degradation occurring over a longer period of time than catastrophic facet damage. The decrease in mirror reflectivity and the increase in nonradiative recombination at the facets owing to facet erosion lower the internal quantum efficiency of the laser and increase the threshold current. In GaAlAs lasers facet erosion arises from oxidation of the mirror surface. It is speculated that the oxidation process is stimulated by the optical radiation emitted from the laser. Facet erosion is minimized by depositing a half-wavelength-thick Al_2O_3 film on the facet. This type of coating acts as a moisture barrier and does not affect the mirror reflectivity or the lasing threshold current.

A comparison[82] of two definitions of failure for laser diodes operating at 70°C is shown in Fig. 4-41. The lower curve shows the time required for the laser output to drop to one-half its initial value when a constant current passes through the device. This is the "3-dB life."

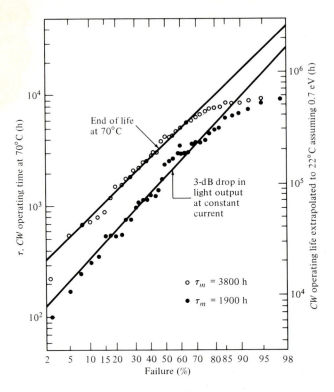

FIGURE 4-41
Time-to-failure plot on log-normal coordinates for 40 low-threshold ($\simeq$ 50 mA) oxide-defined stripe lasers at a 70°C heat sink temperature. τ_m is the time it took for 50 percent of the lasers to fail for the two types of failure mechanisms. (Reproduced with permission from Ettenberg and Kressel,[82] © 1980, IEEE.)

The "end-of-life" failure is given by the top trace in Fig. 4-41. This condition is defined as the time at which the device can no longer emit a fixed power level (1.25 mW in this case) at the 70°C heat sink temperature. The mean operating times (time for 50 percent of the lasers to fail) are 3800 h and 1900 h for the end-of-life and 3-dB-life conditions respectively. The right-hand ordinate of Fig. 4-41 gives an estimate of the operating time at 22°C, assuming an activation energy of 0.7 eV.

4.7 SUMMARY

In this chapter we have examined the basic operating characteristics of hetero-junction-structured semiconductor light-emitting diodes (LEDs) and laser diodes. We first discussed the structures of these light sources, which is a sandwich type construction of different semiconductor materials. These layers serve to confine the electrical and optical carriers to yield optical sources with high output and high efficiencies. The principal materials of which these layers are composed include the ternary alloy GaAlAs for operation in the 800- to 900-nm wavelength region and the quaternary alloy InGaAsP for use between 1100 and 1600 nm.

We then discussed in detail the internal quantum efficiency and the various factors that can reduce this efficiency. The two principal factors are:

1. Nonradiative recombination of charge carriers at the boundaries of the heterostructure layers resulting from crystal lattice mismatches. This effect can be minimized by choosing materials with closely matched lattice spacings for adjacent heterojunction layers.
2. Optical absorption in the active region of the device. This effect, which is known as self-absorption, is minimized by using thin active regions.

In conjunction with the quantum efficiency we addressed the modulation capability of optical sources and their response when subjected to transient current pulses. We noted in particular that the time delay between the application of a current pulse and the onset of optical power output can be reduced by applying a small dc bias to the source. This bias reduces the parasitic diode space charge capacitance, which could cause a delay of the carrier injection into the active region, which in turn could delay the optical output.

We next turned our attention to semiconductor laser diode optical sources. When deciding whether to choose an LED or a laser diode source, a tradeoff must be made between the advantages and drawbacks of each type of device. The advantages that a laser diode has over an LED are:

1. A faster response time, so that much greater modulation rates (data transmission rates) are possible with a laser diode
2. A narrower spectral width of the output, which implies less dispersion-induced signal distortion
3. A much higher optical power level that can be coupled into a fiber with a laser diode, thus allowing greater transmission distances

Some of the drawbacks of laser diodes are:

1. Their construction is more complicated, mainly because of the requirement of current confinement in a small lasing cavity.
2. The optical output power level is strongly dependent on temperature. This increases the complexity of the transmitter circuitry. If a laser diode is to be used over a wide temperature range, then either a cooling mechanism must be used to maintain the laser at a constant temperature or a threshold-sensing circuit can be implemented to adjust the bias current with changes in temperature.
3. They are susceptible to catastrophic facet degradation which greatly reduces the device lifetime. This is mechanical damage of the facets that may arise after short operating times at high optical power densities.

High-radiance LEDs and laser diodes are well suited for wideband analog applications, provided a method is used to compensate for any nonlinearities in the output of these devices. These nonlinearities will create frequency components in the output signal that were not present in the input signal. Two important nonlinear effects are harmonic and intermodulation distortions. For intermodulation distortion, only the third-order terms are usually important. All other intermodulation products can be filtered out in the receiver if the operating bands are less than an octave. Total harmonic distortions in GaAlAs LEDs and laser diodes (which operate in the 800- to 900-nm range) tend to be about 30 to 40 dB below the output of the fundamental modulation frequency for modulation depths of around 0.5. Special circuit design techniques can be implemented for linearization of optical sources in analog communication systems. These designs can reduce total harmonic distortions by 30 to 40 dB.

An important issue in any application is device reliability. Degradation of light sources can be divided into three basic categories: internal damage and contact degradation, which hold for both laser diodes and LEDs, and damage to the facets of laser diodes. Lifetime tests of optical sources are often done at elevated temperatures (e.g., 70°C) to accelerate the degradation process. Since optical sources are adversely affected by high currents and high temperatures, it is recommended that in order to increase the light source lifetime, they be operated at as low a current and temperature as is practical in a system.

PROBLEMS

4-1. Measurements have shown that the band-gap energy E_g for GaAs varies with temperature according to the empirical formula

$$E_g(T) \simeq 1.55 - 4.3 \times 10^{-4}T$$

where E_g is given in electron volts. Using this expression, show that the temperature dependence of the intrinsic electron concentration n_i is

$$n_i = 5 \times 10^{15}T^{3/2}e^{-8991/T}$$

where n_i is measured in cm^{-3}. Use the values of m_e and m_h given in Example 4-1.

4-2. Repeat the steps given in Example 4-2 for a p-type semiconductor. In particular, show that when the net acceptor concentration is much greater than n_i, we have $p_p = N_A$ and $n_p = n_i^2/N_A$.

4-3. An engineer has two Ga$_{1-x}$Al$_x$As LEDs; one has a band-gap energy of 1.540 eV and the other has $x = 0.015$. (a) Find the aluminum mole fraction x and the emission wavelength for the first LED; (b) find the band-gap energy and the emission wavelength of the other LED.

4-4. The lattice spacing of In$_{1-x}$Ga$_x$As$_y$P$_{1-y}$ has been shown to obey Vegard's law.[89] This states that for quaternary alloys of the form A$_{1-x}$B$_x$C$_y$D$_{1-y}$, where A and B are group III elements (e.g., Al, In, and Ga) and C and D are group V elements (e.g., As, P, and Sb), the lattice spacing $a(x, y)$ of the quaternary alloy can be

approximated by
$$a(x, y) = xya(BC) + x(1 - y)a(BD) + (1 - x)ya(AC)$$
$$+ (1 - x)(1 - y)a(AD)$$

where the $a(IJ)$ are the lattice spacings of the binary compounds IJ.

(a) Show that for $In_{1-x}Ga_xAs_yP_{1-y}$ with

$$a(GaAs) = 5.6536 \text{ Å}$$

$$a(GaP) = 5.4512 \text{ Å}$$

$$a(InAs) = 6.0590 \text{ Å}$$

$$a(InP) = 5.8696 \text{ Å}$$

the quaternary lattice spacing becomes

$$a(x, y) = 0.1894y - 0.4184x + 0.0130xy + 5.8696 \text{ Å}$$

(b) For quaternary alloys that are lattice-matched to InP the relation between x and y can be determined by letting $a(x, y) = a(InP)$. Show that since $0 \le x \le 0.47$, the resulting expression can be approximated by $y \simeq 2.20x$.

(c) A simple empirical relation that gives the band-gap energy in terms of x and y is[89]

$$E_g(x, y) = 1.35 + 0.668x - 1.17y + 0.758x^2 + 0.18y^2$$

$$- 0.069xy - 0.322x^2y + 0.03xy^2 \text{ eV}$$

Find the band-gap energy and the peak emission wavelength of $In_{0.74}Ga_{0.26}As_{0.56}P_{0.44}$.

4-5. (a) If the radiative and nonradiative recombination lifetimes of the minority carriers in the active region of an LED are 3 ns and 100 ns, respectively, find the internal efficiency and the bulk recombination lifetime in the absence of self-absorption and recombination at the heterojunction.

(b) If the surface recombination velocity at the heterojunction interfaces is 5000 cm/s, what are the lifetime reductions for 1-μm- and 2-μm-thick active layers? Assume that the condition $L_DS/D \ll 1$ holds.

4-6. By using Eqs. (4-11) and (4-12) derive Eq. (4-13), which gives the density of excess electrons at the position x relative to the pn junction in the active layer of a light source.

4-7. Derive Eq. (4-15) by using Eqs. (4-13) and (4-14).

4-8. Show that Eq. (4-15) reduces to Eq. (4-16) under the conditions that $L_DS \ll D$ and $d \le L_D$.

4-9. Derive Eq. (4-19) from Eq. (4-17).

4-10. Calculate and plot the relative reduction in the internal quantum efficiency, η_i^{dh}/η_0, for a 10^{18}-cm^{-3} active-area doping concentration in a GaAs LED. Let the active-layer widths range from $d = 0.1L_D$ to $10L_D$, and assume the surface recombination velocity $S = 10^5$ cm/s. At a 10^{18}-cm^{-3} doping level $\alpha_\lambda = 4000$ cm^{-1}, $L_D = 5 \times 10^{-4}$ cm, and $D = 80$ cm^2/s for GaAs. Compare the results with the $\alpha_\lambda L_D = 2.0$ curves for $S = 10^3$ cm/s and 10^6 cm/s shown in Fig. 4-15.

4-11. A particular InGaAsP LED emitting at 1.3 μm is found to have a radiative recombination coefficient of 3×10^{-10} cm^3/s, a carrier concentration of $n_0 +$

$p_0 = 10^{17}$ cm^{-3} in its 1-μm-thick active region, and a surface recombination velocity of 10^4 cm/s. Using Eq. (4-24) for the effective carrier lifetime, plot the frequency response of the relative optical output intensity $I(\omega)/I_0$ of this LED for modulation frequencies ranging from 1 to 500 MHz at current densities of 0.5, 2.0, and 10 kA/cm^2. (*Note:* Use three-cycle semilog paper with the frequency assigned to the log scale.) What is the 3-dB modulation bandwidth at each of these current densities?

4-12. A practical surface-emitting LED has a 50-μm-diameter emitting area and operates at a peak modulation current of 100 mA. Assuming Eq. (4-29) holds, what is the bandwidth of a GaAlAs LED having a 2.0-μm active-area thickness? Take $B_r = 10^{-10}$ cm^3/s and $S = 10^4$ cm/s.

4-13. Derive Eq. (4-27). Show that this equation reduces to Eq. (4-28) when $\Delta n \ll n_0 + p_0$ and to Eq. (4-29) when $\Delta n \gg n_0 + p_0$.

4-14. An LED has a 500-pF space charge capacitance, a 1.0-pA saturation current, and a 5-ns minority carrier lifetime. Plot the half-current and 10- to 90-percent rise times as a function of current for drive current amplitudes ranging from 10 to 100 mA. (The fact that $t_{1/2}$ is greater than t_{10-90} in this plot for low drive currents in nonprebiased LEDs is a result of the time delay between the application of a current pulse and the onset of optical output power.)

4-15. Derive Eq. (4-31) by letting the effects of the lifetime τ be modeled as a diffusion capacitance C_d in shunt with the junction of the LED, as Fig. P4-15 shows. Here $R = 2k_BT/(qI_p)$, where I_p is the amplitude of the current step function used to drive the LED, and $\tau = RC_d$. Use Laplace transforms to first find $I_F(t)$ and then solve for t_{90-10}.

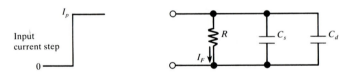

FIGURE P4-15

4-16. (*a*) A GaAlAs laser diode has a 500-μm cavity length which has an effective absorption coefficient of 10 cm^{-1}. For uncoated facets the reflectivities are 0.32 at each end. What is the optical gain at the lasing threshold?
(*b*) If one end of the laser is coated with a dielectric reflector so that its reflectivity is now 90 percent, what is the optical gain at the lasing threshold?
(*c*) If the internal quantum efficiency is 0.65, what is the external quantum efficiency in cases (*a*) and (*b*)?

4-17. Find the external quantum efficiency for a Ga$_{1-x}$Al$_x$As laser diode (with $x = 0.03$) which has the optical-power-versus-drive-current relationship shown in Fig. 4-31.

4-18. A GaAs laser emitting at 800 nm has a 400-μm-long cavity with a refractive index $n = 3.6$. If the gain g exceeds the total loss α_t throughout the range 750 nm $< \lambda <$ 850 nm, how many modes will exist in the laser?

4-19. A laser emitting at $\lambda_0 = 850$ nm has a gain-spectral width of $\sigma = 32$ nm and a peak gain of $g(0) = 50$ cm^{-1}. Plot $g(\lambda)$ from Eq. (4-44). If $\alpha_t = 32.2$ cm^{-1}, show the region where lasing takes place. If the laser is 400 μm long and $n = 3.6$, how many modes will be excited in this laser?

4-20. The derivation of Eq. (4-47) assumes that the refractive index n is independent of wavelength.

(a) Show that when n depends on λ, we have

$$\Delta\lambda = \frac{\lambda^2}{2L(n - \lambda\, dn/d\lambda)}$$

(b) If the group refractive index $(n - \lambda\, dn/d\lambda)$ is 4.5 for GaAs at 850 nm, what is the mode spacing for a 400-μm-long laser?

4-21. For laser structures having strong carrier confinement, the threshold current density for stimulated emission J_{th} can to a good approximation be related to the lasing-threshold optical gain g_{th} by $g_{th} = \beta J_{th}$, where β is a constant that depends on the specific device construction. Consider a GaAs laser with an optical cavity of length 250 μm and width 100 μm. At the normal operating temperature, the gain factor $\beta = 21 \times 10^{-3}$ A/cm^3 and the effective absorption coefficient $\bar{\alpha} = 10$ cm^{-1}.

(a) If the refractive index is 3.6, find the threshold current density and the threshold current I_{th}. Assume the laser end faces are uncoated and the current is restricted to the optical cavity.

(b) What is the threshold current if the laser cavity width is reduced to 10 μm?

4-22. A distributed feedback laser has a Bragg wavelength of 1570 nm, a second-order grating with $\Lambda = 460$ nm, and a 300-μm cavity length. Assuming a perfectly symmetrical DFB laser, find the zeroth-, first-, and second-order lasing wavelengths to a tenth of a nanometer. Draw a relative amplitude-versus-wavelength plot.

4-23. When a current pulse is applied to a laser diode, the injected carrier pair density Δn within the recombination region of width d changes with time according to the relationship

$$\frac{\partial(\Delta n)}{\partial t} = \frac{J}{qd} - \frac{\Delta n}{\tau}$$

(a) Assume τ is the average carrier lifetime in the recombination region when the injected carrier pair density is Δn_{th} near the threshold current density J_{th}. That is, in the steady state we have $\partial(\Delta n)/\partial t = 0$, so that

$$\Delta n_{th} = \frac{J_{th}\tau}{qd}$$

If a current pulse of amplitude I_p is applied to an unbiased laser diode, show that the time needed for the onset of stimulated emission is

$$t_d = \tau \ln \frac{I_p}{I_p - I_{th}}$$

Assume the drive current $I = JA$, where J is the current density and A is the area of the active region.

(b) If the laser is now prebiased to a current density $J_B = I_B/A$, so that the initial excess carrier pair density is $\Delta n_B = J_B\tau/qd$, then the current density in the active region during a current pulse I_p is $J = J_B + J_p$. Show that in this case Eq. (4-50) results.

4-24. Assume we have an LED which operates with a 5-V bias supply and which we want to drive with a 50-mA peak current. Design the following simple drive circuits:

(*a*) A common-emitter transistor configuration.

(*b*) A low-speed driver using a TTL NAND logic gate (e.g., a 7437 positive-NAND buffer).

(*c*) A high-speed ECL-compatible emitter-coupled driver using a commercial ECL device (e.g., a 10210 circuit).

4-25. When designing a driver for a laser diode, one must take into account the temperature dependence of the threshold current as noted in Sec. 4.3.6. Design a laser-diode transmitter having a low-speed rear-facet-detection bias-stabilization circuit and a high-speed drive circuit. How would a thermoelectric cooler be incorporated into this design?

4-26. A laser diode has a maximum average output of 1 mW (0 dBm). The laser is to be amplitude-modulated with a signal $x(t)$ that has a dc component of 0.2 and a periodic component of ± 2.56. If the current-input to optical-output relationship is $P(t) = i(t)/10$, find the values of I_0 and m if the modulating current is $i(t) = I_0[1 + mx(t)]$.

4-27. Consider the following Taylor series expansion of the optical power versus drive current relationship of an optical source about a given bias point:

$$y(t) = a_1 x(t) + a_2 x^2(t) + a_3 x^3(t) + a_4 x^4(t)$$

Let the modulating signal $x(t)$ be the sum of two sinusoidal tones at frequencies ω_1 and ω_2 given by

$$x(t) = b_1 \cos \omega_1 t + b_2 \cos \omega_2 t$$

(*a*) Find the second-, third-, and fourth-order intermodulation distortion coefficients B_{mn} (where m and $n = \pm 1, \pm 2, \pm 3,$ and ± 4) in terms of $b_1, b_2,$ and the a_i.

(*b*) Find the second-, third-, and fourth-order harmonic distortion coefficients $A_2,$ $A_3,$ and A_4 in terms of $b_1, b_2,$ and the a_i.

4-28. An optical source is selected from a batch characterized as having lifetimes which follow a slow internal degradation mode. The -3-dB mean time to failure of these devices at room temperature is specified as 5×10^4 h. If the device emits 1 mW at room temperature, what is the expected optical output power after 1 month of operation? after 1 yr? after 5 yr?

4-29. A group of optical sources is found to have operating lifetimes of 4×10^4 h at 60°C and 6500 h at 90°C. What is the expected lifetime at 20°C if the device lifetime follows an Arrhenius type relationship?

REFERENCES

1. H. Kressel and J. K. Butler, *Semiconductor Lasers and Heterojunction LEDs*, Academic, New York, 1977.

2. H. C. Casey, Jr. and M. B. Panish, *Heterostructure Lasers: Part A—Fundamental Principles: Part B—Materials and Operating Characteristics*, Academic, New York, 1978.

3. G. H. B. Thompson, *Physics of Semiconductor Laser Devices*, Wiley, New York, 1980.

4. (*a*) D. Botez, "Laser diodes are power-packed," *IEEE Spectrum*, vol. 22, pp. 43–53, June 1985.
   (*b*) D. Botez, "Recent developments in high-power single-element fundamental-mode diode lasers," *Laser Focus/E-O*, vol. 23, pp. 68–79, Mar. 1987.

5. D. Botez and M. Ettenberg, "Comparison of surface and edge emitting LEDs for use in fiber optical communications," *IEEE Trans. Electron Devices*, vol. ED-26, pp. 1230–1238, Aug. 1979.

6. (*a*) J. E. Bowers, "High speed semiconductor laser design and performance," *Solid State Electronics*, vol. 30, pp. 1–11, Jan. 1987.
   (*b*) S. Oshiba and Y. Tamura, "Recent progress in high-power InGaAsP lasers," *J. Lightwave Tech.*, vol. 8, pp. 1350–1356, Sept. 1990.
   (*c*) J. Arnaud and M. Esteban, "Circuit theory of laser diode modulation and noise," *IEE Proc.*, vol. 137, pp. 55–63, Feb. 1990.

7. M. Ohtsu, "Tutorial review: Frequency stabilization in semiconductor lasers," *Opt. Quantum Electron.*, vol. 20, no. 4, pp. 283–300, July 1988.

8. (*a*) T. P. Lee, C. A. Burrus, Jr., and R. H. Saul, "Light-emitting diodes for telecommunications," in S. E. Miller and I. P. Kaminow, eds., *Optical Fiber Telecommunications*, *II*, Academic, New York, 1988.
   (*b*) J. E. Bowers and M. A. Pollack, "Semiconductor lasers for telecommunications," in S. E. Miller and I. P. Kaminow, eds., *Optical Fiber Telecommunications*, *II*, Academic, New York, 1988.

9. (*a*) D. M. Fye, R. Olshansky, J. La Course, W. Powazinik, and R. B. Lauer, "Low-current, 1.3-μm edge-emitting LED for single-mode subscriber loop applications," *Electron. Lett.*, vol. 22, pp. 87–88, Jan. 1986.
   (*b*) R. B. Lauer, D. M. Fye, L. W. Ulbricht, and M. J. Teare, "A light source for the fibered local loop," *GTE J. Science and Tech.*, vol. 1, no. 1, pp. 28–39, 1987.

10. (*a*) J. L. Gimlett, M. Stern, R. S. Vodhanel, N. K. Cheung, G. K. Chang, H. P. Leblanc, P. W. Shumate, and A. Suzuki, "Transmission experiments at 560 Mb/s and 140-Mb/s using single-mode fiber and 1300-nm LEDs," *Electron. Lett.*, vol. 21, pp. 1198–1200, Dec. 5, 1985.
    (*b*) G. K. Chang, H. P. Leblanc, and P. W. Shumate, "Novel high-speed LED transmitter for single-mode fiber and wideband loop transmission systems," *Electron. Lett.*, vol. 23, pp. 1338–1340, Dec. 3, 1987.

11. (*a*) L. A. Reith and P. A. Shumate, "Coupling sensitivity of an edge-emitting LED to a single-mode fiber," *J. Lightwave Tech.*, vol. LT-5, pp. 29–34, Jan. 1987.
    (*b*) D. N. Christodoulides, L. A. Reith, and M. A. Saifi, "Theory of LED coupling to single-mode fibers," *J. Lightwave Tech.*, vol. LT-5, pp. 1623–1629, Nov. 1987.

12. B. Hillerich, "New analysis of LED to a single-mode fiber coupling," *Electron. Lett.*, vol. 22, pp. 1176–1177, Oct. 1986; "Efficiency and alignment tolerances of LED to a single-mode fiber coupling—theory and experiment," *Opt. Quantum Electron.*, vol. 19, no. 4, pp. 209–222, July 1987.

13. T. Tsubota, Y. Kashima, H. Takano, and Y. Hirose, "InGaAsP/InP long-wavelength high-efficiency edge-emitting LED for single-mode fiber optic communication," *Fiber Integrated Optics*, vol. 7, no. 4, pp. 353–360, 1988.

14. T. Ohtsuka, N. Fujimoto, K. Yamaguchi, A. Taniguchi, H. Naitou, and Y. Nabeshima, "Gigabit single-mode fiber transmission using 1.3-μm edge-emitting LEDs for broadband subscriber loops," *J. Lightwave Tech.*, vol. LT-5, pp. 1534–1541, Oct. 1987.

15. E. Yang, *Microelectronic Devices*, McGraw-Hill, New York, 1988.

16. M. Zambuto, *Semiconductor Devices*, McGraw-Hill, New York, 1989.

17. S. Wang, *Fundamentals of Semiconductor Theory and Device Physics*, Prentice-Hall, Englewood Cliffs, NJ, 1989.

18. C. M. Wolfe, N. Holonyak, Jr., and G. E. Stillman, *Physical Properties of Semiconductors*, Prentice-Hall, Englewood Cliffs, NJ, 1989.

19. S. M. Sze, *Physics of Semiconductor Devices*, Wiley, New York, 1981.

20. (*a*) J. A. Long, R. A. Logan, and R. F. Karlicek, Jr., "Epitaxial growth methods for lightwave devices," in S. E. Miller and I. P. Kaminow, eds., *Optical Fiber Telecommunications*, *II*, Academic, New York, 1988.

(b) G. H. Olsen, "Long-wavelength components by vapor phase epitaxy," *Laser Focus/Electro-Optics*, vol. 21, pp. 124–132, Jan. 1985.

21. (a) C. A. Burrus and B. I. Miller, "Small-area double heterostructure AlGaAs electroluminescent diode sources for optical fiber transmission lines," *Opt. Commun.*, vol. 4, pp. 307–309, Dec. 1971.

    (b) J. P. Wittke, M. Ettenberg, and H. Kressel, "High radiance LED for single fiber optical links," *RCA Rev.*, vol. 37, pp. 159–183, June 1976.

22. T. P. Lee and A. J. Dentai, "Power and modulation bandwidth of GaAs-AlGaAs high radiance LEDs for optical communication systems," *IEEE J. Quantum Electron.*, vol. QE-14, pp. 150–159, Mar. 1978.

23. W. Harth, J. Heinen, and W. Huber, "Influence of active layer width on the performance of homojunction and single heterojunction GaAs light emitting diodes," *Electron. Lett.*, vol. 11, pp. 23–24, Jan. 1975.

24. (a) H. Namizaki, M. Nagano, and S. Nakahara, "Frequency response of GaAlAs light emitting diodes," *IEEE Trans. Electron. Devices*, vol. ED-21, pp. 688–691, Nov. 1974.

    (b) Y. S. Lin and D. A. Smith, "The frequency response of an amplitude modulated GaAs luminescent diode," *Proc. IEEE*, vol. 63, pp. 542–544, Mar. 1975.

25. T. P. Lee, "Effects of junction capacitance on the rise time of LEDs and the turn-on delay of injection lasers," *Bell Sys. Tech. J.*, vol. 54, pp. 53–68, Jan. 1975.

26. I. Hino and K. Iwamoto, "LED pulse response analysis," *IEEE Trans. Electron. Devices*, vol. ED-26, pp. 1238–1242, Aug. 1979.

27. J. Zucker and R. B. Lauer, "Optimization and characterization of high radiance (Al, Ga)As double heterostructure LEDs for optical communication systems," *IEEE Trans. Electron. Devices*, vol. ED-25, pp. 193–198, Feb. 1978.

28. J. Zucker, "Closed-form calculation of the transient behavior of (Al, Ga)As double-heterojunction LEDs," *J. Appl. Phys.*, vol. 49, pp. 2543–2545, Apr. 1978.

29. K. Kobayashi and I. Mito, "Single frequency and tunable laser diodes," *J. Lightwave Tech.*, vol. 6, pp. 1623–1633, Nov. 1988.

30. H. Imai and T. Kaneda, "High-speed distributed feedback lasers and InGaAs avalanche photodiodes," *J. Lightwave Tech.*, vol. 6, pp. 1634–1642, Nov. 1988.

31. T. L. Koch, "Distributed feedback and distributed Bragg reflector lasers: Comparisons of performance and use," *Tech. Digest OSA/IEEE Optical Fiber Commun. Conf.*, p. 32, Houston, TX, Feb. 1989.

32. M. Ettenberg, "A new dielectric facet reflector for semiconductor lasers," *Appl. Phys. Lett.*, vol. 32, pp. 724–725, June 1978.

33. M. Kume, H. Shimizu, K. Hamada, F. Susa, M. Wada, K. Itoh, G. Kano, and I. Teramoto, "Noise reduction in single longitudinal mode lasers by high-reflectivity coatings," *IEEE J. Quantum Electron.*, vol. QE-21, pp. 707–711, June 1985.

34. M. Yamada, "Transverse and longitudinal mode control in semiconductor injection lasers," *IEEE J. Quantum Electron.*, vol. QE-19, pp. 1365–1380, Sept. 1983.

35. J. K. Butler and D. Botez, "Lateral mode discrimination and control in high-power single-mode diode lasers of the large-optical cavity (LOC) type," *IEEE J. Quantum Electron.*, vol. QE-20, pp. 879–891, Aug. 1984.

36. T. P. Lee, C. A. Burrus, R. A. Linke, and R. J. Nelson, "Short-cavity, single-frequency InGaAsP buried heterostructure lasers," *Electron. Lett.*, vol. 19, pp. 82–84, Feb. 3, 1983.

37. S. Kinoshita and K. Iga, "Circular buried heterostructure (CBH) GaAlAs/GaAs surface-emitting lasers," *IEEE J. Quantum Electron.*, vol. QE-23, pp. 882–888, June 1987.

38. (a) Z. L. Liau and J. N. Walpole, "Surface-emitting GaInAsP/InP laser with low threshold current and high efficiency," *Appl. Phys. Lett.*, vol. 46, pp. 115–117, Jan. 15, 1985.

    (b) Z. L. Liau and J. N. Walpole, "Large monolithic two-dimensional arrays of GaInAsP/InP surface-emitting lasers," *Appl. Phys. Lett.*, vol. 50, pp. 528–530, Mar. 1987.

39. (a) M. Y. A. Raja, S. R. J. Brueck, M. Osinski, C. F. Schaus, J. G. McInerney, T. M. Brennan, and B. E. Hammons, "Resonant periodic gain surface-emitting semiconductor lasers," *IEEE J. Quantum Electron.*, vol. 25, pp. 1500–1512, June 1989.

(*b*) S. W. Corzine, R. S. Geels, J. W. Scott, R.-H. Yan, and L. A. Coldren, "Design of Fabry-Perot surface-emitting lasers with a periodic gain structure," *IEEE J. Quantum Electron.*, vol. 25, pp. 1513–1524, June 1989.

(*c*) A. Hardy, D. F. Welch, and W. Streifer, "Analysis of a dual grating-type surface-emitting laser," *IEEE J. Quantum Electron.*, vol. 26, pp. 50–60, Jan. 1990.

(*d*) J. L. Jewell, Y. H. Lee, A. Scherer, S. L. McCall, N. A. Olsson, J. P. Harbison, and L. T. Florez, "Surface-emitting microlasers for photonic switching and interchip connections," *Optical Engineering*, vol. 29, pp. 210–214, Mar. 1990.

40. H. Kogelnik and C. V. Shank, "Coupled-wave theory of distributed feedback lasers," *J. Appl. Phys.*, vol. 43, pp. 2327–2335, May 1972.

41. W. Streifer, R. D. Burnham, and D. R. Scifres, "Effect of external reflectors on longitudinal modes of distributed feedback lasers," *IEEE J. Quantum Electron.*, vol. QE-11, pp. 154–161, Apr. 1975.

42. S. Akiba, M. Usami, and K. Utaka, "1.5-μm $\lambda/4$ shifted InGaAsP/InP DFB lasers," *J. Lightwave Tech.*, vol. LT-5, pp. 1564–1573, Nov. 1987.

43. J. E. A. Whiteaway, G. H. B. Thompson, A. J. Collar, and C. J. Armistead, "The design and assessment of $\lambda/4$ phase-shifted DFB laser structures," *IEEE J. Quantum Electron.*, vol. 25, pp. 1261–1279, June 1989.

44. Y. Suematsu, S. Arai, and K. Kinoshita, "Dynamic single-mode semiconductor lasers with a distributed reflector," *J. Lightwave Tech.*, vol. LT-1, pp. 161–176, Mar. 1983.

45. K. Komori, S. Arai, Y. Suematsu, I. Arima, and M. Aoki, "Single-mode properties of distributed-reflector lasers," *IEEE J. Quantum Electron.*, vol. 25, pp. 1235–1244, June 1989.

46. M. Stix, M. P. Kesler, and E. Ippen, "Observation of subpicosecond dynamics in GaAlAs laser diodes," *Appl. Phys. Lett.*, vol. 48, no. 25, pp. 1722–1724, June 23, 1986.

47. (*a*) M. Ettenberg, C. J. Neuse, and H. Kressel, "The temperature dependence of threshold current for double-heterojunction lasers," *J. Appl. Phys.*, vol. 50, pp. 2949–2950, Apr. 1979.

(*b*) H. Ghafoori-Shiraz, "Temperature, bandgap-wavelength, and doping dependence of peak-gain coefficient parabolic model parameters for InGaAsP/InP semiconductor laser diodes," *J. Lightwave Tech.*, vol. 6, pp. 500–506, Apr. 1988.

48. R. Chin, N. Holonyak, B. A. Vojak, K. Hess, R. D. Dupuis, and P. D. Dapkus, "Temperature dependence of threshold current for quantum-well heterostructure laser diodes," *Appl. Phys. Lett.*, vol. 36, pp. 19–21, Jan. 1980.

49. G. H. B. Thompson, "Temperature dependence of threshold current in GaInAsP DH lasers at 1.3 and 1.5 μm wavelength," *IEE Proc.*, vol. 128, pp. 37–43, Apr. 1981.

50. P.W. Shumate, Jr., F. S. Chen, and P. W. Dorman, "GaAlAs laser transmitter for lightwave transmission systems," *Bell Sys. Tech. J.*, vol. 57, pp. 1823–1836, July–Aug. 1978.

51. F. S. Chen, "Simultaneous feedback control of bias and modulation currents for injection lasers," *Electron. Lett.*, vol. 16, pp. 7–8, Jan. 1980.

52. D. W. Smith, "Laser level-control circuit for high-bit-rate systems using a slope detector," *Electron. Lett.*, vol. 14, pp. 775–776, Nov. 1978.

53. J. F. Svacek, "Transmitter feedback techniques stabilize laser diode outputs," *EDN*, vol. 25, pp. 107–111, Mar. 1980.

54. M. Ettenberg, D. R. Patterson, and E. J. Denlinger, "A temperature-compensated laser module for optical communications," *RCA Rev.*, vol. 40, pp. 103–114, June 1979.

55. A. Albanese, "An automatic bias control circuit for injection lasers," *Bell Sys. Tech. J.*, vol. 57, pp. 1533–1544, May 1978.

56. (*a*) A. B. Carlson, *Communication Systems*, 3rd ed., McGraw-Hill, New York, 1986.

(*b*) Tri T. Ha, *Solid-State Microwave Amplifier Design*, Wiley, New York, 1981, Chap. 6.

(*c*) H. Taub and D. L. Shilling, *Principles of Communication Systems*, 2nd ed., McGraw-Hill, New York, 1986.

57. K. Asatani and T. Kimura, "Analysis of LED nonlinear distortions," *IEEE Trans. Electron Devices*, vol. ED-25, pp. 199–207, Feb. 1978; "Linearization of LED nonlinearity by predistortions," *ibid.*, pp. 207–212.

58. R. W. Dawson, "Frequency and bias dependence of video distortion in Burrus-type homostructure and heterostructure LEDs," *IEEE Trans. Electron Devices*, vol. ED-25, pp. 550–551, May 1978.

59. F. D. King, J. Straus, O. I. Szentesi, and A. J. Springthorpe, "High-radiance long-lived LEDs for analogue signalling," *Proc. IEE*, vol. 123, pp. 619–622, June 1976.

60. T. Ozeki and E. H. Hara, "Measurement of nonlinear distortion in light emitting diodes," *Electron. Lett.*, vol. 12, pp. 78–80, Feb. 1976.

61. K. Asatani, "Nonlinearity and its compensation of semiconductor laser diodes for analog intensity modulation systems," *IEEE Trans. Commun.*, vol. COM-28, pp. 297–300, Feb. 1980.

62. (a) J. Straus, "Linearized transmitters for analog fiber links," *Laser Focus*, vol. 14, pp. 54–61, Oct. 1978.

    (b) M. Ohtsu, M. Murata, and M. Kourogi, "FM noise reduction and subkilohertz linewidth of an AlGaAs laser by negative electrical feedback," *IEEE J. Quantum Electron.*, vol. 26, pp. 231–241, Feb. 1990.

63. J. Straus, A. J. Springthorpe, and O. I. Szentesi, "Phase shift modulation technique for the linearization of analog transmitters," *Electron. Lett.*, vol. 13, pp. 149–151, Mar. 1977.

64. (a) R. E. Patterson, J. Straus, G. Blenman, and T. Witkowicz, "Linearization of multichannel analog optical transmitters by quasi-feedforward compensation technique," *IEEE Trans. Comm.*, vol. COM-27, pp. 582–588, Mar. 1979.

    (b) J. Straus and O. I. Szentesi, "Linearization of optical transmitters by a quasi-feedforward compensation technique," *Electron. Lett.*, vol. 13, pp. 158–159, Mar. 1977.

65. D. Kato, "High-quality broadband optical communication by TDM-PAM: Nonlinearity in laser diodes," *IEEE J. Quantum Electron.*, vol. QE-14, pp. 343–346, May 1978.

66. K. O. Hill, Y. Tremblay, and B. S. Kawasaki, "Modal noise in multimode fiber links: Theory and experiment," *Opt. Letters*, vol. 5, pp. 270–272, June 1980.

67. K. Sato and K. Asatani, "Speckle noise reduction in fiber optic analog video transmission using semiconductor laser diodes," *IEEE Trans. Comm.*, vol. COM-29, pp. 1017–1024, July 1981.

68. A. R. Michelson and A. Weierholt, "Modal-noise limited signal-to-noise ratios in multimode optical fibers," *Appl. Opt.*, vol. 22, pp. 3084–3089, Oct. 1983.

69. K. Petermann, "Nonlinear distortions and noise in optical communication systems due to fiber connectors," *IEEE J. Quantum Electron.*, vol. QE-16, pp. 761–770, July 1980.

70. T. Kanada, "Evaluation of modal noise in multimode fiber-optic systems," *J. Lightwave Tech.*, vol. LT-2, pp. 11–18, Feb. 1984.

71. P. E. Couch and R. E. Epworth, "Reproducible modal-noise measurements in system design and analysis," *J. Lightwave Tech.*, vol. LT-1, pp. 591–595, Dec. 1983.

72. F. M. Sears, I. A. White, R. B. Kummer, and F. T. Stone, "Probability of modal noise in single-mode lightguide systems," *J. Lightwave Tech.*, vol. LT-4, pp. 652–655, June 1986.

73. K. Petermann and G. Arnold, "Noise and distortion characteristics of semiconductor lasers in optical fiber communication systems," *IEEE J. Quantum Electron.*, vol. QE-18, pp. 543–554, Apr. 1982.

74. N. H. Jensen, H. Olesen, and K. E. Stubkjaer, "Partition noise in semiconductor lasers under CW and pulsed operation," *IEEE J. Quantum Electron.*, vol. QE-18, pp. 71–80, Jan. 1987.

75. M. Ohtsu and Y. Teramachi, "Analyses of mode partition and mode hopping in semiconductor lasers," *IEEE J. Quantum Electron.*, vol. 25, pp. 31–38, Jan. 1989.

76. (a) C. H. Henry, P. S. Henry, and M. Lax, "Partition fluctuations in nearly single longitudinal mode lasers," *J. Lightwave Tech.*, vol. LT-2, pp. 209–216, June 1984.

    (b) S. E. Miller, "On the prediction of the mode-partitioning floor in injection lasers with multiple side modes at 2 and 10 Gb/s," *IEEE J. Quantum Electron.*, vol. 26, pp. 242–249, Feb. 1990.

77. E. E. Basch, R. F. Kearns, and T. G. Brown, "The influence of mode partition fluctuations in nearly single-longitudinal-mode lasers on receiver sensitivity," *J. Lightwave Tech.*, vol. LT-4, pp. 516–519, May 1986.

78. O. Hirota and Y. Suematsu, "Noise properties of injection lasers due to reflected waves," *IEEE J. Quantum Electron.*, vol. QE-15, pp. 142–149, Mar. 1979.

79. Y. C. Chen, "Noise characteristics of semiconductor laser diodes coupled to short optical fibers," *Appl. Phys. Lett.*, vol. 37, pp. 587–589, Oct. 1980.

80. G. P. Agrawal, N. A. Olsson, and N. K. Dutta, "Effect of far-end reflections on intensity and phase noise in InGaAsP semiconductor lasers," *Appl. Phys. Lett.*, vol. 45, pp. 597–599, Sept. 1984.

81. W. I. Way and M. M. Choy, "Optical feedback on linearity performance of 1300 nm DFB and multimode lasers under microwave intensity modulation," *J. Lightwave Tech.*, vol. 6, pp. 100–108, Jan. 1988.

82. M. Ettenberg and H. Kressel, "The reliability of (AlGa)As CW laser diodes," *IEEE J. Quantum Electron.*, vol. QE-16, pp. 186–196, Feb. 1980.

83. S. Yamakoshi, O. Hasegawa, H. Hamaguchi, M. Abe, and T. Yamaoka, "Degradation of high-radiance $Ga_{1-x}Al_xAs$ LEDs," *Appl. Phys. Lett.*, vol. 31, pp. 627–629, Nov. 1977.

84. L. R. Dawson, V. G. Keramidas, and C. L. Zipfel, "Reliable, high-speed LEDs for short-haul optical data links," *Bell Sys. Tech. J.*, vol. 59, pp. 161–168, Feb. 1980.

85. S. Yamakoshi, M. Abe, O. Wada, S. Komiya, and T. Sakurai, "Reliability of high-radiance InGaAsP/InP LEDs operating in the 1.2–1.3 μm wavelength," *IEEE J. Quantum Electron.*, vol. QE-17, pp. 167–173, Feb. 1981.

86. A. R. Goodwin, I. G. A. Davis, R. M. Gibb, and R. H. Murphy, "The design and realization of a high reliability semiconductor laser for single-mode fiber-optical communication links," *J. Lightwave Tech.*, vol. 6, pp. 1424–1434, Sept. 1988.

87. M. Fukuda, O. Fujita, and S. Uehara, "Homogeneous degradation of surface emitting type InGaAsP/InP light emitting diodes," *J. Lightwave Tech.*, vol. 6, pp. 1808–1814, Dec. 1988.

88. M. Fukuda, "Lasers and LED reliability update," *J. Lightwave Tech.*, vol. 6, pp. 1488–1495, Oct. 1988.

89. R. E. Nahory, M. A. Pollack, W. D. Johnston, Jr., and R. L. Barns, "Band gap versus composition and demonstration of Vegard's law for InGaAsP lattice matched to InP," *Appl. Phys. Lett.*, vol. 33, pp. 659–661, Oct. 1978.

CHAPTER
5

POWER
LAUNCHING
AND COUPLING

In implementing an optical fiber link two of the major system questions are how to launch optical power into a particular fiber from some type of luminescent source and how to couple optical power from one fiber into another. Launching optical power from a source into a fiber entails considerations such as the numerical aperture, core size, refractive-index profile, and core-cladding index difference of the fiber, plus the size, radiance, and angular power distribution of the optical source.

A measure of the amount of optical power emitted from a source that can be coupled into a fiber is usually given by the *coupling efficiency* η defined by

$$\eta = \frac{P_F}{P_S}$$

Here P_F is the power coupled into the fiber and P_S is the power emitted from the light source. The launching or coupling efficiency depends on the type of fiber that is attached to the source and on the coupling process, for example, whether or not lenses or other coupling improvement schemes are used.

In practice, many source suppliers offer devices with a short length of optical fiber (1 m or less) already attached in an optimum power-coupling configuration. This section of fiber is generally referred to as a *flylead* or a *pigtail*. The power-launching problem for these pigtailed sources thus reduces to a simpler one of coupling optical power from one fiber into another. The effects to be considered in this case include fiber misalignments, different core sizes, numerical apertures, and core refractive-index profiles, plus the need for clean and smooth fiber end faces that are perpendicular to the fiber axis.

Care must also be exercised when measuring the coupling efficiency between the fiber flylead and the cabled fiber, since the source can launch a significant amount of optical power into the cladding of the flylead. Although this power may be present at the end of the short flylead, it will not be coupled into the core of the following fiber. A true measurement of the power available from the flylead for coupling into a fiber can only be determined by stripping off the cladding modes before measuring the output optical power.

5.1 SOURCE-TO-FIBER POWER LAUNCHING

A convenient and useful measure of the optical output of a luminescent source is its radiance (or brightness) B at a given diode drive current. *Radiance* is the optical power radiated into a unit solid angle per unit emitting surface area and is generally specified in terms of watts per square centimeter per steradian. Since the optical power which can be coupled into a fiber depends on the radiance (that is, on the spatial distribution of the optical power), the radiance of an optical source rather than the total output power is the important parameter when considering source-to-fiber coupling efficiencies.

5.1.1 Source Output Pattern

To determine the optical power-accepting capability of a fiber, the spatial radiation pattern of the source must first be known. This pattern can be fairly complex. Consider Fig. 5-1, which shows a spherical coordinate system characterized by R, θ, and ϕ, with the normal to the emitting surface being the polar axis. The radiance may be a function of both θ and ϕ, and can also vary from point to point on the emitting surface. A reasonable assumption for simplicity of analysis is to take the emission to be uniform across the source area.

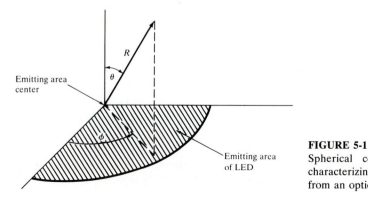

FIGURE 5-1
Spherical coordinate system for characterizing the emission pattern from an optical source.

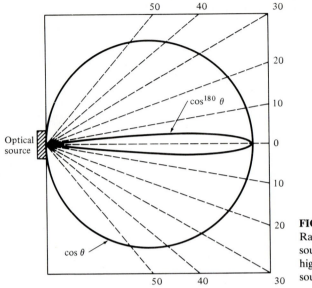

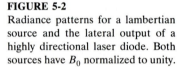

FIGURE 5-2
Radiance patterns for a lambertian source and the lateral output of a highly directional laser diode. Both sources have B_0 normalized to unity.

Surface-emitting LEDs are characterized by their lambertian output pattern, which means the source is equally bright when viewed from any direction. The power delivered at an angle θ measured relative to a normal to the emitting surface varies as $\cos \theta$ because the projected area of the emitting surface varies as $\cos \theta$ with viewing direction. The emission pattern for a lambertian source thus follows the relationship

$$B(\theta, \phi) = B_0 \cos \theta \tag{5-1}$$

where B_0 is the radiance along the normal to the radiating surface. The radiance pattern for this source is shown in Fig. 5-2.

Edge-emitting LEDs and laser diodes have a more complex emission pattern. These devices have different radiances $B(\theta, 0°)$ and $B(\theta, 90°)$ in the planes parallel and normal, respectively, to the emitting-junction plane of the device. These radiances can be approximated by the general form[1]

$$\frac{1}{B(\theta, \phi)} = \frac{\sin^2 \phi}{B_0 \cos^T \theta} + \frac{\cos^2 \phi}{B_0 \cos^L \theta} \tag{5-2}$$

The integers T and L are the transverse and lateral power distribution coefficients, respectively. In general, for edge emitters, $L = 1$ (which is a lambertian distribution with a 120° half-power beam width) and T is significantly larger. For laser diodes, L can take on values over 100.

Example 5-1. Figure 5-2 compares a lambertian pattern with a laser diode having a lateral ($\phi = 0°$) half-power beam width of $2\theta = 10°$. In this case from Eq. (5-2) we have

$$B(\theta = 5°, \phi = 0°) = B_0(\cos 5°)^L = \tfrac{1}{2}B_0$$

Solving for L, we have

$$L = \frac{\log 0.5}{\log(\cos 5°)} = \frac{\log 0.5}{\log 0.9962} = 182$$

The much narrower output beam from a laser diode allows significantly more light to be coupled into an optical fiber.

5.1.2 Power-Coupling Calculation

To calculate the maximum optical power coupled into a fiber, consider first the case shown in Fig. 5-3 for a symmetric source of brightness $B(A_s, \Omega_s)$, where A_s and Ω_s are the area and solid emission angle of the source, respectively. Here the fiber end face is centered over the emitting surface of the source and is positioned as close to it as possible. The coupled power can be found using the relationship

$$P = \int_{A_f} dA_s \int_{\Omega_f} d\Omega_s \, B(A_s, \Omega_s)$$

$$= \int_0^{r_m} \int_0^{2\pi} \left[\int_0^{2\pi} \int_0^{\theta_{0,\max}} B(\theta, \phi) \sin\theta \, d\theta \, d\phi \right] d\theta_s \, r \, dr \tag{5-3}$$

where the area and solid acceptance angle of the fiber define the limits of the integrals. In this expression first the radiance $B(\theta, \phi)$ from an individual radiating point source on the emitting surface is integrated over the solid acceptance angle of the fiber. This is shown by the expression in square brackets, where $\theta_{0,\max}$ is the maximum acceptance angle of the fiber, which is related to the numerical aperture NA through Eq. (2-23). The total coupled power is then determined by summing up the contributions from each individual

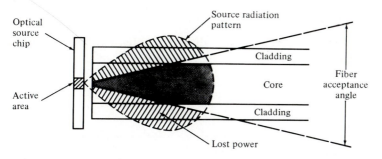

FIGURE 5-3
Schematic diagram of an optical source coupled to an optical fiber. Light outside the acceptance angle is lost.

emitting-point source of incremental area $d\theta_s\, r\, dr$, that is, integrating over the emitting area. For simplicity, here the emitting surface is taken as being circular. If the source radius r_s is less than the fiber core radius a, then the upper integration limit $r_m = r_s$; for source areas larger than the fiber core area, $r_m = a$.

As an example, assume a surface-emitting LED of radius r_s less than the fiber core radius a. Since this is a lambertian emitter, Eq. (5-1) applies and Eq. (5-3) becomes

$$P = \int_0^{r_s}\int_0^{2\pi}\left(2\pi B_0 \int_0^{\theta_{0,\max}} \cos\theta \sin\theta\, d\theta\right) d\theta_s\, r\, dr$$

$$= \pi B_0 \int_0^{r_s}\int_0^{2\pi} \sin^2\theta_{0,\max}\, d\theta_s\, r\, dr$$

$$= \pi B_0 \int_0^{r_s}\int_0^{2\pi} \mathrm{NA}^2\, d\theta_s\, r\, dr \qquad (5\text{-}4)$$

where the numerical aperture NA is defined by Eq. (2-23). For step-index fibers the numerical aperture is independent of the positions θ_s and r on the fiber end face, so that Eq. (5-4) becomes (for $r_s < a$),

$$P_{\mathrm{LED,\,step}} = \pi^2 r_s^2 B_0 (\mathrm{NA})^2 \simeq 2\pi^2 r_s^2 B_0 n_1^2 \Delta \qquad (5\text{-}5)$$

Consider now the total optical power P_s that is emitted from the source of area A_s into a hemisphere (2π sr). This is given by

$$P_s = A_s \int_0^{2\pi}\int_0^{\pi/2} B(\theta,\phi)\sin\theta\, d\theta\, d\phi$$

$$= \pi r_s^2\, 2\pi B_0 \int_0^{\pi/2} \cos\theta \sin\theta\, d\theta$$

$$= \pi^2 r_s^2 B_0 \qquad (5\text{-}6)$$

Equation (5-5) can, therefore, be expressed in terms of P_s:

$$P_{\mathrm{LED,\,step}} = P_s(\mathrm{NA})^2 \qquad \text{for} \quad r_s \le a \qquad (5\text{-}7)$$

When the radius of the emitting area is larger than the radius a of the fiber core area, Eq. (5-7) becomes

$$P_{\mathrm{LED,\,step}} = \left(\frac{a}{r_s}\right)^2 P_s(\mathrm{NA})^2 \qquad \text{for} \quad r_s > a \qquad (5\text{-}8)$$

Example 5-2. Consider an LED having a circular emitting area of radius 35 μm and a lambertian emission pattern with 150 W/(cm$^2 \cdot$ sr) axial radiance at a given drive current. Let us compare the optical powers coupled into two step-index fibers, one of which has a core radius of 25 μm with NA = 0.20 and the other having a core radius of 50 μm with NA = 0.20. For the larger core fiber we use

Eqs. (5-6) and (5-7) to get

$$P_{LED-step} = P_s(NA)^2 = \pi^2 r_s^2 B_0 (NA)^2$$

$$= \pi^2 (0.0035 \text{ cm})^2 \left[150 \text{ W}/(\text{cm}^2 \cdot \text{sr}) \right] (0.20)^2 = 0.725 \text{ mW}$$

For the case when the fiber end face area is smaller than the emitting surface area, we use Eq. (5-8). Thus the coupled power is less than the above case by the ratio of the radii squared:

$$P_{LED-step} = \left(\frac{25 \ \mu m}{35 \ \mu m} \right)^2 P_s(NA)^2 = \left(\frac{25 \ \mu m}{35 \ \mu m} \right)^2 (0.725 \text{ mW}) = 0.37 \text{ mW}$$

In the case of a graded-index fiber, the numerical aperture depends on the distance r from the fiber axis through the relationship defined by Eq. (2-80). Thus using Eqs. (2-80) and (2-81), the power coupled from a surface-emitting LED into a graded-index fiber becomes (for $r_s < a$)

$$P_{LED, graded} = 2\pi^2 B_0 \int_0^{r_s} \left[n^2(r) - n_2^2 \right] r \, dr$$

$$= 2\pi^2 r_s^2 B_0 n_1^2 \Delta \left[1 - \frac{2}{\alpha + 2} \left(\frac{r_s}{a} \right)^\alpha \right]$$

$$= 2 P_s n_1^2 \Delta \left[1 - \frac{2}{\alpha + 2} \left(\frac{r_s}{a} \right)^\alpha \right] \tag{5-9}$$

where the last expression was obtained from Eq. (5-6).

The foregoing analyses assumed perfect coupling conditions between the source and the fiber. This can only be achieved if the refractive index of the medium separating the source and the fiber end matches the index n_1 of the fiber core. If the refractive index n of this medium is different from n_1, then for perpendicular fiber end faces, the power coupled into the fiber reduces by the factor

$$R = \left(\frac{n_1 - n}{n_1 + n} \right)^2 \tag{5-10}$$

where R is the Fresnel reflection coefficient at the fiber core end face.

Example 5-3. The end faces of two optical fibers with core refractive indices of 1.50 are perfectly aligned and have a small gap between them. If this gap is filled with a gel having a refractive index of 1.30, let us find the optical loss in decibels at this joint. From Eq. (5-10) the Fresnel reflection coefficient at the fiber core end face is

$$R = \left(\frac{n_1 - n}{n_1 + n} \right)^2 = \left(\frac{1.50 - 1.30}{1.50 + 1.30} \right)^2 = 0.0051$$

This value of R corresponds to a reflection of 0.5 percent of the transmitted optical power at a single interface. The power loss in decibels is found from

$P_{\text{coupled}} = (1 - R)P_{\text{emitted}}$:

$$L = 10\log\left(\frac{P_{\text{coupled}}}{P_{\text{emitted}}}\right) = 10\log(1 - R) = 10\log(0.995) = -0.022 \text{ dB}$$

From symmetry considerations the power loss at the second interface is also 0.022 dB, so that the total loss resulting from Fresnel reflection at this fiber-to-fiber joint is 0.044 dB.

The calculation of power coupling for nonlambertian emitters following a cylindrical $\cos^m \theta$ distribution is left as an exercise. The power launched into a fiber from an edge-emitting LED having a noncylindrical distribution is rather complex. An example of this has been given by Marcuse,[2] to which the reader is referred for details. Section 5.4 presents a simplified analysis of this in the discussion on coupling LEDs to single-mode fibers.

5.1.3   Power Launching versus Wavelength

It is of interest to note that the optical power launched into a fiber does not depend on the wavelength of the source but only on its brightness, that is, its radiance. Let us explore this a little further. We saw in Eq. (2-97) that the number of modes which can propagate in a graded-index fiber of core size a and index profile α is

$$M = \frac{\alpha}{\alpha + 2}\left(\frac{2\pi a n_1}{\lambda}\right)^2 \Delta \tag{5-11}$$

Thus, for example, twice as many modes propagate in a given fiber at 900 nm than at 1300 nm.

The radiated power per mode, P_s/M, from a source at a particular wavelength is given by the radiance multiplied by the square of the nominal source wavelength,[3]

$$\frac{P_s}{M} = B_0\lambda^2 \tag{5-12}$$

Thus, twice as much power is launched into a given mode at 1300 nm than at 900 *nm*. Hence two identically sized sources operating at different wavelengths but having identical radiances will launch equal amounts of optical power into the same fiber.

5.1.4   Equilibrium Numerical Aperture

As we noted earlier, a light source is often supplied with a short (1- to 2-m) fiber flylead attached to it in order to facilitate coupling the source to a system fiber. To achieve a low coupling loss, this flylead should be connected to a system fiber having a nominally identical NA and core diameter. A certain amount of optical power (ranging from 0.1 to 1 dB) is lost at this junction, the exact loss

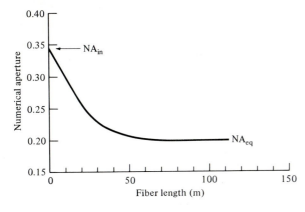

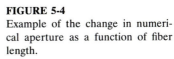

FIGURE 5-4
Example of the change in numerical aperture as a function of fiber length.

depending on the connecting mechanism; this is discussed in Sec. 5.3. In addition to the coupling loss, an excess power loss will occur in the first few tens of meters of the system fiber. This excess loss is a result of nonpropagating modes scattering out of the fiber as the launched modes come to an equilibrium condition (see Sec. 3.4). This is of particular importance for surface-emitting LEDs, which tend to launch power into all modes of the fiber. Fiber-coupled lasers are less prone to this effect, since they tend to excite fewer nonpropagating fiber modes.

The excess power loss must be analyzed carefully in any system design, since it can be significantly higher for some types of fibers than for others.[4,5] An example of the excess power loss is shown in Fig. 5-4 in terms of the fiber numerical aperture. At the input end of the fiber the light acceptance is described in terms of the launch numerical aperture NA_{in}. If the light-emitting area of the LED is less than the cross-sectional area of the fiber core, then at this point the power coupled into the fiber is given by Eq. (5-7), where $NA = NA_{in}$.

However, when the optical power is measured in long fiber lengths after the launched modes have come to equilibrium (which is often taken to occur at 50 m), the effect of the equilibrium numerical aperture NA_{eq} becomes apparent. At this point the optical power in the fiber scales as

$$P_{eq} = P_{50}\left(\frac{NA_{eq}}{NA_{in}}\right)^2 \tag{5-13}$$

where P_{50} is the power expected in the fiber at the 50-m point based on the launch NA. The degree of mode coupling occurring in a fiber is primarily a function of the core-cladding index difference. It can thus vary significantly among different fiber types. Since most optical fibers attain 80 to 90 percent of their equilibrium NA after about 50 m, it is the value of NA_{eq} that is important when calculating launched optical power in telecommunication systems.

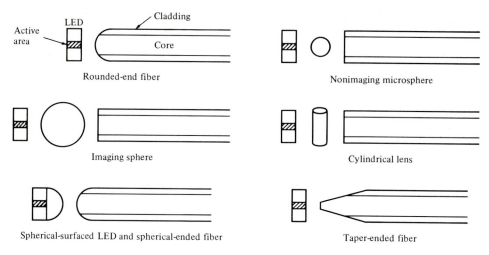

FIGURE 5-5
Examples of possible lensing schemes used to improve optical source-to-fiber coupling efficiency.

5.2 LENSING SCHEMES FOR COUPLING IMPROVEMENT

The optical power-launching analysis given in Sec. 5-1 is based on centering a flat fiber end face directly over the light source as close to it as possible. If the source-emitting area is larger than the fiber core area, when the resulting optical power coupled into the fiber is the maximum that can be achieved. This is a result of fundamental energy and radiance conservation principles[6] (also known as the *law of brightness*). However, if the emitting area of the source is smaller than the core area, a miniature lens may be placed between the source and the fiber to improve the power-coupling efficiency.

The function of the microlens is to magnify the emitting area of the source to match exactly the core area of the fiber end face. If the emitting area is increased by a magnification factor M, the solid angle within which optical power is coupled to the fiber from the LED is increased by the same factor.

Several possible lensing schemes[1, 7–13] are shown in Fig. 5-5. These include a rounded-end fiber, a small glass sphere (nonimaging microsphere) in contact with both the fiber and the source, a larger spherical lens used to image the source on the core area of the fiber end, a cylindrical lens generally formed from a short section of fiber, a system consisting of a spherical-surfaced LED and a spherical-ended fiber, and a taper-ended fiber.

Although these techniques can improve the source-to-fiber coupling efficiency, they also create additional complexities. One problem is that the lens size is similar to the source and fiber core dimensions, which introduces fabrication and handling difficulties. In the case of the taper-ended fiber the mechanical alignment must be done with greater precision since the coupling efficiency becomes a more sharply peaked function of the spatial alignment. However, alignment tolerances are increased for other types of lensing systems.

5.2.1 Nonimaging Microsphere

One of the the most efficient lensing methods is the use of a nonimaging microsphere. Let us first examine its use for a surface emitter, as shown in Fig. 5-6. We first make the following practical assumptions: the spherical lens has a refractive index of about 2.0, the outside medium is air ($n = 1.0$), and the emitting area is circular. To collimate the output from the LED, the emitting surface should be located at the focal point of the lens. The focal point can be found from the gaussian lens formula[14]

$$\frac{n}{s} + \frac{n'}{q} = \frac{n' - n}{r} \tag{5-14}$$

where s and q are the object and image distances, respectively, as measured from the lens surface, n is the refractive index of the lens, n' is the refractive index of the outside medium, and r is the radius of curvature of the lens surface.

The following sign conventions are used with Eq. (5-14):

1. Light travels from left to right.
2. Object distances are measured as positive to the left of a vertex and negative to the right.
3. Image distances are measured as positive to the right of a vertex and negative to the left.
4. All convex surfaces encountered by the light have a positive radius of curvature, and concave surfaces have a negative radius.

With the use of these conventions we shall now find the focal point for the right-hand surface of the lens shown in Fig. 5-6. To find the focal point, we set $q = \infty$ and solve for s in Eq. (5-14), where s is measured from point B. With

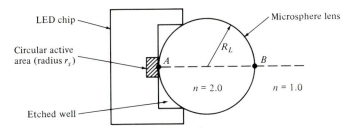

FIGURE 5-6
Schematic diagram of an LED emitter with a microsphere lens.

$n = 2.0$, $n' = 1.0$, $q = \infty$, and $r = -R_L$, Eq. (5-14) yields

$$s = f = 2R_L$$

Thus the focal point is located on the lens surface at point A. (This, of course, changes if the refractive index of the sphere is not equal to 2.0.)

Placing the LED close to the lens surface thus results in a magnification M of the emitting area. This is given by the ratio of the cross-sectional area of the lens to that of the emitting area:

$$M = \frac{\pi R_L^2}{\pi r_s^2} = \left(\frac{R_L}{r_s}\right)^2 \tag{5-15}$$

Using Eq. (5-4) we can show that, with the lens, the optical power P_L that can be coupled into a full aperture angle 2θ is given by

$$P_L = P_s \left(\frac{R_L}{r_s}\right)^2 \sin^2 \theta \tag{5-16}$$

where P_s is the total output power from the LED without the lens.

The theoretical coupling efficiency that can be achieved is based on energy and radiance conservation principles.[15] This efficiency is usually determined by the size of the fiber. For a fiber of radius a and numerical aperture NA, the maximum coupling efficiency η_{max} is given by

$$\eta_{max} = \begin{cases} \left(\dfrac{a}{r_s}\right)^2 (NA)^2 & \text{for} \quad \dfrac{r_s}{a} > NA \\ 1 & \text{for} \quad \dfrac{r_s}{a} \leq NA \end{cases} \tag{5-17}$$

Thus when the radius of the emitting area is larger than the fiber radius, no improvement in coupling efficiency is possible with a lens. In this case the best coupling efficiency is achieved by a direct-butt method.

Based on Eq. (5-17), the theoretical coupling efficiency as a function of the emitting diameter is shown in Fig. 5-7 for a fiber with a numerical aperture of 0.20 and 50-μm core diameter.

5.2.2   Laser Diode-to-Fiber Coupling

As we noted in Chap. 4 laser diodes have an emission pattern which nominally has a full width at half-maximum (FWHM) of 30 to 50° in the plane perpendicular to the active-area junction and an FWHM of 5 to 10° in the plane parallel to the junction. Since the angular output distribution of the laser is greater than the fiber acceptance angle and since the laser emitting area is much smaller than the fiber core, spherical or cylindrical lenses[12, 16, 17] or optical fiber tapers[18, 19] can also be used to improve the coupling efficiency between laser diodes and optical fibers.

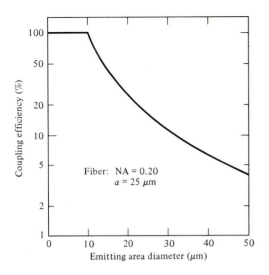

Fiber: NA = 0.20
        $a = 25 \ \mu$m

FIGURE 5-7
Theoretical coupling efficiency for a surface-emitting LED as a function of the emitting diameter. Coupling is to a fiber with NA = 0.20 and radius $a = 25 \ \mu$m.

The use of homogeneous glass microsphere lenses has been tested in a series of several hundred laser diode assemblies by Khoe and Kyut.[16] Spherical glass lenses with a refractive index of 1.9 and diameters ranging between 50 and 60 μm were epoxied to the ends of 50-μm core-diameter graded-index fibers having a numerical aperture of 0.2. The measured FWHM values of the laser output beams were:

1. Between 3 and 9 μm for the near field parallel to the junction
2. Between 30 and 60° for the field perpendicular to the junction
3. Between 15 and 55° for the field parallel to the junction

Coupling efficiencies in these experiments ranged between 50 and 80 percent.

5.3 FIBER-TO-FIBER JOINTS

A significant factor in any fiber optic system installation is the interconnection of fibers in a low-loss manner. These interconnections occur at the optical source, at the photodetector, at intermediate points within a cable where two fibers are joined, and at intermediate points in a link where two cables are connected. The particular technique selected for joining the fibers depends on whether a permanent bond or an easily demountable connection is desired. A permanent bond is generally referred to as a *splice*, whereas a demountable joint is known as a *connector*.

Every jointing technique is subject to certain conditions which can cause various amounts of optical power loss at the joint. These losses depend on parameters such as the input power distribution to the joint, the length of the fiber between the optical source and the joint, the geometrical and waveguide characteristics of the two fiber ends at the joint, and the fiber end face qualities.

The optical power that can be coupled from one fiber to another is limited by the number of modes that can propagate in each fiber. For example if a fiber in which 500 modes can propagate is connected to a fiber in which only 400 modes can propagate, then at most 80 percent of the optical power from the first fiber can be coupled into the second fiber (if we assume that all modes are equally excited). The total number of modes in a fiber can be found from Eq. (2-94). Letting the maximum value of the parameter $\beta = kn_2$, where $k = 2\pi/\lambda$, we have, for a fiber of radius a,

$$M = k^2 \int_0^a \left[n^2(r) - n_2^2 \right] r\, dr \tag{5-18}$$

where $n(r)$ defines the variation in the refractive-index profile of the core. This can be related to a general local numerical aperture $NA(r)$ through Eq. (2-80) to yield

$$M = k^2 \int_0^a NA^2(r) r\, dr$$

$$= k^2 NA^2(0) \int_0^a \left[1 - \left(\frac{r}{a} \right)^\alpha \right] r\, dr \tag{5-19}$$

In general, any two fibers that are to be joined will have varying degrees of differences in their radii a, axial numerical apertures $NA(0)$, and index profiles α. Thus the fraction of energy coupled from one fiber to another is proportional to the common mode volume M_{comm} (if a uniform distribution of energy over the modes is assumed). The fiber-to-fiber coupling efficiency η_F is given by

$$\eta_F = \frac{M_{comm}}{M_E} \tag{5-20}$$

where M_E is the number of modes in the *emitting fiber* (the fiber which launches power into the next fiber).

The fiber-to-fiber coupling loss L_F is given in terms of η_F as

$$L_F = -10 \log \eta_F \tag{5-21}$$

An analytical estimate of the optical power loss at a joint between multimode fibers is difficult to make, since the loss depends on the power distribution among the modes in the fiber.[20-23] For example, consider first the case where all modes in a fiber are equally excited, as shown in Fig. 5-8a. The emerging optical beam thus fills the entire exit numerical aperture of this emitting fiber. Suppose now that a second identical fiber, which we shall call the *receiving fiber*, is to be joined to the emitting fiber. For the receiving fiber to accept all the optical power emitted by the first fiber, there must be perfect mechanical alignment between the two optical waveguides, and their geometric and waveguide characteristics must match precisely.

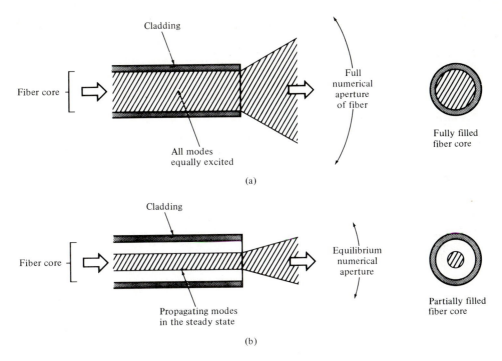

FIGURE 5-8
Different modal distributions of the optical beam emerging from a fiber lead to different degrees of coupling loss. (*a*) When all modes are equally excited, the output beam fills the entire output NA; (*b*) for a steady-state modal distribution only the equilibrium NA is filled by the output beam.

On the other hand, if steady-state modal equilibrium has been established in the emitting fiber, most of the energy is concentrated in the lower-order fiber modes. This means that the optical power is concentrated near the center of the fiber core, as shown in Fig. 5-8*b*. The optical power emerging from the fiber then fills only the equilibrium numerical aperture (see Fig. 5-4). In this case, since the input NA of the receiving fiber is larger than the equilibrium NA of the emitting fiber, slight mechanical misalignments of the two joined fibers and small variations in their geometric characteristics do not contribute significantly to joint loss.

Steady-state modal equilibrium is generally established in long fiber lengths (see Chap. 3). Thus when estimating joint losses between long fibers, calculations based on a uniform modal power distribution tend to lead to results which may be too pessimistic. However, if a steady-state equilibrium modal distribution is assumed, the estimate may be too optimistic, since mechanical misalignments and fiber-to-fiber variations in characteristics cause a redistribution of power among the modes in the second fiber. As the power propagates along the second fiber, an additional loss will thus occur when a steady-state distribution is again established.

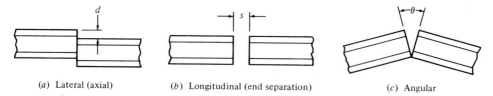

(a) Lateral (axial)   (b) Longitudinal (end separation)   (c) Angular

FIGURE 5-9
Three types of mechanical misalignments that can occur between two joined fibers.

An exact calculation of coupling loss between different optical fibers, which takes into account nonuniform distribution of power among the modes and propagation effects in the second fiber, is lengthy and involved.[24] Here we shall, therefore, make the assumption that all modes in the fiber are equally excited. Although this gives a somewhat pessimistic prediction of joint loss, it will allow an estimate of the relative effects of losses resulting from mechanical misalignments, geometrical mismatches, and variations in the waveguide properties between two joined fibers.

5.3.1   Mechanical Misalignment

Mechanical alignment is a major problem when joining two fibers, because of their microscopic size.[25-29] A standard multimode graded-index fiber core is 50 to 100 μm in diameter, which is roughly the thickness of a human hair, whereas single-mode fibers have diameters on the order of 9 μm. Radiation losses result from mechanical misalignments because the radiation cone of the emitting fiber does not match the acceptance cone of the receiving fiber. The magnitude of the radiation loss depends on the degree of misalignment. The three fundamental types of misalignments between fibers are shown in Fig. 5-9.

Longitudinal separation occurs when the fibers have the same axis but have a gap s between their end faces. *Angular misalignment* results when the two axes form an angle so that the fiber end faces are no longer parallel. *Axial displacement* (which is also often called *lateral displacement*) results when the axes of the two fibers are separated by a distance d.

The most common misalignment occurring in practice, which also causes the greatest power loss, is axial displacement. This axial offset reduces the overlap area of the two fiber core end faces, as illustrated in Fig. 5-10, and consequently reduces the amount of optical power that can be coupled from one fiber into the other.

To illustrate the effects of axial misalignment, let us first consider the simple case of two identical step-index fibers of radii a. Suppose that their axes are offset by a separation d at the common junction, as is shown in Fig. 5-10, and assume there is a uniform modal power distribution in the emitting fiber. Since the numerical aperture is constant across the end faces of the two fibers,

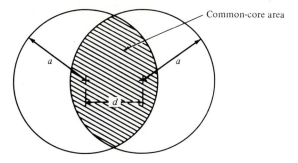

Common-core area

FIGURE 5-10
Axial offset reduces the common core area of the two fiber end faces.

the optical power coupled from one fiber to another is simply proportional to the common area A_{comm} of the two fiber cores. It is straightforward to show that this is (see Prob. 5-9)

$$A_{comm} = 2a^2 \arccos \frac{d}{2a} - d\left(a^2 - \frac{d^2}{4}\right)^{1/2} \qquad (5\text{-}22)$$

For the step-index fiber the coupling efficiency is simply the ratio of the common core area to the core end face area,

$$\eta_{F,\text{step}} = \frac{A_{comm}}{\pi a^2} = \frac{2}{\pi} \arccos \frac{d}{2a} - \frac{d}{\pi a}\left[1 - \left(\frac{d}{2a}\right)^2\right]^{1/2} \qquad (5\text{-}23)$$

The calculation of power coupled from one graded-index fiber into another identical one is more complex, since the numerical aperture varies across the fiber end face. Because of this, the total power coupled into the receiving fiber at a given point in the common core area is limited by the numerical aperture of the transmitting or receiving fiber, depending on which is smaller at that point.

If the end face of a graded-index fiber is uniformly illuminated, the optical power accepted by the core will be that power that falls within the numerical aperture of the fiber. The optical power density $p(r)$ at a point r on the fiber end is proportional to the square of the local numerical aperture $NA(r)$ at that point:[30]

$$p(r) = p(0)\frac{NA^2(r)}{NA^2(0)} \qquad (5\text{-}24)$$

where $NA(r)$ and $NA(0)$ are defined by Eqs. (2-80) and (2-81), respectively. The parameter $p(0)$ is the power density at the core axis, which is related to the total power P in the fiber by

$$P = \int_0^{2\pi}\int_0^a p(r)r\,dr\,d\theta \qquad (5\text{-}25)$$

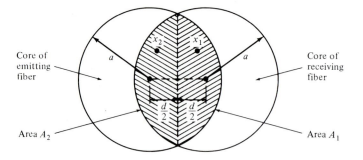

FIGURE 5-11
Core overlap region for two identical parabolic graded-index fibers with an axial separation d.
Points x_1 and x_2 are arbitrary points of symmetry in areas A_1 and A_2.

For an arbitrary index profile the double integral in Eq. (5-25) must be evaluated numerically. However, an analytic expression can be found by using a fiber with a parabolic index profile ($\alpha = 2.0$). Using Eq. (2-80) the power density expression at a point r given by Eq. (5-24) becomes

$$p(r) = p(0)\left[1 - \left(\frac{r}{a}\right)^2\right]$$ (5-26)

Using Eqs. (5-25) and (5-26) the relationship between the axial power density $p(0)$ and the total power P in the emitting fiber is

$$P = \frac{\pi a^2}{2}p(0)$$ (5-27)

Let us now calculate the power transmitted across the butt joint of the two parabolic graded-index fibers with an axial offset d, as shown in Fig. 5-11. The overlap region must be considered separately for the areas A_1 and A_2. In area A_1 the numerical aperture is limited by that of the emitting fiber, whereas in area A_2 the numerical aperture of the receiving fiber is smaller than that of the emitting fiber. The vertical dashed line separating the two areas is the locus of points where the numerical apertures are equal.

To determine the power coupled into the receiving fiber, the power density given by Eq. (5-26) is integrated separately over areas A_1 and A_2. Since the numerical aperture of the emitting fiber is smaller than that of the receiving fiber in area A_1, all of the power emitted in this region will be accepted by the receiving fiber. The received power P_1 in area A_1 is thus

$$P_1 = 2\int_0^{\theta_1}\int_{r_1}^a p(r)r\,dr\,d\theta$$

$$= 2p(0)\int_0^{\theta_1}\int_{r_1}^a\left[1 - \left(\frac{r}{a}\right)^2\right]r\,dr\,d\theta$$ (5-28)

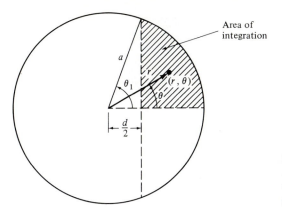

Area of integration

FIGURE 5-12
Area and limits of integration for the common core area of two parabolic graded-index fibers.

where the limits of integration, shown in Fig. 5-12, are

$$r_1 = \frac{d}{2 \cos \theta}$$

and

$$\theta_1 = \arccos \frac{d}{2a}$$

Carrying out the integration yields

$$P_1 = \frac{a^2}{2} p(0) \left\{ \arccos \frac{d}{2a} - \left[1 - \left(\frac{d}{2a} \right)^2 \right]^{1/2} \frac{d}{6a} \left(5 - \frac{d^2}{2a^2} \right) \right\} \quad (5\text{-}29)$$

where $p(0)$ is given by Eq. (5-27). The derivation of Eq. (5-29) is left as an exercise.

In area A_2 the emitting fiber has a larger numerical aperture than the receiving fiber. This means that the receiving fiber will accept only that fraction of the emitted optical power that falls within its own numerical aperture. This power can be found easily from symmetry considerations.[31] The numerical aperture of the receiving fiber at a point x_2 in area A_2 is the same as the numerical aperture of the emitting fiber at the symmetrical point x_1 in area A_1. Thus the optical power accepted by the receiving fiber at any point x_2 in area A_2 is equal to that emitted from the symmetrical point x_1 in area A_1. The total power P_2 coupled across area A_2 is thus equal to the power P_1 coupled across area A_1. Combining these results, we have that the total power P_T accepted by the receiving fiber is

$$P_T = 2P_1$$

$$= \frac{2}{\pi} P \left\{ \arccos \frac{d}{2a} - \left[1 - \left(\frac{d}{2a} \right)^2 \right]^{1/2} \frac{d}{6a} \left(5 - \frac{d^2}{2a^2} \right) \right\} \quad (5\text{-}30)$$

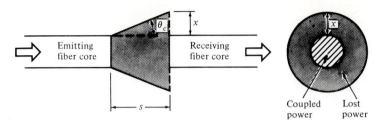

FIGURE 5-13
Loss effect when the fiber ends are separated longitudinally by a gap s.

When the axial misalignment d is small compared to the core radius a, Eq. (5-30) can be approximated by

$$P_T \simeq P\left(1 - \frac{8d}{3\pi a}\right) \qquad (5\text{-}31)$$

This is accurate to within 1 percent for $d/a < 0.4$. The coupling loss for the offsets given by Eqs. (5-30) and (5-31) is

$$L_F = -10 \log \eta_F = -10 \log \frac{P_T}{P} \qquad (5\text{-}32)$$

The effect of separating the two fiber ends longitudinally by a gap s is shown in Fig. 5-13. All the higher-mode optical power emitted in the ring of width x will not be intercepted by the receiving fiber. It is straightforward to show that, for a step-index fiber, the loss occurring in this case is

$$L_F = -10 \log\left(\frac{a}{a + s \tan \theta_c}\right)^2 \qquad (5\text{-}33)$$

where θ_c is the critical acceptance angle of the fiber.

When the axes of two joined fiber are angularly misaligned at the joint, the optical power that leaves the emitting fiber outside of the solid acceptance angle of the receiving fiber will be lost. For two step-index fibers having an angular misalignment θ, the optical power loss at the joint has been shown to be [32, 33]

$$L_F = -10 \log\left(\cos \theta \left\{\frac{1}{2} - \frac{1}{\pi}p(1 - p^2)^{1/2} - \frac{1}{\pi}\arcsin p\right.\right.$$

$$\left.\left. - q\left[\frac{1}{\pi}y(1 - y^2)^{1/2} + \frac{1}{\pi}\arcsin y + \frac{1}{2}\right]\right\}\right) \qquad (5\text{-}34)$$

where

$$p = \frac{\cos \theta_c \, (1 - \cos \theta)}{\sin \theta_c \, \sin \theta}$$

$$q = \frac{\cos^3 \theta_c}{\left(\cos^2 \theta_c - \sin^2 \theta\right)^{3/2}}$$

$$y = \frac{\cos^2 \theta_c \, (1 - \cos \theta) - \sin^2 \theta}{\sin \theta_c \, \cos \theta_c \, \sin \theta}$$

Here n is the refractive index of the material between the fibers ($n = 1.0$ for air). The derivation of Eq. (5-34) again assumes that all modes are uniformly excited.

An experimental comparison[27] of the losses induced by the three types of mechanical misalignments is shown in Fig. 5-14. The measurements were based on two independent experiments using LED sources and graded-index fibers. The core diameters were 50 and 55 μm for the first and second experiments, respectively. A 1.83-m-long fiber was used in the first test and a 20-m length in the second. In either case the power output from the fibers was first optimized. The fibers were then cut at the center, so that the mechanical misalignment loss measurements were done on identical fibers. The axial offset and longitudinal separation losses are plotted as functions of misalignments normalized to the

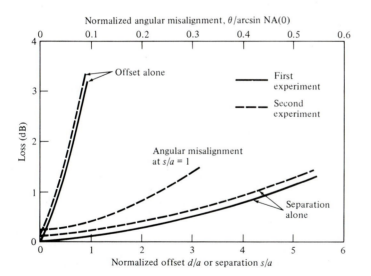

FIGURE 5-14
Experimental comparison of loss (in dB) as a function of mechanical misalignments. (Reproduced with permission from Chu and McCormick,[27] © 1978, AT & T.)

core radius. A normalized angular misalignment of 0.1 corresponds to a 1° angular offset.

> **Example 5-4.** Suppose two graded index fibers are misaligned with an axial offset of $d = 0.3a$. From Eq. (5-30) we have that the fraction of optical power coupled from the first fiber into the second fiber is
>
> $$\frac{P_T}{P} = \frac{2}{\pi}\left\{\arccos(0.15) - \left[1 - (0.15)^2\right]^{1/2}\left(\frac{0.15}{3}\right)\left[5 - \frac{(0.3)^2}{2}\right]\right\} = 0.748$$
>
> or in decibels
>
> $$10\log\frac{P_T}{P} = -1.27 \text{ dB}$$
>
> which compares with the experimental value shown in Fig. 5-14.

Figure 5-14 shows that of the three mechanical misalignments the dominant loss arises from lateral displacement. In practice, angular misalignments of less than 1° are readily achievable in splices and connectors. From the experimental data shown in Fig. 5-14, these misalignments result in losses of less than 0.5 dB.

For splices the separation losses are normally negligible, since the fibers should be in relatively close contact. In most connectors the fiber ends are intentionally separated by a small gap. This prevents them from rubbing against each other and becoming damaged during connector engagement. Typical gaps in these applications range form 0.025 to 0.10 mm, which results in losses of less than 0.8 dB for a 50-μm-diameter fiber.

5.3.2 Fiber-Related Losses

In addition to mechanical misalignments, differences in the geometrical and waveguide characteristics of any two waveguides being joined can have a profound effect on fiber-to-fiber coupling loss. These include variations in core diameter, core area ellipticity, numerical aperture, refractive-index profile, and core-cladding concentricity of each fiber. Since these are manufacturer-related variations, the user generally has little control over them. Theoretical and experimental studies[24, 34-38] of the effects of these variations have shown that, for a given percentage mismatch, differences in core radii and numerical apertures have a significantly larger effect on joint loss than mismatches in the refractive-index profile or core ellipticity.

The joint losses resulting from core diameter, numerical aperture, and core refractive-index-profile mismatches can easily be found from Eqs. (5-19) and (5-20). For simplicity, let the subscripts E and R refer to the emitting and receiving fibers, respectively. If the radii a_E and a_R are not equal but the axial numerical apertures and the index profiles are equal [$NA_E(0) = NA_R(0)$ and

$\alpha_E = \alpha_R$], then the coupling loss is

$$L_F(a) = \begin{cases} -10\log\left(\dfrac{a_R}{a_E}\right)^2 & \text{for} \quad a_R < a_E \\ 0 & \text{for} \quad a_R \geq a_E \end{cases} \tag{5-35}$$

If the radii and the index profiles of the two coupled fibers are identical but their axial numerical apertures are different, then

$$L_F(\text{NA}) = \begin{cases} -10\log\left[\dfrac{\text{NA}_R(0)}{\text{NA}_E(0)}\right]^2 & \text{for} \quad \text{NA}_R(0) < \text{NA}_E(0) \\ 0 & \text{for} \quad \text{NA}_R(0) \geq \text{NA}_E(0) \end{cases} \tag{5-36}$$

Finally, if the radii and the axial numerical apertures are the same but the core refractive-index profiles differ in two joined fibers, then the coupling loss is

$$L_F(\alpha) = \begin{cases} -10\log\dfrac{\alpha_R(\alpha_E + 2)}{\alpha_E(\alpha_R + 2)} & \text{for} \quad \alpha_R < \alpha_E \\ 0 & \text{for} \quad \alpha_R \geq \alpha_E \end{cases} \tag{5-37}$$

This results because for $\alpha_R < \alpha_E$ the number of modes that can be supported by the receiving fiber is less than the number of modes in the emitting fiber. If $\alpha_R > \alpha_E$, then all modes in the emitting fiber can be captured by the receiving fiber. The derivations of Eqs. (5-35) to (5-37) are left as an exercise (see Probs. 5-15 through 5-17).

5.3.3   Fiber End Face Preparation

One of the first steps that must be followed before fibers are connected or spliced to each other is to properly prepare the fiber end faces. In order not to have light deflected or scattered at the joint, the fiber ends must be flat, perpendicular to the fiber axis, and smooth.[39] End-preparation techniques that have been extensively used include sawing, grinding and polishing, and con-trolled fracture.[40–45]

Conventional grinding and polishing techniques can produce a very smooth surface that is perpendicular to the fiber axis. However, this method is quite time-consuming and requires a fair amount of operator skill. Although it is often implemented in a controlled environment such as a laboratory or a factory, it is not readily adaptable for field use. The procedure employed in the grinding and polishing technique is to use successively finer abrasives to polish the fiber end face. The end face is polished with each successive abrasive until the scratches created by the previous abrasive material are replaced by the finer

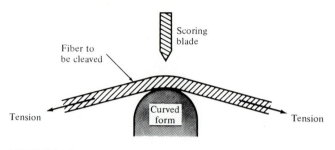

FIGURE 5-15
Controlled-fracture procedure for fiber end preparation.

scratches of the present abrasive. The number of abrasives used depends on the degree of smoothness that is desired.

Controlled-fracture techniques are based on score-and-break methods for cleaving fibers. In this operation the fiber to be cleaved is first scratched to create a stress concentration at the surface. The fiber is then bent over a curved form while tension is simultaneously applied, as shown in Fig. 5-15. This action produces a stress distribution across the fiber. The maximum stress occurs at the scratch point so that a crack starts to propagate through the fiber.

One can produce a highly smooth and perpendicular end face this way. A number of different tools based on the controlled-fracture technique have been developed[46-52] and are being used both in the field and in factory environments. However, the controlled-fracture method requires careful control of the curvature of the fiber and of the amount of tension applied. If the stress distribution across the crack is not properly controlled, the fracture propagating across the fiber can fork into several cracks. This forking produces defects such as a lip or a hackled portion on the fiber end,[53,54] as shown in Fig. 5-16. The EIA Fiber Optic Test Procedures (FOTP) 57 and 179 define these and other common end-face defects as follows:[54]

Lip. This is a sharp protrusion from the edge of a cleaved fiber which prevents the cores from coming in close contact. Excessive lip height can cause fiber damage.

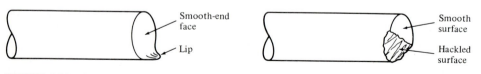

FIGURE 5-16
Two examples of improperly cleaved fiber ends.

Rolloff. This rounding-off of the edge of a fiber is the opposite condition to lipping. It is also known as *breakover* and can cause high insertion or splice loss.

Chip. A chip is a localized fracture or break at the end of a cleaved fiber.

Hackle. Figure 5-16 shows this as severe irregularities across a fiber end face.

Mist. This is similar to hackle but much less severe.

Spiral or step. These are abrupt changes in the end-face surface topology.

Shattering. This is the result of an uncontrolled fracture and has no definable cleave or surface characteristics.

5.4 LED COUPLING TO SINGLE-MODE FIBERS

In the early years of optical fiber applications, LEDs were traditionally only considered for multimode-fiber systems. However, around 1985 researchers recognized that edge-emitting LEDs can launch sufficient optical power into a single-mode fiber for transmission at data rates up to 560 Mb/s over several kilometers.[55-58] The interest in this arose because of the cost and reliability advantages of LEDs over laser diodes. Investigators have used edge-emitting LEDs rather than surface-emitting LEDs because the edge-emitters have a laserlike output pattern in the direction perpendicular to the junction plane.

To rigorously evaluate the coupling between an LED and a single-mode fiber we need to use the formalism of electromagnetic theory rather than geometrical optics, because of the monomode nature of the fiber. However, coupling analyses of the output from an edge-emitting LED to a single-mode fiber can be carried out wherein the results of electromagnetic theory are interpreted from a geometrical point of view,[59-61] which involves defining a numerical aperture for the single-mode fiber. Agreement with experimental measurements and with a more exact theory are quite good.[60-63]

Here we will use the analysis of Reith and Shumate[60] to look at the following two cases: (*a*) direct coupling of an LED into a single-mode fiber, and (*b*) coupling into a single-mode fiber from a multimode flylead attached to the LED. In general edge-emitting LEDs have gaussian near-field output profiles with $1/e^2$ full widths of approximately 0.9 and 22 μm in the directions perpendicular and parallel to the junction plane, respectively. The far-field patterns vary approximately as $\cos^7 \theta$ in the perpendicular direction and as $\cos \theta$ (lambertian) in the parallel direction.

For a source with a circularly asymmetric radiance $B(A_S, \Omega_S)$, Eq. (5-3) is in general not separable into contributions from the perpendicular and parallel directions. However, we can approximate the independent contributions by evaluating Eq. (5-3) as if each component were a circularly symmetric source, and then taking the geometric mean to find the total coupling efficiency. Calling these the x (parallel) and y (perpendicular) directions, and letting τ_x and τ_y be the x and y power transmissivities (directional coupling efficiencies), respectively, we can find the maximum LED-to-fiber coupling efficiency η from the relation

$$\eta = \frac{P_{in}}{P_s} = \tau_x \tau_y \qquad (5\text{-}38)$$

where P_{in} is the optical power launched into the fiber and P_s is the total source output power.

Using a small-angle approximation, we first integrate over the effective solid acceptance angle of the fiber to get πNA_{SM}^2, where the geometrical-optics-based fiber numerical aperture $NA_{SM} = 0.11$. Assuming a gaussian output for the source, for butt coupling of the LED to the single-mode fiber of radius a, the coupling efficiency in the y direction is

$$
\begin{aligned}
\tau_y &= \left(\frac{P_{in;\, y}}{P_s} \right)^{1/2} \\
&= \left[\frac{\int_0^{2\pi} \int_0^a B_0 e^{-2r^2/\omega_y^2} r\, dr\, d\theta_s\ \pi\ NA_{SM}^2}{\int_0^{2\pi} \int_0^\infty B_0 e^{-2r^2/\omega_y^2} y\, dy\, d\theta_s \int_0^{2\pi} \int_0^{\pi/2} \cos^7 \theta \sin \theta\, d\theta\, d\phi} \right]^{1/2} \qquad (5\text{-}39)
\end{aligned}
$$

where $P_{in;\, y}$ is the optical power coupled into the fiber from the y-direction source output, which has a $1/e^2$ LED intensity radius ω_y. One can write a similar set of integrals for τ_x. Letting $a = 4.5\ \mu m$, $\omega_x = 10.8\ \mu m$, and $\omega_y = 0.47$ μm, Reith and Shumate calculated $\tau_x = -12.2$ dB and $\tau_y = -6.6$ dB to yield a total coupling efficiency $\eta = -18.8$ dB. Thus, for example, if the LED emits 200 μW (-7 dBm), then 2.6 μW (-25.8 dBm) gets coupled into the single-mode fiber.

When a 1- to 2-m multimode-fiber flylead is attached to an edge-emitting LED, the near field profile of the multimode fiber has the same asymmetry as the LED. In this case one can assume that the multimode-fiber optical output is a simple gaussian with different beam widths along the x and y directions. Using a similar coupling analysis with effective beam widths of $\omega_x = 19.6\ \mu m$ and $\omega_y = 10.0\ \mu m$, the directional coupling efficiencies are $\tau_x = -7.8$ dB and $\tau_y = -5.2$ dB, yielding a total LED-to-fiber coupling efficiency $\eta = -13.0$ dB.

5.5   FIBER SPLICING

In joining single fibers or multiple groups (arrays) of fibers, one needs to implement some type of splicing mechanism. Initially in the development of fiber optics, the major concern was to make low-loss splices for long-haul applications using multimode fibers. Recent trends toward the use of single-mode fibers for both low-loss, long-haul systems and high-speed localized networks has provided a strong incentive for designing cost-effective, straightforward single-mode splicing techniques. This section first addresses general splicing methods and then examines single-mode splicing.

5.5.1 Splicing Techniques

Many different fiber-splicing techniques have arisen during the evolution of optical fiber technology. Among these are the fusion splice,[64-68] the V-groove and tube mechanical splice,[69-73] the elastic-tube splice,[74-75] and the rotary splice.[36, 76]

Fusion splices are made by thermally bonding together prepared fiber ends, as pictured in Fig. 5-17. In this method the fiber ends are first prealigned and butted together. This is done either in a grooved fiber holder or under a microscope with micromanipulators. The butt joint is then heated with an electric arc or a laser pulse so that the fiber ends are momentarily melted and hence bonded together. This technique can produce very low splice losses (typically averaging less than 0.06 dB). However, care must be exercised in this technique, since surface damage due to handling, surface defect growth created during heating, and residual stresses induced near the joint as a result of changes in chemical composition arising from the material melting can produce a weak splice.[77-78]

In the V-groove splice technique the prepared fiber ends are first butted together in a V-shaped groove, as shown in Fig. 5-18. They are then bonded together with an adhesive or are held in place by means of a cover plate. The V-shaped channel could be either a grooved silicon, plastic, ceramic, or metal substrate. The splice loss in this method depends strongly on the fiber size (outside dimensions and core diameter variations) and eccentricity (the position of the core relative to the center of the fiber).

The elastic-tube splice shown cross-sectionally in Fig. 5-19 is a unique device that automatically performs lateral, longitudinal, and angular alignment. It splices multimode fibers with losses in the same range as commercial fusion splices, but much less equipment and skill are needed. The splice mechanism is

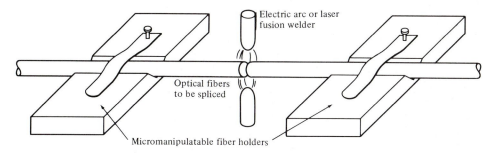

FIGURE 5-17
Fusion splicing of optical fibers.

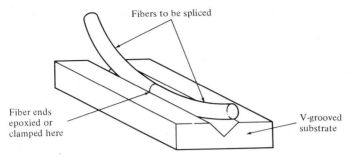

FIGURE 5-18
V-groove optical fiber splicing technique.

basically a tube made of an elastic material. The central hole diameter is slightly smaller than that of the fiber to be spliced and is tapered on each end for easy fiber insertion. When a fiber is inserted it expands the hole diameter so that the elastic material exerts a symmetrical force on the fiber. This symmetry feature allows an accurate and automatic alignment of the axes of the two joined fibers. A wide range of fiber diameters can be inserted into the elastic tube. Thus the fibers to be spliced do not have to be equal in diameter, since each fiber moves into position independently relative to the tube axis.

5.5.2 Splicing Single-Mode Fibers

As is the case in multimode fibers, in single-mode fibers the lateral (axial) offset loss presents the most serious misalignment. This loss depends on the shape of the propagating mode. For gaussian-shaped beams the loss between identical fibers is[79]

$$L_{\text{SM; lat}} = -10 \log \left\{ \exp \left[-\left(\frac{d}{W} \right)^2 \right] \right\} \tag{5-40}$$

where the spot size W is the mode-field radius defined in Eq. (2-74), and d is

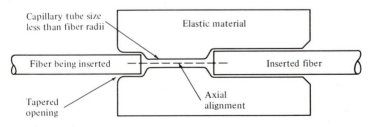

FIGURE 5-19
Schematic of an elastic-tube splice.

the lateral displacement shown in Fig. 5-9. Since the spot size is only a few micrometers in single-mode fibers, low-loss coupling requires a very high degree of mechanical precision in the axial dimension.

Example 5-5. A single-mode fiber has a normalized frequency $V = 2.40$, a core refractive index $n_1 = 1.47$, a cladding refractive index $n_2 = 1.465$, and a core diameter $2a = 9$ μm. Let us find the insertion losses of a fiber joint having a lateral offset of 1 μm.

First, using the expression for the mode-field diameter from Prob. 2-24, we have

$$W_0 = a(0.65 + 1.619V^{-3/2} + 2.879V^{-6})$$

$$= 4.5\left[0.65 + 1.619(2.40)^{-3/2} + 2.879(2.40)^{-6}\right] = 4.95\ \mu\text{m}$$

Then from Eq. (5-40) we have

$$L_{\text{SM; lat}} = -10\log\left\{\exp\left[-(1/4.95)^2\right]\right\} = 0.18\ \text{dB}$$

For angular misalignment in single-mode fibers, the loss at a wavelength gl is[79]

$$L_{\text{SM; ang}} = -10\log\left\{\exp\left[-\left(\frac{\pi n_2 W\theta}{\lambda}\right)^2\right]\right\} \tag{5-41}$$

where n_2 is the refractive index of the cladding, θ is the angular misalignment in radians shown in Fig. 5-9, and W is the mode-field radius.

Example 5-6. Consider the single-mode fiber described in Example 5-5. Let us find the loss at a joint having an angular misalignment of 1° at a 1300-nm wavelength. From Eq. (5-41) we have

$$L_{\text{SM; ang}} = -10\log\left\{\exp\left[-\left(\frac{\pi(1.465)(4.95)(0.0175)}{1.3}\right)^2\right]\right\} = 0.41\ \text{dB}$$

The gap (longitudinal) loss for single-mode fibers is given by[80]

$$L_{\text{SM; gap}} = -10\log\frac{4(4Z^2 + 1)}{(4Z^2 + 2) + 4Z^2} \tag{5-42}$$

where $Z = s\lambda/(2\pi n_2 W^2)$ with s being the end face separation shown in Fig. 5-9.

5.5.3   Splicing Fluoride Glass Fibers

When splicing heavy-metal-fluoride glass fibers, one ideally should have splice losses less than 0.005 dB per splice in order not to negate the advantage of the intrinsic 0.01-dB/km loss of the fiber at 2.5 μm. Unlike silica fibers, microcrystallite formation in fluoride glass fiber splices may cause an additional extrinsic loss. Laboratory tests[81] on single-mode fibers with a mode-field radius of 8.4 μm have yielded losses averaging 0.08 dB per splice when using a fusion splicer in an inert helium atmosphere. From Eq. (5-40) we see that to have an average loss of 0.005 dB for such fibers, one must maintain a lateral core offset of less than 0.28 μm.

5.6   OPTICAL FIBER CONNECTORS

A wide variety of optical fiber connectors has evolved for numerous different applications. Their uses range from simple single-channel fiber-to-fiber connectors in a benign location to multichannel connectors used in harsh military field environments. Some of the principal requirements of a good connector design are as follows:

1. *Low coupling losses.* The connector assembly must maintain stringent alignment tolerances to assure low mating losses. These low losses must not change significantly during operation and after numerous connects and disconnects.
2. *Interchangeability.* Connectors of the same type must be compatible from one manufacturer to another.
3. *Ease of assembly.* A service technician should be able to readily install the connector in a field environment, that is, in a location other than the connector factory. The connector loss should also be fairly insensitive to the assembly skill of the technician.
4. *Low environmental sensitivity.* Conditions such as temperature, dust, and moisture should have a small effect on connector-loss variations.
5. *Low-cost and reliable construction.* The connector must have a precision suitable to the application, but its cost must not be a major factor in the fiber system.
6. *Ease of connection.* Generally one should be able to simply mate and demate the connector by hand.

5.6.1   Connector Types

Connectors are available in screw-on, bayonet-mount, and push-pull configurations.[72, 82-88] These include both single-channel and multichannel assemblies for cable-to-cable and for cable-to-circuit card connections. The basic coupling mechanisms used in these connectors belong to either the *butt-joint* or the *expanded-beam*[89-91] classes.

The majority of connectors today are of the butt-joint type. These connectors employ a metal, ceramic, or molded-plastic ferrule for each fiber and a precision sleeve into which the ferrule fits. The fiber is epoxied into a precision hole which has been drilled into the ferrule. The mechanical challenges of

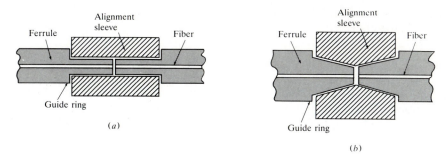

FIGURE 5-20
Examples of two popular alignment schemes used in fiber optic connectors: (*a*) straight sleeve, (*b*) tapered sleeve.

ferrule connectors include maintaining both the dimensions of the hole diameter and its position relative to the ferrule outer surface.

Figure 5-20 shows two popular butt-joint alignment designs used in both multimode and single-mode fiber systems. These are the *straight-sleeve* and the *tapered-sleeve* (or *biconical*) mechanisms. In the straight-sleeve connector the length of the sleeve and a guide ring on the ferrules determine the end separation of the fibers. The biconical connector uses a tapered sleeve to accept and guide tapered ferrules. Again the sleeve length and the guide rings maintain a given fiber end separation.

An expanded-beam connector, illustrated in Fig. 5-21, employs lenses on the ends of the fibers. These lenses either collimate the light emerging from the transmitting fiber, or focus the expanded beam onto the core of the receiving fiber. The fiber-to-lens distance is equal to the focal length of the lens. The advantage of this scheme is that, since the beam is collimated, separation of the fiber ends may take place within the connector. Thus the connector is less dependent on lateral alignments. In addition, optical processing elements, such

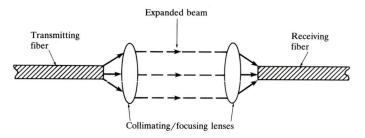

FIGURE 5-21
Schematic representation of an expanded-beam fiber optic connector.

as beam splitters and switches, can easily be inserted into the expanded beam between the fiber ends.

5.6.2   Single-Mode Fiber Connectors

Because of the wide use of single-mode fiber optic links and because of the greater alignment precision required for these systems, this section addresses single-mode connector coupling losses. Based on the gaussian-beam model of single-mode fiber fields,[80] Nemota and Makimoto[92] derived the following coupling loss (in decibels) between single-mode fibers that have unequal mode-field diameters (which is an intrinsic factor) and lateral, longitudinal, and angular offsets plus reflections (which are all extrinsic factors):

$$L_{\text{SM; ff}} = -10 \log \left[\frac{16 n_1^2 n_3^2}{(n_1 + n_3)^4} \frac{4\sigma}{q} \exp\left(-\frac{\rho u}{q} \right) \right] \tag{5-43}$$

where $\rho = (kW_1)^2$
$q = G^2 + (\sigma + 1)^2$
$u = (\sigma + 1)F^2 + 2\sigma FG \sin\theta + \sigma(G^2 + \sigma + 1)\sin^2\theta$
$F = \dfrac{d}{kW_1^2}$
$G = \dfrac{s}{kW_1^2}$
$\sigma = (W_2/W_1)^2$
$k = 2\pi n_3/\lambda$
$n_1 = $ core refractive index of fibers
$n_3 = $ refractive index of medium between fibers
$\lambda = $ wavelength of source
$d = $ lateral offset
$s = $ longitudinal offset
$\theta = $ angular misalignment
$W_1 = 1/e$ mode-field radius of transmitting fiber
$W_2 = 1/e$ mode-field radius of receiving fiber

This general equation gives very good correlation with experimental investigations.[84]

5.7   SUMMARY

In this chapter we have addressed the problem of launching optical power from a light source into a fiber and the factors involved in coupling light from one fiber into another. The coupling of optical power from a luminescent source into a fiber is influenced by the following:

1. The numerical aperture of the fiber, which defines the light acceptance cone of the fiber.

2. The cross-sectional area of the fiber core compared to the source-emitting area. If the emitting area is smaller than the fiber core, then lensing schemes can be used to improve the coupling efficiency.

3. The radiance (or brightness) of the light source, which is the optical power radiated into a unit solid angle per unit emitting area (measured in watts per square centimeter per steradian).

4. The spatial radiation pattern of the source. The incompatibility between the wide beam divergence of LEDs and the narrow acceptance cone of the fiber is a major contributor to coupling loss. This holds to a lesser extent for laser diodes.

In practice, many suppliers offer optical sources which have a short length of optical fiber (1 to 2 m) already attached in an optimum power coupling configuration. This fiber, which is referred to as a flylead or a pigtail, makes it easier for the user to couple the source to a system fiber. The power-launching problem now becomes a simpler one of coupling optical power from one fiber into another. To achieve a low coupling loss, the fiber flylead should be connected to a system fiber having a nominally identical numerical aperture and core diameter.

Fiber-to-fiber joints can exist between the source flylead and the system fiber, at the photodetector, at intermediate points within a cable where two fibers are joined, and at intermediate points in a link where two cable sections are connected. The two principal types of joints are splices, which are permanent bonds between the two fibers, and connectors, which are used when an easily demountable connection between two fibers is desired.

Each jointing technique is subject to certain conditions which can cause varying degrees of optical power loss at the joint. These losses depend on parameters such as:

1. *The geometrical characteristics of the fibers.* For example, optical power will be lost because of area mismatches if an emitting fiber has a larger core diameter than the receiving fiber.

2. *The waveguide characteristics of the fibers.* For example, if an emitting fiber has a larger numerical aperture than the receiving fiber, all optical power falling outside of the acceptance cone of the receiving fiber is lost.

3. *The various mechanical misalignments between the two fiber ends at the joint.* These include longitudinal separation, angular misalignment, and axial (or lateral) displacement. The most common misalignment occurring in practice, which also causes the greatest power loss, is axial misalignment.

4. *The input power distribution to the joint.* If all the modes of an emitting fiber are equally excited, there must be perfect mechanical alignment between the two optical waveguides, and their geometric and waveguide characteristics must match precisely in order for no optical power loss to occur at the joint. On the other hand, if steady-state modal equilibrium has been established in

the emitting fiber (which happens in long fiber lengths), most of the energy is concentrated in the lower-order fiber modes. In this case, slight mechanical misalignments of the two joined fibers and small variations in their geometric and waveguide characteristics do not contribute significantly to joint loss.

5. *The fiber end face quality.* One criterion for low-loss joints is that the fiber end faces be clean, smooth, and perpendicular to the fiber axis. End preparation techniques include sawing, grinding and polishing, and controlled fracture.

PROBLEMS

5-1. Use polar graph paper to compare the emission patterns from a lambertian source and a source with an emission pattern given by $B(\theta) = B_0 \cos^3 \theta$. Assume B_0 is normalized to unity in both cases.

5-2. A laser diode has lateral ($\phi = 0°$) and transverse ($\phi = 90°$) half-power beam widths of $2\theta = 60$ and $30°$, respectively. What are the transverse and lateral power distribution coefficients for this device?

5-3. An LED with a circular emitting area of radius 20 μm has a lambertian emission pattern with a 100-W/(cm² · sr) axial radiance at 100-mA drive current. How much optical power can be coupled into a step-index fiber having a 100-μm core diameter and NA = 0.22? How much optical power can be coupled from this source into a 50-μm core-diameter graded-index fiber having $\alpha = 2.0$, $n_1 = 1.48$, and $\Delta = 0.01$?

5-4. Derive the right-hand side of Eq. (5-9).

5-5. When the refractive index of the medium between an optical source and a fiber end is different from that of the fiber core, part of the optical energy incident on the fiber end face is lost through Fresnel reflection. What is this loss in percent and in decibels when the intervening medium is air ($n = 1.0$) and the fiber core has a refractive index $n_1 = 1.48$?

5-6. Use Eq. (5-3) to derive an expression for the power coupled into a step-index fiber from an LED having a radiant distribution given by

$$B(\theta) = B_0 \cos^m \theta$$

5-7. Use Eq. (5-4) to verify Eq. (5-16).

5-8. On the same graph plot the maximum coupling efficiencies as a function of the source radius r_s for the following fibers:
(*a*) core radius of 25 μm and NA = 0.16
(*b*) core radius of 50 μm and NA = 0.20
Let r_s range from 0 to 50 μm. In what regions can a lens improve the coupling efficiency?

5-9. Verify that Eq. (5-22) gives the common core area of the two axially misaligned step-index fibers shown in Fig. 5-10. If $d = 0.1a$, what is the coupling efficiency in decibels?

5-10. Consider the fibers listed in Table P5-10. Use Eq. (5-23) to complete this table for connector losses (in decibels) due to the indicated axial misalignments.

5-11. Derive Eq. (5-27) by using Eqs. (5-25) and (5-26).

TABLE P5-10

Fiber size: Core diameter (μm) / clad diameter (μm)	Coupling loss (dB) for given axial misalignment (μm)			
	1	3	5	10
50/125			0.590	
62.5/125				
100/140				

5-12. Derive Eq. (5-29) for the received power in area A_1 of Fig. 5-11.

5-13. Show that, when the axial misalignment of d is small compared to the core radius a, Eq. (5-30) can be approximated by Eq. (5-31). Plot Eqs. (5-30) and (5-31) in terms of P_T/P as a function of d/a for $0 \le d/a \le 0.4$.

5-14. Verify that Eq. (5-33) gives the coupling loss between two step-index fibers separated by a gap s, as shown in Fig. 5-13.

5-15. Using Eqs. (5-19) and (5-20), show that Eq. (5-35) gives the coupling loss for two fibers with unequal core radii. Make a plot of the coupling loss as a function of a_R/a_E for $0.5 \le a_R/a_E \le 1.0$.

5-16. Using Eqs. (5-19) and (5-20), show that Eq. (5-36) gives the coupling loss for two fibers with unequal axial numerical apertures. Plot this coupling loss as a function of $\mathrm{NA}_R(0)/\mathrm{NA}_E(0)$ for $0.5 \le \mathrm{NA}_R(0)/\mathrm{NA}_E(0) \le 1.0$.

5-17. Show that Eq. (5-37) gives the coupling loss for two fibers having different core refractive-index profiles. Plot this coupling loss as a function of α_R/α_E for the range $0.75 \le \alpha_R/\alpha_E \le 1.0$. Take $\alpha_E = 2.0$.

5-18. Compare the following single-mode coupling losses: (a) lateral offset up to 4 μm; (b) longitudinal offset up to 40 μm; (c) angular misalignment up to 2°.

5-19. Assuming that a single-mode connector has no losses due to extrinsic factors, show that a 10-percent mismatch in mode-field diameters yields a loss of 0.05 dB.

REFERENCES

1. (a) Y. Uematsu, T. Ozeki, and Y. Unno, "Efficient power coupling between an MH LED and a taper-ended multimode fiber," *IEEE J. Quantum Electron.*, vol. QE-15, pp. 86–92, Feb. 1979.

   (b) H. Kuwahara, M. Sasaki, and N. Tokoyo, "Efficient coupling from semiconductor lasers into single-mode fibers with tapered hemispherical ends," *Appl. Opt.*, vol. 19, pp. 2578–2583, Aug. 1980.

2. D. Marcuse, "Excitation of parabolic-index fibers with incoherent sources," *Bell Sys. Tech. J.*, vol. 54, pp. 1507–1530, Nov. 1975; "LED Fundamentals: Comparison of front and edge-emitting diodes," *IEEE J. Quantum Electron.*, vol. QE-13, pp. 819–827, Oct. 1977.

3. A. Yariv, *Quantum Electronics*, Wiley, New York, 3rd ed., 1988.

4. S. Zemon and D. Fellows, "Characterization of the approach to steady state and the steady state properties of multimode optical fibers using LED excitation," *Optics Commun.*, vol. 13, pp. 198–202, Feb. 1975.

5. R. B. Lauer and J. Schlafer, "LEDs or DLs: Which light source shines brightest in fiber optic telecomm systems?," *Electron. Design*, vol. 8, pp. 131–135, Apr. 12, 1980.

6. M. Born and E. Wolf, *Principles of Optics*, Pergamon, Oxford, 6th ed., 1980.
7. A. Nicia, "Lens coupling in fiber-optic devices: Efficiency limits," *Appl. Opt.*, vol. 20, pp. 3136–3145, Sept. 1981.
8. B. S. Kawasaki and D. C. Johnson, "Bulb-ended fiber coupling to LED sources," *Opt. Quantum Electron.*, vol. 7, pp. 281–288, 1975.
9. J. G. Ackenhusen, "Microlenses to improve LED-to-optical fiber coupling and alignment tolerance," *Appl. Opt.*, vol. 18, pp. 3694–3699, Nov. 1979.
10. M. Sumida and K. Takamoto, "Lens coupling of laser diodes to single-mode fibers," *J. Lightwave Tech.*, vol. LT-2, pp. 305–311, June 1984.
11. B. Hillerich, "Influence of lens imperfections with LD and LED to single-mode fiber coupling," *J. Lightwave Tech.*, vol. 7, pp. 77–86, Jan. 1989.
12. M. L. Dakss and B. Kim, "Effects of fiber propagation loss on diode laser-fiber coupling," *Electron. Lett.*, vol. 16, pp. 421–422, May 1980; "Self-centering technique for mounting microsphere coupling lens on a fiber," *ibid.*, vol. 16, pp. 463–464, June 1980.
13. O. Hasegawa, R. Namazu, M. Abe, and Y. Toyoma, "Coupling of spherical-surfaced LED, and spherical-ended fiber," *J. Appl. Phys.*, vol. 51, pp. 30–36, Jan. 1980.
14. See any general physics book or introductory optics book; for example:
    (*a*) M. V. Klein and T. E. Furtak, *Optics*, Wiley, New York, 2nd ed., 1986.
    (*b*) E. Hecht and A. Zajac, *Optics*, Addition-Wesley, Reading, Mass., 2nd ed., 1987.
    (*c*) F. A. Jenkins and H. E. White, *Fundamentals of Optics*, McGraw-Hill, New York, 4th ed., 1976.
15. M. C. Hudson, "Calculation of the maximum optical coupling efficiency into multimode optical waveguides, *Appl. Opt.*, vol. 13, pp. 1029–1033, May 1974.
16. G. K. Khoe and G. Kuyt, "Realistic efficiency of coupling light from GaAs laser diodes into parabolic-index optical fibers," *Electron. Lett.*, vol. 14, pp. 667–669, Sept. 28, 1978.
17. L. d'Auria, Y. Combemale, and C. Moronvalle, "High index microlenses for GaAlAs laser-fibre coupling," *Electron. Lett.*, vol. 16, pp. 322–324, Apr. 24, 1980.
18. H. M. Presby, N. Amitay, F. V. DiMarcello, and K. T. Nelson, "Optical fiber tapers—a novel approach to self-aligned beam expansion and single-mode hardware," *J. Lightwave Tech.*, vol. LT-5, pp. 70–76, Jan. 1987.
19. H. M. Presby, N. Amitay, R. Scotti, and A. F. Benner, "Laser-to-fiber coupling via optical fiber up-tapers," *J. Lightwave Tech.*, vol. 7, pp. 274–277, Feb. 1989.
20. R. B. Kummer, "Lightguide splice loss—effects of launch beam numerical aperture," *Bell Sys. Tech. J.*, vol. 59, pp. 441–447, Mar. 1980.
21. D. H. Rice and G. E. Keiser, "Short-haul fiber-optic link connector loss," *Int. Wire & Cable Symp. Proc.*, Nov. 13–15, 1984, Reno, NV, pp. 190–192.
22. Y. Daido, E. Miyauchi, and T. Iwama, "Measuring fiber connection loss using steady-state power distribution: A method," *Appl. Opt.*, vol. 20, pp. 451–456, Feb. 1981.
23. M. J. Hackert, "Evolution of power distributions in fiber optic systems: Development of a measurement strategy," *Fiber & Integrated Optics*, vol. 8, pp. 163–167, 1989.
24. P. DiVita and U. Rossi, "Realistic evaluation of coupling loss between different optical fibers," *J. Opt. Commun.*, vol. 1, pp. 26–32, Sept. 1980; "Evaluation of splice losses induced by mismatch in fibre parameters," *Opt. Quantum Electron.*, vol. 13, pp. 91–94, Jan. 1981.
25. M. J. Adams, D. N. Payne, and F. M. E. Sladen, "Splicing tolerances in graded index fibers," *Appl. Phys. Lett.*, vol. 28, pp. 524–526, May 1976.
26. D. Gloge, "Offset and tilt loss in optical fiber splices," *Bell Sys. Tech. J.*, vol. 55, pp. 905–916, Sept. 1976.
27. T. C. Chu and A. R. McCormick, "Measurement of loss due to offset, end separation, and angular misalignment in graded index fibers excited by an incoherent source," *Bell Sys. Tech. J.*, vol. 57, pp. 595–602, Mar. 1978.
28. P. DiVita and U. Rossi, "Theory of power coupling between multimode optical fibers," *Opt. Quantum Electron.*, vol. 10, pp. 107–117, Jan. 1978.
29. C. M. Miller, "Transmission vs. transverse offset for parabolic-profile fiber splices with unequal core diameters," *Bell Sys. Tech. J.*, vol. 55, pp. 917–927, Sept. 1976.

30. D. Gloge and E. A. J. Marcatili, "Multimode theory of graded-core fibers," *Bell Sys. Tech. J.*, vol. 52, pp. 1563–1578, Nov. 1973.
31. H. G. Unger, *Planar Optical Waveguides and Fibres*, Clarendon, Oxford, 1977.
32. F. L. Thiel and R. M. Hawk, "Optical waveguide cable connection," *Appl. Opt.*, vol. 15, pp. 2785–2791, Nov. 1976.
33. F. L. Thiel and D. H. Davis, "Contributions of optical-waveguide manufacturing variations to joint loss," *Electron. Lett.*, vol. 12, pp. 340–341, June 1976.
34. S. C. Mettler, "A general characterization of splice loss for multimode optical fibers," *Bell Sys. Tech. J.*, vol. 58, pp. 2163–2182, Dec. 1979.
35. D. J. Bond and P. Hensel, "The effects on joint losses of tolerances in some geometrical parameters of optical fibres," *Opt. Quantum Electron.*, vol. 13, pp. 11–18, Jan. 1981.
36. S. C. Mettler and C. M. Miller, "Optical fiber splicing," in S. E. Miller and I. P. Kaminow, eds., *Optical Fiber Telecommunications II*, Academic, New York, 1988.
37. V. C. Y. So, R. P. Hughes, J. B. Lamont, and P. J. Vella, "Splice loss measurement using local launch and detect," *J. Lightwave Tech.*, vol. LT-5, pp. 1663–1666, Dec. 1987.
38. D. W. Peckham and C. R. Lovelace, "Multimode optical fiber splice loss: Relating system and laboratory measurements," *J. Lightwave Tech.*, vol. LT-5, pp. 1630–1636, Dec. 1987.
39. M. D. Drake, "Low reflectance terminations and connections for duplex fiber-optic transmission links," *Appl. Opt.*, vol. 20, pp. 1640–1644, May 1981.
40. D. Gloge, A. H. Cherin, C. M. Miller, and P. W. Smith, "Fiber splicing," in S. E. Miller and A. G. Chynoweth, eds., *Optical Fiber Telecommunications*, Academic, New York, 1979.
41. G. D. Khoe, G. Kuyt, and J. A. Luijendijk, "Optical fiber end preparation: A new method for producing perpendicular fractures in glass fibers," *Appl. Opt.*, vol. 20, pp. 707–714, Feb. 1981.
42. H. Taya, K. Ito, T. Yamada, and M. Yoshinuma, "New splicing method for polarization-maintaining fibers," *Tech. Digest OSA/IEEE Optical Fiber Commun. Conf.*, p. 164, Jan. 1989.
43. R. P. Drexel, J. A. Nelson, and J. Schnecker, "New approaches to termination," *Photonics Spectra*, vol. 22, pp. 101–106, Mar. 1988.
44. G. Cancellieri and U. Ravaioli, *Measurements of Optical Fibers and Devices: Theory and Experiment*, Artech House, Dedham, MA, 1984.
45. A. Tomita and N. M. Shen, "Reducing Fresnel reflection in fiber-optic connectors by use of a disposable elastomer," *Tech. Digest OSA/IEEE Optical Fiber Commun. Conf.*, p. 167, Jan. 1989.
46. J. E. Fulenwider and M. L. Dakss, "Hand-held tool for optical fibre end preparation," *Electron. Lett.*, vol. 13, pp. 578–580, Sept. 1977.
47. D. Gloge, P. W. Smith, and E. L. Chinnock, "Apparatus for breaking brittle rods or fibers," U.S. Patent 4,027,817, 1977.
48. A. Albanese and L. Maggi, "New fiber breaking tool," *Appl. Opt.*, vol. 16, pp. 2604–2605, Oct. 1977.
49. C. Belmonte, M. L. Dakss, and J. E. Fulenwider, "Hand-held tool for optical fiber waveguide end preparation," U.S. Patent 4,159,793, July 1979.
50. P. C. Hensel, "Simplified optical-fibre breaking machine," *Electron. Lett.*, vol. 11, pp. 581–582, Nov. 1975; "Spark-induced fracture of optical fibres," *ibid.*, vol. 13, pp. 603–604, Sept. 1977.
51. T. Haibara, M. Matsumoto, and M. Miyauchi, "Design and development of an automatic cutting tool for optical fibers," *J. Lightwave Tech.*, vol. LT-4, pp. 1434–1439, Sept. 1986.
52. M. deJong, "Cleave and crimp fiber optic connector for field installation," in *Technical Digest OFC '90*, p. 139, San Francisco, Jan. 1990.
53. K. Johnson, "High-quality cleaves for better fiber splices," *Photonics Spectra*, vol. 21, pp. 135–142, Oct. 1987.
54. (*a*) Electronic Industries Association, Standard EIA-455-57, *Optical Fiber End Preparation and Examination*, Feb. 1987.
    (*b*) Electronic Industries Association, Standard EIA-455-179, *Inspection of Cleaved Fiber End-faces by Interferometry*, May 1988.
55. (*a*) D. M. Fye, R. Olshansky, J. LaCourse, W. Powazinik, and R. B. Lauer, "Low-current, 1.3-μm edge-emitting LED for single-mode subscriber loop applications," *Electron. Lett.*, vol. 22, pp. 87–88, Jan. 1986.

(*b*) R. B. Lauer, D. M. Fye, L. W. Ulbricht, and M. J. Teare, "A light source for the fibered local loop," *GTE J. Science and Tech.*, vol. 1, no. 1, pp. 28–39, 1987.

56. (*a*) J. L. Gimlett, M. Stern, R. S. Vodhanel, N. K. Cheung, G. K. Chang, H. P. Leblanc, P. W. Shumate, and A. Suzuki, "Transmission experiments at 560 Mb/s and 140 Mb/s using single-mode fiber and 1300-nm LEDs," *Electron. Lett.*, vol. 21, pp. 1198–1200, Dec. 5, 1985.

(*b*) G. K. Chang, H. P. Leblanc, and P. W. Shumate, "Novel high-speed LED transmitter for single-mode fiber and wideband loop transmission systems," *Electron. Lett.*, vol. 23, pp. 1338–1340, Dec. 3, 1987.

57. T. Tsubota, Y. Kashima, H. Takano, and Y. Hirose, "InGaAsP/InP long-wavelength high-efficiency edge-emitting LED for single-mode fiber optic communication," *Fiber Integrated Optics*, vol. 7, no. 4, pp. 353–360, 1988.

58. T. Ohtsuka, N. Fujimoto, K. Yamaguchi, A. Taniguchi, H. Naitou, and Y. Nabeshima, "Gigabit single-mode fiber transmission using 1.3-μm edge-emitting LEDs for broadband subscriber loops," *J. Lightwave Tech.*, vol. LT-5, pp. 1534–1541, Oct. 1987.

59. D. N. Christodoulides, L. A. Reith, and M. A. Saifi, "Coupling efficiency and sensitivity of an LED to a single-mode fiber," *Electron. Lett.*, vol. 22, pp. 1110–1111, Oct. 1986.

60. L. A. Reith and P. A. Shumate, "Coupling sensitivity of an edge-emitting LED to a single-mode fiber," *J. Lightwave Tech.*, vol. LT-5, pp. 29–34, Jan. 1987.

61. B. Hillerich, "New analysis of LED to a single-mode fiber coupling," *Electron. Lett.*, vol. 22, pp. 1176–1177, Oct. 1986; "Efficiency and alignment tolerances of LED to a single-mode fiber coupling—theory and experiment," *Opt. Quantum Electron.*, vol. 19, no. 4, pp. 209–222, July 1987.

62. W. van Etten, "Coupling of LED light into a single-mode fiber," *J. Opt. Commun.*, vol. 9, no. 3, pp. 100–101, Sept. 1988.

63. D. N. Christodoulides, L. A. Reith, and M. A. Saifi, "Theory of LED coupling to single-mode fibers," *J. Lightwave Tech.*, vol. LT-5, pp. 1623–1629, Nov. 1987.

64. D. L. Bisbee, "Splicing silica fibers with an electric arc," *Appl. Opt.*, vol. 15, pp. 796–798, Mar. 1976.

65. J. T. Krause, C. R. Kurkjian, and U. C. Paek, "Strength of fusion splices for fiber lightguides," *Electron. Lett.*, vol. 17, pp. 232–233, Mar. 1981.

66. T. Yamada, Y. Ohsato, M. Yoshinuma, T. Tanaka, and K.-I. Itoh, "Arc fusion splicer with profile alignment system for high-strength low-loss optical submarine cable," *J. Lightwave Tech.*, vol. LT-4, pp. 1204–1210, Aug. 1986.

67. M. Fujise, Y. Iwamoto, and S. Takei, "Self core-alignment arc-fusion splicer based on a simple local monitoring method," *J. Lightwave Tech.*, vol. LT-4, pp. 1211–1218, Aug. 1986.

68. G. D. Khoe, J. A. Luijendijk, and L. J. C. Vroomen, "Arc-welded monomode fiber splices made with the aid of local injection and detection of blue light," *J. Lightwave Tech.*, vol. LT-4, pp. 1219–1222, Aug. 1986.

69. E. E. Basch, R. A. Beaudette, and H. A. Carnes, "Optical transmission for interoffice trunks," *IEEE Trans. Commun.*, vol. COM-26, pp. 1007–1014, July 1978.

70. M. L. Dakss, W. J. Carlsen, and J. E. Benasutti, "Field installable connectors and splices for glass optical fiber communication systems," *12th Annual Connector Symp.*, Cherry Hill, NJ, Oct. 17–18, 1979.

71. C. M. Miller, S. C. Mettler, and I. A. White, *Optical Fiber Splices and Connectors*, Marcel Dekker, New York, 1986.

72. D. B. Keck, A. J. Morrow, D. A. Nolan, and D. A. Thompson, "Passive components in the subscriber loop," *J. Lightwave Tech.*, vol. 7, pp. 1623–1633, Nov. 1989.

73. R. A. Patterson, "A new low-cost high-performance mechanical optical fiber splicing system for construction and restoration in the subscriber loop," *J. Lightwave Tech.*, vol. 7, pp. 1682–1688, Nov. 1989.

74. (*a*) W. J. Carlsen, "An elastic-tube fiber splice," *Laser Focus*, vol. 16, pp. 58–62, Apr. 1980.

(*b*) P. Melman and W. J. Carlsen, "Elastic-tube splice performance with single-mode and multimode fibers," *Electron. Lett.*, vol. 18, no. 8, pp. 320–321, Apr. 1982.

75. D. M. Knecht, W. J. Carlsen, and P. Melman, "Fiber-optic field splice," *Proc. SPIE Intl. Soc. Opt. Eng., Tech. Symp.*, Los Angeles, vol. 326, pp. 57–60, Jan. 25–29, 1982.

76. F. M. Sears, "A passive mechanical splice for polarization-maintaining fibers," *J. Lightwave Tech.*, vol. 7, pp. 1494–1498, Oct. 1989.

77. J. T. Krause, W. A. Reed, and K. L. Walker, "Splice loss of single-mode fiber as related to fusion time, temperature, and index profile alteration," *J. Lightwave Tech.*, vol. LT-4, pp. 837–840, July 1986.

78. E. Serafini, "Statistical approach to the optimization of optical fiber fusion splicing in the field," *J. Lightwave Tech.*, vol. 7, pp. 431–435, Feb. 1989.

79. D. Marcuse, D. Gloge, and E. A. J. Marcatili, "Guiding properties of fibers," in S. E. Miller and A. G. Chynoweth, eds., *Optical Fiber Telecommunications*, Academic, New York, 1979.

80. D. Marcuse, "Loss analysis of single-mode splices," *Bell Sys. Tech. J.*, vol. 56, pp. 703–718, May 1977.

81. B. B. Harbison, W. I. Roberts, and I. D. Aggarwal, "Fusion splicing of heavy metal fluoride glass optical fibers," *Electron. Lett.*, vol. 25, pp. 1214–1216, Aug. 1989.

82. (*a*) R. A. Wey, "Connectors: Trends," *Laser Focus/E-O*, vol. 23, pp. 130–146, June 1987. (*b*) R. Mack, "Fiber optic connectors," *Laser Focus/E-O*, vol. 23, pp. 148–156, June 1987.

83. T. Ormand, "Fiber optic connectors come of age," *EDN*, vol. 35, pp. 89–96, Feb. 15, 1990.

84. W. C. Young and D. R. Frey, "Fiber connectors," in S. E. Miller and I. P. Kaminow, eds., *Optical Fiber Telecommunications II*, Academic, New York, 1988.

85. K. Bowes, "Fiber connectors: A primer for the wary designer," *Photonics Spectra*, vol. 23, pp. 115–120, Oct. 1989.

86. G. M. Alameel and A. W. Carlisle, "The performance of the AT & T single-mode ST connector and its latest enhanced features," *Fiber Integ. Optics*, vol. 8, no. 1, pp. 45–59, 1989.

87. T. Satake, S. Nagasawa, and R. Arioka, "A new type of demountable plastic-molded single-mode multifiber connector," *J. Lightwave Tech.*, vol. LT-4, pp. 1232–1236, Aug. 1986.

88. Y. Tamaki, K. Koyama, H. Furukawa, H. Yokosuka, and O. Watanabe, "Field-installable plastic multifiber connector," *J. Lightwave Tech.*, vol. LT-4, pp. 1248–1254, Aug. 1986.

89. A. Nicia, "Practical low-loss lens connector for optical fibres," *Electron. Lett.*, vol. 14, pp. 511–512, Aug. 1978.

90. D. M. Knecht, W. J. Carlsen, and P. Melman, "Expanded beam fiber optic connectors," *Proc. SPIE Intl. Soc. Opt. Eng., Tech. Symp.*, Los Angeles, vol. 326, pp. 44–50, Jan. 25–29, 1982.

91. J. C. Baker and D. N. Payne, "Expanded beam connector design study," *Appl. Opt.*, vol. 20, pp. 2861–2867, Aug. 1981.

92. S. Nemota and T. Makimoto, "Analysis of splice loss in single-mode fibers using a gaussian field approximation," *Optical Quantum Electron.*, vol. 11, no. 5, pp. 447–457, Sept. 1979.

CHAPTER
6

PHOTODETECTORS

At the output end of an optical transmission line there must be a receiving device which interprets the information contained in the optical signal. The first element of this receiver is a photodetector. The photodetector senses the luminescent power falling upon it and converts the variation of this optical power into a correspondingly varying electric current. Since the optical signal is generally weakened and distorted when it emerges from the end of the fiber, the photodetector must meet very high performance requirements. Among the foremost of these requirements are a high response or sensitivity in the emission wavelength range of the optical source being used, a minimum addition of noise to the system, and a fast response speed or sufficient bandwidth to handle the desired data rate. The photodetector should also be insensitive to variations in temperature, be compatible with the physical dimensions of the optical fiber, have a reasonable cost in relation to the other components of the system, and have a long operating life.

Several different types of photodetectors are in existence. Among these are photomultipliers,[1-3] pyroelectric detectors,[4] and semiconductor-based photoconductors, phototransistors, and photodiodes.[5] However, many of these detectors do not meet one or more of the foregoing requirements. Photomultipliers consisting of a photocathode and an electron multiplier packaged in a vacuum tube are capable of very high gain and very low noise. Unfortunately, their large size and high voltage requirements make them unsuitable for optical fiber systems. Pyroelectric photodetectors involve the conversion of photons to heat. Photon absorption results in a temperature change of the detector material. This gives rise to a variation in the dielectric constant which is usually measured as a capacitance change. The response of this detector is quite flat over a broad spectral band, but its speed is limited by the detector cooling rate after it has been excited. Its principal use is for detecting high-speed laser pulses, and it is not well suited for optical fiber systems.

Of the semiconductor-based photodetectors, the photodiode is used almost exclusively for fiber optic systems because of its small size, suitable material, high sensitivity, and fast response time. The two types of photodiodes used are the *pin* photodetector and the avalanche photodiode (APD). Detailed reviews of these photodiodes have been presented in the literature.[3,5-10] We shall examine the fundamental characteristics of these two device types in the following sections. In describing these components we shall make use of the elementary principles of semiconductor device physics given in Sec. 4.1. The books by Rose[11] and Sze[12] give basic discussions of photodetection processes.

6.1 PHYSICAL PRINCIPLES OF PHOTODIODES

6.1.1 The *pin* Photodetector

The most common semiconductor photodetector is the *pin* photodiode shown schematically in Fig. 6-1. The device structure consists of p and n regions separated by a very lightly n-doped intrinsic (i) region. In normal operation a sufficiently large reverse-bias voltage is applied across the device so that the intrinsic region is fully depleted of carriers. That is, the intrinsic n and p carrier concentrations are negligibly small in comparison with the impurity concentration in this region.

When an incident photon has an energy greater than or equal to the band-gap energy of the semiconductor material, the photon can give up its energy and excite an electron from the valence band to the conduction band. This process generates free electron-hole pairs which are known as *photocarriers*, since they are photon-generated charge carriers, as is shown in Fig. 6-2. The photodetector is normally designed so that these carriers are generated mainly in the depletion region (the depleted intrinsic region) where most of the

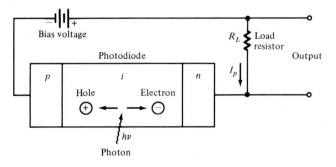

FIGURE 6-1
Schematic representation of a *pin* photodiode circuit with an applied reverse bias.

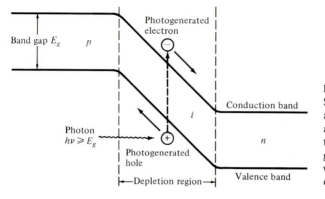

FIGURE 6-2
Simple energy-band diagram for a *pin* photodiode. Photons with an energy greater than or equal to the band-gap energy E_g can generate free electron-hole pairs which act as photocurrent carriers.

incident light is absorbed. The high electric field present in the depletion region causes the carriers to separate and be collected across the reverse-biased junction. This gives rise to a current flow in an external circuit, with one electron flowing for every carrier pair generated. This current flow is known as the *photocurrent*.

As the charge carriers flow through the material, some electron-hole pairs will recombine and hence disappear. On the average, the charge carriers move a distance L_n or L_p for electrons and holes, respectively. This distance is known as the *diffusion length*. The time it takes for an electron or hole to recombine is known as the *carrier lifetime* and is represented by τ_n and τ_p, respectively. The lifetimes and the diffusion lengths are related by the expressions

$$L_n = (D_n \tau_n)^{1/2} \quad \text{and} \quad L_p = (D_p \tau_p)^{1/2}$$

where D_n and D_p are the electron and hole diffusion coefficients (or constants), respectively, which are expressed in units of centimeters squared per second.

Optical radiation is absorbed in the semiconductor material according to the exponential law

$$P(x) = P_0(1 - e^{-\alpha_s(\lambda)x}) \tag{6-1}$$

Here $\alpha_s(\lambda)$ is the *absorption coefficient* at a wavelength λ, P_0 is the incident optical power level, and $P(x)$ is the optical power absorbed in a distance x.

The dependence of the optical absorption coefficient on wavelength is shown in Fig. 6-3 for several photodiode materials.[13] As the curves clearly show, α_s depends strongly on the wavelength. Thus a particular semiconductor material can only be used over a limited wavelength range. The upper wavelength cutoff λ_c is determined by the band-gap energy E_g of the material. If E_g is expressed in units of electron volts (eV), then λ_c is given in units of micrometers (μm) by

$$\lambda_c (\mu\text{m}) = \frac{hc}{E_g} = \frac{1.24}{E_g (\text{eV})} \tag{6-2}$$

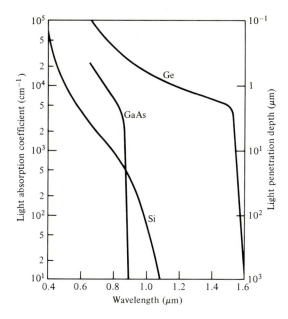

FIGURE 6-3
Optical absorption coefficient as a function of wavelength for silicon, germanium, and gallium arsenide. (Reproduced with permission from Miller, Marcatili, and Li,[13] © 1973, IEEE.)

The cutoff wavelength is about 1.06 μm for Si and 1.6 μm for Ge. For longer wavelengths the photon energy is not sufficient to excite an electron from the valence to the conduction band.

Example 6-1. A photodiode is constructed of GaAs, which has a band-gap energy of 1.43 eV at 300 K. From Eq. (6-2) the long-wavelength cutoff is

$$\lambda_c = \frac{hc}{E_g} = \frac{(6.625 \times 10^{-34} \text{ J} \cdot \text{s})(3 \times 10^8 \text{ m/s})}{(1.43 \text{ eV})(1.6 \times 10^{-19} \text{ J/eV})} = 869 \text{ nm}$$

This GaAs photodiode will not operate for photons of wavelength greater than 869 nm.

At the lower-wavelength end, the photoresponse cuts off as a result of the very large values of α_s at the shorter wavelengths. In this case the photons are absorbed very close to the photodetector surface where the recombination time of the generated electron-hole pairs is very short. The generated carriers thus recombine before they can be collected by the photodetector circuitry.

If the depletion region has a width w, then, from Eq. (6-1), the total power absorbed in the distance w is

$$P(w) = P_0(1 - e^{-\alpha_s w}) \tag{6-3}$$

If we take into account a reflectivity R_f at the entrance face of the photodiode, then the primary photocurrent I_p resulting from the power absorption of Eq.

(6-3) is given by

$$I_p = \frac{q}{h\nu} P_0 (1 - e^{-\alpha_s w})(1 - R_f) \tag{6-4}$$

where P_0 is the optical power incident on the photodetector, q is the electron charge, and $h\nu$ is the photon energy.

Two important characteristics of a photodetector are its quantum efficiency and its response speed. These parameters depend on the material band gap, the operating wavelength, and the doping and thickness of the p, i, and n regions of the device. The *quantum efficiency* η is the number of electron-hole carrier pairs generated per incident photon of energy $h\nu$ and is given by

$$\eta = \frac{\text{number of electron-hole pairs generated}}{\text{number of incident photons}} = \frac{I_p/q}{P_0/h\nu} \tag{6-5}$$

Here I_p is the average photocurrent generated by a steady-state average optical power P_0 incident on the photodetector.

> **Example 6-2.** In a 100-ns pulse, 6×10^6 photons at a wavelength of 1300 nm fall on an InGaAs photodetector. On the average 3.9×10^6 electron-hole (e-h) pairs are generated. The quantum efficiency is found from Eq. (6-5) as
>
> $$\eta = \frac{\text{number of e-h pairs generated}}{\text{number of incident photons}} = \frac{3.9 \times 10^6}{6 \times 10^6} = 0.65$$
>
> Thus the quantum efficiency at 1300 nm is 65 percent.

In a practical photodiode, 100 photons will create between 30 and 95 electron-hole pairs, thus giving a detector quantum efficiency ranging from 30 to 95 percent. To achieve a high quantum efficiency, the depletion layer must be thick enough to permit a large fraction of the incident light to be absorbed. However, the thicker the depletion layer, the longer it takes for the photogenerated carriers to drift across the reverse-biased junction. Since the carrier drift time determines the response speed of the photodiode, a compromise has to be made between response speed and quantum efficiency. We shall discuss this further in Sec. 6.3.

The performance of a photodiode is often characterized by the *responsivity* $\mathcal{R}$. This is related to the quantum efficiency by

$$\mathcal{R} = \frac{I_p}{P_0} = \frac{\eta q}{h\nu} \tag{6-6}$$

This parameter is quite useful, since it specifies the photocurrent generated per unit optical power. Typical *pin* photodiode responsivities as a function of wavelength are shown in Fig. 6-4. Representative values are 0.65 μA/μW for silicon at 900 nm, 0.45 μA/μW for germanium at 1.3 μm, and 0.6 μA/μW for InGaAs at 1.3 μm.

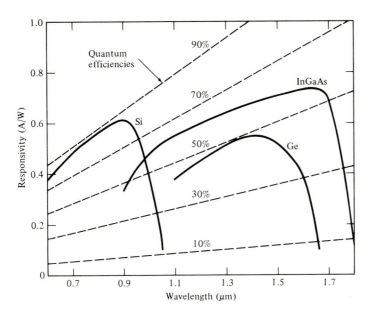

FIGURE 6-4

Comparison of the responsivity and quantum efficiency as a function of wavelength for *pin* photodiodes constructed of different materials.

Example 6-3. Photons of energy 1.53×10^{-19} J are incident on a photodiode which has a responsivity of 0.65 A/W. If the optical power level is 10 μW, then from Eq. (6-6) the photocurrent generated is

$$I_p = \mathscr{R}P_0 = (0.65 \text{ A/W})(10 \ \mu\text{W}) = 6.5 \ \mu\text{A}$$

In most photodiodes the quantum efficiency is independent of the power level falling on the detector at a given photon energy. Thus the responsivity is a linear function of the optical power. That is, the photocurrent I_p is directly proportional to the optical power P_0 incident upon the photodetector, so that the responsivity $\mathscr{R}$ is constant at a given wavelength (a given value of $h\nu$). Note, however, that the quantum efficiency is not a constant at all wavelengths, since it varies according to the photon energy. Consequently, the responsivity is a function of the wavelength and of the photodiode material (since different materials have different band-gap energies). For a given material, as the wavelength of the incident photon becomes longer, the photon energy becomes less than that required to excite an electron from the valence band to the conduction band. The responsivity thus falls off rapidly beyond the cutoff wavelength, as can be seen in Fig. 6-4.

Example 6-4. As shown in Fig. 6-4, for the wavelength range 1100 nm $< \lambda <$ 1600 nm the quantum efficiency for InGaAs is about 60 percent. Thus in this wavelength range the responsivity is

$$\mathscr{R} = \frac{\eta q}{h\nu} = \frac{\eta q\lambda}{hc} = \frac{(0.60)(1.6 \times 10^{-19}\ \text{C})\lambda}{(6.625 \times 10^{-34}\ \text{J} \cdot \text{s})(3 \times 10^8\ \text{m/s})} = 4.83 \times 10^5 \lambda$$

For example, at 1300 nm we have

$$\mathscr{R} = \left[4.83 \times 10^5\ (\text{A/W})/\text{m}\right](1.30 \times 10^{-6}\ \text{m}) = 0.63\ \text{A/W}$$

At wavelengths higher than 1600 nm, the photon energy is not sufficient to excite an electron from the valence band to the conduction band. For example, $In_{0.53}Ga_{0.47}As$ has an energy gap $E_g = 0.73$ eV, so that from Eq. (6-2) the cutoff wavelength is

$$\lambda_c = \frac{1.24}{E_g} = \frac{1.24}{0.73} = 1.70\ \mu\text{m}$$

At wavelengths less than 1100 nm, the photons are absorbed very close to the photodetector surface where the recombination rate of the generated electron-hole pairs is very short. The responsivity thus decreases rapidly for smaller wavelengths, since many of the generated carriers do not contribute to the photocurrent.

6.1.2   Avalanche Photodiodes

Avalanche photodiodes (APDs) internally multiply the primary signal photocurrent before it enters the input circuitry of the following amplifier. This increases receiver sensitivity, since the photocurrent is multiplied before encountering the thermal noise associated with the receiver circuit. In order for carrier multiplication to take place, the photogenerated carriers must traverse a region where a very high electric field is present. In this high-field region a photogenerated electron or hole can gain enough energy so that it ionizes bound electrons in the valence band upon colliding with them. This carrier multiplication mechanism is known as *impact ionization*. The newly created carriers are also accelerated by the high electric field, thus gaining enough energy to cause further impact ionization. This phenomenon is the *avalanche effect*. Below the diode breakdown voltage a finite total number of carriers are created, whereas above breakdown the number can be infinite.

A commonly used structure for achieving carrier multiplication with very little excess noise is the *reach-through* construction[7, 13-16] shown in Fig. 6-5. The reach-through avalanche photodiode (RAPD) is composed of a high-resistivity p-type material deposited as an epitaxial layer on a p^+ (heavily doped p-type) substrate. A p-type diffusion or ion implant is then made in the high-resistivity material followed by the construction of an n^+ (heavily doped n-type) layer. For silicon the dopants used to form these layers are normally boron and phosphorus, respectively. This configuration is referred to as a $p^+\pi pn^+$ *reach-through* structure. The π layer is basically an intrinsic material that inadvertently has some p doping because of imperfect purification.

The term "reach-through" arises from the photodiode operation. When a low reverse-bias voltage is applied, most of the potential drop is across the pn^+

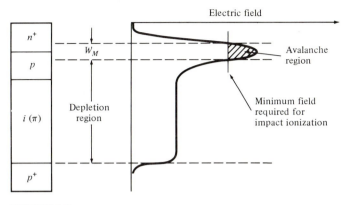

FIGURE 6-5
Reach-through avalanche photodiode structure and the electric fields in the depletion and multiplication regions.

junction. The depletion layer widens with increasing bias until a certain voltage is reached at which the peak electric field at the pn^+ junction is about 5 to 10 percent below that needed to cause avalanche breakdown. At this point the depletion layer just "reaches through" to the nearly intrinsic π region.

In normal usage the RAPD is operated in the fully depleted mode. Light enters the device through the p^+ region and is absorbed in the π material, which acts as the collection region for the photogenerated carriers. Upon being absorbed the photon gives up its energy, thereby creating electron-hole pairs, which are then separated by the electric field in the π region. The photogenerated electrons drift through the π region to the pn^+ junction where a high electric field exists. It is in this high-field region that carrier multiplication takes place.

The average number of electron-hole pairs created by a carrier per unit distance traveled is called the *ionization rate*. Most materials exhibit different *electron ionization rates* α and *hole ionization rates* β. Experimentally obtained values of α and β for five different semiconductor materials are shown in Fig. 6-6. The ratio $k = \beta/\alpha$ of the two ionization rates is a measure of the photodetector performance. As we shall see in Sec. 6.4, avalanche photodiodes constructed of materials in which one type of carrier largely dominates impact ionization exhibit low noise and large gain-bandwidth products. As shown in Fig. 6-6, of all the materials examined so far,[16-34] only silicon has a significant difference between electron and hole ionization rates.

The multiplication M for all carriers generated in the photodiode is defined by

$$M = \frac{I_M}{I_p} \tag{6-7}$$

where I_M is the average value of the total multiplied output current and I_p is

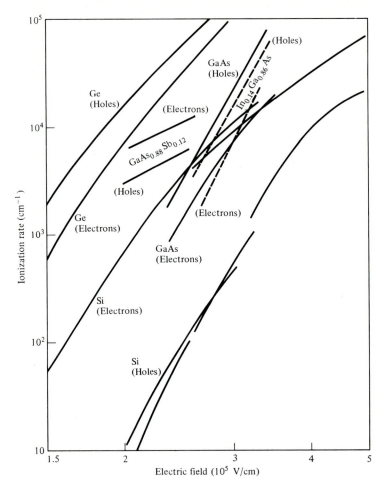

FIGURE 6-6
Carrier ionization rates obtained experimentally for silicon, germanium, gallium arsenide, gallium arsenide antimonide, and indium gallium arsenide. (Reproduced with permission from Melchior.[2])

the primary unmultiplied photocurrent defined in Eq. (6-4). In practice, the avalanche mechanism is a statistical process, since not every carrier pair generated in the diode experiences the same multiplication. Thus the measured value of M is expressed as an average quantity.

Example 6-5. A given silicon avalanche photodiode has a quantum efficiency of 65 percent at a wavelength of 900 nm. Suppose 0.50 μW of optical power produces a multiplied photocurrent of 10 μA. Let us find the multiplication M. From Eq.

(6-6) the primary photocurrent is

$$I_p = \mathscr{R}P_0 = \frac{\eta q}{h\nu}P_0 = \frac{\eta q\lambda}{hc}P_0$$

$$= \frac{(0.65)(1.6 \times 10^{-19} \text{ C})(9 \times 10^{-7} \text{ m})}{(6.625 \times 10^{-34} \text{ J} \cdot \text{s})(3 \times 10^8 \text{ m/s})} 5 \times 10^{-7} \text{ W} = 0.235 \ \mu\text{A}$$

From Eq. (6-7) the multiplication is

$$M = \frac{I_M}{I_p} = \frac{10 \ \mu\text{A}}{0.235 \ \mu\text{A}} = 43$$

Thus the primary photocurrent is multiplied by a factor of 43.

Typical current gains for different wavelengths[15] as a function of bias voltage for a silicon reach-through avalanche photodiode are shown in Fig. 6-7. The dependence of the gain on the excitation wavelength is attributable to mixed initiation of the avalanche process by electrons and holes when most of the light is absorbed in the n^+p region close to the detector surface. This is especially noticeable at short wavelengths, where a larger portion of the optical power is absorbed close to the surface than at longer wavelengths. Since the ionization coefficient for holes is smaller than that for electrons in silicon, the total current gain is reduced at the short wavelengths.

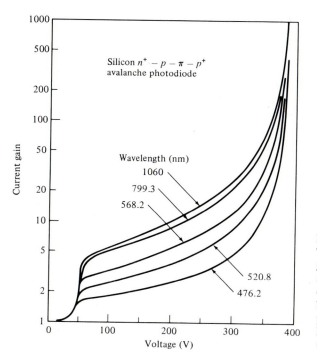

FIGURE 6-7

Typical room-temperature current gains of a silicon reach-through avalanche photodiode for different wavelengths as a function of bias voltage. (Reproduced with permission from Melchior, Hartman, Schinke, and Seidel,[15] © 1978, AT & T.)

Analogous to the *pin* photodiode, the performance of an APD is characterized by its responsivity $\mathscr{R}_{APD}$, which is given by

$$\mathscr{R}_{APD} = \frac{\eta q}{h\nu}M = \mathscr{R}_0 M \qquad (6\text{-}8)$$

where $\mathscr{R}_0$ is the unity gain responsivity.

6.2 PHOTODETECTOR NOISE

In fiber optic communication systems the photodiode is generally required to detect very weak optical signals. Detection of the weakest possible optical signals requires that the photodetector and its following amplification circuitry be optimized so that a given signal-to-noise ratio is maintained. The power signal-to-noise ratio S/N at the output of an optical receiver is defined by

$$\frac{S}{N} = \frac{\text{signal power from photocurrent}}{\text{photodetector noise power} + \text{amplifier noise power}} \qquad (6\text{-}9)$$

The noise sources in the receiver arise from the photodetector noises resulting from the statistical nature of the photon-to-electron conversion process and the thermal noises associated with the amplifier circuitry.

To achieve a high signal-to-noise ratio, the following conditions should be met:

1. The photodetector must have a high quantum efficiency to generate a large signal power.
2. The photodetector and amplifier noises should be kept as low as possible.

For most applications it is the noise currents which determine the minimum optical power level that can be detected, since the photodiode quantum efficiency is normally close to its maximum possible value.

The sensitivity of a photodetector in an optical fiber communication system is describable in terms of the *minimum detectable optical power*. This is the optical power necessary to produce a photocurrent of the same magnitude as the root mean square of the total noise current or, equivalently, a signal-to-noise ratio of one. A thorough understanding of the source, characteristics, and interrelationships of the various noises in a photodetector is therefore necessary to make a reliable design and to evaluate optical receivers.

6.2.1 Noise Sources

To see the interrelationship of the different types of noises affecting the signal-to-noise ratio, let us examine the simple receiver model and its equivalent circuit shown in Fig. 6-8. The photodiode has a small series resistance R_s, a total capacitance C_d consisting of junction and packaging capacitances, and a bias (or load) resistor R_L. The amplifier following the photodiode has an input

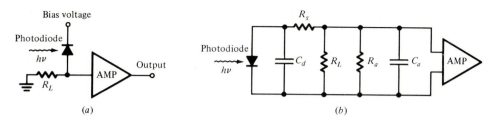

FIGURE 6-8
(*a*) Simple model of a photodetector receiver, and (*b*) its equivalent circuit.

capacitance C_a and a resistance R_a. For practical purposes, R_s is much smaller than the load resistance R_L and can be neglected.

If a modulated signal of optical power $P(t)$ falls on the detector, the primary photocurrent $i_{ph}(t)$ generated is

$$i_{ph}(t) = \frac{\eta q}{h\nu}P(t) \tag{6-10}$$

This primary current consists of a dc value I_p, which is the average photocurrent due to the signal power, and a signal component $i_p(t)$. For *pin* photodiodes the mean square signal current $\langle i_s^2 \rangle$ is

$$\langle i_s^2 \rangle = \langle i_p^2(t) \rangle \tag{6-11a}$$

whereas for avalanche photodetectors,

$$\langle i_s^2 \rangle = \langle i_p^2(t) \rangle M^2 \tag{6-11b}$$

where M is the average of the statistically varying avalanche gain as defined in Eq. (6-7). For a sinusoidally varying input signal of modulation index m, the signal component $\langle i_p^2 \rangle$ is of the form (see Prob. 6-5)

$$\langle i_p^2(t) \rangle = \frac{m^2}{2}I_p^2 \tag{6-12}$$

where m is defined in Eq. (4-54).

The principal noises associated with photodetectors having no internal gain are quantum noise, dark-current noise generated in the bulk material of the photodiode, and surface leakage current noise. The *quantum* or *shot noise* arises from the statistical nature of the production and collection of photoelectrons when an optical signal is incident on a photodetector. It has been demonstrated[35] that these statistics follow a Poisson process. Since the fluctuations in the number of photocarriers created from the photoelectric effect are a fundamental property of the photodetection process, they set the lower limit on the receiver sensitivity when all other conditions are optimized. The quantum noise current has a mean square value in a bandwidth B which is proportional to the average value of the photocurrent I_p:

$$\langle i_Q^2 \rangle = 2qI_p BM^2 F(M) \tag{6-13}$$

where $F(M)$ is a noise figure associated with the random nature of the avalanche process. From experimental results it has been found that to a reasonable approximation $F(M) \simeq M^x$, where x (with $0 \le x \le 1.0$) depends on the material. This is discussed in more detail in Sec. 6.4. For *pin* photodiodes M and $F(M)$ are unity.

The photodiode dark current is the current that continues to flow through the bias circuit of the device when no light is incident on the photodiode. This is a combination of bulk and surface currents. The *bulk dark current* i_{DB} arises from electrons and/or holes which are thermally generated in the *pn* junction of the photodiode. In an APD these liberated carriers also get accelerated by the high electric field present at the *pn* junction, and are therefore multiplied by the avalanche gain mechanism. The mean square value of this current is given by

$$\langle i_{DB}^2 \rangle = 2qI_D M^2 F(M) B \tag{6-14}$$

where I_D is the primary (unmultiplied) detector bulk dark current.

The *surface dark current* is also referred to as a *surface leakage current* or simply the leakage current. It is dependent on surface defects, cleanliness, bias voltage, and surface area. An effective way of reducing surface dark current is through the use of a guard ring structure which shunts surface leakage currents away from the load resistor. The mean square value of the surface dark current is given by

$$\langle i_{DS}^2 \rangle = 2qI_L B \tag{6-15}$$

where I_L is the surface leakage current. Note that since avalanche multiplication is a bulk effect, the surface dark current is not affected by the avalanche gain.

A comparison[30] of typical dark currents for Si, Ge, GaAs, and $In_x Ga_{1-x} As$ photodiodes is given in Fig. 6-9 as a function of applied voltage normalized to the breakdown voltage V_B. Note that for $In_x Ga_{1-x} As$ photodiodes the dark current increases with the composition x. Under a reverse bias, both dark currents also increase with area. The surface dark current increases in proportion to the square root of the active area, and the bulk dark current is directly proportional to the area.

Since the dark currents and the signal current are uncorrelated, the total mean square photodetector noise current $\langle i_N^2 \rangle$ can be written as

$$\langle i_N^2 \rangle = \langle i_Q^2 \rangle + \langle i_{DB}^2 \rangle + \langle i_{DS}^2 \rangle$$
$$= 2q(I_p + I_D)M^2 F(M) B + 2qI_L B \tag{6-16}$$

To simplify the analysis of the receiver circuitry, we shall assume here that the amplifier input impedance is much greater than the load resistance so that

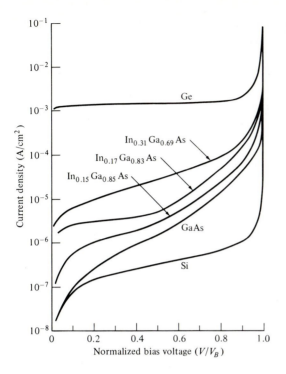

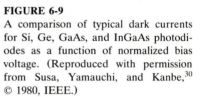

FIGURE 6-9

A comparison of typical dark currents for Si, Ge, GaAs, and InGaAs photodiodes as a function of normalized bias voltage. (Reproduced with permission from Susa, Yamauchi, and Kanbe,[30] © 1980, IEEE.)

its thermal noise is much smaller than that of R_L. The photodetector load resistor contributes a mean square thermal (Johnson) noise current

$$\langle i_T^2 \rangle = \frac{4k_BT}{R_L}B \qquad (6\text{-}17)$$

where k_B is Boltzmann's constant and T is the absolute temperature. This noise can be reduced by using a load resistor which is large but still consistent with the receiver bandwidth requirements. Further details on this are given in Chap. 7 along with a detailed discussion of the amplifier noise current i_{amp}.

Example 6-6. An InGaAs *pin* photodiode has the following parameters at a wavelength of 1300 nm: $I_D = 4$ nA, $\eta = 0.65$, $R_L = 1000$ Ω, and the surface leakage current is negligible. The incident optical power is 300 nW (-35 dBm), and the receiver bandwidth is 20 MHz. Let us find the various noise terms of the receiver.

First we need to find the primary photocurrent. From Eq. (6-6)

$$I_p = \mathscr{R}P_0 = \frac{\eta q}{h\nu}P_0 = \frac{\eta q\lambda}{hc}P_0$$

$$= \frac{(0.65)(1.6 \times 10^{-19}\text{ C})(1.3 \times 10^{-6}\text{ m})}{(6.625 \times 10^{-34}\text{ J} \cdot \text{s})(3 \times 10^8\text{ m/s})} 3 \times 10^{-7}\text{ W} = 0.204\ \mu\text{A}$$

From Eq. (6-13) the mean square quantum noise current for a *pin* photodiode is

$$\langle I_Q^2 \rangle = 2qI_p B = 2(1.6 \times 10^{-19} \text{ C})(0.204 \times 10^{-6} \text{ A})(20 \times 10^6 \text{ Hz})$$

$$= 1.3 \times 10^{-18} \text{ A}^2$$

or

$$\langle I_Q^2 \rangle^{1/2} = 1.1 \text{ nA}$$

From Eq. (6-14) the mean square dark current is

$$\langle I_{DB}^2 \rangle = 2qI_D B = 2(1.6 \times 10^{-19} \text{ C})(4 \times 10^{-9} \text{ A})(20 \times 10^6 \text{ Hz})$$

$$= 2.56 \times 10^{-20} \text{ A}^2$$

or

$$\langle I_{DB}^2 \rangle^{1/2} = 0.16 \text{ nA}$$

The mean square thermal noise current for the receiver is found from Eq. (6-17) as

$$\langle I_T^2 \rangle = \frac{4k_B T}{R_L} B = \frac{4(1.38 \times 10^{-23} \text{ J/K})(293 \text{ K})}{1 \text{ k}\Omega} 20 \times 10^6 \text{ Hz}$$

$$= 323 \times 10^{-18} \text{ A}^2$$

or

$$\langle I_T^2 \rangle^{1/2} = 18 \text{ nA}$$

Thus for this receiver the rms thermal noise current is about 16 times greater than the rms shot noise current and about 100 times greater than the rms dark current.

6.2.2   Signal-to-Noise Ratio

Substituting Eqs. (6-11), (6-16), and (6-17) into Eq. (6-9) for the signal-to-noise ratio at the input of the amplifier, we have

$$\frac{S}{N} = \frac{\langle i_p^2 \rangle M^2}{2q(I_p + I_D)M^2 F(M)B + 2qI_L B + 4k_B TB/R_L} \tag{6-18}$$

In general, when *pin* photodiodes are used, the dominating noise currents are those of the detector load resistor (the thermal current i_T) and the active elements of the amplifier circuitry (i_{amp}). For avalanche photodiodes the thermal noise is of lesser importance, and the photodetector noises usually dominate.[36] From Eq. (6-18) it can be seen that the signal power is multiplied by M^2 and the quantum noise plus bulk dark current is multiplied by $M^2 F(M)$. The surface leakage current is not altered by the avalanche gain mechanism. Since the noise figure $F(M)$ increases with M, there always exists an optimum value of M that maximizes the signal-to-noise ratio. The optimum gain at the maximum signal-to-noise ratio can be found by differentiating Eq. (6-18) with respect to M, setting the result equal to zero, and solving for M. Doing so for a sinusoidally modulated signal, with $m = 1$ and $F(M)$ approximated by M^x,

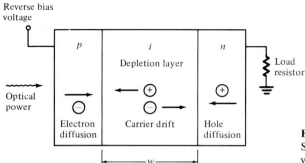

FIGURE 6-10

Schematic representation of a reverse-biased *pin* photodiode.

yields

$$M_{opt}^{x+2} = \frac{2qI_L + 4k_BT/R_L}{xq(I_p + I_D)} \tag{6-19}$$

6.3 DETECTOR RESPONSE TIME

6.3.1 Depletion Layer Photocurrent

To understand the frequency response of photodiodes, let us first consider the schematic representation of a reverse-biased *pin* photodiode shown in Fig. 6-10. Light enters the device through the *p* layer and produces electron-hole pairs as it is absorbed in the semiconductor material. Those electron-hole pairs that are generated in the depletion region or within a diffusion length of it will be separated by the reverse-bias-voltage-induced electric field, thereby leading to a current flow in the external circuit as the carriers drift across the depletion layer.

Under steady-state conditions the total current density J_{tot} flowing through the reverse-biased depletion layer is[37]

$$J_{tot} = J_{dr} + J_{diff} \tag{6-20}$$

Here J_{dr} is the drift current density resulting from carriers generated inside the depletion region, and J_{diff} is the diffusion current density arising from the carriers that are produced outside of the depletion layer in the bulk of the semiconductor (that is, in the *n* and *p* regions) and diffuse into the reverse-biased junction. The drift current density can be found from Eq. (6-4),

$$J_{dr} = \frac{I_p}{A} = q\Phi_0(1 - e^{-\alpha_s w}) \tag{6-21}$$

where A is the photodiode area and Φ_0 is the incident photon flux per unit area given by

$$\Phi_0 = \frac{P_0(1 - R_f)}{Ah\nu} \tag{6-22}$$

The surface p layer of a pin photodiode is normally very thin. The diffusion current is thus principally determined by hole diffusion from the bulk n region. The hole diffusion in this material can be determined by the one-dimensional diffusion equation[12]

$$D_p \frac{\partial^2 p_n}{\partial x^2} - \frac{p_n - p_{n0}}{\tau_p} + G(x) = 0 \tag{6-23}$$

where D_p is the hole diffusion coefficient, p_n is the hole concentration in the n-type material, τ_p is the excess hole lifetime, p_{n0} is the equilibrium hole density, and $G(x)$ is the electron-hole generation rate given by

$$G(x) = \Phi_0 \alpha_s e^{-\alpha_s x} \tag{6-24}$$

From Eq. (6-23) the diffusion current density is found to be (see Prob. 6-9)

$$J_{\text{diff}} = q\Phi_0 \frac{\alpha_s L_p}{1 + \alpha_s L_p} e^{-\alpha_s w} + q p_{n0} \frac{D_p}{L_p} \tag{6-25}$$

Substituting Eqs. (6-21) and (6-25) into Eq. (6-20) we have that the total current density through the reverse-biased depletion layer is

$$J_{\text{tot}} = q\Phi_0 \left(1 - \frac{e^{-\alpha_s w}}{1 + \alpha_s l_p} \right) + q p_{n0} \frac{D_p}{L_p} \tag{6-26}$$

The term involving p_{n0} is normally small, so that the total photogenerated current is proportional to the photon flux Φ_0.

6.3.2   Response Time

The response time of a photodiode together with its output circuit (see Fig. 6-8) depends mainly on the following three factors:

1. The transit time of the photocarriers in the depletion region
2. The diffusion time of the photocarriers generated outside the depletion region
3. The RC time constant of the photodiode and its associated circuit

The photodiode parameters responsible for these three factors are the absorption coefficient α_s, the depletion region width w, the photodiode junction and package capacitances, the amplifier capacitance, the detector load resistance, the amplifier input resistance, and the photodiode series resistance. The photodiode series resistance is generally only a few ohms and can be neglected in comparison with the large load resistance and the amplifier input resistance. Let us first look at the transit time of the photocarriers in the depletion region. The response speed of a photodiode is fundamentally limited by the time it takes photogenerated carriers to travel across the depletion region. This transit time t_d depends on the carrier drift velocity v_d and the depletion layer width w, and is given by

$$t_d = \frac{w}{v_d} \tag{6-27}$$

In general, the electric field in the depletion region is large enough so that the carriers have reached their scattering-limited velocity. For silicon the maximum velocities for electrons and holes are 8.4×10^6 and 4.4×10^6 cm/s, respectively, when the field strength is on the order of 2×10^4 V/cm. A typical high-speed silicon photodiode with a 10-μm depletion layer width thus has a response time limit of about 0.1 ns.

The diffusion processes are slow compared to the drift of carriers in the high-field region. Therefore, to have a high-speed photodiode, the photocarriers should be generated in the depletion region or so close to it that the diffusion times are less than or equal to the carrier drift times. The effect of long diffusion times can be seen by considering the photodiode response time. This response time is described by the rise time and fall time of the detector output when the detector is illuminated by a step input of optical radiation. The rise time τ_r is typically measured from the 10- to the 90-percent points of the leading edge of the output pulse, as is shown in Fig. 6-11. For fully depleted photodiodes the rise time τ_r and fall time τ_f are generally the same. However, they can be different at low bias levels where the photodiode is not fully depleted, since the photon collection time then starts to become a significant contributor to the rise time. In this case, charge carriers produced in the depletion region are separated and collected quickly. On the other hand, electron-hole pairs generated in the n and p regions must slowly diffuse to the depletion region before they can be separated and collected. A typical response

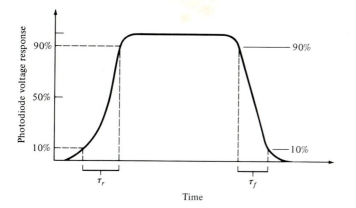

FIGURE 6-11
Photodiode response to an optical input pulse showing the 10- to 90-percent rise time and the 10- to 90-percent fall time.

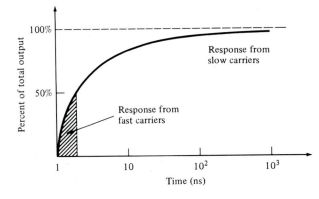

FIGURE 6-12
Typical response time of a photodiode that is not fully depleted.

time of a partially depleted photodiode is shown in Fig. 6-12. The fast carriers
allow the device output to rise to 50 percent of its maximum value in approximately 1 ns, but the slow carriers cause a relatively long delay before the output
reaches its maximum value.

To achieve a high quantum efficiency the depletion layer width must be
much larger than $1/\alpha_s$ (the inverse of the absorption coefficient), so that most
of the light will be absorbed. The response to a rectangular input pulse of a
low-capacitance photodiode having $w \gg 1/\alpha_s$ is shown in Fig. 6-13b. The rise
and fall times of the photodiode follow the input pulse quite well. If the
photodiode capacitance is larger, the response time becomes limited by the RC
time constant of the load resistor R_L and the photodiode capacitance. The
photodetector response then begins to appear as that shown in Fig. 6-13c.

If the depletion layer is too narrow, any carriers created in the undepleted
material would have to diffuse back into the depletion region before they could
be collected. Devices with very thin depletion regions thus tend to show distinct
slow- and fast-response components, as shown in Fig. 6-13d. The fast component in the rise time is due to carriers generated in the depletion region,
whereas the slow component arises from the diffusion of carriers that are

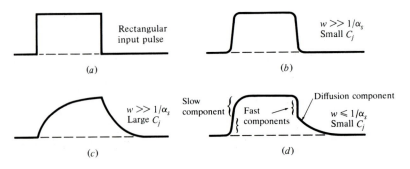

FIGURE 6-13
Photodiode pulse responses under various detector parameters.

created within a distance L_n from the edge of the depletion region. At the end of the optical pulse, the carriers in the depletion region are collected quickly, which results in the fast-detector-response component in the fall time. The diffusion of carriers which are within a distance L_n of the depletion region edge appears as the slowly decaying tail at the end of the pulse. Also, if w is too thin, the junction capacitance will become excessive. The junction capacitance C_j is

$$C_j = \frac{\epsilon_s A}{w} \tag{6-28}$$

where ϵ_s = the permittivity of the semiconductor material = $\epsilon_0 K_s$
  K_s = the semiconductor dielectric constant
  ϵ_0 = 8.8542×10^{-12} F/m is the free-space permittivity
  A = the diffusion layer area

This excessiveness will then give rise to a large RC time constant which limits the detector response time. A reasonable compromise between high-frequency response and high quantum efficiency is found for absorption region thicknesses between $1/\alpha_s$ and $2/\alpha_s$.

If R_T is the combination of the load and amplifier input resistances and C_T is the sum of the photodiode and amplifier capacitances, as shown in Fig. 6-8, the detector behaves approximately like a simple RC low-pass filter with a passband given by

$$B = \frac{1}{2\pi R_T C_T} \tag{6-29}$$

Example 6-7. If the photodiode capacitance is 3 pF, the amplifier capacitance is 4 pF, the load resistor is 1 kΩ, and the amplifier input resistance is 1 MΩ, then $C_T = 7$ pF and $R_T \simeq 1$ kΩ, so that the circuit bandwidth is

$$B = \frac{1}{2\pi R_T C_T} = 23 \text{ MHz} \tag{6-30}$$

If we reduce the photodetector load resistance to 50 Ω, then the circuit bandwidth becomes $B = 455$ MHz.

6.4 AVALANCHE MULTIPLICATION NOISE

As we noted earlier, the avalanche process is statistical in nature, since not every photogenerated carrier pair undergoes the same multiplication.[38-41] The probability distribution of possible gains that any particular electron-hole pair might experience is sufficiently wide so that the mean square gain is greater than the average gain squared. That is, if m denotes the statistically varying gain then

$$\langle m^2 \rangle > \langle m \rangle^2 = M^2 \tag{6-31}$$

where the symbols $\langle\ \rangle$ denote an ensemble average and $\langle m \rangle = M$ is the average carrier gain defined in Eq. (6-7). Since the noise created by the avalanche process depends on the mean square gain $\langle m^2 \rangle$, the noise in an avalanche photodiode can be relatively high. From experimental observations it has been found that, in general, $\langle m^2 \rangle$ can be approximated by

$$\langle m^2 \rangle \simeq M^{2+x} \tag{6-32}$$

where the exponent x varies between 0 and 1.0 depending on the photodiode material and structure.

The ratio of the actual noise generated in an avalanche photodiode to the noise that would exist if all carrier pairs were multiplied by exactly M is called the *excess noise factor F* and defined by

$$F = \frac{\langle m^2 \rangle}{\langle m \rangle^2} = \frac{\langle m^2 \rangle}{M^2} \tag{6-33}$$

This excess noise factor is a measure of the increase in detector noise resulting from the randomness of the multiplication process. It depends on the ratio of the electron and hole ionization rates and on the carrier multiplication.

The derivation of an expression for F is complex, since the electric field in the avalanche region (of width W_M, as shown in Fig. 6-5) is not uniform, and both holes and electrons produce impact ionization. McIntyre[40] has shown that, for injected electrons and holes, the excess noise factors are

$$F_e = \frac{k_2 - k_1^2}{1 - k_2} M_e + 2\left[1 - \frac{k_1(1 - k_1)}{1 - k_2}\right] - \frac{(1 - k_1)^2}{M_e(1 - k_2)} \tag{6-34}$$

$$F_h = \frac{k_2 - k_1^2}{k_1^2(1 - k_2)} M_h - 2\left[\frac{k_2(1 - k_1)}{k_1^2(1 - k_2)} - 1\right] + \frac{(1 - k_1)^2 k_2}{k_1^2(1 - k_2) M_h} \tag{6-35}$$

where the subscripts e and h refer to electrons and holes, respectively. The weighted ionization rate ratios k_1 and k_2 take into account the nonuniformity of the gain and the carrier ionization rates in the avalanche region. They are given by

$$k_1 = \frac{\int_0^{W_M} \beta(x) M(x)\, dx}{\int_0^{W_M} \alpha(x) M(x)\, dx} \tag{6-36}$$

$$k_2 = \frac{\int_0^{W_M} \beta(x) M^2(x)\, dx}{\int_0^{W_M} \alpha(x) M^2(x)\, dx} \tag{6-37}$$

where $\alpha(x)$ and $\beta(x)$ are the electron and hole ionization rates, respectively.

Normally to a first approximation k_1 and k_2 do not change much with variations in gain and can be considered as constant and equal. Thus Eqs. (6-34) and (6-35) can be simplified as[7]

$$F_e = M_e \left[1 - (1 - k_{\text{eff}}) \left(1 - \frac{1}{M_e} \right)^2 \right]$$

$$= k_{\text{eff}} M_e + \left(2 - \frac{1}{M_e} \right) (1 - k_{\text{eff}}) \tag{6-38}$$

for electron injection, and

$$F_h = M_h \left[1 - \left(1 - \frac{1}{k'_{\text{eff}}} \right) \left(1 - \frac{1}{M_h} \right)^2 \right]$$

$$= k'_{\text{eff}} M_h - \left(2 - \frac{1}{M_h} \right) (k'_{\text{eff}} - 1) \tag{6-39}$$

for hole injection, where the effective ionization rate ratios are

$$k_{\text{eff}} = \frac{k_2 - k_1^2}{1 - k_2} \simeq k_2$$

$$k'_{\text{eff}} = \frac{k_{\text{eff}}}{k_1^2} \simeq \frac{k_2}{k_1^2} \tag{6-40}$$

Figure 6-14 shows F_e as a function of the average electron gain M_e for various values of the effective ionization rate ratio k_{eff}. If the ionization rates are equal, the excess noise is at its maximum so that F_e is at its upper limit of M_e. As the ratio β/α decreases from unity, the electron ionization rate starts to be the dominant contributor to impact ionization, and the excess noise factor becomes smaller. If only electrons cause ionization, $\beta = 0$ and F_e reaches its lower limit of 2.

This shows that, to keep the excess noise factor at a minimum, it is desirable to have small values of k_{eff}. Referring back to Fig. 6-6, we thus see the superiority of silicon over other materials for making avalanche photodiodes. The effective ionization rate ratio k_{eff} varies between 0.015 and 0.035 for silicon, between 0.3 and 0.5 for InGaAs, and between 0.6 and 1.0 for germanium.

From the empirical relationship for the mean square gain given by Eq. (6-32), the excess noise factor can be approximated by

$$F = M^x \tag{6-41}$$

The parameter x takes on values of 0.3 for silicon, 0.7 for InGaAs, and 1.0 for germanium avalanche photodiodes.

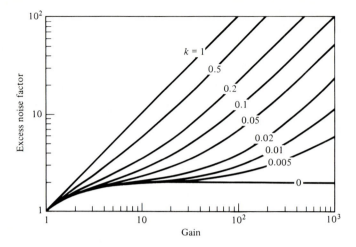

Excess noise factor

$k = 1$
0.5
0.2
0.1
0.05
0.02
0.01
0.005
0

Gain

FIGURE 6-14
Variation of the electron excess noise factor F_e as a function of the electron gain for various values of the effective ionization rate ratio k_{eff}. (Reproduced with permission from Webb, McIntyre, and Conradi.[7])

6.5 TEMPERATURE EFFECT ON AVALANCHE GAIN

The gain mechanism of an avalanche photodiode is very temperature-sensitive because of the temperature dependence of the electron and hole ionization rates.[42-44] This temperature dependence is particularly critical at high bias voltages, where small changes in temperature can cause large variations in gain. An example of this is shown in Fig. 6-15 for a silicon avalanche photodiode. For example, if the operating temperature decreases and the applied bias voltage is kept constant, the ionization rates for electrons and holes will increase and so will the avalanche gain.

To maintain a constant gain as the temperature changes, the electric field in the multiplying region of the *pn* junction must also be changed. This requires that the receiver incorporate a compensation circuit which adjusts the applied bias voltage on the photodetector when the temperature changes.

The dependence of gain on temperature has been studied in detail by Conradi.[44] In that work the gain curves were described by using the explicit temperature dependence of the ionization rates α and β together with a detailed knowledge of the device structure. Although excellent agreement was found between the theoretically computed and the experimentally measured gains, the calculations are rather involved. However, a simple temperature-dependent expression can be obtained from the empirical relationship[45]

$$M = \frac{1}{1 - (V/V_B)^n} \tag{6-42}$$

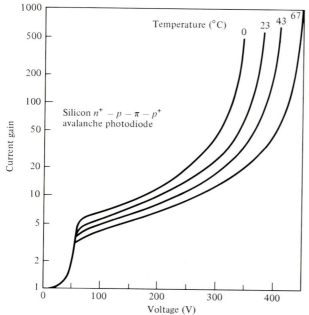

FIGURE 6-15
Example of how the gain mechanism of a silicon avalanche photodiode depends on temperature. The measurements for this device were done at 825 nm. (Reproduced with permission from Melchior, Hartman, Schinke, and Seidel,[15] © 1978, AT & T.)

where V_B is the breakdown voltage at which M goes to infinity, the parameter n varies between 2.5 and 7, depending on the material, and $V = V_a - I_M R_M$, with V_a being the reverse-bias voltage applied to the detector, I_M is the multiplied photocurrent, and R_M accounts for the photodiode series resistance and the detector load resistance. Since the breakdown voltage is known to vary with temperature as[46, 47]

$$V_B(T) = V_B(T_0)[1 + a(T - T_0)] \qquad (6\text{-}43)$$

the temperature dependence of the avalanche gain can be approximated by substituting Eq. (6-43) into Eq. (6-42) together with the expression

$$n(T) = n(T_0)[1 + b(T - T_0)] \qquad (6\text{-}44)$$

The constants a and b are positive for reach-through avalanche photodiodes and can be determined from experimental curves of gain versus temperature.

6.6 PHOTODIODE MATERIALS

The responsivity of a photodetector is principally determined by the construction of the detector and the type of material used. The absorption coefficient α_s of semiconductor materials varies greatly with wavelength, as is shown in Fig. 6-3. For a practical photodiode, the best responsivity and the highest quantum efficiency are obtained in a material having a band-gap energy slightly less than the energy of the photons at the longest wavelength of interest. In addition to ensuring good detector quantum efficiency and response speed, this condition simultaneously keeps the dark current low.

Any of a number of different materials, including Si, Ge, GaAs, InGaAs, and InGaAsP, could be used for photodiode operation in the 800- to 900-nm spectral region. However, silicon is used almost exclusively because, in addition to its highly developed technology, it exhibits the lowest avalanche carrier multiplication noise, thus permitting high receiver sensitivity.

For operation at wavelengths above 1.0 μm the responsivity of Si is too low for it to be used as a photodiode, since photons at these wavelengths do not have enough energy to excite an electron across the 1.17-eV silicon band gap. Various high-sensitivity photodetectors have been developed for the 1.0- to 1.65-μm range. The materials examined have included Ge,[17] InP,[18] InGaAsP,[19-22] GaSb,[23,24] GaAlSb,[25] HgCdTe,[26] and InGaAs.[27-33] Of these the most widely used compound for both *pin* and avalanche photodiodes is InGaAs. This material can absorb light with wavelengths as long as 1650 nm, and has been used in high-speed experimental systems with over 200 km between repeaters. Germanium is an alternative long-wavelength detector material. It has a large absorption coefficient of approximately 10^4 cm^{-1} over the wavelength range of 1.0 to 1.55 μm, which should make it an ideal photodetector for long-wavelength applications. A number of Ge photodetectors with reasonable sensitivity and fast response times have been fabricated, but the material exhibits a number of shortcomings. For example, Ge has a high excess noise factor for avalanche multiplication owing to a carrier ionization rate ratio of only 2. Furthermore, since the band gap of Ge is narrower than that of Si, the bulk dark current is much higher, thus limiting the usable avalanche gain. Despite these limitations, Ge avalanche photodiodes have been successfully used in high-data-transmission experiments, one example being an 800-Mb/s link operating at 1.3 μm over an 11-km distance.

In addition to Ge a variety of III-V semiconductor alloys such as InGaAsP, GaAlSb, InGaAs, GaSb, and GaAsSb have been investigated for long-wavelength applications.[34] There are several reasons for examining these materials. First, since the band gaps of these alloys depend on their molecular composition, the absorption edge can be selected to be just above the longest wavelength of operation by varying the molecular concentrations of the constituent elements of the alloys. This results in detectors with high quantum efficiency, fast response speed, and low dark current. Another reason for studying these alloys is to search for a material having a large difference in the electron and hole ionization rates. Unfortunately, the ionization rate ratios in all III-V materials measured to date are inferior to silicon. This tends to limit operation of these devices to moderate avalanche gains of between 10 and 30.

6.7 SUMMARY

Semiconductor *pin* and avalanche photodiodes are the principal devices used as photon detectors in optical fiber links because of their size compatibility with fibers, their high sensitivities at the desired optical wavelengths, and their fast response times. In addition to these two photodiode structures, considerable attention is being given to the heterojunction phototransistor,[48–51] which is capable of satisfying many of the detector requirements of a fiber communication system.

When light having photon energies greater than or equal to the band-gap energy of the semiconductor material is incident on a photodetector, the photons can give up their energy and excite electrons from the valence band to the conduction band. This process generates free electron-hole pairs, which are known as photocarriers. When a reverse-bias voltage is applied across the photodetector, the resultant electric field in the device causes the carriers to separate. This gives rise to a current flow in an external circuit, which is known as the photocurrent.

The quantum efficiency η is an important photodetector performance parameter. This is defined as the number of electron-hole carrier pairs generated per incident photon of energy $h\nu$. In practice, quantum efficiencies range from 30 to 95 percent. Another important factor is the responsivity. This is related to the quantum efficiency by

$$\mathcal{R} = \frac{\eta q}{h\nu}$$

This parameter is quite useful in that it specifies the photocurrent generated per unit optical power. Representative responsivities for *pin* photodiodes are 0.65 $\mu A/\mu W$ for Si at 800 nm, 0.45 $\mu A/\mu W$ for Ge at 1300 nm, and 0.6 $\mu A/\mu W$ for InGaAs at 1300 nm.

Avalanche photodiodes (APDs) internally multiply the primary signal photocurrent. This increases receiver sensitivity, since the photocurrent is multiplied before encountering the thermal noise associated with the receiver circuitry. The carrier multiplication M is a result of impact ionization. Since the avalanche mechanism is a statistical process, not every carrier pair generated in the photodiode experiences the same multiplication. Thus the measured value of M is expressed as an average quantity. Analogous to the *pin* photodiode, the performance of an APD is characterized by its responsivity

$$\mathcal{R}_{APD} = \frac{\eta q}{h\nu} M = \mathcal{R}_0 M$$

where $\mathcal{R}_0$ is the unity gain responsivity.

The sensitivity of a photodetector and its associated receiver is essentially determined by the photodetector noises resulting from the statistical nature of the photon-to-electron conversion process and the thermal noises in the amplifier circuitry. The main noise currents of photodetectors are:

1. Quantum or shot noise current arising from the statistical nature of the production and collection of photoelectrons
2. Bulk dark current arising from electrons and/or holes which are thermally generated in the *pn* junction of the photodiode

3. Surface dark current (or leakage current) which depends on surface defects, cleanliness, bias voltage, and surface area

In general, for *pin* photodiodes the thermal noise currents of the detector load resistor and the active elements of the amplifier circuitry are the dominant noise sources. For avalanche photodiodes the thermal noise is of lesser importance and the photodetector noises usually dominate.

The usefulness of a given photodiode in a particular application depends on the required response time. To reproduce faithfully the incoming signal, the photodiode must be able to track accurately the variations in this signal. As we showed in Sec. 6.3, this depends on the absorption coefficient of the material at the desired operating wavelength, the photodiode depletion layer width, and the various capacitances and resistances of the photodiode and its associated receiver circuitry.

Since the multiplication process in an avalanche photodiode is statistical in nature, an additional noise factor is introduced which is not present in a *pin* photodiode. A measure of this noise increase is given by the excess noise factor, which we described in Sec. 6.4. This noise factor depends on the electron and hole ionization rates and on the carrier multiplication. The ionization rates also depend on the temperature, so that there is a strong variation in avalanche gain with changes in temperature, as is shown in Sec. 6.5.

We concluded the chapter by looking at various semiconductor materials used for making photodiodes and the wavelengths over which they are suitable. Silicon is the main material used for the 800- to 900-nm region. For wavelengths above 1000 nm the responsivity of Si is too low for it to be used as a photodetector. Photodetectors exhibiting high quantum efficiency and fast response times for the 1.0- to 1.65-μm region have been made from various materials, in particular Ge and InGaAs. Although these materials show excellent performance as *pin* photodiodes, the large ionization rate ratios of these materials limit avalanche gains to values that are much lower than those achievable in silicon.

PROBLEMS

6-1. Consider the absorption coefficient of silicon as a function of wavelength, as shown in Fig. P6-1. Ignoring reflections at the photodiode surface, plot the quantum efficiency for depletion layer widths of 1, 5, 10, 20, and 50 μm over the wavelength range 0.6 to 1.0 μm.

6-2. If an optical power level P_0 is incident on a photodiode, the electron-hole generation rate $G(x)$ in the photodetector is given by

$$G(x) = \Phi_0 \alpha_s e^{-\alpha_s x}$$

Here Φ_0 is the incident photon flux per unit area given by

$$\Phi_0 = \frac{P_0(1 - R_f)}{Ah\nu}$$

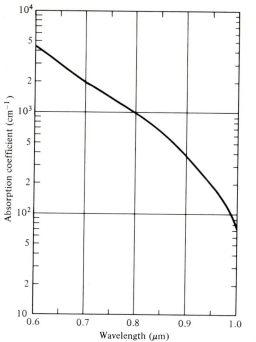

FIGURE P6-1
Absorption coefficient of Si as a function of wavelength.

where A is the detector area. From this show that the primary photocurrent generated in the depletion region of width w is given by Eq. (6-4).

6-3. Using the data from Fig. P6-1, plot the responsivity over the wavelength range 0.6 to 1.0 μm for a silicon *pin* photodiode having a 20-μm-thick depletion layer. Assume $R_f = 0$.

6-4. The low-frequency gain M_0 of an avalanche photodiode depends on the carrier ionization rate and on the width of the multiplication region, both of which depend on the applied reverse-bias voltage V_a. This gain can be described by the empirical relationship[45,52]

$$M_0 = \frac{I_M}{I_p} = \frac{1}{1 - \left(\dfrac{V_a - I_M R_M}{V_B}\right)^n} \tag{P6-4}$$

where V_B is the breakdown voltage at which M_0 goes to infinity ($M_0 \to \infty$), I_M is the total multiplied current, and R_M accounts for the photodiode series resistance and the detector load resistance. The exponential factor n depends on the semiconductor material and its doping profile. Its value varies between about 2.5 and 7.

(*a*) Show that, for applied voltages near the breakdown voltage, at which point $V_B \gg I_M R_M$, Eq. (P6-4) can be approximated by

$$M_0 = \frac{I_M}{I_p} \approx \frac{V_B}{n(V_B - V_a + I_M R_M)} \approx \frac{V_B}{n I_M R_M}$$

(b) The maximum value of M_0 occurs when $V_a = V_B$. Show that at this point

$$M_{0,\,max} = \left(\frac{V_B}{nR_M I_p}\right)^{1/2}$$

6-5. Consider a sinusoidally modulated optical signal $P(t)$ of frequency ω, modulation index m, and average power P_0 given by

$$P(t) = P_0(1 + m \cos \omega t)^2$$

Show that, when this optical signal falls on a photodetector, the mean square signal current $\langle i_s^2 \rangle$ generated consists of a dc (average) component I_p and a signal current i_p given by

$$\langle i_s^2 \rangle = I_p^2 + \langle i_p^2 \rangle = (\mathcal{R}_0 P_0)^2 + \tfrac{1}{2}(m\mathcal{R}_0 P_0)^2$$

where the responsivity $\mathcal{R}_0$ is given by Eq. (6-6).

6-6. Consider an avalanche photodiode receiver having the following parameters: dark current $I_D = 1$ nA, leakage current $I_L = 1$ nA, quantum efficiency $\eta = 0.85$, gain $M = 100$, excess noise factor $F = M^{1/2}$, load resistor $R_L = 10^4 \ \Omega$, and bandwidth $B = 10$ kHz. Suppose a sinusoidally varying 850-nm signal having a modulation index $m = 0.85$ falls on the photodiode, which is at room temperature ($T = 300$ K). To compare the contributions from the various noise terms to the signal-to-noise ratio for this particular set of parameters, plot the following terms in decibels [that is, $10 \log(S/N)$] as a function of the average received optical power P_0. Let P_0 range from -70 to 0 dBm, that is, from 0.1 nW to 1.0 mW:

(a) $\left(\dfrac{S}{N}\right)_Q = \dfrac{\langle i_s^2 \rangle}{\langle i_Q^2 \rangle}$

(b) $\left(\dfrac{S}{N}\right)_{DB} = \dfrac{\langle i_s^2 \rangle}{\langle i_{DB}^2 \rangle}$

(c) $\left(\dfrac{S}{N}\right)_{DS} = \dfrac{\langle i_s^2 \rangle}{\langle i_{DS}^2 \rangle}$

(d) $\left(\dfrac{S}{N}\right)_T = \dfrac{\langle i_s^2 \rangle}{\langle i_T^2 \rangle}$

What happens to these curves if either the load resistor, the gain, the dark current, or the bandwidth is changed?

6-7. Suppose an avalanche photodiode has the following parameters: $I_L = 1$ nA, $I_D = 1$ nA, $\eta = 0.85$, $F = M^{1/2}$, $R_L = 10^3 \ \Omega$, and $B = 1$ kHz. Consider a sinusoidally varying 850-nm signal, which has a modulation index $m = 0.85$ and an average power level $P_0 = -50$ dBm, to fall on the detector at room temperature. Plot the signal-to-noise ratio as a function of M for gains ranging from 20 to 100. At what value of M does the maximum signal-to-noise ratio occur?

6-8. Derive Eq. (6-19).

6-9. (a) Show that under the boundary conditions

$$p_n = p_{n0} \qquad \text{for} \quad x = \infty$$

and

$$p_n = 0 \quad \text{for} \quad x = w$$

the solution to Eq. (6-23) is given by

$$p_n = p_{n0} - (p_{n0} + Be^{-\alpha_s w})e^{(w-x)/L_p} + Be^{-\alpha_s x}$$

where $L_p = (D_p \tau_p)^{1/2}$ is the diffusion length and

$$B = \left(\frac{\Phi_0}{D_p}\right)\frac{\alpha_s L_p^2}{1 - \alpha_s^2 L_p^2}$$

(b) Derive Eq. (6-25) using the relationship

$$J_{\text{diff}} = qD_p\left(\frac{\partial p_n}{\partial x}\right)_{x=w}$$

(c) Verify that J_{tot} is given by Eq. (6-26).

6-10. Consider a modulated photon flux density

$$\Phi = \Phi_0 e^{j\omega t} \text{ photons}/(\text{s} \cdot \text{cm}^2)$$

to fall on a photodetector, where ω is the modulation frequency. The total current through the depletion region generated by this photon flux can be shown to be[37]

$$J_{\text{tot}} = \left(\frac{j\omega\epsilon_s V}{w} + q\Phi_0\frac{1 - e^{-j\omega t_d}}{j\omega t_d}\right)e^{j\omega t}$$

where ϵ_s is the material permittivity, V is the voltage across the depletion layer, and t_d is the transit time of carriers through the depletion region.

(a) From the short-circuit current density ($V = 0$), find the value of ωt_d at which the photocurrent amplitude is reduced by $\sqrt{2}$.

(b) If the depletion region thickness is assumed to be $1/\alpha_s$, what is the 3-dB modulation frequency in terms of α_s and v_d (the drift velocity)?

6-11. Suppose we have a silicon *pin* photodiode which has a depletion layer width $w = 20$ μm, an area $A = 0.05$ mm^2, and a dielectric constant $K_s = 11.7$. If the photodiode is to operate with a 10-kΩ load resistor at 800 nm, where the absorption coefficient $\alpha_s = 10^3$ cm^{-1}, compare the *RC* time constant and the carrier drift time of this device. Is carrier diffusion time of importance in this photodiode?

6-12. Verify that, when the weighted ionization rate ratios k_1 and k_2 are assumed to be approximately equal, Eqs. (6-34) and (6-35) can be simplified to yield Eqs. (6-38) and (6-39).

6-13. Derive the limits of F_e given by Eq. (6-38) when (a) only electrons cause ionization; (b) the ionization rates α and β are equal.

REFERENCES

1. H. Melchior, "Sensitive high speed photodetectors for the demodulation of visible and near infra-red light," *J. Luminescence*, vol. 7, pp. 390–414, 1973.
2. H. Melchior, "Detectors for lightwave communications," *Phys. Today*, vol. 30, pp. 32–39, Nov. 1977.

3. (a) F. Capasso, "Multilayer avalanche photodiodes and solid-state photomultipliers," *Laser Focus/Electro-Optics*, vol. 20, pp. 84–101, July 1984.

   (b) Y. Wang, D. H. Park, and K. F. Brennan, "Theoretical analysis of confined quantum state GaAs/AlGaAs solid-state photomultipliers," *IEEE J. Quantum Electron.*, vol. 26, pp. 285–295, Feb. 1990.

4. E. H. Putley, "The pyro-electric detector," in R. K. Willardson and A. C. Beer, eds., *Semiconductors and Semimetals*, vol. 5, Academic, New York, 1970; vol. 12, Academic, New York, 1977.

5. S. R. Forrest, "Optical detectors: Three contenders," *IEEE Spectrum*, vol. 23, pp. 76–84, May 1986.

6. G. E. Stillman and C. M. Wolfe, "Avalanche photodiodes," in R. K. Willardson and A. C. Beer, eds., *Semiconductors and Semimetals*, vol. 12, Academic, New York, 1977.

7. P. P. Webb, R. J. McIntyre, and J. Conradi, "Properties of avalanche photodiodes," *RCA Review*, vol. 35, pp. 234–278, June 1974.

8. J. E. Bowers and C. A. Burrus, "Ultrawide-band long-wavelength *p-i-n* photodetectors," *J. Lightwave Tech.*, vol. LT-5, pp. 1339–1350, Oct. 1987.

9. J. N. Hollenhorst, "Fabrication and performance of high-speed InGaAs APDs," *Tech Digest OSA/IEEE Opt. Fiber Comm. Conf.*, p. 148, Jan. 1990.

10. (a) T. P. Lee and T. Li, "Photodetectors," in S. E. Miller and A. C. Chynoweth, eds., *Optical Fiber Telecommunications*, Academic, New York, 1979.

    (b) S. R. Forrest, "Optical detectors for lightwave communications," in S. E. Miller and I. P. Kaminow, eds., *Optical Fiber Telecommunications II*, Academic, New York, 1988.

11. A. Rose, *Concepts in Photoconductivity and Allied Problems*, Wiley, New York, 1963.

12. S. M. Sze, *Physics of Semiconductor Devices*, 2nd ed., chap. 13, Wiley, New York, 1981.

13. S. E. Miller, E. A. J. Marcatili, and T. Li, "Research toward optical-fiber transmission systems," *Proc. IEEE*, vol. 61, pp. 1703–1751, Dec. 1973.

14. J.-W. Hong, Y.-W. Chen, W.-L. Laih, Y.-K. Fanf, C.-Y. Chang, and C. Gong, "The hydrogenated amorphous silicon reach-through avalanche photodiode," *IEEE J. Quantum Electron.*, vol. 26, pp. 280–284, Feb. 1990.

15. H. Melchior, A. R. Hartman, D. P. Schinke, and T. E. Seidel, "Planar epitaxial silicon avalanche photodiode," *Bell Sys. Tech. J.*, vol. 57, pp. 1791–1807, July–Aug. 1978.

16. K. Berchtold, O. Krumpholz, and J. Suri, "Avalanche photodiodes with a gain-bandwidth product of more than 200 GHz," *Appl. Phys. Lett.*, vol. 26, pp. 585–587, May 1975.

17. H. Sudo, Y. Nakano, and G. Iwanea, "Reliability of germanium avalanche photodiodes for optical transmission systems," *IEEE Trans. Electron. Devices*, vol. 33, pp. 98–103, Jan. 1986.

18. T. P. Lee, C. A. Burrus, A. G. Dentai, A. A. Ballman, and W. A. Bonner, "High avalanche gain in small-area InP photodiodes," *Appl. Phys. Lett.*, vol. 35, pp. 511–513, Oct. 1979.

19. R. Yeats and S. H. Chiao, "Long-wavelength InGaAsP avalanche photodiodes," *Appl. Phys. Lett.*, vol. 34, pp. 581–583, May 1979; "Leakage current in InGaAsP avalanche photodiodes," *ibid.*, vol. 36, pp. 160–170, Jan. 1980.

20. T. Shirai, S. Yamasaki, F. Osaka, K. Nakajima, and T. Kaneda, "Multiplication noise in planar InP/InGaAsP heterostructure avalanche photodiodes," *Appl. Phys. Lett.*, vol. 40, pp. 532–533, Mar. 1982.

21. B. L. Casper and J. C. Campbell, "Multigigabit-per-second avalanche photodiode lightwave receivers," *J. Lightwave Tech.*, vol. LT-5, pp. 1351–1364, Oct. 1987.

22. F. Osaka, T. Mikawa, and T. Kaneda, "Impact ionization coefficients for electrons and holes in (100)-oriented GaInAsP," *IEEE J. Quantum Electron.*, vol. QE-21, pp. 1326–1338, Sept. 1985.

23. F. Capasso, M. B. Panish, S. Sumski, and P. W. Foy, "Very high quantum efficiency GaSb mesa photodetectors between 1.3 and 1.6 μm," *Appl. Phys. Lett.*, vol. 36, pp. 165–167, Jan. 1980.

24. Y. Nagao, T. Hariu, and Y. Shibata, "GaSb Schottky diodes for infrared detectors," *IEEE Trans. Electron. Devices*, vol. ED-28, pp. 407–411, Apr. 1981.

25. L. R. Tomasetta, H. D. Law, R. C. Eden, I. Deyhimy, and K. Nakano, "High sensitivity optical receivers for 1.0–1.4 μm fiber optic systems," *IEEE J. Quantum Electron.*, vol. QE-14, pp. 800–804, Nov. 1978.

26. (*a*) R. Alabedra, B. Orsal, G. Lecoy, G. Pichard, J. Meslage, and P. Fragnon, "An HgCdTe avalanche photodiode for optical-fiber transmission systems at 1.3 μm," *IEEE Trans. Electron. Devices*, vol. 32, pp. 1302–1306, July 1985.

    (*b*) B. Orsal, R. Alabedra, M. Valenza, G. Lecoy, J. Meslage, and C. Y. Boisrobert, "HgCdTe 1.55-μm avalanche photodiode noise analysis in the vicinity of resonant impact ionization connected with the spin-orbit split-off band," *IEEE Trans. Electron. Devices*, vol. 35, pp. 101–107, Jan. 1988.

27. J. G. Bauer and R. Trommer, "Long-term operation of planar InGaAs/InP *p-i-n* photodiodes," *IEEE Trans. Electron. Devices*, vol. 35, pp. 2349–2354, Dec. 1988.

28. Y. Liu, S. R. Forrest, V. S. Ban, K. M. Woodruff, J. Colosi, G. C. Erikson, M. J. Lange, and G. H. Olsen, "Simple, very low dark current, planar long-wavelength avalanche photodiode," *Tech. Digest OSA/IEEE Opt. Fiber Comm. Conf.*, p. 171, Feb. 1989.

29. M. C. Brain and T. P. Lee, "Optical receivers for lightwave communication systems," *J. Lightwave Tech.*, vol. LT-3, pp. 1281–1300, Dec. 1985.

30. N. Susa, Y. Yamauchi, and H. Kanbe, "Vapor phase epitaxially grown InGaAs photodiodes," *IEEE Trans. Electron. Devices*, vol. ED-27, pp. 92–98, Jan. 1980.

31. C. P. Skrimshire, J. R. Farr, D. F. Sloan, M. J. Robertson, P. A. Putland, J. C. D. Stokoe, and R. R. Sutherland, "Reliability of mesa and planar InGaAs *pin* photodiodes," *IEE Proc.*, vol. 137, pp. 74–78, Feb. 1990.

32. M. C. Brain, "Comparison of available detectors for digital optical fiber systems for the 1.2–1.55 μm wavelength region," *IEEE J. Quantum Electron.*, vol. QE-18, pp. 219–224, Feb. 1982.

33. S. R. Sloan, "Processing and passivation techniques for fabrication of high-speed InP/InGaAs/InP mesa photodetectors," *Hewlett-Packard J.*, vol. 40, no. 5, pp. 69–75, Oct. 1989.

34. H. D. Law, K. Nakano, and L. R. Tomasetta, "III-V alloy heterostructure high speed avalanche photodiodes," *IEEE J. Quantum Electron.*, vol. QE-15, pp. 549–558, July 1979.

35. B. M. Oliver, "Thermal and quantum noise," *Proc. IEEE*, vol. 53, pp. 436–454, May 1965.

36. W. M. Hubbard, "Utilization of optical-frequency carriers for low and moderate bandwidth channels," *Bell Sys. Tech. J.*, vol. 52, pp. 731–765, May–June 1973.

37. W. W. Gaertnar, "Depletion-layer photoeffects in semiconductors," *Phys. Rev.*, vol. 116, pp. 84–87, Oct. 1959.

38. R. S. Fyath and J. J. O'Reilly, "Performance degradation of APD-optical receivers due to dark current generated within the multiplication region," *J. Lightwave Tech.*, vol. 7, pp. 62–67, Jan. 1989.

39. S. D. Personick, "Statistics of a general class of avalanche detectors with applications to optical communications," *Bell Sys. Tech. J.*, vol. 50, pp. 3075–3096, Dec. 1971.

40. R. J. McIntyre, "The distribution of gains in uniformly multiplying avalanche photodiodes: Theory," *IEEE Trans. Electron. Devices*, vol. ED-19, pp. 703–713, June 1972.

41. J. Conradi, "The distribution of gains in uniformly multiplying avalanche photodiodes: Experimental," *IEEE Trans. Electron. Devices*, vol. ED-19, pp. 713–718, June 1972.

42. T. Mikawa, S. Kagawa, T. Kaneda, Y. Toyama, and O. Mikami, "Crystal orientation dependence of ionization rates in germanium," *Appl. Phys. Lett.*, vol. 37, pp. 387–389, Aug. 1980.

43. C. R. Crowell and S. M. Sze, "Temperature dependence of avalanche multiplication in semiconductors," *Appl. Phys. Lett.*, vol. 9, pp. 242–244, Sept. 1966.

44. J. Conradi, "Temperature effects in silicon avalanche photodiodes," *Solid State Electron.*, vol. 17, pp. 99–106, Jan. 1974.

45. S. L. Miller, "Avalanche breakdown in germanium," *Phys. Rev.*, vol. 99, pp. 1234–1241, Aug. 1955.

46. M. S. Tyagi, "Zener and avalanche breakdown in silicon alloyed *p-n* junction," *Solid State Electron.*, vol. 11, pp. 99–115, Feb. 1968.

47. N. Susa, H. Nakagome, H. Ando, and H. Kanbe, "Characteristics in InGaAs/InP avalanche photodiodes with separated absorption and multiplication regions," *IEEE J. Quantum Electron.*, vol. QE-17, pp. 243–250, Feb. 1981.

48. J. C. Campbell, A. G. Dentai, C. A. Burrus, Jr., and J. F. Ferguson, "InP/InGaAs heterojunction phototransistors," *IEEE J. Quantum Electron.*, vol. QE-17, pp. 264–269, Feb. 1981.
49. K. Tabatabaie-Alavi and C. G. Fonstad, "Performance comparison of heterojunction phototransistors, *pin* FETs, and APD-FETs for optical fiber communication systems," *IEEE J. Quantum Electron.*, vol. QE-17, pp. 2259–2261, Dec. 1981.
50. R. A. Milano, D. P. Dapkus, and G. E. Stillman, "An analysis of the performance of heterojunction phototransistors for fiber optic communications," *IEEE Trans. Electron. Devices*, vol. ED-29, pp. 266–274, Feb. 1982.
51. M. C. Brain and D. R. Smith, "Phototransistors in digital optical communication systems," *IEEE J. Quantum Electron.*, vol. QE-19, pp. 1139–1148, June 1983.
52. H. Melcior and W. T. Lynch, "Signal and noise response of high speed germanium avalanche photodiodes," *IEEE Trans. Electron. Devices*, vol. ED-13, pp. 829–838, Dec. 1966.

CHAPTER
7

OPTICAL
RECEIVER
OPERATION

Having discussed the characteristics and operation of photodetectors in the previous chapter, we now turn our attention to the optical receiver. An optical receiver consists of a photodetector, an amplifier, and signal-processing circuitry. It has the task of first converting the optical energy emerging from the end of a fiber into an electric signal, and then amplifying this signal to a large enough level so that it can be processed by the electronics following the receiver amplifier.

In these processes various noises and distortions will unavoidably be introduced which can lead to errors in the interpretation of the received signal. As we saw in the previous chapter, the current generated by the photodetector is generally very weak and is adversely affected by the random noises associated with the photodetection process. When this electric signal output from the photodiode is amplified, additional noises arising from the amplifier electronics will further corrupt the signal. Noise considerations are thus important in the design of optical receivers, since the noise sources operating in the receiver generally set the lowest limit for the signals that can be processed.

In designing a receiver it is desirable to predict its performance based on mathematical models of the various receiver stages. These models must take into account the noises and distortions added to the signal by the components in each stage, and they must show the designer which components to choose so that the desired performance criteria of the receiver are met.

The most meaningful criterion for measuring the performance of a digital communication system is the average error probability. In an analog system the fidelity criterion is usually specified in terms of a peak signal-to-rms-noise ratio.

The calculation of the error probability for a digital optical communication receiver differs from that of conventional electric systems. This is because of the discrete quantum nature of the optical signal and also because of the probabilistic character of the gain process when an avalanche photodiode is used. Various authors[1-8] have used different numerical methods to derive approximate predictions for receiver performance. In carrying out these predictions a tradeoff results between simplicity of the analysis and accuracy of the approximation. General reviews and concepts of optical receiver designs are given in Refs. 9–18.

In this chapter we first present an overview of the fundamental operational characteristics of the various stages of an optical receiver. This consists of tracing the path of a digital signal through the receiver and showing what happens at each step along the way. This is followed in Sec. 7.2 by mathematical models for predicting the performance of a digital receiver under various noise and distortion conditions. Practical receiver design examples and their performance predictions based on these models are discussed next. The chapter concludes with an analysis of analog receivers.

7.1 FUNDAMENTAL RECEIVER OPERATION

The design of an optical receiver is much more complicated than that of an optical transmitter because the receiver must first detect weak, distorted signals and then make decisions on what type of data was sent based on amplified version of this distorted signal. To get an appreciation of the function of the optical receiver, we first examine what happens to a signal as it is sent through the optical data link shown in Fig. 7-1. Since most fiber optic systems use a two-level binary digital signal, we shall analyze receiver performance by using this signal form first. Analog receivers are discussed in Sec. 7.4.

7.1.1 Digital Signal Transmission

A typical digital fiber transmission link is shown in Fig. 7-1. The transmitted signal is a two-level binary data stream consisting of either a 0 or a 1 in a time slot of duration T_b. This time slot is referred to as a *bit period*. Electrically there are many ways of sending a given digital message.[19-21] One of the simplest (but not necessarily the most efficient) techniques for sending binary data is amplitude-shift keying, wherein a voltage level is switched between two values, which are usually *on* and *off*. The resultant signal wave thus consists of a voltage pulse of amplitude V relative to the zero voltage level when a binary 1 occurs and a zero-voltage-level space when a binary 0 occurs. Depending on the coding scheme to be used, a 1 may or may not fill the time slot T_b. For simplicity here we assume that when a 1 is sent, a voltage pulse of duration T_b occurs, whereas for a 0 the voltage remains at its zero level. A discussion of more efficient transmission code is given in Chap. 8.

The function of the optical transmitter is to convert the electric signal to an optical signal. As shown in Chap. 4, an electric current $i(t)$ can be used to modulate directly an optical source (either an LED or a laser diode) to produce

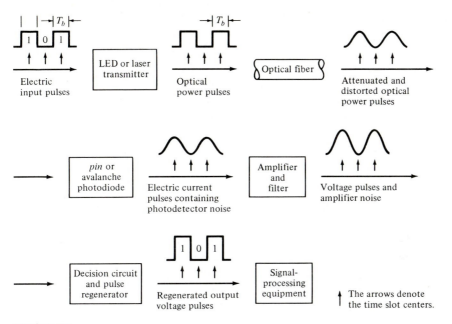

FIGURE 7-1
Signal path through an optical data link. (Adapted with permission from Personick et al.,[4] © 1977, IEEE.)

an optical output power $P(t)$. Thus, in the optical signal emerging from the transmitter, a 1 is represented by a pulse of optical power (light) of duration T_b, whereas a 0 is the absence of any light.

The optical signal that gets coupled from the light source to the fiber becomes attenuated and distorted as it propagates along the fiber waveguide. Upon reaching the receiver either a *pin* or an avalanche photodiode converts the optical signal back to an electrical format. After the electric signal produced by the photodetector is amplified and filtered, a decision circuit compares the signal in each time slot with a certain reference voltage known as the *threshold level*. If the received signal level is greater than the threshold level, a 1 is said to have been received. If the voltage is below the threshold level, a 0 is assumed to have been received.

7.1.2 Error Sources

Errors in the detection mechanism can arise from various noises and disturbances associated with the signal detection system, as shown in Fig. 7-2. The term *noise* is used customarily to describe unwanted components of an electric signal that tend to disturb the transmission and processing of the signal in a physical system, and over which we have incomplete control. The noise sources can be either external or the system (for example, atmospheric noise, equipment-generated noise) or internal to the system. Here we shall be concerned

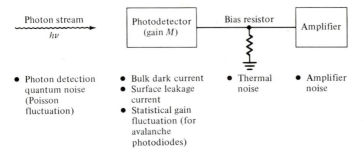

FIGURE 7-2
Noise sources and disturbances in the optical pulse detection mechanism.

mainly with internal noise, which is present in every communication system and represents a basic limitation on the transmission or detection of signals. This noise is caused by the spontaneous fluctuations of current or voltage in electric circuits. The two most common examples of these spontaneous fluctuations are shot noise and thermal noise. Shot noise arises in electronic devices because of the discrete nature of current flow in the device. Thermal noise arises from the random motion of electrons in a conductor. Detailed treatments of electric noise may be found in Ref. 21.

As discussed in Chap. 6, the random arrival rate of signal photons produces a quantum (or shot) noise at the photodetector. Since this noise depends on the signal level, it is of particular importance for *pin* receivers having large optical input levels and for avalanche photodiode receivers. When using an avalanche photodiode an additional shot noise arises from the statistical nature of the multiplication process. This noise level increases with increasing avalanche gain *M*. Additional photodetector noises come from the dark current and leakage current. These are independent of the photodiode illumination and can generally be made very small in relation to other noise currents by a judicious choice of components.

Thermal noises arising from the detector load resistor and from the amplifier electronics tend to dominate in applications with low signal-to-noise ratio when a *pin* photodiode is used. When an avalanche photodiode is used in low-optical-signal-level applications, the optimum avalanche gain is determined by a design tradeoff between the thermal noise and the gain-dependent quantum noise.

Since the thermal noises are of a gaussian nature, they can be readily treated by standard techniques. This is shown in Sec. 7.2. The analysis of the noises and the resulting error probabilities associated with the primary photocurrent generation and the avalanche multiplication are complicated, since neither of these processes is gaussian. The primary photocurrent generated by the photodiode is a time-varying Poisson process resulting from the random arrival of photons at the detector. If the detector is illuminated by an optical signal $P(t)$, then the average number of electron-hole pairs $\overline{N}$ generated in a time τ is

$$\overline{N} = \frac{\eta}{h\nu} \int_0^\tau P(t)\, dt = \frac{\eta E}{h\nu} \tag{7-1}$$

where η is the detector quantum efficiency, $h\nu$ is the photon energy, and E is the energy received in a time interval τ. The actual number of electron-hole pairs n that are generated fluctuates from the average according to the Poisson distribution

$$P_r(n) = \overline{N}^n \frac{e^{-\overline{N}}}{n!} \tag{7-2}$$

where $P_r(n)$ is the probability that n electrons are emitted in an interval τ. The fact that it is not possible to predict exactly how many electron-hole pairs are generated by a known optical power incident on the detector is the origin of the type of shot noise called *quantum noise*. The random nature of the avalanche multiplication process gives rise to another type of shot noise. Recall from Chap. 6 that, for a detector with a mean avalanche gain M and an ionization rate ratio k, the excess noise factor $F(M)$ for electron injection is

$$F(M) = kM + \left(2 - \frac{1}{M}\right)(1 - k)$$

This equation is often approximated by the empirical expression

$$F(M) \simeq M^x \tag{7-3}$$

where the factor x ranges between 0 and 1.0 depending on the photodiode material.

A further error source is attributed to *intersymbol interference* (ISI), which results from pulse spreading in the optical fiber. When a pulse is transmitted in a given time slot, most of the pulse energy will arrive in the corresponding time slot at the receiver, as shown in Fig. 7-3. However, because of pulse spreading induced by the fiber, some of the transmitted energy will progressively spread into neighboring time slots as the pulse propagates along the fiber. The presence of this energy in adjacent time slots results in an interfering signal,

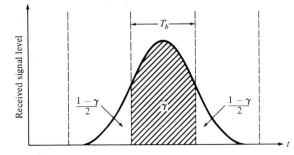

FIGURE 7-3
Pulse spreading in an optical signal which leads to intersymbol interference.

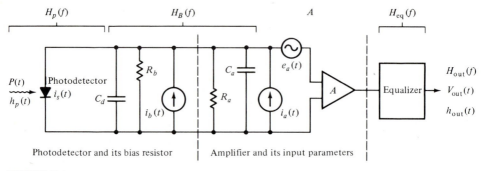

Photodetector and its bias resistor | Amplifier and its input parameters

FIGURE 7-4
Schematic diagram of a typical optical receiver.

hence the term *intersymbol interference*. In Fig. 7-3 the fraction of energy remaining in the appropriate time slot is designated by γ, so that $1 - \gamma$ is the fraction of energy that has spread into adjacent time slots.

7.1.3 Receiver Configuration

A schematic diagram of a typical optical receiver is shown in Fig. 7-4. The three basic stages of the receiver are a photodetector, an amplifier, and an equalizer. The photodetector can be either an avalanche photodiode with a mean gain M or a *pin* photodiode for which $M = 1$. The photodiode has a quantum efficiency η and a capacitance C_d. The detector bias resistor has a resistance R_b which generates a thermal noise current $i_b(t)$.

The amplifier has an input impedance represented by the parallel combination of a resistance R_a and a shunt capacitance C_a. Voltages appearing across this impedance cause current to flow in the amplifier output. This amplifying function is represented by the voltage-controlled current source which is characterized by a *transconductance* g_m (given in amperes/volt, or *siemens*). There are two amplifier noise sources. The input noise current source $i_a(t)$ arises from the thermal noise of the amplifier input resistance R_a, whereas the noise voltage source $e_a(t)$ represents the thermal noise of the amplifier channel. These noise sources are assumed gaussian in statistics, flat in spectrum (which characterizes *white* noise), and uncorrelated (statistically independent). They are thus completely described by their noise spectral densities[19] S_I and S_E (see App. E).

The equalizer that follows the amplifier is normally a linear frequency-shaping filter that is used to mitigate the effects of signal distortion and intersymbol interference. Ideally[20] it accepts the combined frequency response of the transmitter, the transmission medium, and the receiver, and transforms it into a signal response that is suitable for the following signal-processing electronics. In some cases, the equalizer may be used only to correct for the electric frequency response of the photodetector and the amplifier.

To account for the fact that the rectangular digital pulses that were sent out by the transmitter arrive rounded and distorted at the receiver, the binary digital pulse train incident on the photodetector can be described by

$$P(t) = \sum_{n=-\infty}^{\infty} b_n h_p(t - nT_b) \tag{7-4}$$

Here $P(t)$ is the received optical power, T_b is the bit period, b_n is an amplitude parameter representing the nth message digit, and $h_p(t)$ is the received pulse shape, which is positive for all t. For binary data the parameter b_n can take on the two values b_{on} and b_{off} corresponding to a binary 1 and 0, respectively. If we let the nonnegative photodiode input pulse $h_p(t)$ be normalized to have unit area

$$\int_{-\infty}^{\infty} h_p(t)\, dt = 1 \tag{7-5}$$

then b_n represents the energy in the nth pulse.

The mean output current from the photodiode at time t resulting from the pulse train given in Eq. (7-4) is (neglecting dc components arising from dark noise currents)

$$\langle i(t) \rangle = \frac{\eta q}{h\nu} MP(t) = \mathcal{R}_0 M \sum_{n=-\infty}^{\infty} b_n h_p(t - nT_b) \tag{7-6}$$

where $\mathcal{R}_0 = \eta q/h\nu$ is the photodiode responsivity as given in Eq. (6-6). This current is then amplified and filtered to produce a mean voltage at the output of the equalizer given by the convolution of the current with the amplifier impulse response (see App. E):

$$\langle v_{out}(t) \rangle = A\mathcal{R}_0 MP(t) * h_B(t) * h_{eq}(t)$$
$$= \mathcal{R}_0 GP(t) * h_B(t) * h_{eq}(t) \tag{7-7}$$

Here A is the amplifier gain, we define $G = AM$ for brevity, $h_B(t)$ is the impulse response of the bias circuit, $h_{eq}(t)$ is the equalizer impulse response, and $*$ denotes convolution.

From Fig. 7-4 $h_B(t)$ is given by the inverse Fourier transform of the bias circuit transfer function $H_B(f)$:

$$h_B(t) = F^{-1}[H_B(f)] = \int_{-\infty}^{\infty} H_B(f) e^{j2\pi ft}\, df \tag{7-8}$$

where F denotes the Fourier transform operation. The bias current transfer function $H_B(f)$ is simply the impedance of the parallel combination of R_b, R_a, C_d, and C_a:

$$H_B(f) = \frac{1}{1/R + j2\pi fC} \tag{7-9}$$

where

$$\frac{1}{R} = \frac{1}{R_a} + \frac{1}{R_b} \tag{7-10}$$

and

$$C = C_a + C_d \tag{7-11}$$

Analogous to Eq. (7-4), the mean voltage output from the equalizer can be written in the form

$$\langle v_{\text{out}}(t) \rangle = \sum_{n=-\infty}^{\infty} b_n h_{\text{out}}(t - nT_b) \tag{7-12}$$

where

$$h_{\text{out}}(t) = \mathcal{R}_0 G h_p(t) * h_B(t) * h_{\text{eq}}(t) \tag{7-13}$$

is the shape of an isolated amplified and filtered pulse. The Fourier transform of Eq. (7-13) can be written as[19] (see App. E)

$$H_{\text{out}}(f) = \int_{-\infty}^{\infty} h_{\text{out}}(t) e^{-j2\pi ft} \, dt = \mathcal{R}_0 G H_p(f) H_B(f) H_{\text{eq}}(f) \tag{7-14}$$

Here $H_p(f)$ is the Fourier transform of the received pulse shape $h_p(t)$, and $H_{\text{eq}}(f)$ is the transfer function of the equalizer.

7.2 DIGITAL RECEIVER PERFORMANCE CALCULATION

In a digital receiver the amplified and filtered signal emerging from the equalizer is compared with a threshold level once per time slot to determine whether or not a pulse is present at the photodetector in that time slot. Ideally the output signal $v_{\text{out}}(t)$ would always exceed the threshold voltage when a 1 is present and would be less than the threshold when no pulse (a 0) was sent. In actual systems, deviations from the average value of $v_{\text{out}}(t)$ are caused by various noises, interference from adjacent pulses, and conditions wherein the light source is not completely extinguished during a zero pulse.

7.2.1 Probability of Error

In practice there are several standard ways of measuring the rate of error occurrences in a digital data stream.[22] One common approach is to divide the number N_e of errors occurring over a certain time interval t by the number N_t of pulses (ones and zeros) transmitted during this interval. This is called either the *error rate* or the *bit error rate*, which is commonly abbreviated BER. Thus we have

$$\text{BER} = \frac{N_e}{N_t} = \frac{N_e}{Bt} \tag{7-15}$$

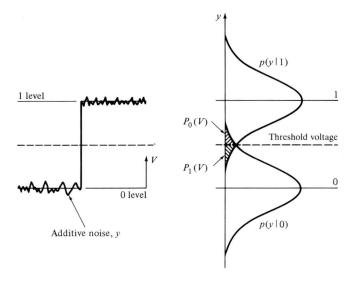

FIGURE 7-5
Probability distributions for 0 and 1 signal levels.

where $b = 1/T_b$ is the bit rate (that is, the pulse transmission rate). The error rate is expressed by a number such as 10^{-6}, for example, which states that on the average one error occurs for every million pulses sent. Typical error rates for optical fiber telecommunication systems range from 10^{-6} to 10^{-10}. This error rate depends on the signal-to-noise ratio at the receiver (the ratio of signal power to noise power). The system error rate requirements and the receiver noise levels thus set a lower limit on the optical signal power level that is required at the photodetector.

To compute the bit error rate at the receiver, we have to know the probability distribution[23] of the signal at the equalizer output. Knowing the signal probability distribution at this point is important because it is here that the decision is made as to whether a 0 or a 1 was sent. The shapes of two signal probability distributions are shown in Fig. 7-5. These are

$$P_1(v) = \int_{-\infty}^{v} p(y|1)\, dy \qquad (7\text{-}16)$$

which is the probability that the equalizer output voltage is less than v when a 1 pulse was sent, and

$$P_0(v) = \int_{v}^{\infty} p(y|0)\, dy \qquad (7\text{-}17)$$

which is the probability that the output voltage exceeds v when a 0 was transmitted. The functions $p(y|1)$ and $p(y|0)$ are the conditional probability distribution functions,[9,23] that is, $p(y|x)$ is the probability that the output voltage is y, given that an x was transmitted.

If the threshold voltage is v_{th} then the error probability P_e is defined as

$$P_e = aP_1(v_{th}) + bP_0(v_{th}) \tag{7-18}$$

The weighting factors a and b are determined by the *a priori* distribution of the data. That is, a and b are the probabilities that either a 1 or 0 occurs, respectively. For unbiased data with equal probability of 1 and 0 occurrences, $a = b = 0.5$. The problem to be solved now is to select the decision threshold at that point where P_e is minimum.

To calculate the error probability we require a knowledge of the mean square noise voltage $\langle v_N^2 \rangle$ which is superimposed on the signal voltage at the decision time. The statistics of the output voltage at the sampling time are very complicated, so that an exact calculation is rather tedious to perform. A number of different approximations[1-18] have therefore been used to calculate the performance of a binary optical fiber receiver. In applying these approximations, we have to make a tradeoff between computational simplicity and accuracy of the results. The simplest method is based on a gaussian approximation. In this method it is assumed that, when the sequence of optical input pulses is known, the equalizer output voltage $v_{out}(t)$ is a gaussian random variable. Thus, to calculate the error probability, we only need to know the mean and standard deviation of $v_{out}(t)$. Other approximations which have been investigated are more involved[4-7,16-18] and will not be discussed here.

Thus let us assume that the noise has a gaussian probability density function with zero mean. If we sample the noise voltage $n(t)$ at any arbitrary time t_1, the probability that the measured sample $n(t_1)$ falls in the range n to $n + dn$ is given by

$$f(n)\,dn = \frac{1}{\sqrt{2\pi\sigma^2}} e^{-n^2/2\sigma^2}\,dn \tag{7-19}$$

where σ^2 is the noise variance and $f(n)$ is the probability density function.

We can now use this function to determine the probability of error for a data stream in which the pulses are all of amplitude V. Let us first consider the case of a 0 being sent, so that no pulse is present at the time of decoding. Assuming that we have unbiased data, the probability of error in this case is the probability that the noise will exceed the threshold voltage $v_{th} = V/2$ and be mistaken for a 1 pulse. The probability of error $P_0(v)$ is then simply the chance that the equalizer output voltage $v(t)$ will fall somewhere between $V/2$ and ∞. Using Eq. (7-17) we have

$$P_0(v_{th}) = \int_{V/2}^{\infty} p(y|0)\,dy = \int_{V/2}^{\infty} f_0(y)\,dy$$

$$= \int_{V/2}^{\infty} \frac{1}{\sqrt{2\pi\sigma^2}} e^{-v^2/2\sigma^2}\,dv \tag{7-20}$$

where the subscript 0 denotes the presence of a 0 bit.

Similarly we can find the probability of error that a transmitted 1 is misinterpreted as a 0 by the decoder electronics following the equalizer. When a 1 is transmitted, the decoder sees a pulse of amplitude V volts plus superimposed noise. In this case the equalizer output voltage $v(t)$ will fluctuate around V, so that the probability density function of Eq. (7-19) becomes

$$f_1(v) = \frac{1}{\sqrt{2\pi\sigma^2}} e^{-(v-V)^2/2\sigma^2}. \tag{7-21}$$

where the subscript 1 denotes the presence of a 1 bit. The probability of error that a 1 is decoded as a 0 is the likelihood that the sampled signal-plus-noise pulse falls below $V/2$. This is simply given by

$$P_1(v_{\text{th}}) = \int_{-\infty}^{V/2} p(y|1)\, dy = \int_{-\infty}^{V/2} f_1(v)\, dv$$

$$= \frac{1}{\sqrt{2\pi\sigma^2}} \int_{-\infty}^{V/2} e^{-(v-V)^2/2\sigma^2}\, dv \tag{7-22}$$

Since we assumed unbiased data, $a = b = 0.5$ in Eq. (7-18), so that we have, substituting Eqs. (7-20) and (7-22) into Eq. (7-18), the probability of error P_e in the decoding of any digit:

$$P_e = \frac{1}{2}\left[1 - \text{erf}\left(\frac{V}{2\sqrt{2}\,\sigma}\right)\right] \tag{7-23}$$

where

$$\text{erf } x = \frac{2}{\sqrt{\pi}} \int_0^x e^{-y^2}\, dy \tag{7-24}$$

is the error function, which is tabulated in various mathematical handbooks.[24] An important point in Eq. (7-23) is that P_e depends only on the parameter V/σ, which is the ratio of the signal amplitude V to the standard deviation σ of the noise. Since σ is usually called the *rms noise*, the ratio V/σ is then the *peak signal-to-rms-noise ratio*. The relationship given in Eq. (7-23) is thus a very fundamental one in communication theory, since it relates the bit error probability (or bit error rate, denoted by BER) to the signal-to-noise ratio, which is often designated by S/N.

A plot of BER versus V/σ is given in Fig. 7-6. To change V/σ to decibels, recall that (see App. D)

$$\left(\frac{S}{N}\right)_{\text{dB}} = 20 \log \frac{V}{\sigma} \tag{7-25}$$

Example 7-1. As an example, Fig. 7-6 shows that for a signal-to-noise ratio of 8.5 (18.6 dB) we have $P_e = 10^{-5}$. In this case on the average 1 out of 10^5 transmitted bits will be interpreted wrong. If this signal is being sent at a standard T1 telephone-line rate (1.544 Mb/s), this BER results in a misinterpreted bit every 0.065 s, which is highly unsatisfactory. However, by increasing the signal strength

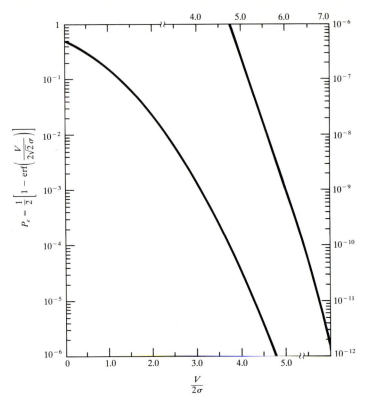

FIGURE 7-6
Bit error rate as a function of signal-to-noise ratio.

so that $V/\sigma = 12.0$ (21.6 dB), the BER decreases to $P_e = 10^{-9}$. For the T1 case this means a bit is misinterpreted every 650 s, or 11 min, on the average, which in general is tolerable.

The above example demonstrates the exponential behavior of the probability of error as a function of the signal-to-noise ratio. Here we saw that by increasing V/σ by $\sqrt{2}$, that is, doubling S/N (a 3-dB power increase), the BER decreased by 10^4. Thus, there exists a narrow range of signal-to-noise ratios above which the error rate is tolerable and below which a highly unacceptable number of errors occur. The signal-to-noise ratio at which this transition occurs is called the *threshold level*. In general, a performance safety margin of 3 to 6 dB is included in the transmission link design to ensure that this threshold level is not exceeded when system parameters such as transmitter output, line attenuation, or noise floor vary with time.

For simplicity we shall first assume that the optical power is completely extinguished in those time slots where a 0 occurs. In practice there usually is some optical signal in the 0 time slot of a baseband binary signal, since the optical source is often biased slightly on at all times. As we saw in Chap. 4, this is done to increase the response speed of the light source. Biasing the light source slightly on during a 0 time slot results in a nonzero *extinction ratio* ϵ. This is defined as the ratio of the optical power in a 0 pulse to the power in a 1 pulse. Its effect on receiver performance is discussed in Sec. 7.2.7.

7.2.2 The Quantum Limit

Suppose we have an ideal photodetector which has unity quantum efficiency and which produces no dark current, that is, no electron-hole pairs are generated in the absence of an optical pulse. Given this condition, it is possible to find the minimum received optical power required for a specific bit-error-rate performance in a digital system. This minimum received power level is known as the *quantum limit*, since all system parameters are assumed ideal and the performance is only limited by the photodetection statistics.

Assume that an optical pulse of energy E falls on the photodetector in a time interval τ. This can only be interpreted by the receiver as a 0 pulse if no electron-hole pairs are generated with the pulse present. From Eq. (7-2) the probability that $n = 0$ electrons are emitted in a time interval τ is

$$P_r(0) = e^{-\bar{N}} \tag{7-26}$$

where the average number of electron-hole pairs, $\bar{N}$, is given by Eq. (7-1). Thus for a given error probability $P_r(0)$, we can find the minimum energy E required at a specific wavelength λ.

> **Example 7-2.** A digital fiber optic link operating at 850 nm requires a maximum BER of 10^{-9}.
>
> (*a*) Let us first find the quantum limit in terms of the quantum efficiency of the detector and the energy of the incident photon. From Eq. (7-26) the probability of error is
>
> $$P_r(0) = e^{-\bar{N}} = 10^{-9}$$
>
> Solving for $\bar{N}$, we have $\bar{N} = 9 \ln 10 = 20.7 \simeq 21$. Hence an average of 21 photons per pulse is required for this BER. Using Eq. (7-1) and solving for E, we get
>
> $$E = 20.7 \frac{h\nu}{\eta}$$
>
> (*b*) Now let us find the minimum incident optical power P_0 that must fall on the photodetector to achieve a 10^{-9} BER at a data rate of 10 Mb/s for a simple binary-level signaling scheme. If the detector quantum efficiency $\eta = 1$, then
>
> $$E = P_0 \tau = 20.7 \, h\nu = 20.7 \frac{hc}{\lambda}$$

where $1/\tau$ is one-half the data rate B, that is, $1/\tau = B/2$. (*Note:* This assumes an equal number of 0 and 1 pulses.) Solving for P_0,

$$P_0 = 20.7 \, \frac{hcB}{2\lambda}$$

$$= \frac{20.7(6.626 \times 10^{-34} \text{J} \cdot \text{s})(3.0 \times 10^8 \text{ m/s})(10 \times 10^6 \text{ bits/s})}{2(0.85 \times 10^{-6} \text{ m})}$$

$$= 24.2 \text{ pW}$$

or, when the reference power level is one milliwatt,

$$P_0 = -76.2 \text{ dBm}$$

7.2.3 Receiver Noises

We now turn our attention to calculating the noise voltages or, equivalently, the noise currents. If $v_N(t)$ is the noise voltage causing $v_{\text{out}}(t)$ to deviate from its average value, then the actual equalizer output voltage is of the form

$$v_{\text{out}}(t) = \langle v_{\text{out}}(t) \rangle + v_N(t) \tag{7-27a}$$

The noise voltage at the equalizer output for the receiver shown in Fig. 7-4 can be represented by

$$v_N^2(t) = v_s^2(t) + v_R^2(t) + v_I^2(t) + v_E^2(t) \tag{7-27b}$$

where

$v_s(t)$ is the quantum (or shot) noise resulting from the random multiplied Poisson nature of the photocurrent $i_s(t)$ produced by the photodetector,

$v_R(t)$ is the thermal (or Johnson) noise associated with the bias resistor R_b,

$v_I(t)$ results from the amplifier input noise current source $i_a(t)$,

$v_E(t)$ results from the amplifier input voltage noise source $e_a(t)$.

The amplifier noise sources will be assumed independent of each other and gaussian in their statistics. The amplifier input current and voltage noise sources are also referred to as a shunt noise current and a series noise voltage, respectively.

Here we are interested in the mean square noise voltage $\langle v_N^2 \rangle$, which is given by

$$\langle v_N^2 \rangle = \left\langle \left[v_{\text{out}}(t) - \langle v_{\text{out}}(t) \rangle \right]^2 \right\rangle$$

$$= \left\langle v_{\text{out}}^2(t) \right\rangle - \left\langle v_{\text{out}}(t) \right\rangle^2$$

$$= \left\langle v_s^2(t) \right\rangle + \left\langle v_R^2(t) \right\rangle + \left\langle v_I^2(t) \right\rangle + \left\langle v_E^2(t) \right\rangle \tag{7-28}$$

We shall first evaluate the last three thermal noise terms of Eq. (7-28) at the output of the equalizer. The thermal noise of the load resistor R_b is[19]

$$\langle v_R^2(t) \rangle = \frac{4k_BT}{R_b} B_{bae} R^2 A^2 \tag{7-29}$$

Here k_BT is Boltzmann's constant times the absolute temperature, R is given by Eq. (7-10), A is the amplifier gain, and B_{bae} is the noise equivalent bandwidth of the bias circuit, amplifier, and equalizer defined for positive frequencies only:[19]

$$B_{bae} = \frac{1}{|H_B(0)H_{eq}(0)|^2} \int_0^\infty |H_B(f)H_{eq}(f)|^2 \, df$$

$$= \frac{1}{|H_{out}(0)/H_p(0)|^2} \int_0^\infty \left| \frac{H_{out}(f)}{H_p(f)} \right|^2 df \tag{7-30}$$

where we have used Eq. (7-14) for the last equality.

Since the thermal noise contributions from the amplifier input noise current source $i_a(t)$ and from the amplifier input noise voltage source $e_a(t)$ are assumed to be gaussian and independent, they are completely characterized by their noise spectral densities.[19] Thus,

$$\langle v_I^2(t) \rangle = S_I B_{bae} R^2 A^2 \tag{7-31}$$

and

$$\langle v_E^2(t) \rangle = S_E B_e A^2 \tag{7-32}$$

where S_I is the spectral density of the amplifier input noise current source (measured in amperes squared per hertz), S_E is the spectral density of the amplifier noise voltage source (measured in volts squared per hertz), and

$$B_e = \frac{1}{|H_{eq}(0)|^2} \int_0^\infty |H_{eq}(f)|^2 \, df$$

$$= \frac{R^2}{|H_{out}(0)/H_p(0)|^2} \int_0^\infty \left| \frac{H_{out}(f)}{H_p(f)} \left(\frac{1}{R} + j2\pi fC \right) \right|^2 df \tag{7-33}$$

is the noise equivalent bandwidth of the equalizer. The last equality comes from Eq. (7-14). The noise spectral densities are described further in Sec. 7.3.

7.2.4 Shot Noise

The nongaussian nature of the photodetection process and the avalanche multiplication noise makes the evaluation of the shot noise term $\langle v_s^2(t) \rangle$ more difficult than that of the thermal noise[1, 25]. Personick[1] carried out a detailed analysis that evaluated the shot noise as a function of time within the bit slot.

This results in an accurate estimate of the shot noise contribution to the equalizer output noise voltage, but at the expense of computational difficulty.

Smith and Garrett[8] subsequently proposed a simplification of Personick's expressions by relating the mean square shot noise voltage $\langle v_s^2(t) \rangle$ at the decision time to the average unity gain photocurrent $\langle i_0 \rangle$ over the bit time T_b through the shot noise expression[11, 21]

$$\langle v_s^2(t) \rangle = 2q\langle i_0 \rangle\langle m^2 \rangle B_{bae} R^2 A^2 \qquad (7\text{-}34)$$

Here $\langle m^2 \rangle$ is the mean square avalanche gain [Eq. (6-32)], which we shall assume takes the form M^{2+x} with $0 < x \leq 1.0$. The other terms are as defined in Eq. (7-29). The factor of 2 arises because there is an equal contribution to the noise from the generation and from the recombination of carriers traversing the photoconductive channel.

We now calculate $\langle i_0 \rangle$ at the decision time within a particular bit slot. For this we must take into account not only the shot noise contribution from a pulse within this particular time slot but also the shot noises resulting from all other pulses that overlap into this bit period. The shot noise within a time slot will thus depend on the shape of the received pulse (that is, how much of it has spread into adjacent time slots as shown in Fig. 7-3) and on the data sequence (the distribution of 1 and 0 pulses in the data stream). The worst case of shot noise in any particular time slot occurs when all neighboring pulses are 1, since this causes the greatest amount of intersymbol interference. For this case the mean unity gain photocurrent over a bit time T_b for a 1 pulse is

$$\langle i_0 \rangle_1 = \sum_{n=-\infty}^{\infty} \frac{\eta q}{h\nu} b_{\text{on}} \frac{1}{T_b} \int_{-T_b/2}^{T_b/2} h_p(t - nT_b)\, dt$$

$$= \frac{\eta q}{h\nu} \frac{b_{\text{on}}}{T_b} \int_{-\infty}^{\infty} h_p(t)\, dt = \frac{\eta q}{h\nu} \frac{b_{\text{on}}}{T_b} \qquad (7\text{-}35)$$

where we have made use of Eq. (7-5).

For a 0 pulse (with all adjacent pulses being 1), we assume $b_{\text{off}} = 0$, so that

$$\langle i_0 \rangle_0 = \sum_{n \neq 0} \frac{\eta q}{h\nu} b_{\text{on}} \frac{1}{T_b} \int_{-T_b/2}^{T_b/2} h_p(t - nT_b)\, dt$$

$$= \frac{\eta q}{h\nu} \frac{b_{\text{on}}}{T_b} \left[\sum_{n=-\infty}^{\infty} \int_{-T_b/2}^{T_b/2} h_p(t - nT_b)\, dt - \int_{-T_b/2}^{T_b/2} h_p(t)\, dt \right]$$

$$= \frac{\eta q}{h\nu} \frac{b_{\text{on}}}{T_b} (1 - \gamma) \qquad (7\text{-}36)$$

The parameter

$$\gamma = \int_{-T_b/2}^{T_b/2} h_p(t)\, dt \qquad (7\text{-}37)$$

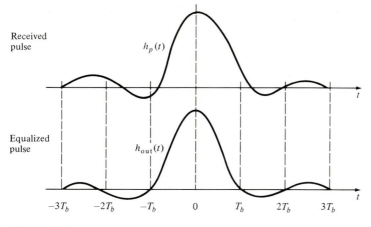

FIGURE 7-7
Equalized output pulse with no intersymbol interference at the decision time.

is the fractional energy of a 1 pulse that is contained within its bit period, as shown by the shaded area in Fig. 7-3. The factor $1 - \gamma$ is thus the fractional energy of a pulse that has spread outside of its bit period as it traveled through the optical fiber.

Equations (7-35) and (7-36) can now be substituted back into Eq. (7-34) to find the worst-case shot noise for a 1 and 0 pulse, respectively.

7.2.5 Receiver Sensitivity Calculation

To calculate the sensitivity of an optimal receiver, we first simplify the noise voltage expressions by using the notation of Personick.[1] We begin by assuming that the equalized pulse stream has no intersymbol interference at the sampling times nT_b, as shown in Fig. 7-7, and that the maximum value of $h_{out}(t)$ at $t = 0$ is unity. This means that

$$
\begin{aligned}
h_{out}(t=0) &= 1 \\
h_{out}(t=nT_b) &= 0 \qquad \text{for} \quad n \neq 0
\end{aligned}
\tag{7-38}
$$

Substituting this into Eq. (7-12), we then have from Eq. (7-27a) that the actual equalizer output voltage at the sampling times $t = nT_b$ is

$$
v_{out} = b_n h_{out}(0) + v_N(nT_b)
\tag{7-39}
$$

This shows that the noise $v_N(t)$ depends on all the b_n values and on the time t.

Furthermore, we introduce the dimensionless time and frequency variables $\tau = t/T_b$ and $\phi = fT_b$ to make the bandwidth integrals of Eqs. (7-30) and (7-33) independent of the bit period T_b. This means that the numerical value will depend only on the shapes of the received and equalized pulses and not on their scale. Using these values it can be shown by considering the Fourier transforms of $h_p(\tau)$ and $h_{out}(\tau)$ that the normalized transforms, denoted by $H'_p(\phi)$ and $H'_{out}(\phi)$, are related to $H_p(f)$ and $H_{out}(f)$ by

$$H'_p(\phi) = H_p(f)$$

$$H'_{out}(\phi) = \frac{1}{T_b} H_{out}(f) \tag{7-40}$$

Thus it follows that the *normalized bandwidth integrals* are[18]

$$I_2 = \frac{1}{T_b} \int_0^\infty \left| \frac{H_{out}(f)}{H_p(f)} \right|^2 df = \int_0^\infty \left| \frac{H'_{out}(\phi)}{H'_p(\phi)} \right|^2 d\phi \tag{7-41}$$

and

$$I_3 = \frac{1}{T_b} \int_0^\infty \left| \frac{H_{out}(f)}{H_p(f)} \right|^2 f^2 df = \int_0^\infty \left| \frac{H'_{out}(\phi)}{H'_p(\phi)} \right|^2 \phi^2 d\phi \tag{7-42}$$

Thus, using Eqs. (7-41) and (7-42), the bandwidth integrals in Eqs. (7-30) and (7-33) become

$$B_{bae} = \frac{I_2}{T_b} = I_2 B \tag{7-43}$$

and

$$B_e = \frac{I_2}{T_b} + \frac{(2\pi RC)^2}{T_b^3} I_3 = I_2 B + (2\pi RC)^2 I_3 B^3 \tag{7-44}$$

In evaluating Eqs. (7-30) and (7-33) we used the normalization conditions on $h_p(t)$ and $h_{out}(t)$ given by Eqs. (7-5) and (7-38), respectively, to derive the relationships

$$H_p(0) = 1 = H'_p(0) \quad \text{and} \quad H_{out}(0) = T_b$$

so that $H'_{out}(0) = 1$.

Using these expressions for B_{bae} and B_e and from Eqs. (7-29), (7-31), (7-32), and (7-34), the total mean square noise voltage in Eq. (7-28) becomes

$$\langle v_N^2 \rangle = R^2 A^2 \left(2q\langle i_0 \rangle M^{2+x} + \frac{4k_B T}{R_b} + S_I + \frac{S_E}{R^2} \right) I_2 B + (2\pi RC)^2 A^2 S_E I_3 B^3$$

$$= (qRAB)^2 \left(\frac{2\langle i_0 \rangle}{q} M^{2+x} T_b I_2 + W \right) \tag{7-45}$$

where

$$W = \frac{1}{q^2 B} \left(S_I + \frac{4k_B T}{R_b} + \frac{S_E}{R^2} \right) I_2 + \frac{(2\pi C)^2}{q^2} S_E I_3 B \tag{7-46}$$

TABLE 7-1
Input signal and noise currents in an optical receiver

Shot noise	$\langle i_s^2 \rangle = 2q\langle i_0 \rangle \langle m^2 \rangle A^2 I_2 B$
Thermal noise	$\langle i_R^2 \rangle = \dfrac{4k_B T}{R_b} A^2 I_2 B$
Shunt noise	$\langle i_I^2 \rangle = S_I A^2 I_2 B$
Series noise	$\langle i_E^2 \rangle = S_E A^2 \left[\dfrac{I_2 B}{R^2} + (2\pi C)^2 I_3 B^3 \right]$
Total noise	$\langle i_N^2 \rangle = \langle i_s^2 \rangle + \langle i_R^2 \rangle + \langle i_I^2 \rangle + \langle i_E^2 \rangle$
	$= A^2 \left(2q\langle i_0 \rangle \langle m^2 \rangle I_2 B + q^2 W B^2 \right)$

is a dimensionless parameter characterizing the thermal noise of the receiver. We shall call this parameter the *thermal noise characteristic* of the receiver amplifier.

Since the signal and noise voltages in Eq. (7-45) are each proportional to the resistance R, we can rewrite all of the expressions making up Eq. (7-45) in terms of input signal and noise *currents*. Doing so yields the results shown in Table 7-1.

Our task now is to find the minimum energy per pulse that is required to achieve a prescribed maximum bit error rate. For this we shall assume that the output voltage is approximately a gaussian variable. This is the *signal-to-noise ratio approximation*. Although the shot noise has a Poisson distribution, the inaccuracy resulting from the gaussian approximation is small.[7] The mean and variance of the gaussian output for a 1 pulse are b_{on} and σ_{on}^2, whereas for a 0 pulse they are b_{off} and σ_{off}^2. This is illustrated in Fig. 7-8. The variances σ_{on}^2 and σ_{off}^2 are defined as the worst-case values of $\langle v_N^2 \rangle$, which are obtained by

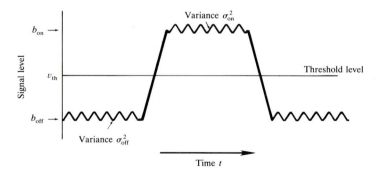

FIGURE 7-8
Gaussian noise statistics of a binary signal.

substituting Eq. (7-35) and (7-36), respectively, for $\langle i_0 \rangle$ into Eq. (7-34):

$$\sigma_{\text{on}}^2 = \left(\frac{h\nu}{\eta}\right)^2 \left(\frac{\eta M^x}{h\nu} b_{\text{on}} I_2 + \frac{W}{M^2}\right) \tag{7-47}$$

$$\sigma_{\text{off}}^2 = \left(\frac{h\nu}{\eta}\right)^2 \left[\frac{\eta M^x}{h\nu} b_{\text{on}} I_2 (1 - \gamma) + \frac{W}{M^2}\right] \tag{7-48}$$

If the decision threshold voltage v_{th} is set so that there is an equal error probability for 0 and 1 pulses and if we assume that there are an equal number of 0 and 1 pulses [that is, $a = b = \frac{1}{2}$ in Eq. (7-18)], then from Eqs. (7-16) and (7-17),

$$P_0(v_{\text{th}}) = P_1(v_{\text{th}}) = \tfrac{1}{2} P_e$$

Assuming that the equalizer output is a gaussian variable, the error probability P_e is given by Eqs. (7-20) and (7-22):

$$P_e = \frac{1}{\sqrt{2\pi}\,\sigma_{\text{off}}} \int_{v_{\text{th}}}^{\infty} \exp\left[-\frac{(v - b_{\text{off}})^2}{2\sigma_{\text{off}}^2}\right] dv$$

$$= \frac{1}{\sqrt{2\pi}\,\sigma_{\text{on}}} \int_{-\infty}^{v_{\text{th}}} \exp\left[\frac{-(-v + b_{\text{on}})^2}{2\sigma_{\text{on}}^2}\right] dv \tag{7-49}$$

Defining the parameter Q as

$$Q = \frac{v_{\text{th}} - b_{\text{off}}}{\sigma_{\text{off}}} = \frac{b_{\text{on}} - v_{\text{th}}}{\sigma_{\text{on}}} \tag{7-50}$$

then Eq. (7-49) becomes

$$P_e(Q) = \frac{1}{\sqrt{\pi}} \int_{Q/\sqrt{2}}^{\infty} e^{-x^2} dx$$

$$= \frac{1}{2}\left[1 - \text{erf}\left(\frac{Q}{\sqrt{2}}\right)\right] \tag{7-51}$$

where $\text{erf}(x)$ is the *error function* which is defined in Eq. (7-24). Analogous to Fig. 7-6, a plot of q versus P_e is given in Fig. 7-9. To an excellent approximation Eq. (7-51) can be replaced by[26]

$$P_e(Q) = \frac{1}{\sqrt{2\pi}} \frac{e^{-Q^2/2}}{Q} \tag{7-52}$$

The parameter Q is related to the signal-to-noise ratio required to achieve the desired bit error rate.[26] Equation (7-51) states that relative to the noise at b_{off} the threshold voltage v_{th} must be at least Q standard deviations above b_{off}, or equivalently, relative to the noise at b_{on} the threshold voltage must be no more than Q standard deviations below b_{on} to have the desired error rate.

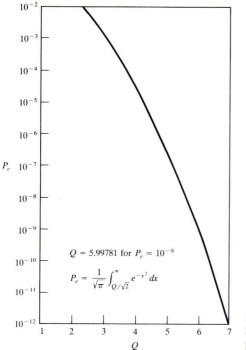

$Q = 5.99781$ for $P_e = 10^{-9}$

$$P_e = \frac{1}{\sqrt{\pi}} \int_{Q/\sqrt{2}}^{\infty} e^{-x^2} dx$$

FIGURE 7-9
Signal-to-noise ratio factor Q versus error probability P_e.

Example 7-3. When there is little intersymbol interference, γ is small, so that $\sigma_{on}^2 \simeq \sigma_{off}^2$. Then, by $b_{off} = 0$, we have from Eq. (7-50) that

$$Q = \frac{b_{on}}{2\sigma_{on}} = \frac{1}{2}\frac{S}{N}$$

which is one-half the signal-to-noise ratio. In this case $v_{th} = b_{on}/2$, so that the optimum decision threshold is midway between the 0 and 1 signal levels.

Example 7-4. For an error rate of 10^{-9} we have from Eq. (7-51) that

$$P_e(Q) = 10^{-9} = \frac{1}{2}\left[1 - \mathrm{erf}\left(\frac{Q}{\sqrt{2}}\right)\right]$$

From Fig. 7-6 we have that $Q \simeq 6$ (an exact evaluation yields $Q = 5.99781$), which gives a signal-to-noise ratio of 12, or 10.8 dB (that is, $10\log(S/N) = 10\log 12 = 10.8$ dB).

Using the expression in Eq. (7-50), the receiver sensitivity is given by

$$b_{on} - b_{off} = Q(\sigma_{on} + \sigma_{off}) \qquad (7\text{-}53)$$

If b_{off} is zero, the required energy per pulse that is needed to achieve a desired bit error rate characterized by the parameter Q is

$$b_{on} = \frac{Q}{M} \frac{h\nu}{\eta} \left\{ \left(M^{2+x} \frac{\eta}{h\nu} b_{on} I_2 + W \right)^{1/2} + \left[M^{2+x} \frac{\eta}{h\nu} b_{on} I_2 (1 - \gamma) + W \right]^{1/2} \right\}$$

(7-54)

We can now determine the optimum value of the avalanche gain, M_{opt}, by differentiating Eq. (7-54) with respect to M and putting $db_{on}/dM = 0$. After going through some lengthy but straightforward algebra, we obtain[8]

$$M_{opt}^{2+x} b_{on} = \frac{h\nu}{\eta} \frac{W}{2 I_2} \left(\frac{2 - \gamma}{1 - \gamma} \right) K$$

(7-55)

where

$$K = -1 + \left[1 + 16 \frac{1 + x}{x^2} \frac{1 - \gamma}{(2 - \gamma)^2} \right]^{1/2}$$

(7-56)

The minimum energy per pulse necessary to achieve a bit error rate characterized by Q can then be found by substituting Eq. (7-55) into Eq. (7-54) and solving for b_{on}. Doing so yields[8]

$$b_{on, min} = Q^{(2+x)/(1+x)} \frac{h\nu}{\eta} W^{x/(2+2x)} I_2^{1/(1+x)} L$$

(7-57)

where

$$L = \left[\frac{2(1 - \gamma)}{K(2 - \gamma)} \right]^{1/(1+x)} \left\{ \left[\frac{(2 - \gamma)K}{2(1 - \gamma)} + 1 \right]^{1/2} + \left[\tfrac{1}{2}(2 - \gamma)K + 1 \right]^{1/2} \right\}^{(2+x)/(1+x)}$$

(7-58)

The parameter L has a somewhat involved expression, but it has the feature of depending only on the fraction γ of the pulse energy contained within a bit period T_b and on the avalanche photodiode factor x. Values for L are typically between 2 and 3. Recalling from Chap. 6 that x takes on values between 0 and 1.0 (for example, 0 for *pin* photodiodes, 0.3 for silicon APDs, 0.7 for InGaAs APDs, and 1.0 for Ge APDs), we plot L as a function of γ in Fig. 7-10 for three different values of x. Note that these curves give L for any received pulse shape, since L depends only on x and γ.

The optimum gain at the desired bit error rate characterized by Q can be found by substituting Eq. (7-57) into Eq. (7-55) to obtain

$$M_{opt}^{1+x} = \frac{W^{1/2}}{Q I_2} \left[\frac{(2 - \gamma)K}{2(1 - \gamma)L} \right]^{(1+x)/(2+x)}$$

(7-59)

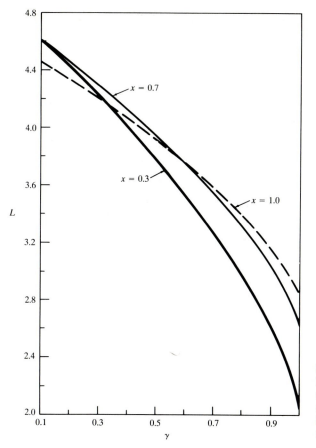

FIGURE 7-10
Relationship between the parameter L and the fraction γ of received optical energy of a pulse in a time slot T_b for $x = 0.3$ (Si), 0.7 (InGaAs), and 1.0 (Ge).

The optimum avalanche gain becomes, when $\gamma = 1$ (no intersymbol interference),

$$M_{\text{opt}}^{1+x} = \frac{2W^{1/2}}{xQI_2} \qquad (7\text{-}60)$$

The proof of this is left as an exercise.

7.2.6   Performance Curves

Using Eq. (7-57) we can calculate the effect of intersymbol interference on the required energy per pulse at the optimum gain for any received and equalized pulse shape. The minimum optical power required occurs for very narrow optical input pulses.[1] Ideally this is a unit impulse or delta function. More power is necessary for other received pulse shapes. The additional or *excess power* ΔP required for pulse shapes other than impulses is normally defined as a *power penalty* measured in decibels. Thus,

$$\Delta P = 10 \log \frac{b_{\text{on, nonimpulse}}}{b_{\text{on, impulse}}} \qquad (7\text{-}61)$$

As an example we shall calculate the case for which the amplifier resistance R given by Eq. (7-10) is sufficiently large so that the term

$$\frac{(2\pi C)^2}{T_b q^2} S_E I_3$$

dominates the thermal noise in Eq. (7-46). In this case,

$$\Delta P = 10 \log \frac{I_{3,n}^{x/(2+2x)} I_{2,n}^{1/(1+x)} L_n}{I_{3,i}^{x/(2+2x)} I_{2,i}^{1/(1+x)} L_i} \qquad (7\text{-}62)$$

where the subscripts n and i refer to *nonimpulse* and *impulse*, respectively.

For the input pulse shape $h_p(t)$ to the receiver we shall choose a gaussian pulse,

$$h_p(t) = \frac{1}{\sqrt{2\pi}\,\alpha T_b} e^{-t^2/2\alpha^2 T_b^2} \qquad (7\text{-}63)$$

the normalized Fourier transform of which is

$$H_p'(\phi) = e^{-(2\pi\alpha\phi)^2/2} \qquad (7\text{-}64)$$

As shown in Fig. 7-11, the parameter αT_b, where T_b is the bit period, defines the variance or spread of the pulse.

For the equalizer output waveform $h_{\text{out}}(t)$ we choose the commonly used raised-cosine pulse[1, 9, 20]

$$h_{\text{out}}(t) = \frac{\sin \pi\tau}{\pi\tau} \frac{\cos \pi\beta\tau}{1 - (2\beta\tau)^2} \qquad (7\text{-}65)$$

where $\tau = t/T_b$. A plot of $h_{\text{out}}(t)$ with $\beta = 0, 0.5,$ and 1.0 is shown in Fig. 7-12. The normalized Fourier transform is

$$H_{\text{out}}'(\phi) = \begin{cases} 1 & \text{for } 0 < |\phi| \leq \dfrac{1-\beta}{2} \\[2ex] \dfrac{1}{2}\left[1 - \sin\left(\dfrac{\pi\phi}{\beta} - \dfrac{\pi}{2\beta}\right)\right] & \text{for } \dfrac{1-\beta}{2} < |\phi| \leq \dfrac{1+\beta}{2} \\[2ex] 0 & \text{otherwise} \end{cases} \qquad (7\text{-}66)$$

The parameter β varies between 0 and 1 and determines the bandwidth used by the pulse, as shown in Fig. 7-12. A β value of unity indicates the bandwidth is $2/T_b$, whereas $\beta = 0$ means that the minimum bandwidth of $1/T_b$ is used. Although less bandwidth is used as β decreases, the tails of $h_{\text{out}}(t)$ become

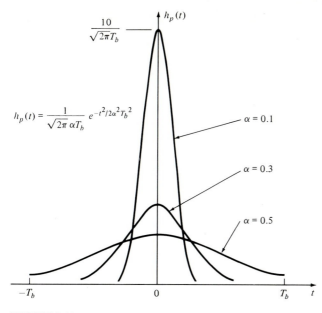

FIGURE 7-11
Shape of a gaussian pulse as a function of the parameter α.

larger, and signal timing and equalization become more difficult. For simplicity, we shall choose $\beta = 1$ in the examples given here.

The unit impulse or Dirac delta function is characterized by

$$h_{pi}(t) = \delta(t) \tag{7-67}$$

where

$$\delta(t) = 0 \qquad \text{for} \quad t \neq 0 \tag{7-68}$$

and

$$\int_{-\infty}^{\infty} \delta(t)\, dt = 1$$

Thus the Fourier transform of the impulse function is

$$H'_{pi}(\phi) = F[\delta(t)] = \int_{-\infty}^{\infty} \delta(t) e^{j2\pi ft}\, dt = 1 \tag{7-69}$$

Using Eqs. (7-41), (7-66), and (7-69) we have for an impulse input $H'_p(\phi)$ and a raised-cosine output,

$$I_{2i} = \int_{0}^{\infty} |H'_{\text{out}}(\phi)|^2\, d\phi = \frac{1}{2}\left(1 - \frac{\beta}{4}\right) \tag{7-70}$$

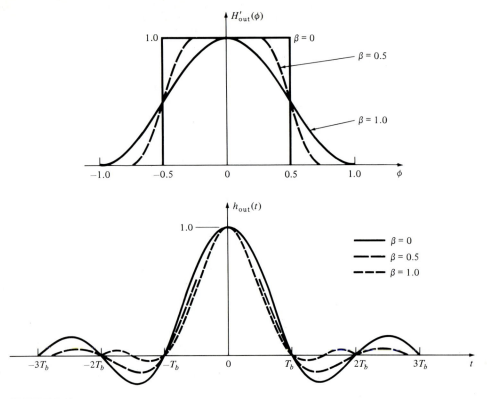

FIGURE 7-12
Shape of a raised-cosine pulse and its Fourier transform for three different values of the parameter β.

which, for $\beta = 1$, becomes

$$I_{2i} = \tfrac{3}{8} \tag{7-71}$$

From Eq. (7-42)

$$I_{3i} = \int_0^\infty |H'_{\text{out}}(\phi)|^2 \phi^2 \, d\phi$$

$$= \frac{\beta^3}{16}\left(\frac{1}{\pi^2} - \frac{1}{6}\right) + \beta^2\left(\frac{1}{8} - \frac{1}{\pi^2}\right) - \frac{\beta}{32} + \frac{1}{24} \tag{7-72}$$

For $\beta = 1$, we have

$$I_{3i} = 0.03001$$

We now evaluate I_{2n} and I_{3n} in Eq. (7-62) for the gaussian input pulse shape given by eq. (7-64) and for a raised-cosine output. With $\beta = 1$ in Eq.

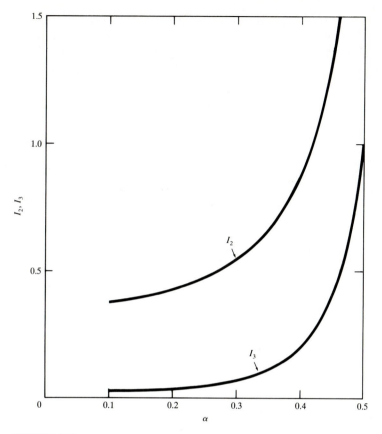

FIGURE 7-13
Plots of the normalized dimensionless bandwidth variables I_2 and I_3 as a function of α for a gaussian input pulse. The output pulse is a raised cosine with $\beta = 1.0$.

(7-66), we have

$$I_{2n} = \int_0^\infty \left| \frac{H'_{\text{out}}(\phi)}{H'_p(\phi)} \right|^2 d\phi$$

$$= \frac{2}{\pi} \int_0^{\pi/2} e^{16\alpha^2 \dot{x}^2} \cos^4 x \, dx \tag{7-73}$$

The results of a numerical evaluation of Eq. (7-73) as a function of α are given in Fig. 7-13. With the same assumptions,

$$I_{3n} = \left(\frac{2}{\pi} \right)^3 \int_0^{\pi/2} x^2 e^{16\alpha^2 x^2} \cos^4 x \, dx \tag{7-74}$$

The numerical evaluation of I_{3n} as a function of α is shown in Fig. 7-13.

Two more parameters (L_i and L_n) now remain to be evaluated in Eq. (7-62) in order to determine the receiver power penalty. Since all the pulse energy is contained within the bit period for a unit impulse input, L_i is determined by taking the limit of L in Eq. (7-58) as γ goes on unity. Thus,

$$L_i = \lim_{\gamma \to 1} L = (1 + x)\left(\frac{2}{x}\right)^{x/(1+x)} \tag{7-75}$$

The proof of this is left as an exercise.

The parameter L_n is given by Eq. (7-58). It depends on the pulse energy per bit period (γ) and on the excess noise coefficient x of the avalanche process. For a gaussian input pulse, γ, in turn, depends on the parameter α in Eq. (7-63). The relation between γ and α is found from Eqs. (7-37) and (7-63):

$$\gamma = \int_{-T_b/2}^{T_b/2} h_p(t)\, dt = \frac{2}{\sqrt{\pi}} \int_0^{1/(2\sqrt{2}\,\alpha)} e^{-x^2}\, dx$$

$$= \operatorname{erf}\left(\frac{1}{2\sqrt{2}\,\alpha}\right) \tag{7-76}$$

where the error function $\operatorname{erf}(x)$ is defined in Eq. (7-24). The relationship between γ and α given by Eq. (7-76) is shown in Fig. 7-14.

We are now finally ready to evaluate Eq. (7-62). Choosing $x = 0.3$, which is characteristic of silicon avalanche photodiodes, Eq. (7-75) yields $L_i = 2.38$. For InGaAs we have $x = 0.7$ and $L_i = 2.620$. What we wish to plot is the penalty in minimum received power ΔP (required for a certain bit error rate) as a function of the fraction of pulse energy $1 - \gamma$ that has spread outside of the bit period T_b. For a given value of γ we read the corresponding value of L_n from Fig. 7-10. To find I_{2n} and I_{3n}, we first find the value of α corresponding to γ from Fig. 7-14, and then we find the values of I_{2n} and I_{3n} corresponding to

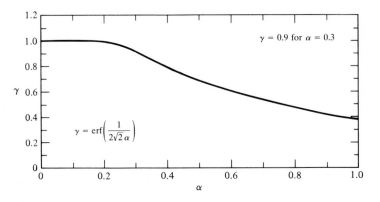

FIGURE 7-14
A plot of the fraction of pulse energy, γ, as a function of the gaussian pulse shape parameter α.

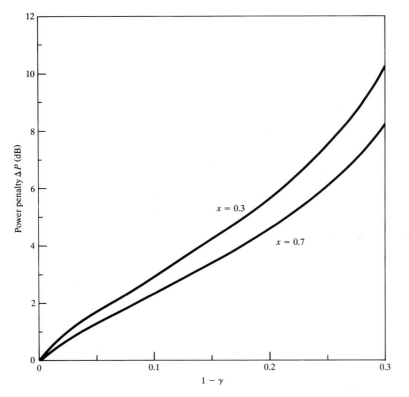

FIGURE 7-15
Plots of Eq. (7-62) which show the penalty in minimum received optical power (receiver sensitivity) arising from pulse spreading outside of the bit period for gaussian received pulses.

this value of α from Fig. 7-13. Substituting all these values into Eq. (7-62) for various values of γ and letting $x = 0.3$ for Si and 0.7 for InGaAs we obtain the results shown in Fig. 7-15.

The effect of intersymbol interference (or bandwidth limitation) on the receiver power penalty is readily deduced from Fig. 7-15. As the fraction of pulse energy outside the bit period increases, there is a steep rise in the power penalty curve. This curve gives a clear implication as to the effects of attempting to operate an optical fiber system at such high data rates that bandwidth limitations arise. Intersymbol interference becomes more pronounced at higher data rates, since the individual data pulses start to overlap significantly as the data rate approaches the system bandwidth limit. Since the receiver power penalty increases rapidly for larger pulse overlaps, operating a fiber optic system much beyond its bandwidth is generally not worthwhile, even though it may be possible to correct for intersymbol interference through the use of an equalization circuit.

7.2.7   Nonzero Extinction Ratio

In the previous section we assumed that there is no optical power incident on the photodetector during a 0 pulse, so that $b_{off} = 0$. In actual systems the light source may be biased slightly on at all times in order to obtain a shorter light source turnon time (see Sec. 4.3.5). Thus some optical power is also emitted during a 0 pulse. This is of particular importance for laser diodes, since it is generally desirable to bias them on to just below the lasing threshold. This means that during a zero pulse the laser acts like an LED and can launch a significant amount of optical power into a fiber.

The ratio ϵ of the optical energy emitted in the 0 pulse state to that emitted during a 1 pulse is called the *extinction ratio*

$$\epsilon = \frac{b_{off}}{b_{on}} \tag{7-77}$$

The extinction ratio as defined here thus varies between 0 and 1. *Note:* Equation (7-77) is not a universal definition for the extinction ratio. In many cases in the literature the reciprocal of Eq. (7-77) is used, that is, the extinction ratio is also defined as b_{on}/b_{off}. In that case the extinction ratio ranges between 1 and ∞. Thus when looking at the literature, the reader can easily see which definition is being used. A receiver performance calculation[27] identical to that of Sec. 7.2.6 can be carried out for a nonzero extinction ratio by simply using b_{off} from Eq. (7-77) and by replacing γ by $\gamma' = \gamma(1 - \epsilon)$ in Eq. (7-36). With these replacements Eq. (7-53) yields

$$b_{on}(1 - \epsilon) = \frac{Q}{M}\frac{h\nu}{\eta}\left\{\left(M^{2+x}\frac{\eta}{h\nu}b_{on}I_2 + W\right)^{1/2}\right.$$
$$\left. + \left[M^{2+x}\frac{\eta}{h\nu}b_{on}I_2(1 - \gamma') + W\right]^{1/2}\right\} \tag{7-78}$$

Analogous to the derivation of Eq. (7-57), we differentiate Eq. (7-78) for b_{on} with respect to M to find the minimum energy $b_{on,min}(\epsilon)$ per pulse required at the optimum gain M_{opt}, which results in

$$b_{on,min}(\epsilon) = Q^{(2+x)/(1+x)}\frac{h\nu}{\eta}W^{x/(2+2x)}I_2^{1/(1+x)}L'\left(\frac{1}{1-\epsilon}\right)^{(2+x)/(1+x)} \tag{7-79}$$

where

$$L'^{(1+x)} = \frac{2(1 - \gamma')}{K'(2 - \gamma')}\left\{\left(\frac{1}{2}\frac{2 - \gamma'}{1 - \gamma'}K' + 1\right)^{1/2} + \left[\tfrac{1}{2}(2 - \gamma')K' + 1\right]^{1/2}\right\}^{2+x} \tag{7-80}$$

with

$$K' = -1 + \left[1 + 16\frac{1 + x}{x^2}\frac{1 - \gamma'}{(2 - \gamma')^2}\right]^{1/2} \tag{7-81}$$

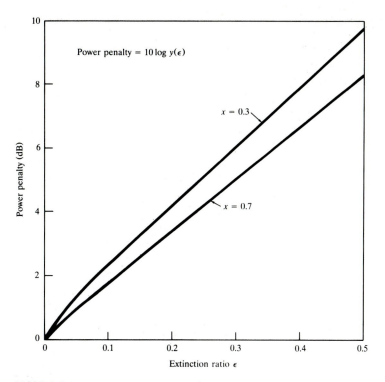

FIGURE 7-16
Plots of Eq. (7-83) which give the penalty in receiver sensitivity as a function of the extinction ratio ϵ for $\gamma = 1.0$ and $x = 0.3$ (Si) and 0.7 (InGaAs).

If a data stream has an equal probability of 1 and 0 pulses, then the minimum received power (or the receiver sensitivity), $P_{r,\,\mathrm{min}}$, is given by the average energy detected per pulse times the pulse rate $1/T_b$.

$$P_{r,\,\mathrm{min}} = \frac{b_{\mathrm{on}} + b_{\mathrm{off}}}{2T_b} = \frac{1 + \epsilon}{2T_b} b_{\mathrm{on}} \qquad (7\text{-}82)$$

where the last equality was obtained by using Eq. (7-77). The extinction ratio penalty, that is, the penalty in receiver sensitivity as a function of the extinction ratio, is

$$y(\epsilon) = \frac{P_{r,\,\mathrm{min}}(\epsilon)}{P_{r,\,\mathrm{min}}(0)} = (1 + \epsilon)\left(\frac{1}{1 - \epsilon}\right)^{(2+x)/(1+x)} \frac{L'}{L} \qquad (7\text{-}83)$$

Both L and L' are given by Fig. 7-10 on using γ and γ', respectively, for the abscissa. Plots of Eq. (7-83) in the form $10 \log y(\epsilon)$ are given in Fig. 7-16 for $x = 0.3$ (Si) and $x = 0.7$ (InGaAs) in the special case where $\gamma = 1.0$.

7.3 PREAMPLIFIER TYPES

Having examined the performance characteristics of a general class of receivers, we now turn our attention to some specific types of receiver preamplifiers. Since the sensitivity and bandwidth of a receiver are dominated by the noise sources at the front end (the preamplifier stage), the major emphasis in the literature has been on the design of a low-noise preamplifier. The goals generally are to maximize the receiver sensitivity while maintaining a suitable bandwidth.

Preamplifiers used in optical fiber communication receivers can be classified into three broad categories. These categories are not actually distinct, since a continuum of intermediate designs are possible, but they serve to illustrate the design approaches. The three categories encompass the low-impedance, the high-impedance, and the transimpedance preamplifiers.

The *low-impedance (LZ) preamplifier* is the most straightforward, but not necessarily the optimum preamplifier design. In this design, a photodiode operates into a low-impedance amplifier (for example, 50 Ω). Here a bias or load resistor R_b is used to match the amplifier impedance (to suppress standing waves for uniform frequency response). The value of the bias resistor, in conjunction with the amplifier input capacitance, is such that the preamplifier bandwidth is equal to or greater than the signal bandwidth. Although low-impedance preamplifiers can operate over a wide bandwidth, they do not provide a high receiver sensitivity, because only a small signal voltage can be developed across the amplifier input impedance and the resistor R_b. This limits their use to special short-distance applications where high sensitivity is not a major concern.

In the *high-impedance (HZ) preamplifier* design shown in Fig. 7-4, the goal is to reduce all sources of noise to the absolute minimum. This is accomplished by reducing the input capacitance through the selection of low-capacitance, high-frequency devices, by selecting a detector with low dark currents, and by minimizing the thermal noise contributed by the biasing resistors. The thermal noise can be reduced by using a high-impedance amplifier (such as a bipolar transistor or an FET) together with a large photodetector bias resistor R_b, which is why this design is referred to as a high-impedance preamplifier. Since the high impedance produces a large input RC time constant, the front-end bandwidth is less than the signal bandwidth. Thus the input signal is integrated, and equalization techniques must be employed to compensate for this.

The *transimpedance preamplifier* design largely overcomes the drawbacks of the high-impedance preamplifier. This is done by utilizing a low-noise, high-impedance amplifier with a negative-feedback resistor R_f with an equivalent thermal noise current $i_f(t)$ shunting the input as shown in Fig. 7-17. The amplifier has an input equivalent series voltage noise source $e_a(t)$, an equivalent shunt current noise $i_a(t)$, and an input impedance given by the parallel combination of R_a and C_a.

What we are interested in here is to evaluate the noise current given by Eq. (7-45) for different amplifier designs. To this end we shall examine only the

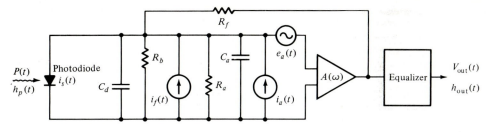

FIGURE 7-17
An equivalent circuit of a transimpedance receiver design. (Reproduced with permission from Personick, Rhodes, Hanson, and Chan, *Proc. IEEE*, vol. 68, p. 1260, Oct. 1980, © 1980, IEEE.)

parameter W of Eq. (7-46). This is a very useful figure of merit for a receiver, since it measures the noisiness of the amplifier. From Eq. (7-46) it can be seen that the noise is minimized if the amplifier and bias resistances R_a and R_b are large, the total capacitance C at the amplifier input is small, and the noise current and voltage spectral heights S_I and S_E are small. In general, these parameters are not independent, so that tradeoffs have to be made among them to minimize the noise. In addition, the freedom of the designer to optimize the device parameters is often restricted by the limited variety of components available. We shall examine several high-impedance and transimpedance amplifier configurations to illustrate some of the design considerations that must be taken into account.

7.3.1 High-Impedance FET Amplifiers

A number of different FETs (field effect transistors) can be used for front-end receiver designs.[12, 15, 17, 28–31] For gigabit-per-second data rates, for example, the lowest-noise receivers are made using GaAs MESFET preamplifiers. At lower frequencies silicon MOSFETs or JFETs are generally used. The circuit of a simple FET amplifier is shown in Fig. 7-18. Typical FETs have very large input

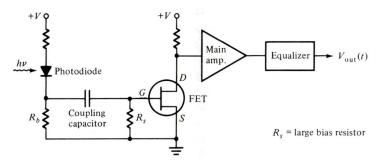

FIGURE 7-18
Simple high-impedance preamplifier design using a FET.

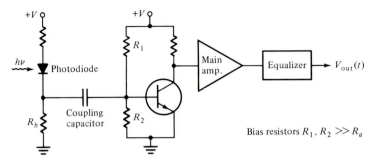

FIGURE 7-19
Simple high-impedance preamplifier design using a bipolar transistor.

resistances R_a (usually greater than 10^6 Ω), so for practical purposes $R_a = \infty$. The total resistance R given by Eq. (7-10) then reduces to the value of the detector bias resistor R_b.

The principal noise sources are thermal noise associated with the FET channel conductance, thermal noise from the load or feedback resistor, and shot noise arising from gate leakage current. A fourth noise source is FET $1/f$ noise. This was not included in the above analyses because it only contributes to the overall noise at very low bit rates[11] (see Prob. 7-20). Since the amplifier input resistance is very large, the input current noise spectral density S_I is

$$S_{I, \text{FET}} = \frac{4k_B T}{R_a} + 2qI_{\text{gate}} \tag{7-84}$$

$$\simeq 2qI_{\text{gate}}$$

where I_{gate} is the gate-leakage current of the FET. In an FET the thermal noise of the conducting channel resistance is characterized by the transconductance g_m. The voltage noise spectral density is[32]

$$S_E = \frac{4k_B T\Gamma}{g_m} \tag{7-85}$$

where the FET channel-noise factor Γ is a numerical constant that accounts for thermal noise and gate-induced noise plus the correlation between these two noises. The thermal noise characteristic W [Eq. (7-46)] at the equalizer output is then

$$W = \frac{1}{q^2 B}\left(2qI_{\text{gate}} + \frac{4k_B T}{R_b} + \frac{4k_B T\Gamma}{g_m R_b^2}\right)I_2 + \left(\frac{2\pi C}{q}\right)^2 \frac{4k_B T\Gamma}{g_m} I_3 B \tag{7-86}$$

Some typical values of the various parameters for GaAs MESFETs, Si MOSFETs, and Si JFETs are given in Table 7-2. Here C_{gs} and C_{gd} are the FET

TABLE 7-2
**Typical values of various parameters for GaAs MESFETs,
Si MOSFETs, and Si JFETs**

Parameter	Si JFET	Si MOSFET	GaAs MESFET
g_m (mS)	5—10	20—40	15—50
C_{gs} (pF)	3—6	0.5—1.0	0.2—0.5
C_{gd} (pF)	0.5—1.0	0.05—0.1	0.01—0.05
Γ	0.7	1.5—3.0	1.1—1.75
I_{gate} (nA)	0.01—0.1	0	1—1000
f_c (MHz)	< 0.1	1—10	10—100

gate-source and gate-drain capacitances, respectively. For a typical FET and a good photodiode we can expect values of $C = C_a + C_d + C_{gs} + C_{gd} = 10$ pF. The $1/f$-noise corner frequency f_c is defined as the frequency at which $1/f$ noise, which dominates the FET noise at low frequencies and has a $1/f$ power spectrum, becomes equal to the high-frequency channel noise described by Γ.

To minimize the noise in a high-impedance design, the bias resistor should be very large. The effect of this is that the detector output signal is integrated by the amplifier input resistance. We can compensate for this by differentiation in the equalizing filter. This integration-differentiation approach is known as the *high-impedance amplifier design* technique. It yields low noise, but also results in a low dynamic range (the range of signal levels that can be processed with high quality). An alternative method to deal with this is described in Sec. 7.3.3.

As the signal frequency reaches values of about 25 to 50 MHz, the gain of a silicon FET approaches unity. Much higher frequencies (4 Gb/s and above) can be achieved with either a GaAs MESFET or a silicon bipolar transistor.[29, 33]

7.3.2 High-Impedance Bipolar Transistor Amplifiers

The circuit of a simple bipolar grounded-emitter transistor amplifier is shown in Fig. 7-19. The input resistance of a bipolar transistor is given by[32, 34]

$$R_{in} = \frac{k_B T}{q I_{BB}} \qquad (7\text{-}87)$$

where I_{BB} is the base bias current. For a bipolar transistor amplifier the input resistance R_a is given by the parallel combination of the bias resistors R_1 and R_2 and the transistor input resistance R_{in}. For a low-noise design R_1 and R_2 are chosen to be much greater than R_{in}, so that $R_a \simeq R_{in}$. Thus, in contrast to the FET amplifier, R_a for a transistor amplifier is adjustable by the designer.

The spectral density (in A^2/Hz) of the input noise current source results from the shot noise of the base current:[32, 34]

$$S_I = 2qI_{BB} = \frac{2k_BT}{R_{in}} \tag{7-88}$$

where the last equality comes from Eq. (7-87). The spectral height (in V^2/Hz) of the noise voltage source is[32, 34]

$$S_E = \frac{2k_BT}{g_m} \tag{7-89}$$

Here the transconductance g_m is related to the shot noise by virtue of the collector current I_c:

$$g_m = \frac{qI_c}{k_BT} = \frac{\beta}{R_{in}} \tag{7-90}$$

where Eq. (7-87) has been used in the last equality to express g_m in terms of the current gain $\beta = I_c/I_{BB}$ and the input resistance R_{in}.

Substituting Eqs. (7-87) through (7-89) into Eq. (7-46), we have

$$W = \frac{T_b}{q^2} 2k_BT \left[\left(\frac{1}{R_{in}} + \frac{2}{R_b} + \frac{R_{in}}{\beta R^2} \right) I_2 + \frac{(2\pi C)^2}{T_b^2} \frac{R_{in}}{\beta} I_3 \right] \tag{7-91}$$

The contribution C_a to C from the bipolar transistor is a few picofarads. If the photodetector bias resistor R_b is much larger than the amplifier resistance R_a, then from Eq. (7-10) $R \simeq R_a \simeq R_{in}$, so that

$$W = \frac{2k_BT}{q^2} \left[\frac{T_b}{R_{in}} \frac{\beta + 1}{\beta} I_2 + \frac{(2\pi C)^2}{\beta T_b} R_{in} I_3 \right] \tag{7-92}$$

As in the case with a high-impedance FET preamplifier, the impedance loading the photodetector integrates the detector output signal. Again, to compensate for this, the amplified signal is differentiated in the equalizing filter.

7.3.3 Transimpedance Amplifier

Although a high-impedance design produces the lowest-noise amplifier, it has two limitations: (*a*) for broadband applications equalization is required and (*b*) it has a limited dynamic range. An alternative design is the *transimpedance amplifier*[17, 29, 35–37] shown in Fig. 7-17. This is basically a high-gain high-impedance amplifier with feedback provided to the amplifier input through the feedback resistor R_f. This design yields both low noise and a large dynamic range.

To compare the nonfeedback and the feedback designs, we make the restriction that both have the same transfer function $H_{out}(f)/H_p(f)$. For the transimpedance amplifier the thermal noise characteristic W_{TZ} at the equalizer output is therefore simply found by replacing R_b in Eq. (7-46) with R_b', where[35]

$$R_b' = \frac{R_b R_f}{R_b + R_f} \tag{7-93}$$

is the parallel combination of R_b and R_f. Thus, from Eq. (7-46),

$$W_{TZ} = \frac{T_b}{q^2}\left(S_I + \frac{4k_BT}{R_b'} + \frac{S_E}{(R')^2}\right)I_2 + \frac{(2\pi C)^2}{q^2 T_b}S_E I_3 \tag{7-94}$$

where, from Eq. (7-10),

$$\frac{1}{R'} = \frac{1}{R} + \frac{1}{R_f} = \frac{1}{R_a} + \frac{1}{R_b} + \frac{1}{R_f} \tag{7-95}$$

In practice, the feedback resistance R_f is much greater than the amplifier input resistance R_a. Consequently, $R' \simeq R$ in Eq. (7-95), so that

$$W_{TZ} = W_{HZ} + \frac{T_b}{q^2}\frac{4k_BT}{R_f}I_2 \tag{7-96}$$

where W_{HZ} is the high-impedance amplifier noise characteristic given by either Eq. (7-86) for FET designs or by Eq. (7-92) for the bipolar transistor case. The thermal noise of the transimpedance amplifier is thus modeled as the sum of the output noise of a nonfeedback amplifier plus the thermal noise associated with the feedback resistance. In practice, the noise considerations tend to be more involved, since R_f has an effect on the frequency response of the amplifier. More details are given by Smith and Personick.[18]

We now compare the bandwidths of the two designs. From Eq. (7-9) the transfer function of the nonfeedback amplifier is (in V/A)

$$H(f) = \frac{AR}{1 + j2\pi RCf} \tag{7-97}$$

where R and C are given by Eqs. (7-10) and (7-11), respectively, and A is the frequency-independent gain of the amplifier. Using Eq. (E-10), this yields a bandwidth of $(4RC)^{-1}$. For the transimpedance amplifier the transfer function $H_{TZ}(f)$ is

$$H_{TZ} = \frac{1}{1 + j2\pi RCf/A} \tag{7-98}$$

which yields a bandwidth of

$$B_{TZ} = \frac{A}{4RC} \tag{7-99}$$

which is A times that of the high-impedance design. This makes the equalization task simpler in the feedback amplifier case.

In summary, the benefits of a transimpedance amplifier are as follows:

1. It has a wide dynamic range compared to the high-impedance amplifier.
2. Usually little or no equalization is required because the combination of R_{in} and the feedback resistor R_f is very small, which means the time constant of the detector is also small.
3. The output resistance is small, so that the amplifier is less susceptible to pickup noise, cross talk, electromagnetic interference (EMI), etc.
4. The transfer characteristic of the amplifier is actually its transimpedance, which is the feedback resistor. Therefore, the transimpedance amplifier is very easily controlled and stable.
5. Although the transimpedance amplifier is less sensitive than the high-impedance amplifier (since $W_{TZ} > W_{HZ}$), this difference is usually only about 2 to 3 dB for most practical wideband designs.

7.3.4   High-Speed Circuits

Improvements in component performance, cost, and reliability in the 1980s led to major applications of fiber optic technology for long-distance carriers, local telephone services, and local area networks. To utilize the wide bandwidth available, there was an increased implementation of high-speed systems for both digital and analog links.[38-42] Along with this came the miniaturization of transmitters and receivers into integrated circuit formats. Many different types of receivers with operating speeds up to multigigahertz rates are thus commercially available from a wide variety of vendors.

7.4   ANALOG RECEIVERS

In addition to the wide usage of fiber optics for the transmission of digital signals, there are many potential applications for analog links. These range from individual 4-kHz voice channels to microwave links operating in the multigigahertz region.[38-40] In the previous sections we discussed digital receiver performance in terms of error probability. For an analog receiver the performance fidelity is measured in terms of a *signal-to-noise ratio*. This is defined as the ratio of the mean square signal current to the mean square noise current.

The simplest analog technique is to use amplitude modulation of the source.[2] In this scheme the time-varying electric signal $s(t)$ is used to modulate directly an optical source about some bias point defined by the bias current I_B, as shown in Fig. 7-20. The transmitted optical power $P(t)$ is thus of the form

$$P(t) = P_t[1 + ms(t)] \tag{7-100}$$

where P_t is the average transmitted optical power, $s(t)$ is the analog modulation

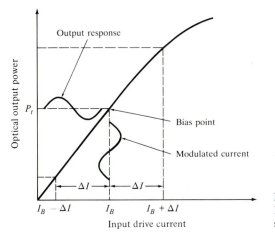

Optical output power

Output response

P_t

Bias point

Modulated current

$\leftarrow\Delta I\rightarrow$ $\leftarrow\Delta I\rightarrow$

$I_B - \Delta I$ I_B $I_B + \Delta I$

Input drive current

FIGURE 7-20
Direct analog modulation of an LED source.

signal, and m is the modulation index defined by (see Sec. 4.4)

$$m = \frac{\Delta I}{I_B} \qquad (7\text{-}101)$$

Here ΔI is the variation in current about the bias point. In order not to introduce distortion into the optical signal, the modulation must be confined to the linear region of the light source output curve shown in Fig. 7-20. Also, if $\Delta I > I_B$, the lower portion of the signal gets cut off and severe distortion results.

At the receiver end the photocurrent generated by the analog optical signal is

$$i_s(t) = \mathcal{R}_0 M P_r[1 + ms(t)]$$
$$= I_p M[1 + ms(t)] \qquad (7\text{-}102)$$

where $\mathcal{R}_0$ is the detector responsivity, P_r is the average received optical power, $I_p = \mathcal{R}_0 P_r$ is the primary photocurrent, and M is the photodetector gain. If $s(t)$ is a sinusoidally modulated signal, then the mean square signal current at the photodetector output is (ignoring a dc term)

$$\langle i_s^2 \rangle = \tfrac{1}{2}(\mathcal{R}_0 M m P_r)^2 = \tfrac{1}{2}(M m I_p)^2 \qquad (7\text{-}103)$$

Recalling from Eq. (6-18) that the mean square noise current for a photodiode receiver is the sum of the mean square quantum noise current, the equivalent-resistance thermal noise current, the dark noise current, and the surface-leakage noise current, we have

$$\langle i_N^2 \rangle = 2q(I_p + I_D)M^2 F(M)B + 2qI_L B + \frac{4k_B TB}{R_{eq}}F_t \qquad (7\text{-}104)$$

where I_p = primary (unmultiplied) photocurrent = $\mathscr{R}_0 P_r$
$\quad\quad I_D$ = primary bulk dark current
$\quad\quad I_L$ = surface leakage current
$\quad F(M)$ = excess photodiode noise factor $\simeq M^x$ $(0 < x \le 1)$
$\quad\quad B$ = effective noise bandwidth
$\quad R_{\mathrm{eq}}$ = equivalent resistance of photodetector load and amplifier
$\quad\quad F_t$ = noise figure of the baseband amplifier

By a suitable choice of the photodetector the leakage current can be rendered negligible. With this assumption the signal-to-noise ratio S/N is

$$\frac{S}{N} = \frac{\langle i_s^2 \rangle}{\langle i_N^2 \rangle} = \frac{\frac{1}{2}(\mathscr{R}_0 M m P_r)^2}{2q(\mathscr{R}_0 P_r + I_D)M^2 F(M)B + (4k_B TB/R_{\mathrm{eq}})F_t}$$

$$= \frac{\frac{1}{2}(I_p M m)^2}{2q(I_p + I_D)M^2 F(M)B + (4k_B TB/R_{\mathrm{eq}})F_t} \tag{7-105}$$

For a *pin* photodiode we have $M = 1$. When the optical power incident on the photodiode is small, the circuit noise term dominates the noise current, so that

$$\frac{S}{N} \simeq \frac{\frac{1}{2}m^2 I_p^2}{(4k_B TB/R_{\mathrm{eq}})F_t} = \frac{\frac{1}{2}m^2 \mathscr{R}_0^2 P_r^2}{(4k_B TB/R_{\mathrm{eq}})F_t} \tag{7-106}$$

Here the signal-to-noise ratio is directly proportional to the square of the photodiode output current and inversely proportional to the thermal noise of the circuit.

For large optical signals incident on a *pin* photodiode, the quantum noise associated with the signal detection process dominates, so that

$$\frac{S}{N} \simeq \frac{m^2 I_p}{4qB} = \frac{m^2 \mathscr{R}_0 P_r}{4qB} \tag{7-107}$$

Since the signal-to-noise ratio in this case is independent of the circuit noise, it represents the fundamental or quantum limit for analog receiver sensitivity.

When an avalanche photodiode is employed at low signal levels and with low values of gain M, the circuit noise term dominates. At a fixed low signal level, as the gain is increased from a low value, the signal-to-noise ratio increases with gain until the quantum noise term becomes comparable to the circuit noise term. As the gain is increased further beyond this point, the signal-to-noise ratio *decreases* as $F(M)^{-1}$. Thus for a given set of operating conditions, there exists an optimum value of the avalanche gain for which the signal-to-noise ratio is a maximum. Since an avalanche photodiode increases the signal-to-noise ratio for small optical signal levels, it is the preferred photodetector for this situation.

For very large optical signal levels the quantum noise term dominates the receiver noise. In this case, an avalanche photodiode serves no advantage, since the detector noise increases more rapidly with increasing gain M than the signal

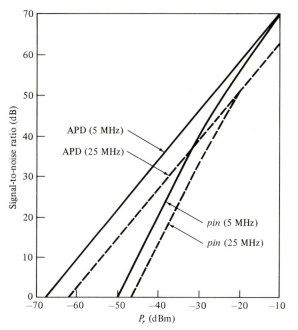

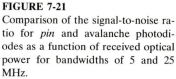

FIGURE 7-21
Comparison of the signal-to-noise ratio for *pin* and avalanche photodiodes as a function of received optical power for bandwidths of 5 and 25 MHz.

level. This is shown in Fig. 7-21, where we compare the signal-to-noise ratio for a *pin* and an avalanche photodiode receiver as a function of the received optical power. The signal-to-noise ratio for the avalanche photodetector is at the optimum gain (see Probs. 7-28 and 7-29). The parameter values chosen for this example are $B = 5$ MHz and 25 MHz, $x = 0.5$ for the avalanche photodiode and 0 for the *pin* diode, $m = 80$ percent, $\mathcal{R}_0 = 0.5$ A/W, and $R_{eq}/F_t = 10^4$ Ω. We see that for low signal levels an avalanche photodiode yields a higher signal-to-noise ratio, whereas at large received optical power levels a *pin* photodiode results in better performance.

7.5 SUMMARY

The task of an optical receiver is first to convert the optical energy emerging from the end of a fiber into an electric signal, and then to amplify this signal to a large enough level so that it can be processed by the electronics following the receiver amplifier. In these processes various noises and distortions will unavoidably be introduced, which can lead to errors in the interpretation of the received signal. The three basic stages of a receiver are a photodetector, an amplifier, and an equalizer. The design of the amplifier which follows the photodiode is of critical importance, because it is in the amplifier where the major noise sources are expected to arise. The equalizer that follows the amplifier is normally a linear frequency-shaping filter that is used to mitigate the effects of signal distortion and intersymbol interference.

In a digital receiver the amplified and filtered signal emerging from the equalizer is compared with a threshold level once per time slot to determine whether or not a pulse is present at the photodetector in that time slot. Various noises, interference from adjacent pulses, and conditions wherein the light source is not completely extinguished during a zero pulse can cause errors in the decision-making process. To calculate the error probability, we require a knowledge of the mean square noise voltage which is superimposed on the signal voltage at the decision time. Since the statistics of the output voltage at the sampling time are very complicated, approximations are used to calculate the performance of a binary optical fiber receiver. In applying these approximations, we have to make a tradeoff between computational simplicity and accuracy of the results. The simplest method is based on a gaussian approximation. In this method it is assumed that when the sequence of optical input pulses is known, the equalizer output voltage is a gaussian random variable. Thus, to calculate the error probability, we only need to know the mean and standard deviation of the output voltage.

The two basic approaches to the design of preamplifiers for fiber optic receivers are the high-impedance and the transimpedance preamplifiers. The high-impedance design produces the lowest noise, but it has two limitations: (*a*) for broadband applications equalization is required and (*b*) it has a limited dynamic range. The transimpedance amplifier is less sensitive, but it has the benefits of a wide dynamic range and little need for equalization.

For the reader who is interested in constructing some actual receivers and employing them in a data link, some detailed receiver designs and their uses and limitations are discussed in Refs. 18, 43, and 44.

PROBLEMS

7-1. In avalanche photodiodes the ionization ratio k is approximately 0.02 for silicon, 0.35 for InGaAs, and 1.0 for germanium. Show that, for gains up to 100 in Si and up to 25 in InGaAs and Ge, the excess noise factor $F(M)$ can be approximated to within 10% by M^x, where x is 0.3 for Si, 0.7 for InGaAs, and 1.0 for Ge.

7-2. Find the Fourier transform $h_B(t)$ of the bias circuit transfer function $H_B(f)$ given by Eq. (7-9).

7-3. Show that the following pulse shapes satisfy the normalization condition

$$\int_{-\infty}^{\infty} h_p(t) \, dt = 1$$

(*a*) Rectangular pulse (α = constant)

$$h_p(t) = \begin{cases} \dfrac{1}{\alpha T_b} & \text{for} \quad \dfrac{-\alpha T_b}{2} < t < \dfrac{\alpha T_b}{2} \\ 0 & \text{otherwise} \end{cases}$$

(*b*) Gaussian pulse

$$h_p(t) = \frac{1}{\sqrt{2\pi}} \frac{1}{\alpha T_b} e^{-t^2/2(\alpha T_b)^2}$$

(*c*) Exponential pulse

$$h_p(t) = \begin{cases} \dfrac{1}{\alpha T_b} e^{-t/\alpha T_b} & \text{for } 0 \le t < \infty \\ 0 & \text{otherwise} \end{cases}$$

7-4. The mathematical operation of convolving two real-valued functions of the same variable is defined as

$$p(t) * q(t) = \int_{-\infty}^{\infty} p(x)q(t-x)\,dx$$

$$= \int_{-\infty}^{\infty} q(x)p(t-x)\,dx$$

where $*$ denotes convolution. If $P(f)$ and $Q(f)$ are the Fourier transforms of $p(t)$ and $q(t)$, respectively, show that

$$F[p(t) * q(t)] = P(f)Q(f) = F[p(t)]F[q(t)]$$

That is, the convolution of two signals in the time domain corresponds to the multiplication of their Fourier transforms in the frequency domain.

7-5. Derive Eq. (7-23) from Eq. (7-18).

7-6. A transmission system sends out information at 200,000 b/s. During the transmission process, fluctuation noise is added to the signal so that at the decoder output the signal pulses are 1 V in amplitude and the rms noise voltage is 0.2 V.

(*a*) Assuming that ones and zeros are equally likely to be transmitted, what is the average time in which an error occurs?

(*b*) How is this time changed if the voltage amplitude is doubled with the rms noise voltage remaining the same?

7-7. Consider the probability distributions shown in Fig. 7-5, where the signal voltage for a binary 1 is V_1 and $v_{th} = V_1/2$.

(*a*) If $\sigma = 0.20V_1$ for $p(y|0)$ and $\sigma = 0.24V_1$ for $p(y|1)$, find the error probabilities $P_0(v_{th})$ and $P_1(v_{th})$.

(*b*) If $a = 0.65$ and $b = 0.35$, find P_e.

(*c*) If $a = b = 0.5$, find P_e.

7-8. An LED operating at 1300 nm injects 25 μW of optical power into a fiber. If the attenuation between the LED and the photodetector is 40 dB and the photodetector quantum efficiency is 0.65, what is the probability that less than 5 electron-hole pairs will be generated at the detector in a 1-ns interval?

7-9. Derive the expression for the mean square noise voltage at the equalizer output given by Eq. (7-28).

7-10. (*a*) Shows that Eqs. (7-30) and (7-33) can be rewritten as Eqs. (7-43) and (7-44).

(*b*) Show that Eq. (7-28) can be rewritten as Eq. (7-45).

7-11. Show that, by using Eq. (7-50), the error-probability expressions given by Eq. (7-49) both reduce to Eq. (7-51).

7-12. A useful approximation to $\frac{1}{2}(1 - \text{erf } x)$ for values of x greater than 3 is given by

$$\frac{1}{2}(1 - \text{erf } x) \simeq \frac{\exp(-x^2)}{2\sqrt{\pi}\, x} \qquad x > 3$$

Using this approximation, consider an on-off binary system that transmits the signal levels 0 and A with equal probability in the presence of gaussian noise. Let the signal amplitude A be K times the standard deviation of the noise.
(a) Calculate the net probability of error if $K = 10$.
(b) Find the value of K required to give a net error probability of 10^{-5}.

7-13. Derive Eq. (7-55) by differentiating both sides of Eq. (7-54) with respect to M and setting $db_{\text{on}}/dM = 0$.

7-14. Verify the expression given by Eq. (7-57) for the minimum energy per pulse needed to achieve a bit error rate characterized by Q.

7-15. Show that, when there is no intersymbol interference $(\gamma = 1)$, the optimum avalanche gain given by Eq. (7-59) reduces to Eq. (7-60).

7-16. Derive Eq. (7-70) for I_2 and Eq. (7-72) for I_3 (impulse input and raised-cosine output).

7-17. Verify the expressions given by Eqs. (7-73) and (7-74) for I_2 and I_3, respectively, for a gaussian input pulse and a raised-cosine output.

7-18. Plot I_2 versus α for a gaussian input pulse for values of $\beta = 0.1$, 0.5, and 1.0.

7-19. Plot I_3 versus α for a gaussian input pulse for values of $\beta = 0.1$, 0.5, and 1.0.

7-20. Derive Eq. (7-75).

7-21. Compare the values of $y(\epsilon)$ in Eq. (7-83) in decibels for 10-percent extinction ratio $(\epsilon = 0.10)$ when (a) $\gamma = 0.90$, $x = 0.5$; (b) $\gamma = 0.90$, $x = 1.0$.

7-22. (a) Plot the values of the thermal noise characteristic W for a high-impedance FET amplifier for data rates $1/T_b$ ranging from 1 to 50 Mb/s. Let $T = 300$ K, $g_m = 0.005$ S, $R_b = 10^5$ Ω, $C = 10$ pF, and $\gamma = 0.90$. Use Fig. 7-13 to find I_2 and I_3. Recall that I_2 and I_3 depend on α, which in turn depends on γ.
(b) Plot the values of W for a high-impedance bipolar transistor preamplifier for data rates ranging from 20 to 100 Mb/s. Let $T = 300$ K, $\beta = 100$, $I_{BB} = 5$ μA, $C = 10$ pF, and $\gamma = 0.90$.

7-23. The receiver sensitivity P_r is given by the average energy b_{on} detected per pulse times the pulse rate $1/T_b$ (if $b_{\text{off}} = 0$):

$$P_r = \frac{b_{\text{on}}}{T_b}$$

Find the sensitivity in dBm (see App. D) of an avalanche photodiode receiver with an FET preamplifier at a 10-Mb/s data rate. Let the required bit error rate be 10^{-9} and take $T = 300$ K, $x = 0.5$, $\gamma = 0.9$, $\eta q/h\nu = 0.7$ A/W (the detector responsivity), $g_m = 0.005$ S, $R_b = 10^5$ Ω, and $C = 10$ pF.

7-24. If a transimpedance amplifier has a feedback resistance of 5000 Ω, by how much does W_{TZ} differ from W_{HZ} at a 10-Mb/s data rate? Assume $\gamma = 0.9$ and $T = 300$ K. Compared to a high-impedance amplifier, what is the decrease in receiver sensitivity (in dB) for this transimpedance amplifier at 10 Mb/s if $x = 0.5$ and $W_{HZ} = 1 \times 10^6$?

7-25. (*a*) To calculate the receiver sensitivity P_r as a function of gain M, we need to solve Eq. (7-54) for b_{on}. Show that, for $\gamma = 1.0$ and $b_{\text{off}} = 0$, this becomes

$$b_{\text{on}} = \frac{h\nu}{\eta}\left(M^x Q^2 I_2 + \frac{2QW^{1/2}}{M}\right)$$

(*b*) Consider a receiver operating at 50 Mb/s. Let the receiver have an avalanche photodiode with $x = 0.5$ and a bipolar transistor front end (preamplifier). Assume $W = 2 \times 10^6$, $Q = 6$ for a 10^{-9} bit error rate, $I_2 = 1.08$, and $\eta q/h\nu = 0.7$ A/W. Using the foregoing expression for b_{on}, plot $P_r = b_{\text{on}}/T_b$ in dBm as a function of gain M for values of M ranging from 30 to 120.

7-26. For $1/f$ noise the noise current $\langle i_f^2 \rangle$ is given by

$$\langle i_f^2 \rangle = \frac{8k_B T \Gamma (2\pi C)^2 f_C I_f B^2}{g_m}$$

where

$$I_f = \frac{1}{T_b}\int_0^\infty \left|\frac{H_{\text{out}}(f)}{H_p(f)}\right|^2 f\,df = \int_0^\infty \left|\frac{H'_{\text{out}}(\phi)}{H'_p(\phi)}\right|^2 \phi\,d\phi$$

and f_c is the corner frequency.

(*a*) Show that

$$I_f = \left(\frac{2}{\pi}\right)^2 \int_0^{\pi/2} xe^{16\alpha^2 x^2}\cos^4 x\,dx$$

for a gaussian input pulse and a raised-cosine output.

(*b*) For a GaAs FET at 20°C, compare the values of $\langle i_R^2 \rangle$, $\langle i_I^2 \rangle$, $\langle i_E^2 \rangle$, and $\langle i_f^2 \rangle$ for bit rates B ranging from 50 Mb/s to 5 Gb/s. Letting the amplifier gain $A = 1$, use the performance parameter values in Table P7-26.

TABLE P7-26

Parameter	Value
FET gate leakage I_{gate}	50 nA
Transconductance g_m	30 mS
Noise figure Γ	1.1
Capacitance C	1.5 pF
$1/f$ corner frequency f_c	20 MHz
Resistance $R_b = R$	400 Ω
I_f	0.12
I_2	0.50
I_3	0.085

7-27. Using Eq. (E-10) for the bandwidth definition, show that the bandwidths of the transfer functions given by Eqs. (7-97) and (7-98) are $1/4RC$ and $A/4RC$.

7-28. Show that the signal-to-noise ratio given by Eq. (7-105) is a maximum when the gain is optimized at

$$M_{opt}^{2+x} = \frac{4k_B TF_t/R_{eq}}{q(I_p + I_D)x}$$

7-29. (*a*) Show that, when the gain M is given by the expression in Prob. 7-28, the signal-to-noise ratio given by Eq. (7-105) can be written as

$$\frac{S}{N} = \frac{xm^2}{2B(2+x)} \frac{I_p^2}{\left[q(I_p + I_D)x\right]^{2/(2+x)}} \left(\frac{R_{eq}}{4k_B TF_t}\right)^{x/(2+x)}$$

(*b*) Show that, when I_p is much larger than I_D, the foregoing expression becomes

$$\frac{S}{N} = \frac{m^2}{2Bx(2+x)} \left[\frac{(xI_p)^{2(1+x)}}{q^2(4k_B TF_t/R_{eq})^x}\right]^{1/(2+x)}$$

7-30. Consider the signal-to-noise ratio expression given in Prob. 7-29*a*. Analogous to Figure 7-21, plot S/N in dB [that is, $10\log(S/N)$] as a function of the received power level P_r in dBm when the dark current $I_D = 10$ nA and $x = 1.0$. Let $B = 5$ MHz, $m = 0.8$, $\mathcal{R}_0 = 0.5$ A/W, $T = 300$ K, and $R_{eq}/F_t = 10^4 \Omega$. Recall that $I_p = \mathcal{R}_0 P_r$.

REFERENCES

1. S.D. Personick, "Receiver design for digital fiber optic communication systems," *Bell Sys. Tech. J.*, vol. 52, pp. 843–886, July–Aug. 1973.
2. W. M. Hubbard, "Utilization of optical-frequency carriers for low and moderate bandwidth channels," *Bell Sys. Tech. J.*, vol. 52, pp. 731–765, May–June 1973.
3. J. E. Mazo and J. Salz, "On optical data communication via direct detection of light pulses," *Bell Sys. Tech. J.*, vol. 55, pp. 347–369, Mar. 1976.
4. S. D. Personick, P. Balaban, J. Bobsin, and P. Kumer, "A detailed comparison of four approaches to the calculation of the sensitivity of optical fiber receivers," *IEEE Trans. Commun.*, vol. COM-25, pp. 541–548, May 1977.
5. (*a*) G. L. Cariolaro, "Error probability in digital optical fiber communication systems," *IEEE Trans. Inform. Theory*, vol. IT-24, pp. 213–221, Mar. 1978.
   (*b*) R. Dogliotti, A. Luvison, and G. Pirani, "Error probability in optical fiber transmission systems," *IEEE Trans. Inform. Theory*, vol.IT-25, pp. 170–178, Mar. 1979.
   (*c*) D. G. Messerschmitt, "Minimum MSE equalization of digital fiber optic systems," *IEEE Trans. Commun.*, vol. COM-26, pp. 1110–1118, July 1978.
   (*d*) W. Hauk, F. Bross, and M. Ottka, "The calculation of error rates for optical fiber systems," *IEEE Trans. Commun.*, vol. COM-26, pp. 1119–1126, July 1978.
6. (*a*) M. Mansuripur, J. W. Goodman, E. G. Rawson, and R. E. Norton, "Fiber optics receiver error rate prediction using the Gram-Charlier method," *IEEE Trans. Commun.*, vol. COM-28, pp. 402–407, Mar. 1980.
   (*b*) J. C. Cartledge and L. W. Coathup, "A Gram-Charlier series method of calculating the probability of error in lightwave transmission systems," *J. Lightwave Tech.*, vol. LT-4, pp. 1736–1739, Dec. 1986.
   (*c*) J. J. O'Reilly, J. R. F. da Rocha, and K. Schumacher, "Optical fiber direct detection receivers optimally tolerant to jitter," *IEEE Trans. Commun.*, vol. COM-34, pp. 1141–1147, Nov. 1986.

7. P. Balaban, "Statistical evaluation of the error rate of the fiberguide repeater using importance sampling," *Bell Sys. Tech. J.*, vol. 55, pp. 745–766, July–Aug. 1976.

8. D. R. Smith and I. Garrett, "A simplified approach to digital optical receiver design," *Opt. Quantum Electron.*, vol. 10, pp. 211–221, 1978.

9. S. D. Personick, "Receiver design for optical fiber systems," *Proc. IEEE*, vol. 65, pp. 1670–1678, Dec. 1977.

10. T. V. Muoi, "Receiver design for high-speed optical-fiber systems," *J. Lightwave Tech.*, vol. LT-2, pp. 243–267, June 1984.

11. S. R. Forrest, "The sensitivity of photoconductive receivers for long-wavelength optical communications," *J. Lightwave Tech.*, vol. LT-3, pp. 347–360, Apr. 1985.

12. M. Brain and T.-P. Lee, "Optical receivers for lightwave communication systems," *J. Lightwave Tech.*, vol. LT-3, pp. 1281–1300, Dec. 1985.

13. B. L. Casper and J. C. Campbell, "Multigigabit-per-second avalanche photodiode lightwave receivers," *J. Lightwave Tech.*, vol. LT-5, pp. 1351–1364, Oct. 1987.

14. C. W. Helstrom, "Computing the performance of optical receivers with avalanche diode detectors," *IEEE Trans. Commun.*, vol. 36, pp. 61–66, Jan. 1988.

15. J. L. Gimlett, "Ultrahigh speed optical receivers," *Tech. Digest Opt. Fiber Commun. Conf.*, p. 81, Feb. 1989.

16. S. D. Personick, "Receiver design," in S. E. Miller and A. G. Chynoweth, eds., *Optical Fiber Telecommunications*, Academic, New York, 1979.

17. B. L. Kasper, "Receiver design," in S. E. Miller and I. P. Kaminow, eds., *Optical Fiber Telecommunications II*, Academic, New York, 1988.

18. R. G. Smith and S. D. Personick, "Receiver design for optical fiber communication systems," in H. Kressel, ed., *Semiconductor Devices for Optical Communications*, 2nd ed., Springer-Verlag, New York, 1982.

19. See any basic book on communication systems, for example:
    (*a*) A. B. Carlson, *Communication Systems*, 3rd ed., McGraw-Hill, New York, 1986.
    (*b*) K. S. Shanmugan, *Digital and Analog Communication Systems*, Wiley, New York, 1979.
    (*c*) J. G. Proakis, *Digital Communications*, 2nd ed., McGraw-Hill, New York, 1989.

20. E. A. Lee and D. G. Messerschmitt, *Digital Communication*, Kluwer Academic, Boston, 1988.

21. (*a*) H. Taub and D. L. Schilling, *Principles of Communication Systems*, 2nd ed., McGraw-Hill, New York, 1986.
    (*b*) M. J. Buckingham, *Noise in Electronic Devices and Systems*, Wiley, New York, 1983.
    (*c*) A. van der Ziel, *Noise in Solid State Devices and Circuits*, Wiley, New York, 1986.

22. E. A. Newcombe and S. Pasupathy, "Error rate monitoring for digital communications," *Proc. IEEE*, vol. 70, pp. 805–828, Aug. 1982.

23. (*a*) A. Papoulis, *Probability, Random Variables, and Stochastic Processes*, 2nd ed., McGraw-Hill, New York, 1984.
    (*b*) P. Z. Peebles, Jr., *Probability, Random Variables, and Random Signal Principles*, 2nd ed., McGraw-Hill, New York, 1987.

24. M. Abramowitz and I. A. Stegun, *Handbook of Mathematical Functions*, Dover, New York, 1965.

25. R. S. Fyath and J. J. O'Reilly, "Performance degradation of APD-optical receivers due to dark current generated within the multiplication region," *J. Lightwave Tech.*, vol. 7, pp. 62–67, Jan. 1989.

26. J. M. Wozencroft and I. M. Jacobs, *Principles of Communication Engineering*, Wiley, New York, 1965.

27. R. C. Hooper and R. B. White, "Digital optical receiver design for non-zero extinction ratio using a simplified approach," *Opt. Quantum Electron.*, vol. 10, pp. 279–282, 1978.

28. J. E. Goell, "An optical repeat with high impedance input amplifier," *Bell Sys. Tech. J.*, vol. 53, pp. 629–643, Apr. 1974.

29. R. A. Minasian, "Optimum design of a 4-Gb/s GaAs MESFET optical preamplifier," *J. Lightwave Tech.*, vol. LT-5, pp. 373–379, Mar. 1987.

30. A. A. Abidi, "On the choice of optimum FET size in wide-band transimpedance amplifiers," *J. Lightwave Tech.*, vol. 6, pp. 64–66, Jan. 1988.

31. R. M. Jopson, A. H. Gnauck, B. L. Kasper, R. E. Tench, N. A. Olsson, C. A. Burrus, and A. R. Chraplyvy, "8 Gbit/s 1.3 μm receiver using optical preamplifier," *Electron. Lett.*, vol. 25, pp. 233–235, Feb. 2, 1989.

32. (*a*) A. van der Ziel, *Introductory Electronics*, Prentice-Hall, Englewood Cliffs, NJ, 1974.
    (*b*) D. Schilling and C. Belove, *Electronic Circuits: Discrete and Integrated*, 3rd ed., McGraw-Hill, New York, 1989.

33. T. T. Ha, *Solid State Microwave Amplifier Design*, Wiley, New York, 1981.

34. (*a*) E. Yang, *Microelectronic Devices*, McGraw-Hill, New York, 1988.
    (*b*) M. Zambuto, *Semiconductor Devices*, McGraw-Hill, New York, 1989.

35. J. L. Hullett and T. V. Muoi, "A feedback receiver amplifier for optical transmission systems," *IEEE Trans. Comm.*, vol. COM-24, pp. 1180–1185, Oct. 1976.

36. B. L. Kasper, A. R. McCormick, C. A. Burrus, and J. R. Talman, "An optical-feedback transimpedance receiver for high sensitivity and wide dynamic range at low bit rates," *J. Lightwave Tech.*, vol. 6, pp. 329–338, Feb. 1988.

37. G. F. Williams and H. P. LeBlanc, "Active feedback lightwave receivers," *J. Lightwave Tech.*, vol. LT-4, pp. 1502–1508, Oct. 1986.

38. R. G. Swartz, "High performance integrated circuits for lightwave systems," in S. E. Miller and I. P. Kaminow, eds., *Optical Fiber Telecommunications II*, Academic, New York, 1988.

39. See special issue on "Microwave Aspects and Applications of GHz/Gbit Technology," *J. Lightwave Tech.*, vol. LT-5, Mar. 1987.

40. See special issue on "Applications of RF and Microwave Subcarriers to Optical Fiber Transmission in Present and Future Broadband Networks," *J. Lightwave Tech.*, vol. 8, 1990.

41. R. G. Swartz, "Electronics for high bit rate systems," *Tech. Digest Opt. Fiber Commun. Conf.*, p. 82, Feb. 1989.

42. B. Wedding, D. Schlump, E. Schlag, W. Pöhlmann, and B. Franz, "2.24-Gb/s 151-km optical transmission system using high-speed integrated silicon chips," *J. Lightwave Tech.*, vol. 8, pp. 227–234, Feb. 1990.

43. V. L. Mirtich, "Designer's guide to fiber optic data links—parts 1, 2, 3," *EDN*, vol. 25, pp. 133–140, June 20, 1980; pp. 113–117, Aug. 5, 1980; pp. 103–110, Aug. 20, 1980.

44. "High speed fiber optic link design with discrete components," *Hewlett Packard Application Note 1022*, Jan. 1985.

CHAPTER
8

DIGITAL TRANSMISSION SYSTEMS

The preceding chapters have presented the fundamental characteristics of the individual building blocks of an optical fiber transmission link. These include the optical fiber transmission medium, the optical source, the photodetector and its associated receiver, and the connectors used to join individual fiber cables to each other and to the source and detector. Now we shall examine how these individual parts can be put together to form a complete optical fiber transmission link. In particular, we shall study digital links in this chapter, and analog links in Chap. 9.

The first discussion involves the simplest case of a point-to-point link. This will include examining the components that are available for a particular application and seeing how these components relate to the system performance criteria (such as dispersion and bit error rate). For a given set of components and a given set of system requirements, we then carry out a power budget analysis to determine whether the fiber optic link meets the attenuation requirements or if repeaters are needed to boost the power level. The final step is to perform a system rise-time analysis to verify that the overall system performance requirements are met.

We next turn our attention to line-coding schemes that are suitable for digital data transmission over optical fibers. These coding schemes are used to introduce randomness and redundancy into the digital information stream to ensure efficient timing recovery and to facilitate error monitoring at the receiver.

315

Once a digital optical fiber link has been installed, a measurement of its overall performance is usually necessary. A simple but powerful method for assessing the data transmission characteristics of the link is the eye-pattern technique. This method has been extensively used for wire systems, and its application to optical fiber links is described in Sec. 8.3.

As one moves to very high-speed (over 400 Mb/s) single-mode applications, a variety of system and component noise factors effect the fiber optic transmission quality. These include modal noise, mode-partition noise, laser chirping, and reflection noise, which are the topic of Sec. 8.4.

8.1   POINT-TO-POINT LINKS

The simplest transmission link is a point-to-point line having a transmitter on one end and a receiver on the other, as is shown in Fig. 8-1. This type of link places the least demand on optical fiber technology and thus sets the basis for examining more complex system architectures.[1-8]

The design of an optical link involves many interrelated variables among the fiber, source, and photodetector operating characteristics, so that the actual link design and analysis may require several iterations before they are completed satisfactorily. Since performance and cost constraints are very important factors in fiber optic communication links, the designer must carefully choose the components to ensure that the desired performance level can be maintained over the expected system lifetime without overspecifying the component characteristics.

The key system requirements needed in analyzing a link are:

1. The desired (or possible) transmission distance
2. The data rate or channel bandwidth
3. The bit error rate (BER)

To fulfill these requirements the designer has a choice of the following components and their associated characteristics:

1. Multimode or single-mode optical fiber
   (*a*) Core size
   (*b*) Core refractive-index profile
   (*c*) Bandwidth or dispersion

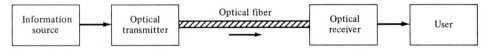

FIGURE 8-1
Simplex point-to-point link.

(*d*) Attenuation

(*e*) Numerical aperture or mode-field diameter

2. LED or laser diode optical source

(*a*) Emission wavelength

(*b*) Spectral line width

(*c*) Output power

(*d*) Effective radiating area

(*e*) Emission pattern

(*f*) Number of emitting modes

3. *pin* or avalanche photodiode

(*a*) Responsivity

(*b*) Operating wavelength

(*c*) Speed

(*d*) Sensitivity

Two analyses are usually carried out to ensure that the desired system performance can be met; these are the *link power budget* and the system *rise-time budget* analyses. In the link power budget analysis one first determines the power margin between the optical transmitter output and the minimum receiver sensitivity needed to establish a specified BER. This margin can then be allocated to connector, splice, and fiber losses, plus any additional margins required for expected component degradation or temperature effects. If the choice of components did not allow the desired transmission distance to be achieved, the components might have to be changed or repeaters might have to be incorporated into the link.

Once the link power budget has been established, the designer can perform a system rise-time analysis to ensure that the desired overall system performance has been met. We shall now examine these two analyses in more detail.

8.1.1 System Considerations

In carrying out a link power budget, we first decide at which wavelength to transmit and then choose components operating in this region. If the distance over which the data are to be transmitted is not too far, we may decide to operate in the 800- to 900-nm region. On the other hand, if the transmission distance is relatively long, we may want to take advantage of the lower attenuation and dispersion that occurs at wavelengths around 1300 or 1550 nm.

Having decided on a wavelength, we next interrelate the system performances of the three major optical link building blocks, that is, the receiver, transmitter, and optical fiber. Normally the designer chooses the characteristics of two of these elements and then computes those of the third to see if the system performance requirements are met. If the components have been over- or underspecified, a design iteration may be needed. The procedure we shall follow here is first to select the photodetector. We then choose an optical source and see how far data can be transmitted over a particular fiber before a repeater is needed in the line to boost up the power level of the optical signal.

In choosing a particular photodetector, we mainly need to determine the minimum optical power that must fall on the photodetector to satisfy the bit error rate (BER) requirement at the specified data rate. In making this choice, the designer also needs to take into account any design cost and complexity constraints. As noted in Chaps. 6 and 7, a *pin* photodiode receiver is simpler, more stable with changes in temperature, and less expensive than an avalanche photodiode receiver. In addition, *pin* photodiode bias voltages are normally less than 50 V, whereas those of avalanche photodiodes are several hundred volts. However, the advantages of *pin* photodiodes may be overruled by the increased sensitivity of the avalanche photodiode if very low optical power levels are to be detected.

The system parameters involved in deciding between the use of an LED and a laser diode are signal dispersion, data rate, transmission distance, and cost. As shown in Chap. 4, the spectral width of the laser output is much narrower than that of an LED. This is of importance in the 800- to 900-nm region, where the spectral width of an LED and the dispersion characteristics of silica fibers limit the data-rate–distance product to around 150 (Mb/s) · km. For higher values [up to 2500 (Mb/s) · km] a laser must be used at these wavelengths. At wavelengths around 1.3 μm, where signal dispersion is very low, bit-rate–distance products of at least 1500 (Mb/s) · km are achievable with LEDs. For InGaAsP lasers this figure is in excess of 25 (Gb/s) · km.

Since laser diodes typically couple from 10 to 15 dB more optical power into a fiber than an LED, greater repeaterless transmission distances are possible with a laser. This advantage and the lower dispersion capability of laser diodes may be offset by cost constraints. Not only is a laser diode itself more expensive than an LED, but also the laser transmitter circuitry is much more complex, since the lasing threshold has to be dynamically controlled as a function of temperature and device aging.

For the optical fiber we have a choice between single-mode and multi-mode fiber, either of which could have a step- or a graded-index core. This choice depends on the type of light source used and on the amount of dispersion that can be tolerated. Light-emitting diodes (LEDs) tend to be used with multimode fibers, although, as we saw in Chap. 5, edge-emitting LEDs can launch sufficient optical power into a single-mode fiber for transmission at data rates up to 560 Mb/s over several kilometers. The optical power that can be coupled into a fiber from an LED depends on the core-cladding index differ-ence Δ, which, in turn, is related to the numerical aperture of the fiber (for $\Delta = 0.01$ the numerical aperture NA $\simeq 0.21$). As Δ increases, the fiber-coupled power increases correspondingly. However, since dispersion also becomes greater with increasing Δ, a tradeoff must be made between the optical power that can be launched into the fiber and the maximum tolerable dispersion.

Either a single-mode or a multimode fiber can be used with a laser diode. A single-mode fiber can provide the ultimate bit-rate–distance product, with values of 30 (Gb/s) · km being achievable. A disadvantage of single-mode fibers is that the small core size (5 to 16 μm in diameter) makes fiber splicing more difficult and critical than for multimode fibers having 50-μm core diameters.

When choosing the attenuation characteristics of a cabled fiber, the excess loss that results from the cabling process must also be considered in addition to the attenuation of the fiber itself. This must also include connector and splice losses as well as environmental-induced losses that could arise from temperature variations, radiation effects, and dust and moisture on the connectors.

8.1.2 Link Power Budget

An optical power loss model for a point-to-point link is shown in Fig. 8-2. The optical power received at the photodetector depends on the amount of light coupled into the fiber and the losses occurring in the fiber and at the connectors and splices. The link loss budget is derived from the sequential loss contributions of each element in the link. Each of these loss elements is expressed in decibels (dB) as

$$\text{loss} = 10 \log \frac{P_{\text{out}}}{P_{\text{in}}} \tag{8-1}$$

where P_{in} and P_{out} are the optical powers emanating into and out of the loss element, respectively.

In addition to the link loss contributors shown in Fig. 8-2, a link power margin is normally provided in the analysis to allow for component aging, temperature fluctuations, and losses arising from components that might be added at future dates. A link margin of 6 to 8 dB is generally used for systems that are not expected to have additional components incorporated into the link in the future.

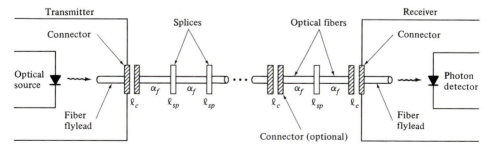

FIGURE 8-2

Optical power loss model for a point-to-point link. The losses occur at connectors (l_c), at splices (l_{sp}), and in the fiber (α_f).

The link loss budget simply considers the total optical power loss P_T that is allowed between the light source and the photodetector, and allocates this loss to cable attenuation, connector loss, splice loss, and system margin. Thus if P_S is the optical power emerging from the end of a fiber flylead attached to the light source, and if P_R is the receiver sensitivity, then

$$P_T = P_S - P_R$$

$$= 2l_c + \alpha_f L + \text{system margin} \qquad (8\text{-}2)$$

Here l_c is the connector loss, α_f is the fiber attenuation (dB/km), L is the transmission distance, and the system margin is nominally taken as 6 dB. Here we assume that the cable of length L has connectors only on the ends and none in between. The splice loss is incorporated into the cable loss for simplicity.

Example 8-1. To illustrate how a link loss budget is set up, let us carry out a specific design example. We shall begin by specifying a data rate of 20 Mb/s and a bit error rate of 10^{-9} (that is, at most one error can occur for every 10^9 bits sent). For the receiver we shall choose a silicon *pin* photodiode operating at 850 nm. Figure 8-3 shows that the required receiver input signal is -42 dBm (42 dB below 1 mW). We next select a GaAlAs LED which can couple a 50-μW (-13-dBm) average optical power level into a fiber flylead with a 50-μm core diameter. We thus have a 29-dB allowable power loss. Assume further that a 1-dB loss occurs

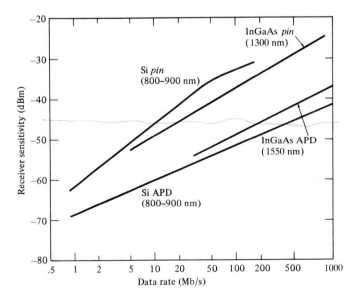

FIGURE 8-3
Receiver sensitivities as a function of bit rate. The Si *pin*, Si APD, and InGaAs *pin* curves are for a 10^{-9} BER. The InGaAs APD curve is for a 10^{-11} BER.

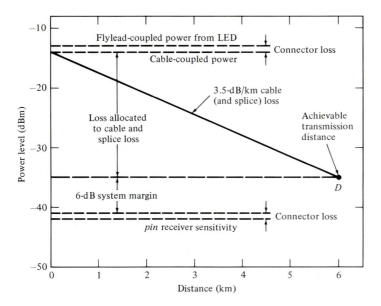

FIGURE 8-4
Graphical representation of a link-loss budget for an 800-nm LED/*pin* system operating at 20 Mb/s.

when the fiber flylead is connected to the cable and another 1-dB connector loss occurs at the cable-to-photodetector interface. Including a 6-dB system margin, the possible transmission distance for a cable with an attenuation of α_f dB/km can be found from Eq. (8-2):

$$P_T = P_S - P_R = 29 \text{ dB}$$

$$= 2(1 \text{ dB}) + \alpha_f L + 6 \text{ dB}$$

If $\alpha_f = 3.5$ dB/km, then a 6.0-km transmission path is possible.

The link power budget can be represented graphically as is shown in Fig. 8-4. The vertical axis represents the optical power loss allowed between the transmitter and the receiver. The horizontal axis gives the transmission distance. Here we show a silicon *pin* receiver with a sensitivity of -42 dBm (at 20 Mb/s) and an LED with an output power of -13 dBm coupled into a fiber flylead. We subtract a 1-dB connector loss at each end, which leaves a total margin of 27 dB. Subtracting a 6-dB system safety margin leaves us with a tolerable loss of 21 dB that can be allocated to cable and splice loss. The slope of the line shown in Fig. 8-4 is the 3.5-dB/km cable (and splice, in this case) loss. This line starts at the -14-dBm point (which is the optical power coupled into the cabled fiber) and ends at the -35-dBm level (the receiver sensitivity minus a 1-dB connector loss and a 6-dB system margin). The intersection point D then defines the maximum possible transmission path length.

8.1.3 Rise-Time Budget

A rise-time budget analysis is a convenient method for determining the dispersion limitation of an optical fiber link. This is particularly useful for digital systems. In this approach the total rise time t_{sys} of the link is the root-sum-square of the rise times from each contributor t_i to the pulse rise-time degradation:

$$t_{sys} = \left(\sum_{i=1}^{N} t_i^2 \right)^{1/2}$$ (8-3)

The four basic elements that may significantly limit system speed are the transmitter rise time t_{tx}, the material dispersion rise time t_{mat} of the fiber, the modal dispersion rise time t_{mod} of the fiber, and the receiver rise time t_{rx}. Generally, the total transition-time degradation of a digital link should not exceed 70 percent of an NRZ (non-return-to-zero) bit period or 35 percent of a bit period for RZ (return-to-zero) data, where one bit period is defined as the reciprocal of the data rate (NRZ and RZ data formats are discussed in more detail in Sec. 8.2).

The rise times of transmitters and receivers are generally known to the designer. The transmitter rise time is attributable primarily to the light source and its drive circuitry. The receiver rise time results from the photodetector response and the 3-dB electrical bandwidth of the receiver front end. The response of the receiver front end can be modeled by a first-order lowpass filter having a step response[9]

$$g(t) = [1 - \exp(-2\pi B_{rx}t)]u(t)$$

where B_{rx} is the 3-dB electrical bandwidth of the receiver and $u(t)$ is the unit step function which is 1 for $t \geq 0$ and 0 for $t < 0$. The rise time t_{rx} of the receiver is usually defined as the time interval between $g(t) = 0.1$ and $g(t) = 0.9$. This is known as the *10- to 90-percent rise time*. Thus, if B_{rx} is given in megahertz, then the receiver front-end rise time in nanoseconds is

$$t_{rx} = \frac{350}{B_{rx}}$$ (8-4)

For multimode fibers the rise time depends on modal and material dispersions. Its analysis is more complicated, since it is a function of the length of the fiber, the type of optical source used, and the operating wavelength. Material dispersion effects can be neglected for laser sources at both short and long wavelengths, and for LEDs at long wavelengths. Using Eq. (3-28), we see that in the 800- to 900-nm region material dispersion adds about 0.07 ns/ (nm · km) to the rise time.

For cables shorter than the modal equilibrium length, the fiber bandwidth resulting from modal dispersion is inversely proportional to the cable length. As shown in Sec. 3.4, in a long, continuous fiber which has no joints, the fiber bandwidth decreases linearly with increasing distance for lengths less than the modal equilibrium length L_c. For lengths greater than L_c, a steady-state equilibrium condition has been established and the bandwidth then decreases as $L^{1/2}$.

In practice, an optical fiber link seldom consists of a continuous, jointless fiber. Instead, several fibers are concatenated (tandemly joined) to form a long link. The situation then becomes more complex because a modal redistribution occurs at fiber-to-fiber joints in the cable. This is a result of misaligned joints, different core index profiles in each fiber, and/or different degrees of mode mixing in individual fibers. The most important of these factors is the effect of different core index profiles in adjacent fibers. As shown in Sec. 3.3, the value of the core index profile α influences the degree of modal-dispersion-induced pulse spreading in a fiber. The value of the index-grading parameter α that minimizes pulse dispersion depends strongly on the wavelength, so that fibers optimized for operation at different wavelengths have different values of α. Variations in α at the same wavelength thus result in overcompensated or undercompensated core index profiles (see Fig. 3-18).

The difficulty in predicting the bandwidth of a series of concatenated fibers arises from the observation that the total route bandwidth can be a function of the order in which fibers are joined. For example, instead of randomly joining together arbitrary (but very similar) fibers, an improved total link bandwidth can be obtained by selecting adjoining fibers with alternating over- and undercompensated refractive-index profiles to provide some modal delay equalization. Although the ultimate concatenated fiber bandwidth can be obtained by judiciously selecting adjoining fibers for optimum modal delay equalization, in practice this is unwieldy and time-consuming, particularly since the initial fiber in the link appears to control the final link characteristics.

A variety of empirical expressions for modal dispersion have thus been developed.[10-15] From practical field experience it has been found that the bandwidth B_M in a link of length L can be expressed to a reasonable approximation by the empirical relation

$$B_M(L) = \frac{B_0}{L^q} \tag{8-5}$$

where the parameter q ranges between 0.5 and 1, and B_0 is the bandwidth of a 1-km length of cable. A value of $q = 0.5$ indicates that a steady-state modal equilibrium has been reached, whereas $q = 1$ indicates little mode mixing. Based on field experience, a reasonable estimate is $q = 0.7$.

Another expression that has been proposed for B_M, based on curve fitting of experimental data, is

$$\frac{1}{B_M} = \left[\sum_{n=1}^{N} \left(\frac{1}{B_n} \right)^{1/q} \right]^q \tag{8-6}$$

where the parameter q ranges between 0.5 (quadrature addition) and 1.0 (linear addition), and B_n is the bandwidth of the nth fiber section. Alternatively,

Eq. (8-6) can be written as

$$t_M(N) = \left[\sum_{n=1}^{N} (t_n)^{1/q} \right]^q \tag{8-7}$$

where $t_M(N)$ is the pulse broadening occurring over N cable sections in which the individual pulse broadenings are given by t_n.

A third empirical expression proposed by Eve[10] for pulse broadening in a jointed link of N fibers is

$$t_M^2(N) = \sum_{k=1}^{N} t_k^2 + \sum_{\substack{1 \\ p \neq k}}^{N} t_p t_k r_{pk} \tag{8-8}$$

where r_{pk} is a correlation coefficient between the pth and the kth fiber. Its magnitude is expected to range between 0 and 1 for strong and little mode mixing, respectively.

We now need to find the relation between the fiber rise time and the 3-dB bandwidth. For this we use a variation of the expression derived by Midwinter.[16] We assume that the optical power emerging from the fiber has a gaussian temporal response described by

$$g(t) = \frac{1}{\sqrt{2\pi}\,\sigma} e^{-t^2/2\sigma^2} \tag{8-9}$$

where σ is the rms pulse width.

The Fourier transform of this function is

$$G(\omega) = \frac{1}{\sqrt{2\pi}} e^{-\omega^2\sigma^2/2} \tag{8-10}$$

From Eq. (8-9) the time $t_{1/2}$ required for the pulse to reach its half-maximum value, that is, the time required to have

$$g(t_{1/2}) = 0.5g(0) \tag{8-11}$$

is given by

$$t_{1/2} = (2\ln 2)^{1/2}\sigma \tag{8-12}$$

If we define the time t_{FWHM} as the full width of the pulse at its half-maximum value, then

$$t_{\mathrm{FWHM}} = 2t_{1/2} = 2\sigma(2\ln 2)^{1/2} \tag{8-13}$$

The 3-dB optical bandwidth $B_{3\,\mathrm{dB}}$ is defined as the modulation frequency $f_{3\,\mathrm{dB}}$ at which the received optical power has fallen to 0.5 of the zero frequency value. Thus from Eqs. (8-10) and (8-13) we find that the relation between the full-width half-maximum rise time t_{FWHM} and the 3-dB optical bandwidth is

$$f_{3\,dB} = B_{3\,dB} = \frac{0.44}{t_{FWHM}} \tag{8-14}$$

Using Eq. (8-5) for the 3-dB optical bandwidth of the fiber link and letting t_{FWHM} be the rise time resulting from modal dispersion, then from Eq. (8-14),

$$t_{mod} = \frac{0.44}{B_M} = \frac{0.44 L^q}{B_0} \tag{8-15}$$

If t_{mod} is expressed in nanoseconds and B_M is given in megahertz, then

$$t_{mod} = \frac{440}{B_M} = \frac{440 L^q}{B_0} \tag{8-16}$$

Substituting Eqs. (3-20), (8-4), and (8-16) into Eq. (8-3) gives a total system rise time of

$$t_{sys} = \left[t_{tx}^2 + D_{mat}^2 \sigma_\lambda^2 L^2 + \left(\frac{440 L^q}{B_0} \right)^2 + \left(\frac{350}{B_{rx}} \right)^2 \right]^{1/2} \tag{8-17}$$

where all the times are given in nanoseconds, σ_λ is the spectral width of the optical source, and D_{mat} is the material dispersion factor of the fiber (given in nanoseconds per nanometer per kilometer). In the 800- to 900-nm region, D_{mat} is about 0.07 ns/(nm · km), but is negligible around 1300 nm (see Fig. 3-13).

Example 8-2. As an example of a rise-time budget, let us continue the analysis of the link we started to examine in Sec. 8.1.2. We shall assume that the LED together with its drive circuit has a rise time of 15 ns. Taking a typical LED spectral width of 40 nm, we have a material-dispersion-related rise-time degradation of 21 ns over the 6-km link. Assuming the receiver has a 25-MHz bandwidth, then from Eq. (8-4) the contribution to the rise-time degradation from the receiver is 14 ns. If the fiber we select has a 400-MHz · km bandwidth-distance product and with $q = 0.7$ in Eq. (8-5), then from Eq. (8-15) the modal-dispersion-induced fiber rise time is 3.9 ns. Substituting all these values back into Eq. (8-17) results in a link rise time of

$$t_{sys} = \left(t_{tx}^2 + t_{mat}^2 + t_{mod}^2 + t_{rx}^2 \right)^{1/2}$$

$$= \left[(15\ ns)^2 + (21\ ns)^2 + (3.9\ ns)^2 + (14\ ns)^2 \right]^{1/2}$$

$$= 30\ ns$$

This value falls below the maximum allowable 35-ns rise-time degradation for our 20-Mb/s NRZ data stream. The choice of components was thus adequate to meet our system design criteria.

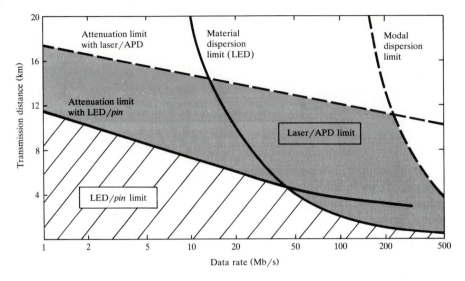

FIGURE 8-5
Transmission-distance limits as a function of data rate for an 800-MHz · km fiber, an 800-nm LED source with a Si *pin* photodiode combination, and an 800-nm laser diode with a Si APD.

8.1.4 First-Window Transmission Distance

Figure 8-5 shows the attenuation and dispersion limitation on the repeaterless transmission distance as a function of data rate for the short-wavelength (800- to 900-nm) LED/*pin* combination. The BER was taken as 10^{-9} for all data rates. The fiber-coupled LED output power was assumed to be a constant -13 dBm for all data rates up to 200 Mb/s. The attenuation limit curve was then derived by using a fiber loss of 3.5 dB/km and the receiver sensitivities shown in Fig. 8-3. Since the minimum optical power required at the receiver for a given BER becomes higher for increasing data rates, the attenuation limit curve slopes downward to the right. We have also included a 1-dB connector-coupling loss at each end and a 6-dB system operating margin.

The dispersion limit depends on material and modal dispersion. Material dispersion at 800 nm is taken as 0.07 ns/(nm · km) or 3.5 ns/km for an LED with a 50-nm spectral width. The curve shown is the material dispersion limit in the absence of modal dispersion. This limit was taken to be the distance at which t_{mat} is 70 percent of a bit period. The modal dispersion was derived from Eq. (8-15) for a fiber with an 800-MHz · km bandwidth-distance product and with $q = 0.7$. The modal dispersion limit was then taken to be the distance at which t_{mod} is 70 percent of a bit period. The achievable repeaterless transmission distances are those that fall below the attenuation limit curve and to the left of the dispersion line, as indicated by the hatched area. The transmission distance is attenuation-limited up to about 40 Mb/s, after which it becomes material-dispersion-limited.

Greater transmission distances are possible when a laser diode is used in conjunction with an avalanche photodiode. Let us consider an AlGaAs laser emitting at 850 nm with a spectral width of 1 nm which couples 0 dBm (1 mW) into a fiber flylead. The receiver uses an APD with a sensitivity depicted in Fig. 8-3. The fiber is the same as described in Sec. 8.1.4. In this case the material-dispersion-limited curve lies off the graph to the right of the modal dispersion curve, and the attenuation limit (with an 8-dB system margin) is as shown in Fig. 8-5. The achievable transmission distances now include those indicated by the shaded area.

8.1.5 Transmission Distance for Single-Mode Links

At the other extreme from that shown in Fig. 8-5, let us examine a single-mode link operating at 1550 nm. In this case the dispersion in the fiber is due only to material and waveguide effects, since there is no modal dispersion. We take the dispersion to be $D = 2.5$ ps/(nm $\cdot$ km) and the attenuation to be 0.30 dB/km at 1550 nm. For the source we choose a distributed-feedback (DFB) laser which couples 0 dBm of optical power into the fiber and which has a spectral width $\sigma_\lambda = 3.5$ nm. The receiver can use either an InGaAs avalanche photodiode (APD) with a sensitivity of $P_r = 11.5 \log B - 71.0$ dBm or an InGaAs *pin*

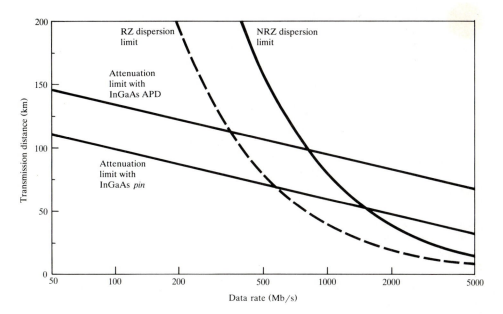

FIGURE 8-6
Transmission-distance limits as a function of data rate for a 1550-nm DFB laser diode, an InGaAs APD, and a single-mode fiber with $\dot{D} = 2.5$ ps/(nm $\cdot$ km) and a 0.3-dB/km attenuation.

photodiode with a sensitivity of $P_r = 11.5 \log B - 60.5$ dBm, where B is the data rate in Mb/s. The attenuation-limited transmission distances for these two photodiodes are shown in Fig. 8-6 with the inclusion of an 8-dB system margin.

For the dispersion limit we examine two cases. First, Fig. 8-6 shows the limit for NRZ data where the product $D\sigma_\lambda L$ is equal to 70 percent of the bit period. Second, for RZ data the product $D\sigma_\lambda L$ is equal to 35 percent of the bit period. These curves are for the ideal case. In reality various noise effects due to laser instabilities coupled with chromatic dispersion in the fiber can greatly decrease the dispersion-limited distance. Section 8.4 discusses these factors and their effect on system performance.

8.2   LINE CODING

In designing an optical fiber link, an important consideration is the format of the transmitted optical signal. This is of importance because, in any practical digital optical fiber data link, the decision circuitry in the receiver must be able to extract precise timing information from the incoming optical signal. The three main purposes of timing are to allow the signal to be sampled by the receiver at the time the signal-to-noise ratio is a maximum, to maintain the proper pulse spacing, and to indicate the start and end of each timing interval. In addition, since errors resulting from channel noise and distortion mechanisms can occur in the signal detection process, it may be desirable for the optical signal to have an inherent error-detecting capability. These features can be incorporated into the data stream by restructuring (or encoding) the signal. This is generally done by introducing extra bits into the raw data stream at the transmitter on a regular and logical basis and extracting them again at the receiver.

Signal encoding uses a set of rules for arranging the signal symbols in a particular pattern. This process is called *channel* or *line coding*. The purpose of this section is to examine the various types of line codes that are well suited for digital transmission on an optical fiber link. The discussion here is limited to binary codes, because they are the most widely used electrical codes and also because they are the most advantageous codes for optical systems.

One of the principal functions of a line code is to introduce redundancy into the data stream for the purpose of minimizing errors resulting from channel interference effects. Depending on the amount of redundancy introduced, any degree of error-free transmission of digital data can be achieved, provided that the data rate that includes this redundancy is less than the channel capacity. This is a result of the well-known Shannon channel-coding theory.[17, 18]

Although large system bandwidths are attainable with optical fibers, the signal-to-noise considerations of the receiver discussed in Chap. 7 show that larger bandwidths result in larger noise contributions. Thus from noise considerations, minimum bandwidths are desirable. However, a larger bandwidth may be needed to have timing data available from the bit stream. In selecting a particular line code, a tradeoff must therefore be made between timing and noise bandwidth.[19] Normally these are largely determined by the expected characteristics of the raw data stream.

The three basic types of two-level binary line codes that can be used for optical fiber transmission links are the non-return-to-zero (NRZ) format, the return-to-zero (RZ) format, and the phase-encoded (PE) format. In NRZ codes a transmitted data bit occupies a full bit period. For RZ formats the pulse width is less than a full bit period. In the PE format both full-width and half-width data bits are present. Multilevel binary (MLB) signaling[20] is also possible, but it is used much less frequently than the popular NRZ and RZ codes. A brief description of some NRZ and RZ codes will be given here. Additional details can be found in numerous communications books.[9, 21-24]

8.2.1 NRZ Codes

A number of different NRZ codes are widely used, and their bandwidths serve as references for all other code groups. The simplest NRZ code is NRZ-level (or NRZ-L), shown in Fig. 8-7. For a serial data stream an on-off (or unipolar) signal represents a 1 by a pulse of current or light filling an entire bit period, whereas for a 0 no pulse is transmitted. These codes are simple to generate and decode, but they possess no inherent error-monitoring or correcting capabilities and they have no self-clocking (timing) features.

The minimum bandwidth is needed with NRZ coding, but the average power input to the receiver is dependent on the data pattern. For example, the high level of received power occurring in a long string of consecutive 1 bits can result in a *baseline wander* effect, as shown in Fig. 8-8. This effect results from the accumulation of pulse tails that arise from the low-frequency characteristics of the ac-coupling filter in the receiver.[25] If the receiver recovery to the original threshold is slow after the long string of 1 bits has ended, an error may occur if the next 1 bit has a low amplitude.

In addition a long string of NRZ ones or zeros contains no timing information, since there are no level transitions. Thus, unless the timing clocks in the system are extremely stable, a long string of N identical bits could be misinterpreted as either $N - 1$ or $N + 1$ bits. However, the use of highly stable clocks increases system costs and requires a long system startup time to achieve synchronization. Two common techniques for restricting the longest time interval in which no level transitions occur are the use of block codes (see Sec. 8.2.3)

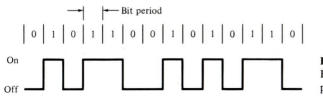

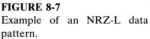

FIGURE 8-7
Example of an NRZ-L data pattern.

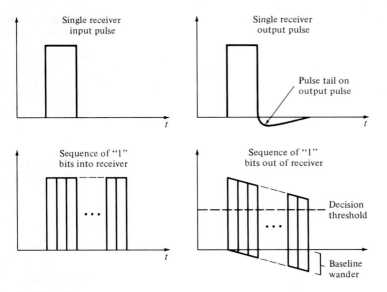

FIGURE 8-8
Baseline wander at the receiver resulting from the transmission of long strings of NRZ 1 bits.

and scrambling.[26-27] *Scrambling* produces a random data pattern by the modulo-2 addition of a known bit sequence with the data stream. At the receiver the same known bit sequence is again modulo-2 added to the received data, and the original bit sequence is recovered. Although the randomness of scrambled NRZ data ensures an adequate amount of timing information, the penalty for its use is an increase in the complexity of the NRZ encoding and decoding circuitry. Transmission experiments at 11 Gb/s over 81 km have been demonstrated with NRZ coding.[28]

8.2.2 RZ Codes

If an adequate bandwidth margin exists, each data bit can be encoded as two optical line code bits. This is the basis of RZ codes. In these codes a signal level transition occurs during either some or all of the bit periods to provide timing information. A variety of RZ code types exist, some of which are shown in Fig. 8-9. The baseband (NRZ-L) data are shown in Fig. 8-9a. In the unipolar RZ data a 1 bit is represented by a half-period optical pulse that can occur in either the first or second half of the bit period. A 0 is represented by no signal during the bit period.

A disadvantage of the unipolar RZ format is that long strings of 0 bits can cause loss of timing synchronization. A common data format not having this limitation is the *biphase* or *optical Manchester* code shown in Fig. 8-9d. Note that this is a unipolar code, which is in contrast to the conventional bipolar

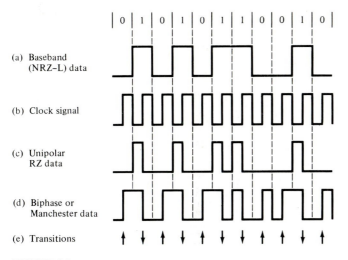

| 0 | 1 | 0 | 1 | 0 | 1 | 1 | 0 | 0 | 1 | 0 |

(a) Baseband (NRZ–L) data

(b) Clock signal

(c) Unipolar RZ data

(d) Biphase or Manchester data

(e) Transitions

FIGURE 8-9

Examples of RZ data formats. (*a*) NRZ-L baseband data; (*b*) clock signal; (*c*) unipolar RZ data; (*d*) biphase or optical Manchester; (*e*) transitions occurring within a bit period for Manchester data.

Manchester code used in wire lines. The optical Manchester signal is obtained by direct modulo-2 addition of the baseband (NRZ-L) signal and a clock signal (Fig. 8-9*b*). In this code there is a transition at the center of each bit interval. A negative-going transition indicates a 1 bit, whereas a positive-going transition means a 0 bit was sent. The Manchester code is simple to generate and decode. Since it is an RZ-type code, it requires twice the bandwidth of an NRZ code. In addition, it has no inherent error-detecting or correcting capability.

Coaxial or wire-pair cable systems commonly use the bipolar RZ or alternate mark inversion (AMI) coding scheme. These wire line codes have also been adapted to unipolar optical systems.[29, 30] The two-level AMI optical pulse formats require twice the transmission bandwidth of NRZ codes, but they provide timing information in the data stream, and the redundancy of the encoded information (which is inherent in these codes) allows for direct in-service error monitoring.[31] Other more complex schemes have been tried for high-speed links.[32–34]

8.2.3 Block Codes

An efficient category of redundant binary codes is the mBnB block code class.[35–38] In this class of codes, blocks of m binary bits are converted to longer blocks of $n > m$ binary bits. These new blocks are then transmitted in NRZ or RZ format. As a result of the additional redundant bits, the increase in bandwidth using this scheme is given by the ratio n/m. At the expense of this increased bandwidth, the mBnB block codes provide adequate timing and

TABLE 8-1
A comparison of several $mBnB$ codes

Code	n / m	N_{max}	D	W (%)
3B4B	1.33	4	± 3	25
6B8B	1.33	6	± 3	75
5B6B	1.20	6	± 4	28
7B8B	1.14	9	± 7	27
9B10B	1.11	11	± 8	24

error-monitoring information, and they do not have baseline wander problems, since long strings of ones and zeros are eliminated.

A convenient concept used for block codes is the *accumulated* or *running disparity*, which is the cumulative difference between the numbers of 1 and 0 bits. A simple means of measuring this is with an up-down counter. The key factors in selecting a particular block code are low disparity and a limit in the disparity variation (the difference between the maximum and minimum values of the accumulated disparity). A low disparity allows the dc component of the signal to be canceled. A bound on the accumulated disparity avoids the low-frequency spectral content of the signal and facilitates error monitoring by detecting the disparity overflow. Generally, one chooses codes that have n even, since for odd values of n there are no coded words with zero disparity.

A comparison of several $mBnB$ codes is given in Table 8-1. The parameters shown in this table are:

1. The ratio n/m, which gives the bandwidth increase
2. The longest number N_{max} of consecutive identical symbols (small values of N_{max} allow for easier clock recovery)
3. The bounds on the accumulated disparity D
4. The percentage W of n-bit words that are not used (the detection of invalid words at the receiver permits character reframing)

The most suitable codes for high data rates are the 3B4B, 5B6B, and 6B8B codes. If simplicity of the encoder and decoder circuits is the main criterion, the 3B4B is the most convenient code. The 5B6B code is the most advantageous if bandwidth reduction is the major concern.

8.3 EYE PATTERN

The eye-pattern technique[9, 21, 23, 39, 40] is a simple but powerful measurement method for assessing the data-handling ability of a digital transmission system. This method has been used extensively for evaluating the performance of wire systems and can also be applied to optical fiber data links. The eye-pattern

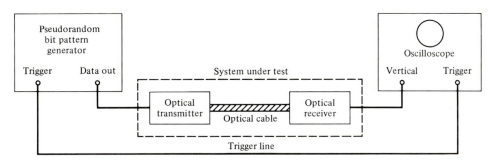

FIGURE 8-10
Basic equipment used for eye-pattern generation.

measurements are made in the time domain and allow the effects of waveform distortion to be shown immediately on an oscilloscope.

An eye-pattern measurement can be made with the basic equipment shown in Fig. 8-10. The output from a pseudorandom data pattern generator is applied to the vertical input of an oscilloscope and the data rate is used to trigger the horizontal sweep. This results in the type of pattern shown in Fig. 8-11, which is called the *eye pattern* because the display shape resembles a human eye. To see how the display pattern is formed, consider the eight possible 3-bit-long NRZ combinations shown in Fig. 8-12. When these eight combinations are superimposed simultaneously, an eye pattern as shown in Fig. 8-11 is formed.

To measure system performance with the eye-pattern method, a variety of word patterns should be provided. A convenient approach is to generate a *random data signal*, because this is the characteristic of data streams found in practice. This type of signal generates ones and zeros at a uniform rate but in a random manner. A variety of pseudorandom pattern generators are available for this purpose. The word *pseudorandom* means that the generated combina-

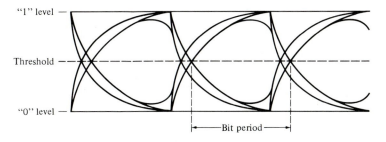

FIGURE 8-11
Sample of an eye-pattern diagram.

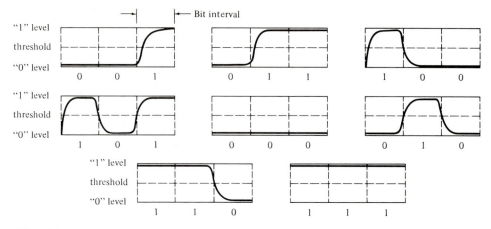

FIGURE 8-12
Eight possible 3-bit-long NRZ combinations.

tion or sequence of ones and zeros will eventually repeat but that it is sufficiently random for test purposes. A pseudorandom bit sequence comprises four different 2-bit-long combinations, eight different 3-bit-long combinations, sixteen different 4-bit-long combinations, and so on (that is, sequences of 2^N different N-bit-long combinations) up to a limit set by the instrument. After this limit has been generated, the data sequence will repeat.

A great deal of system performance information can be deduced from the eye-pattern display. To interpret the eye pattern, consider the simplified drawing shown in Fig. 8-13. The following information regarding the signal ampli-

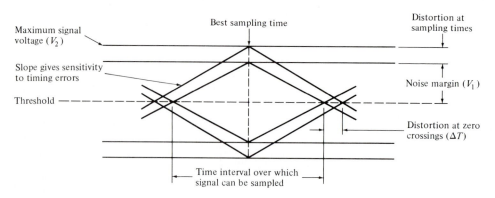

FIGURE 8-13
Simplified eye-pattern diagram and its interpretation.

tude distortion, timing jitter, and system rise time can be derived:

1. The width of the eye opening defines the time interval over which the received signal can be sampled without error from intersymbol interference. The best time to sample the received waveform is when the height of the eye opening is largest.

2. The height of the eye opening is reduced as a result of amplitude distortion in the data signal. The maximum distortion is given by the vertical distance between the top of the eye opening and the maximum signal level. The greater the eye closure becomes, the more difficult it is to detect the signal.

3. The height of the eye opening at the specified sampling time shows the noise margin or immunity to noise. *Noise margin* is the percentage ratio of the peak signal voltage V_1 for an alternating bit sequence (defined by the height of the eye opening) to the maximum signal voltage V_2 as measured from the threshold level, as shown in Fig. 8-13. That is,

$$\text{noise margin (percent)} = \frac{V_1}{V_2} \times 100 \text{ percent} \qquad (8\text{-}18)$$

4. The rate at which the eye closes as the sampling time is varied (that is, the slope of the eye-pattern sides) determines the sensitivity of the system to timing errors. The possibility of timing errors increases as the slope becomes more horizontal.

5. Timing jitter (also referred to as *edge jitter* or *phase distortion*) in an optical fiber system arises from noise in the receiver and pulse distortion in the optical fiber. If the signal is sampled in the middle of the time interval (that is, midway between the times when the signal crosses the threshold level), then the amount of distortion ΔT at the threshold level indicates the amount of jitter. Timing jitter is thus given by

$$\text{timing jitter (percent)} = \frac{\Delta T}{T_b} \times 100 \text{ percent} \qquad (8\text{-}19)$$

where T_b is one bit interval.

6. The 10- to 90-percent rise and fall times of the signal can be easily measured by using the 0- and 100-percent reference levels produced by long strings of zeros and ones, respectively.

7. Any nonlinearities of the channel transfer characteristics will create an asymmetry in the eye pattern. If a purely random data stream is passed through a truly linear system, all the eye openings will be identical and symmetrical.

8.4   NOISE EFFECTS ON SYSTEM PERFORMANCE

In the analysis in Sec. 8.1 we assumed that the optical power falling on the photodetector is a clearly defined function of time within the statistical nature of the quantum detection process. In reality, as noted in Sec. 4.5, various interactions between spectral imperfections in the propagating optical power and the dispersive waveguide give rise to variations in the optical power level falling on the photodetector. These variations create receiver output noises and hence give rise to optical power penalties, which are particularly serious for high-speed links. The main penalties are due to modal noise, wavelength chirp, spectral broadening induced by optical reflections back into the laser, and mode-partition noise. Modal noise is not present in single-mode links; however, mode-partition noise, chirping, and reflection noise are critical in these systems.

8.4.1   Modal Noise

Modal noise arises when the light from a coherent laser is coupled into a multimode fiber.[41-50] This is generally not a problem for links operating below 100 Mb/s, but becomes disastrous at speeds around 400 Mb/s and higher. The following factors can produce modal noise in an optical fiber link:

1. Mechanical disturbances along the link, such as vibrations, connectors, splices, microbends, and source or detector coupling, can result in differential mode delay or modal and spatial filtering of the optical power. This produces temporal fluctuations in the speckle pattern at the receiving end, thus creating modal noise in the receiver.
2. Fluctuations in the frequency of an optical source can also give rise to intermodal delays. A coherent source forms speckle patterns when its coherence time is greater than the intermodal dispersion time δT within the fiber. If the source has a frequency width δf, then its coherence time is $1/\delta f$. Modal noise occurs when the speckle pattern fluctuates, that is, when the source coherence time becomes much less than the intermodal dispersion time. The modal distortion resulting from interference between a single pair of modes will appear as a sinusoidal ripple of frequency

$$f = \delta T \frac{df_{\text{source}}}{dt} \tag{8-20}$$

where df_{source}/dt is the rate of change of optical frequency.

Several researchers have examined how modal noise degrades the bit error rate (BER) performance of a digital link.[48-50] As an example, Fig. 8-14 illustrates the error rates with the addition of modal noise to an avalanche-photodiode receiver system.[49] The analysis is for 280 Mb/s at 1200 nm with a gaussian-shaped received pulse. The factor M' in this figure is related to the

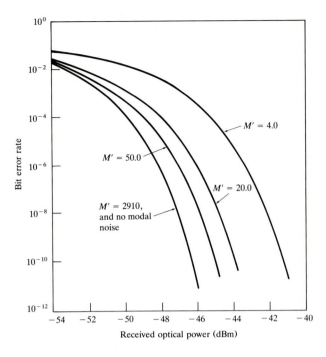

FIGURE 8-14
Error-rate curves for a 280-Mb/s avalanche-photodiode-based system with the addition of modal noise. The factor M' corresponds to the number of speckles. (Reproduced with permission from Chan and Tjhung,[49] © 1989, IEEE.)

number of speckles falling on the photodetector. For a very large number of speckles ($M' = 2910$), the error rate curve is very close to the case when there is no modal noise. As the number of speckles decreases, the performance degrades. When $M' = 50$, one needs an additional 1.0 dB of received optical power to maintain an error rate of 10^{-6}. When $M' = 20$, one must have 2.0 dB more power to achieve a 10^{-6} BER than in the case of no modal noise. This number becomes 4.9 dB when $M' = 4$.

The performance of a high-speed, laser-based multimode fiber link is difficult to predict, since the degree of modal noise which can appear depends greatly on the particular installation. Thus the best policy is to take steps to avoid it. This can be done by the following measures:

1. Use LEDs (which are incoherent sources). This totally avoids modal noise.

2. Use a laser which has a large number of longitudinal modes (10 or more). This increases the graininess of the speckle pattern, thus reducing intensity fluctuations at mechanical disruptions in the link.

3. Use a fiber with a large numerical aperture, since it supports a large number of modes and hence gives a greater number of speckles.

4. Use a single-mode fiber, since it only supports one mode and thus has no modal interference.

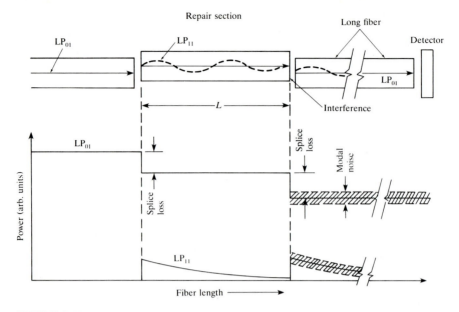

FIGURE 8-15
Repair sections can produce modal noise in a single-mode fiber link. This arises through the interchange of optical power between the LP_{01} and the LP_{11} modes at splice or connector joints. (Reproduced with permission from Sears, White, Kummer, and Stone,[46] © 1986, IEEE.)

The last point needs some further explanation. If a connector or splice point couples some of the optical power from the fundamental mode into the first higher-order mode (the LP_{11} mode), then a significant amount of power could exist in the LP_{11} mode in a short section of fiber between two connectors or at a repair splice.[46, 50] Figure 8-15 illustrates this effect. In a single-mode system modal noise could occur in short connectorized patch-cords, in laser-diode flyleads, or when two high-loss splices are a very short distance apart. To circumvent this problem, one should specify that the effective cutoff wavelength of short patch-cord and flylead fiber lengths is well below the system operating wavelength. Thus mode coupling is not a problem in links having long fiber lengths between connectors and splices, since the LP_{11} mode is usually sufficiently attenuated over the link length.

8.4.2 Mode-Partition Noise

As noted in Sec. 4.5, *mode-partition noise* is associated with intensity fluctuations in the longitudinal modes of a laser diode,[51-59] that is, the side modes are not sufficiently suppressed. This is the dominant noise in single-mode fibers. Intensity fluctuations can occur among the various modes in a multimode laser

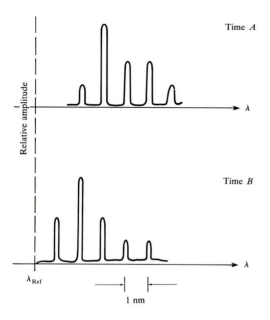

FIGURE 8-16
Time-resolved dynamic spectra of a laser diode. Different modes or groups of modes dominate the optical output at different times. The modes are approximately 1 nm apart.

even when the total optical output is constant as exhibited in Fig. 8-16. This power distribution can vary significantly both within a pulse and from pulse to pulse.

Because the output pattern of a laser diode is highly directional, the light from these fluctuating modes can be coupled into a single-mode fiber with high efficiency. Each of the longitudinal modes that is coupled into the fiber has a different attenuation and time delay, because each is associated with a slightly different wavelength (see Sec. 3.3). Since the power fluctuations among the dominant modes can be quite large, significant variations in signal levels can occur at the receiver in systems with high fiber dispersion.

The signal-to-noise ratio due to mode-partition noise is independent of signal power, so that the overall system error rate cannot be improved beyond the limit set by this noise. This is an important difference from the degradation of receiver sensitivity normally associated with chromatic dispersion, which one can compensate for by increasing the signal power.

Mode-partition noise becomes more pronounced for higher bit rates. The errors due to mode-partition noise can be reduced and sometimes eliminated by setting the bias point of the laser above threshold. However, raising the bias power level reduces the available signal-pulse power, thereby reducing the achievable signal-to-thermal-noise ratio.

In attempts to describe the effects of mode-partition noise, researchers have tried to identify a figure of merit for the laser-diode spectrum that could be measured experimentally, yet give an accurate theoretical prediction of system performance. One approach applies to lasers having many lasing modes,[56]

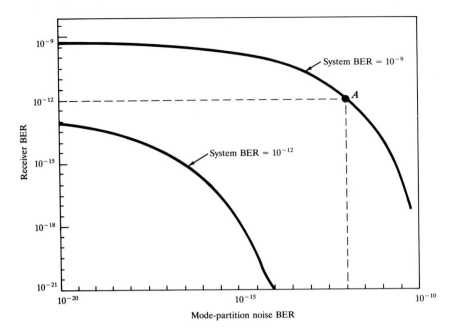

FIGURE 8-17
Example of a tradeoff analysis between the mode-partition-noise BER and the system BER in the absence of mode-partition noise. (Reproduced with permission from Basch, Kearns, and Brown,[54] © 1986, IEEE.)

whereas another addresses two-mode lasers where the side mode is below the lasing threshold.[53, 54] The second case is of interest in practice, since the distribution of mode-partition fluctuations is exponential rather than gaussian. This means that the fluctuations can cause very high error rates in all lasers except those wherein the nonlasing modes are greatly suppressed.

Figured 8-17 shows the result of a tradeoff analysis[54] between the mode-partition-noise BER and the system BER in the absence of mode-partition noise. The curves represent the total system performance for error probabilities of 10^{-9} and 10^{-12}. As an example, to maintain a total system BER of 10^{-9} and also have a receiver error probability of 10^{-12}, the required error rate for mode partitioning is less than 10^{-12}, as shown by point A. This is equivalent to a mode-intensity ratio of $I_0/J_0 \simeq 50$, where I_0 is the average intensity of the main lasing mode and J_0 is the average intensity of the strongest nonlasing mode.

To prevent the occurrence of a high system bit error rate due to power partitioning among insufficiently suppressed side modes, one must select lasers carefully. To evaluate the dynamics of the side modes, one can measure either the time-resolved photon statistics of the laser output, or the bit error rate characteristics under realistic biasing conditions.

8.4.3 Chirping

A laser which oscillates in a single longitudinal mode under CW operation may experience dynamic line broadening when the injection current is directly modulated.[59-66] This line broadening is a frequency "chirp" associated with modulation-induced changes in the carrier density. Laser chirping can lead to significant dispersion effects for intensity-modulated pulses when the laser emission wavelength is displaced from the zero-dispersion wavelength of the fiber. This is particularly true in systems operating at 1550 nm, where fiber dispersion is much greater than at 1300 nm.

To a good approximation, the time-dependent frequency change $\Delta\nu(t)$ of the laser can be given in terms of the output optical power $P(t)$ as[63]

$$\Delta\nu(t) = \frac{-\alpha}{4\pi}\left[\frac{d}{dt}\ln P(t) + \kappa P(t)\right] \tag{8-21}$$

where α is the *linewidth enhancement factor*[66] and κ is a frequency-independent factor which depends on the laser structure.[63] The factor α ranges from -3.5 to -5.5 for AlGaAs lasers[67] and from -6 to -8 for InGaAsP lasers.[68]

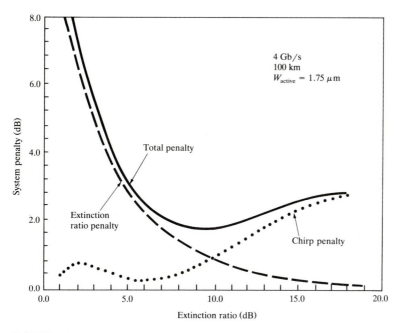

FIGURE 8-18
Extinction-ratio, chirping, and total-system power penalties at 1550 nm for a 4-Gb/s, 100-km-long single-mode link having a fiber with a dispersion $D = 17$ ps/(nm · km) and a DFB laser with an active layer width of 1.75 μm. (Reproduced with permission from Corvini and Koch,[63] © 1987, IEEE.)

One approach to minimize chirp is to increase the bias level of the laser so that the modulation current does not drive it below threshold where $\ln P$ and P change rapidly. However, this results in a lower extinction ratio (ratio of on-state power to off-state power), which leads to an extinction-ratio power penalty at the receiver because of a reduced signal-to-background ratio. This penalty is typically several decibels. Figure 8-18 gives examples of this for two types of laser structures. For higher extinction ratios (bias points progressively lower than threshold), the extinction-ratio power penalty decreases. However, the chirping-induced power penalty increases with lower bias levels.

The best approach to minimizing chirp effects is to choose the laser emission wavelength close to the zero-dispersion wavelength of the fiber. Experiments of this type[28] have shown no degradation in receiver sensitivity due to chromatic dispersion.

Figure 8-19 illustrates the effects of chirping at a 5-Gb/s transmission rate in different single-mode fiber links.[59] Here the laser side-mode suppression is greater than 30 dB, the back-reflected optical power is more than 30 dB below the transmitted signal, and the extinction ratio is about 8 dB. At 1536 nm the standard fiber has a dispersion $D = 17.3$ ps/(nm · km) and the dispersion-shifted fiber has $D = -1.0$ ps/(nm · km). The combined-fiber link consists of concatenated standard positive-dispersion and negative-dispersion fibers. This

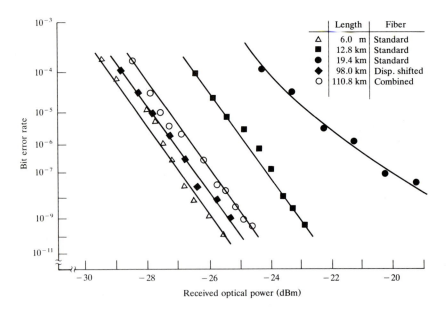

FIGURE 8-19

The effects of chirping at 5 Gb/s in different single-mode fiber links. The laser side-mode suppression is > 30 dB, the reflected power is more than 30 dB below the transmitted signal, and the extinction ratio is ≃ 8 dB. At 1536 nm the standard fiber has $D = 17.3$ ps/(nm · km) and the dispersion-shifted fiber has $D = -1.0$ ps/(nm · km). (Reproduced with permission from Heidemann,[59] © 1988, IEEE.)

leads to a spectral compression of the signal so that dispersion compensation occurs. Thus Fig. 8-19 shows the dramatic reduction in chirping penalty when using a dispersion-shifted fiber, or when combining fibers with positive and negative dispersion.

8.4.4 Reflection Noise

When light travels through a fiber link, some optical power gets reflected at refractive-index discontinuities such as in splices, couplers, and filters, or at air-to-glass interfaces in connectors. The reflected signals can degrade both transmitter and receiver performance.[41, 59, 69-71] In high-speed systems, this reflected power causes optical feedback which can induce laser instabilities. These instabilities show up as either intensity noise (output power fluctuations), jitter (pulse distortion), or phase noise in the laser, and they can change its wavelength, linewidth, and threshold current. Since they reduce the signal-to-noise ratio, these effects cause two types of power penalties in receiver sensitivities. First, as shown in Fig. 8-20a, multiple reflection points set up an interferometric cavity that feeds back into the laser cavity, thereby converting phase noise into intensity noise. A second effect created by multiple optical paths is the appearance of spurious signals arriving at the receiver with variable delays, thereby causing intersymbol interference. Figure 8-20b illustrates this.

Unfortunately these effects are signal-dependent, so that increasing the transmitted or received optical power does not improve the bit error rate performance. Thus one has to find ways to eliminate reflections. Let us first look at their magnitudes. As we saw from Eq. (5-10), a cleaved silica-fiber end face in air typically will reflect about

$$R = \left(\frac{1.47 - 1.0}{1.47 + 1.0} \right)^2 = 3.6 \text{ percent}$$

This corresponds to an optical return loss of 14.4 dB down from the incident signal. Polishing the fiber ends can create a thin surface layer with an increased refractive index of about 1.6. This increases the reflectance to 5.3 percent (a 12.7-dB optical return loss). A further increase in the optical feedback level occurs when the distance between multiple reflection points equals an integral number of half wavelengths of the transmitted wavelength. In this case all roundtrip distances equal an integral number of in-phase wavelengths, so that constructive interference arises. This quadruples the reflection to 14 percent or 8.5 dB for unpolished end faces and to over 22 percent (a 6.6-dB optical return loss) for polished end faces.

The power penalties can be reduced to a few tenths of a decibel by keeping the return losses below values ranging from -15 to -32 dB for data rates varying from 500 Mb/s to 4 Gb/s, respectively.[69] Techniques for reducing

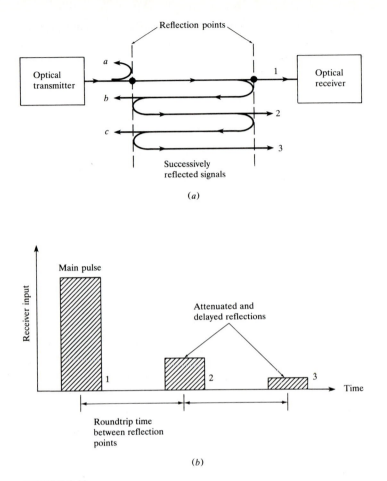

FIGURE 8-20
(*a*) Two refractive-index discontinuities can set up multiple reflections in a fiber link. (*b*) Each roundtrip between reflection points of the back-transmitted portion of a light pulse creates another attenuated and delayed pulse which can cause intersymbol interference.

optical feedback include the following:

1. Prepare fiber end faces with a curved surface or an angle relative to the laser emitting facet. This directs reflected light away from the fiber axis, so it does not reenter the waveguide. Return losses of 45 dB or higher can be achieved with end face angles of 5 to 15°. However, this increases both the insertion loss and the complexity of the connector.

2. Use index-matching oil or gel at air-to-glass interfaces. The return loss with this technique is usually greater than 30 dB. However, this may not be

practical or recommended if connectors need to be remated often, since contaminants could collect on the interface.

3. Use connectors in which the end faces make physical contact (the so-called *PC connectors*). Return losses of 25 to 40 dB have been measured with these connectors.

4. Use optical isolators within the laser transmitter module. These devices easily achieve 25-dB return losses, but they also can introduce up to 1 dB of forward loss in the link.

8.5 COMPUTER-AIDED MODELING

In designing an optical fiber link, one wishes to maximize the transmission distance while maintaining a specified bit error rate and keeping a sufficient power margin to account for possible degradation effects. As a first approximation, one can set up an optical power budget as described in Sec. 8.1.

The discussion in Sec. 8.1 formulated an optical power budget assuming the worst-case values of fiber loss, laser output power, receiver sensitivity, and degradations due to aging and temperature effects. A drawback of the optical power budget is that it does not show the relationships between loss and penalty terms. This is best accomplished with a computer simulation model which takes into account overall performance measures such as error rate, power margin, power penalties due to dispersion and noise, modulation techniques, and noise levels.

In addition to incorporating performance factors, the model should account for the fact that in a real system the individual component parameters can vary considerably. Since most components will not perform at their worst-case values but instead will operate around a certain average value, the worst-case power-budget approach can involve significant cost penalties because the links tend to be overdesigned. Engineers can realize substantial cost savings in transmission equipment if they design the system to exceed its power budget a small percentage of the time. With this method, repeater sections in long-distance links can be increased well above the worst-case value, even if this reduces system margins to the point where a small number of repeaters may fail. This approach is called *statistical design*.[72-77]

To carry out a computer simulation, one should first set up a block diagram such as shown in Fig. 8-21. The input to the optical source can be chosen as the coded output of a pseudorandom data generator. The data is a random binary sequence which has word lengths sufficiently long to analyze the individual effect or combination of effects being studied. For example, to account for intersymbol interference, the input bit pattern must contain all possible combinations of M bits, where M is the memory length of the system. A maximal-length pseudorandom sequence of $N = 2^M$ contains all combinations of length M. For characterizing the group delay or the dispersive effect of

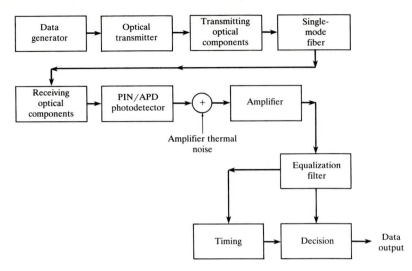

FIGURE 8-21
Block diagram used for computer simulation of a digital optical fiber transmission system. (Reproduced with permission from Elrefaie, Townsend, Romeiser, and Shanmugan,[73] © 1988, IEEE.)

single-mode fibers and typical filters, a bit sequence of 64 to 1024 (M = 6 to 10) is adequate. Line coding can be used to minimize the occurrence of long sequences of ones and zeros, and to shape the spectral density of transmitted waveforms, thus simplifying the transmitter and receiver circuits. The source output is coupled to optical components such as connectors, couplers, or filters. The important characteristics of the transmitter are the optical spectrum and the optical output waveform.

In the next block, the dispersion and loss characteristics of the fiber affect the signal. One can model the fiber either in the optical power domain or in the optical field domain. The first model assumes that the fiber is a linear system in the optical power domain. This is a valid assumption over the range of input bit rates and fiber lengths found in direct detection systems. The power-domain model becomes inaccurate when the source spectral width is comparable to the modulation signal bandwidth.[78] In this case one uses the field-domain model in which the fiber is described as a bandpass filter with flat amplitude and linear group delay over the modulation baseband signal bandwidth.[73]

At the receiver one has the mirror image of the optical components that are at the transmitter. Here either a *pin* or an avalanche photodetector converts the optical power to an electric current with an additional noise component. The thermal noise in the receiver electronics further perturbs the signal, which is then filtered, sampled, and compared against a threshold to determine the transmitted level. Timing information can be extracted from the received waveform and used in the decision process.

Detailed discussions of a variety of analytical expressions for computer simulations can be found in Refs. 72 to 77. The actual simulations can be carried out using commercially available software packages.[79-81]

8.6 SUMMARY

The design of an optical link involves many interrelated variables among the fiber, source, and photodetector operating characteristics. In carrying out an optical fiber link analysis, several iterations with different device characteristics may be required before it is satisfactorily completed. The key requirements needed in analyzing a link are:

1. The desired (or possible) transmission distance
2. The data rate or channel bandwidth
3. The bit error rate (BER)

Two analyses are usually carried out to ensure that the desired system performance can be met. These are the link power budget and the system rise-time analysis. In the link power budget analysis one first determines the power margin between the optical transmitter output and the minimum receiver sensitivity needed to establish a specified BER. This margin can then be allocated to connector, splice, and fiber losses, plus any additional margins required for expected component degradation or temperature effects. If the choice of components did not allow the desired transmission distance to be achieved, the components might have to be changed or repeaters might have to be put into the link.

Once the link power budget has been established, the designer makes a system rise-time analysis to ensure that the dispersion limit of the link has not been exceeded. The four basic elements which may significantly limit the system speed are the transmitter rise time, the material dispersion rise time of the fiber, the modal dispersion rise time of the fiber, and the receiver rise time.

In designing an optical fiber link, line coding schemes are often used to introduce randomness and redundancy into the digital information stream. This is done to ensure efficient timing recovery and to facilitate error monitoring at the receiver. In selecting a particular line code, a tradeoff must be made between timing and noise bandwidth. That is, to have timing data available from the bit stream, a larger bandwidth may be needed. However, larger bandwidths result in larger noise contributions. Popular line codes for optical fiber links are the non-return-to-zero (NRZ) format and the return-to-zero (RZ) format.

Once an optical link has been installed and assembled, a measurement of its overall performance is usually required. The eye-pattern technique is a simple method that provides a great deal of information for assessing the data-handling capability of a digital transmission system. The eye-pattern measurements are made in the time domain and allow the effects of waveform distortion to be seen immediately on an oscilloscope.

When analyzing the performance of an optical link, one needs to take into account the various interactions between spectral imperfections in the propagating optical power and the dispersive waveguide. These effects give rise to variations in the optical power level falling on the photodetector, thereby creating receiver output noises. The resulting optical power penalties are particularly serious for high-speed links (over 400 Mb/s). The main penalties are due to modal noise, wavelength chirp, spectral broadening induced by optical reflections back into the laser, and mode-partition noise. Modal noise is not present in single-mode links; however, the other noise contributors are critical in these systems.

In designing an optical-fiber system, one should take into account that the individual component parameters can vary considerably. Since most components will not perform at their worst-case values but instead will operate around a certain average value, engineers can realize substantial savings in equipment cost if they design the system to exceed its power budget a small percentage of the time. This approach, which is called statistical design, allows distances between repeaters to be increased well above the worst-case values while assuming that a small number of repeater failures may occur.

PROBLEMS

8-1. Make a graphical comparison, as in Fig. 8-4, of the maximum attenuation-limited transmission distance of the following two systems operating at 100 Mb/s:

System 1 operating at 850 nm

(*a*) GaAlAs laser diode: 0-dBm (1-mW) fiber-coupled power

(*b*) Silicon avalanche photodiode: −50-dBm sensitivity

(*c*) Graded-index fiber: 3.5-dB/km attenuation at 850 nm

(*d*) Connector loss: 1 dB/connector

System 2 operating at 1300 nm

(*a*) InGaAsP LED: −13-dBm fiber-coupled power

(*b*) InGaAs *pin* photodiode: −38-dBm sensitivity

(*c*) Graded-index fiber: 1.5-dB/km attenuation at 1300 nm

(*d*) Connector loss: 1 dB/connector

Allow a 6-dB system operating margin in each case.

8-2. An engineer has the following components available:

(*a*) GaAlAs laser diode operating at 850 nm and capable of coupling 1 mW (0 dBm) into a fiber

(*b*) Ten sections of cable each of which is 500 m long, has a 4-dB/km attenuation, and has connectors on both ends

(*c*) Connector loss of 2 dB/connector

(*d*) A *pin* photodiode receiver

(*e*) An avalanche photodiode receiver

Using these components, the engineer wishes to construct a 5-km link operating at

20 Mb/s. If the sensitivities of the *pin* and APD receivers are -45 and -56 dBm, respectively, which receiver should be used if a 6-dB system operating margin is required?

8-3. Using the step response $g(t)$, show that the 10 to 90-percent receiver rise time is given by Eq. (8-4).

8-4. (*a*) Verify Eq. (8-12).

(*b*) Show that Eq. (8-14) follows from Eqs. (8-10) and (8-13).

8-5. Show that, if t_e is the full width of the gaussian pulse in Eq. (8-9) at the $1/e$ points, then the relationship between the 3-dB optical bandwidth and t_e is given by

$$f_{3\,\text{dB}} = \frac{0.53}{t_e}$$

8-6. A 90-Mb/s NRZ data transmission system uses a GaAlAs laser diode having a 1-nm spectral width. The rise time of the laser transmitter output is 2 ns. The transmission distance is 7 km over a graded-index fiber having an 800-MHz · km bandwidth-distance product.

(*a*) If the receiver bandwidth is 90 MHz and the mode-mixing factor $q = 0.7$, what is the system rise time? Does this rise time meet the NRZ data requirement of being less than 70 percent of a pulse width?

(*b*) What is the system rise time if there is no mode mixing in the 7-km link, that is, $q = 1.0$?

8-7. Verify the plot in Fig. 8-5 of the transmission distance versus data rate of the following system. The transmitter is a GaAlAs laser diode operating at 850 nm. The laser power coupled into a fiber flylead is 0 dBm (1 mW), and the source spectral width is 1 nm. The fiber has a 3.5-dB/km attenuation at 850 nm and a bandwidth of 800 MHz · km. The receiver uses a silicon avalanche photodiode which has the sensitivity versus data rate shown in Fig. 8-3. For simplicity, the receiver sensitivity (in dBm) can be approximated from curve fitting by

$$P_R = 9 \log B - 68.5$$

where B is the data rate in Mb/s. For the data rate range of 1 to 1000 Mb/s, plot the attenuation-limited transmission distance (including a 1-dB connector loss at each end and a 6-dB system margin), the modal dispersion limit for full mode mixing ($q = 0.5$), the modal dispersion limit for no mode mixing ($q = 1.0$), and the material dispersion limit.

8-8. Make a plot analogous to Fig. 8-5 of the transmission distance versus data rate of the following system. The transmitter is an InGaAsP LED operating at 1300 nm. The fiber-coupled power from this source is -13 dBm (50 μW), and the source spectral width is 40 nm. The fiber has a 1.5-dB/km attenuation at 1300 nm and a bandwidth of 800 MHz · km. The receiver uses an InGaAs *pin* photodiode which has the sensitivity versus data rate shown in Fig. 8-3. For simplicity, this receiver sensitivity P_R (in dBm) can be approximated from curve fitting by

$$P_R = 11.5 \log B - 60.5$$

where B is the data rate in Mb/s. For the data rate range of 10 to 1000 Mb/s, plot the attenuation-limited transmission distance (including a 1-dB connector loss at each end and a 6-dB system margin), the modal dispersion limit for no mode mixing ($q = 1.0$), and the modal dispersion limit for full mode mixing ($q = 0.5$). Note that the material dispersion is negligible in this case, as can be seen from Fig. 3-13.

8-9. A 1550-nm single-mode digital fiber optic link needs to operate at 565 Mb/s over 50 km without repeaters. A single-mode InGaAsP laser launches an average optical power of -13 dBm into the fiber. The fiber has a loss of 0.35 dB/km, and there is a splice with a loss of 0.1 dB every kilometer. The coupling loss at the receiver is 0.5 dB, and the receiver uses an InGaAs APD with a sensitivity of -39 dBm. Excess-noise penalties are predicted to be 1.5 dB. Set up an optical power budget for this link and find the system margin.

8-10. A popular RZ code used in fiber optic systems is the optical Manchester code. This is formed by direct modulo-2 addition of the baseband (NRZ-L) signal and a double-frequency clock signal as is shown in Fig. 8-9. Using this scheme, draw the pulse train for the data string 001101111001.

8-11. Design the encoder logic for an NRZ-to-optical Manchester converter.

8-12. Consider the encoder shown in Fig. P8-12 which changes NRZ data into a PSK (phase-shift-keyed) waveform. Using this encoder, draw the NRZ and PSK waveforms for the data sequence 0001011101001101.

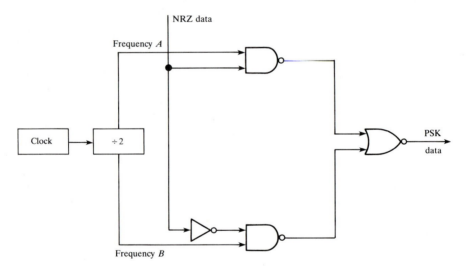

FIGURE P8-12

8-13. A 3B4B code converts blocks of three binary symbols into blocks of four binary digits according to the translation rules shown in Table P8-13. When there are two or more consecutive blocks of three zeros, the coded binary blocks 0010 and 1101 are used alternately. Likewise the coded blocks 1011 and 0100 are used alternately for consecutive blocks of three ones.
(*a*) Using these translation rules, find the coded bit stream for the data input

$$010001111111101000000001111110$$

TABLE P8-13

Original code	3B4B Code	
	Mode 1	Mode 2
000	0010	1101
001	0011	
010	0101	
011	0110	
100	1001	
101	1010	
110	1100	
111	1011	0100

(*b*) What is the maximum number of consecutive identical bits in the coded pattern?

8-14. The power penalty in decibels caused by laser mode-partition noise can be approximated by[56, 82]

$$P_{mpn} = -5\frac{x+2}{x+1}\log\left[1 - \frac{k^2Q^2}{2}(\pi BZD\sigma_\lambda)^4\right]$$

where x is the excess noise factor of an APD, Q is the signal-to-noise ratio (6 at 10^{-9} BER), B is the bit rate in Gb/s, Z is the fiber length in km, D is the fiber chromatic dispersion in ps/(nm · km), σ_λ is the rms spectral width of the source in nm, and k is the mode-partition-noise factor.

(*a*) Plot the power penalty (in dB) as a function of the factor $BZD\sigma_\lambda$ (ranging from 0 to 0.20) at a 10^{-9} BER for mode-partition-noise factors $k = 0.3$ and $k = 0.6$ when using an InGaAs APD with $x = 0.7$.

(*b*) Given that a laser has a spectral width of 3.5 nm, what are the minimum dispersions required for 100-km repeaterless spans operating at 140, 280, and 560 Mb/s with a power penalty of 0.5 dB?

8-15. (*a*) When the effect of laser chirp is small, the eye closure Δ can be approximated by[62]

$$\Delta \simeq \left(\tfrac{4}{3}\pi^2 - 8\right)t_c KB\left[1 + \tfrac{2}{3}(K - t_c B)\right]$$

where $K = DZ\,\delta\lambda\,B$
  t_c = chirp duration time
  B = bit rate
  D = fiber chromatic dispersion
  $\delta\lambda$ = chirp-induced wavelength excursion
  Z = fiber length

The power penalty for an APD system can be estimated by the signal-to-noise ratio degradation (in dB) due to the signal amplitude decrease as

$$P = -10\frac{x+2}{x+1}\log(1 - \Delta)$$

where x is the excess noise factor of the APD. Plot the power penalty in decibels as a function of the factor $DZ\,\delta\lambda$ (product of the total dispersion and wavelength excursion) for the following parameter values:

    (1) $t_c = 0.1$ ns and $B = 1.2$ Gb/s
    (2) $t_c = 0.1$ ns and $B = 565$ Mb/s
    (3) $t_c = 0.05$ ns and $B = 1.2$ Gb/s
    (4) $t_c = 0.05$ ns and $B = 565$ Mb/s

    Let $DZ \, \delta\lambda$ range from zero to 1.5 ns.

  (*b*) Find the distance limitation at 1.2 Gb/s if a 0.5-dB power penalty is allowed with $D = 1.0$ ps/(nm · km) and $\delta\lambda = 0.5$ nm.

The following problems are term exercises which require some research work in the literature and access to a computer for numerical modeling and evaluation.

8-16. Work through the bit-error-rate-degradation analysis of Koonen[48] to derive the optimum decision threshold which minimizes the influence of modal noise (Eq. 33 in Ref. 48). Set up and run a computer program to plot the system BER degradation shown in Fig. 4*a* of Ref. 48.

8-17. Based on the discussions of hybrid modeling given by Elrefaie, Townsend, Romeiser, and Shanmugan,[73] set up the analytical details required for the computer simulation of a digital fiber link. If simulation software is available,[79-81] compute curves of bit error rate versus received power such as those given in Ref. 73.

8-18. Shen and Agrawal[75] show how to account for various degradations present in a 1550-nm link. These include circuit noise, shot noise, laser intensity noise, mode-partition noise, reflections, and frequency chirp. Using these analyses, carry out a computer simulation to duplicate their results for possible transmission distances as a function of data rate under different degradation conditions.

8-19. Examine the statistical-design strategy of Batten, Gibbs, and Nicholson[76] for long optical fiber systems. Expand the example given for a 565-Mb/s, 1300-nm, land-based system to a long-span undersea link operating at 1550 nm. See Ref. 82 (Yamamoto et al.) for some design ideas.

8-20. Kikushima and Hogari[83] have applied the statistical-design approach to dispersion budgeting in single-mode spans containing many splices. Using their analyses, find the allowed increase in repeaterless distances at 10 Gb/s for 1300-nm and 1550-nm links when using the statistical approach rather than the worst-case design method. Alternatively, using the statistical dispersion-budgeting approach, show the range over which the laser wavelength and the fiber zero-dispersion wavelength specification can be relaxed.

REFERENCES

1. J. Bliss and D. W. Stevenson, "Fiber optics: A designer's guide to system budgeting," *Electro-Opt. Sys. Design*, vol. 13, pp. 23–32, Aug. 1981.
2. C. Kleekamp and B. Metcalf, "Designer's guide to fiber optics," *EDN*, vol. 23, pp. 44–51, Jan. 5, 1978 (pt. 1); pp. 46–54, Jan. 20, 1978 (pt. 2); pp. 61–68, Feb. 20, 1978 (pt. 3); pp. 51–62, Mar. 5, 1978 (pt. 4).
3. D. H. Rice and G. E. Keiser, "Applications of fiber optics to tactical communication systems," *IEEE Commun. Mag.*, vol. 23, pp. 46–57, May 1985.
4. (*a*) S. E. Miller, "Transmission system design," in S. E. Miller and A. G. Chynoweth, eds., *Optical Fiber Telecommunications*, Academic, New York, 1979.

(*b*) P. S. Henry, R. A. Linke, and A. H. Gnauck, "Introduction to lightwave systems," in S. E. Miller and I. P. Kaminow, eds., *Optical Fiber Telecommunications II*, Academic, New York, 1988.

5. V. L. Mirtich, "Designer's guide to fiber-optic data links," *EDN*, vol. 25, pp. 133–140, June 20, 1980 (pt. 1); pp. 113–117, Aug. 5, 1980 (pt. 2); pp. 103–110, Aug. 20, 1980 (pt. 3).

6. S. L. Strorozum, "Fiber optic systems: Practical design," *Photonics Spectra*, vol. 19, pp. 61–66, Sept. 1985 (pt. 1); pp. 77–88, Oct. 1985 (pt. 2); pp. 57–66, Nov. 1985 (pt. 3).

7. I. Jacobs, "Design considerations for long-haul lightwave systems," *IEEE J. Sel. Areas Commun.*, vol. 4, pp. 1389–1395, Dec. 1986.

8. T. Kimura, "Factors affecting fiber-optic transmission quality," *J. Lightwave Tech.*, vol. 6, pp. 611–619, May 1988.

9. A. B. Carlson, *Communication Systems*, McGraw-Hill, New York, 3rd ed., 1986.

10. M. Eve, "Multipath time dispersion theory of an optical network," *Opt. Quantum Electron.*, vol. 10, pp. 45–51, Jan. 1978.

11. J. V. Wright and B. P. Nelson, "Bandwidth studies of concatenated multimode fiber links," *Tech. Dig.—Symp. on Optical Fiber Measurements*, NBS Special Publ. 641, pp. 9–12, Oct. 1982.

12. T. C. Olson, F. P. Kapron, and T. W. Geyer, "Describing dispersion in concatenated single-mode fiber cables," *Proc. Intl. Wire and Cable Symp.*, Reno, NV, pp. 276–282, Nov. 1984.

13. D. A. Nolan, R. M. Hawk, and D. B. Keck, "Multimode concatenation modal group analysis," *J. Lightwave Tech.*, vol. LT-5, pp. 1727–1732, Dec. 1987.

14. R. D. de la Iglesia and E. T. Azpitarte, "Dispersion statistics in concatenated single-mode fibers," *J. Lightwave Tech.*, vol. LT-5, pp. 1768–1772, Dec. 1987.

15. P. M. Rodhe, "The bandwidth of a multimode fiber chain," *J. Lightwave Tech.*, vol. LT-3, pp. 145–154, Feb. 1985.

16. J. Midwinter, *Optical Fibers for Transmission*, Wiley, New York, 1979.

17. C. E. Shannon, "A mathematical theory of communication," *Bell Sys. Tech. J.*, vol. 27, pp. 379–423, July 1948 and vol. 27, pp. 623–656, Oct. 1948; "Communication in the presence of noise," *Proc. IRE*, vol. 37, pp. 10–21, Jan. 1949.

18. R. C. Houts and T. A. Green, "Comparing bandwidth requirements for binary baseband signals," *IEEE Trans. Commun.*, vol. COM-21, pp. 776–781, June 1973.

19. A. J. Jerri, "The Shannon sampling theorem—its various extensions and applications: A tutorial review," *Proc. IEEE*, vol. 65, pp. 1565–1596, Nov. 1977.

20. T. V. Muoi and J. L. Hullett, "Receiver design for multilevel digital fiber optic systems," *IEEE Trans. Commun.*, vol. COM-23, pp. 987–994, Sept. 1975.

21. D. J. Morris, *Pulse Code Formats for Fiber Optical Data Communication*, Dekker, New York, 1983.

22. E. A. Lee and D. G. Messerschmitt, *Digital Communication*, Kluwer Academic, Boston, 1988.

23. H. Taub and D. L. Schilling, *Principles of Communication Systems*, 2nd ed., McGraw-Hill, New York, 1986.

24. S. Haykin, *Digital Communication*, Wiley, New York, 1988.

25. S. D. Personick, *Optical Fiber Transmission Systems*, Plenum, New York, 1981.

26. J. E. Savage, "Some simple self-synchronizing digital data assemblers," *Bell Sys. Tech. J.*, vol. 246, pp. 449–487, Feb. 1967.

27. R. D. Gitlin and J. F. Hayes, "Timing recovery and scramblers in data transmission," *Bell Sys. Tech. J.*, vol. 54, pp. 569–593, Mar. 1975.

28. J. L. Gimlett, M. Z. Iqbal, J. Young, L. Curtis, R. Spicer, N. K. Cheung, and S. Tsuji, "11 Gbits/s optical transmission experiment using 1540 nm DFB laser with non-return-to-zero modulation and *pin*/HEMT receiver," *Electron. Lett.*, vol. 25, pp. 596–597, Apr. 1989.

29. Y. Takasaki, M. Tanaka, N. Maeda, K. Yamashita, and K. Nagano, "Optical pulse formats for fiber optic digital communications," *IEEE Trans. Commun.*, vol. COM-24, pp. 404–413, Apr. 1976.

30. R. Petrovic, "New transmission code for digital optical communications," *Electron. Lett.*, vol. 14, pp. 541–542, Aug. 1978.

31. E. A. Newcombe and S. Pasupathy, "Error rate monitoring for digital communications," *Proc. IEEE*, vol. 70, pp. 805–828, Aug. 1982.

32. S. Kawanishi, N. Yoshikai, J.-I. Yamada, and K. Nakagawa, "D*m*B1M code and its performance in a very high-speed optical transmission system," *IEEE Trans. Commun.*, vol. 36, pp. 951–956, Aug. 1988.

33. W. A. Kryzmien, "Transmission performance analysis of a new class of line codes for optical fiber systems," *IEEE Trans. Commun.*, vol. 37, pp. 402–404, Apr. 1989.

34. A. R. Calderbank, M. A. Herro, and V. Telang, "A multilevel approach to the design of DC-free line codes," *IEEE Trans. Inform. Th.*, vol. 35, pp. 579–583, May 1989.

35. M. Rousseau, "Block codes for optical fibre communications," *Electron. Lett.*, vol. 12, pp. 478–479, Sept. 1976.

36. A. X. Widmer and P. A. Franaszek, "Transmission code for high-speed fibre-optic data networks," *Electron. Lett.*, vol. 19, pp. 202–203, Mar. 1983.

37. R. Petrovic, "5B6B optical-fibre line code bearing auxiliary signals," *Electron. Lett.*, vol. 24, pp. 274–275, Mar. 1988.

38. R. Petrovic, "Low redundancy optical-fibre line code," *J. Opt. Commun.*, vol. 9, pp. 108–111, Sept. 1988.

39. H. Nishimoto, T. Okiyama, N. Kuwata, Y. Arai, A. Miyauchi, and T. Touge, "New method of analyzing eye patterns and its application to high-speed optical transmission systems," *J. Lightwave Tech.*, vol. 6, pp. 678–685, May 1988.

40. K. S. Shanmugan, *Digital and Analog Communication Systems*, Wiley, New York, 1979.

41. N. K. Cheung, "Reflection and modal noise associated with connectors in single-mode fibers," *SPIE Proc. on Fiber Optic Couplers, Connectors, and Splice Technology*, vol. 479, pp. 56–63, May 1984.

42. A. R. Michelson and A. Weierholt, "Modal-noise limited signal-to-noise ratios in multimode optical fibers," *Appl. Opt.*, vol. 22, pp. 3084–3089, Oct. 1983.

43. K. Petermann, "Nonlinear distortions and noise in optical communication systems due to fiber connectors," *IEEE J. Quantum Electron.*, vol. QE-16, pp. 761–770, July 1980.

44. T. Kanada, "Evaluation of modal noise in multimode fiber-optic systems," *J. Lightwave Tech.*, vol. LT-2, pp. 11–18, Feb. 1984.

45. P. E. Couch and R. E. Epworth, "Reproducible modal-noise measurements in system design and analysis," *J. Lightwave Tech.*, vol. LT-1, pp. 591–595, Dec. 1983.

46. F. M. Sears, I. A. White, R. B. Kummer, and F. T. Stone, "Probability of modal noise in single-mode lightguide systems," *J. Lightwave Tech.*, vol. LT-4, pp. 652–655, June 1986.

47. K. Petermann and G. Arnold, "Noise and distortion characteristics of semiconductor lasers in optical fiber communication systems," *IEEE J. Quantum Electron.*, vol. QE-18, pp. 543–554, Apr. 1982.

48. A. M. J. Koonen, "Bit-error-rate degradation in a multimode fiber optic transmission link due to modal noise," *IEEE J. Sel. Areas Commun.*, vol. SAC-4, pp. 1515–1522, Dec. 1986.

49. P. Chan and T. T. Tjhung, "Bit-error-rate performance for optical fiber systems with modal noise," *J. Lightwave Tech.*, vol. 7, pp. 1285–1289, Sept. 1989.

50. P. M. Shankar, "Bit-error-rate degradation due to modal noise in single-mode fiber optic communication systems," *J. Opt. Commun.*, vol. 10, pp. 19–23, Mar. 1989.

51. N. H. Jensen, H. Olesen, and K. E. Stubkjaer, "Partition noise in semiconductor lasers under CW and pulsed operation," *IEEE J. Quantum Electron.*, vol. QE-18, pp. 71–80, Jan. 1987.

52. M. Ohtsu and Y. Teramachi, "Analyses of mode partition and mode hopping in semiconductor lasers," *IEEE J. Quantum Electron.*, vol. 25, pp. 31–38, Jan. 1989.

53. C. H. Henry, P. S. Henry, and M. Lax, "Partition fluctuations in nearly single longitudinal mode lasers," *J. Lightwave Tech.*, vol. LT-2, pp. 209–216, June 1984.

54. E. E. Basch, R. F. Kearns, and T. G. Brown, "The influence of mode partition fluctuations in nearly single-longitudinal-mode lasers on receiver sensitivity," *J. Lightwave Tech.*, vol. LT-4, pp. 516–519, May 1986.

55. J. C. Cartledge, "Performance implications of mode partition fluctuations in nearly single longitudinal mode lasers," *J. Lightwave Tech.*, vol. 6, pp. 626–635, May 1988.

56. K. Ogawa, "Analysis of mode partition noise in laser transmission systems," *IEEE J. Quantum Electron.*, vol. QE-18, pp. 849–855, May 1982.

57. N. A. Olsson, W. T. Tsang, H. Temkin, N. K. Dutta, and R. A. Logan, "Bit-error-rate saturation due to mode-partition noise induced by optical feedback in 1.5 μm single longitudinal-mode C^3 and DFB semiconductor lasers," *J. Lightwave Tech.*, vol. LT-3, pp. 215–218, Apr. 1985.

58. S. E. Miller, "On the injection laser contribution to mode partition noise in fiber telecommunication systems," *IEEE J. Quantum Electron.*, vol. 25, pp. 1771–1781, Aug. 1989.

59. R. Heidemann, "Investigations on the dominant dispersion penalties occurring in multigigabit direct detection systems," *J. Lightwave Tech.*, vol. 6, pp. 1693–1697, Nov. 1988.

60. R. A. Linke, "Modulation induced transient chirping in single frequency lasers," *IEEE J. Quantum Electron.*, vol. QE-21, pp. 593–597, June 1985.

61. Y. Yoshikuni and G. Motosugi, "Multielectrode distributed feedback laser for pure frequency modulation and chirping suppressed amplitude modulation," *J. Lightwave Tech.*, vol. LT-5, pp. 516–522, Apr. 1987.

62. S. Yamamoto, M. Kuwazuru, H. Wakabayashi, and Y. Iwamoto, "Analysis of chirp power penalty in 1.55-μm DFB-LD high-speed optical fiber transmission systems," *J. Lightwave Tech.*, vol. LT-5, pp. 1518–1524, Oct. 1987.

63. P. J. Corvini and T. L. Koch, "Computer simulation of high-bit-rate optical fiber transmission using single-frequency lasers," *J. Lightwave Tech.*, vol. LT-5, pp. 1591–1595, Nov. 1987.

64. J. C. Cartledge and G. S. Burley, "The effect of laser chirping on lightwave system performance," *J. Lightwave Tech.*, vol. 7, pp. 568–573, Mar. 1989.

65. J. C. Cartledge and M. Z. Iqbal, "The effect of chirping-induced pulse shaping on the performance of 11 Gb/s lightwave systems," *IEEE Photonics Tech. Lett.*, vol. 1, pp. 346–348, Oct. 1989.

66. C. H. Henry, "Theory of the linewidth of semiconductor lasers," *IEEE J. Quantum Electron.*, vol. QE-18, pp. 259–264, Feb. 1982.

67. C. H. Harder, K. Vahala, and A. Yariv, "Measurement of the linewidth enhancement factor of semiconductor lasers," *Appl. Phys. Lett.*, vol. 42, pp. 328–330, Apr. 1983.

68. R. Schimpe, J. E. Bowers, and T. L. Koch, "Characterization of frequency response of 1.5-μm InGaAsP DFB laser diode and InGaAs *PIN* photodiode by heterodyne measurement technique," *Electron. Lett.*, vol. 22, pp. 453–454, Apr. 24, 1986.

69. M. Shikada, S. Takano, S. Fujita, I. Mito, and K. Minemura, "Evaluation of power penalties caused by feedback noise of distributed feedback laser diodes," *J. Lightwave Tech.*, vol. 6, pp. 655–659, May 1988.

70. M. Nakazawa, "Rayleigh backscattering theory for single-mode fibers," *J. Opt. Soc. Amer.*, vol. 73, pp. 1175–1180, Sept. 1983.

71. F. P. Kapron, S. G. Joshi, J. W. Peters, and R. A. Swanson, "Lightwave reflections degrade end points," *Bellcore Digest of Tech. Information*, vol. 5, pp. 1–5, Oct. 1988.

72. D. G. Duff, "Computer-aided design of digital lightwave systems," *IEEE J. Sel. Areas Commun.*, vol. SAC-2, pp. 171–185, Jan. 1984.

73. A. F. Elrefaie, J. K. Townsend, M. B. Romeiser, and K. S. Shanmugan, "Computer simulation of digital lightwave links," *IEEE J. Sel. Areas Commun.*, vol. 6, pp. 94–105, Jan. 1988.

74. (*a*) D. O. Harris and J. R. Jones, "Baud rate response: Characterizing modal dispersion for digital fiber optic systems," *J. Lightwave Tech.*, vol. 6, pp. 668–677, May 1988.

    (*b*) R. D. de la Iglesia, "Statistical approach to end-to-end dispersion budgeting in single-mode fiber links using multilongitudinal-mode lasers," *J. Lightwave Tech.*, vol. LT-4, pp. 767–771, July 1986.

75. T.-M. Shen and G. P. Agrawal, "Computer simulation and noise analysis of the system performance of 1.55-μm single-frequency semiconductor lasers," *J. Lightwave Tech.*, vol. LT-5, pp. 653–659, May 1987.

76. T. J. Batten, A. J. Gibbs, and G. Nicholson, "Statistical design of long optical fiber systems," *J. Lightwave Tech.*, vol. 7, pp. 209–217, Jan. 1989.

77. M. K. Moaveni and M. Shafi, "A statistical design approach for gigabit-rate fiber optic transmission systems," *J. Lightwave Tech.*, vol. 8, pp. 1064–1072, July 1990.
78. J. Gimlett and N. Cheung, "Dispersion penalty analysis for LED/single-mode fiber transmission systems," *J. Lightwave Tech.*, vol. LT-4, pp. 1381–1392, Sept. 1986.
79. Special Issue on Computer-Aided Modeling, Analysis, and Design of Communication Systems, *IEEE J. Sel. Areas Commun.*, vol. SAC-2, Jan. 1984.
80. K. S. Shanmugan, "An update on software packages for simulation of communication systems (links)," *IEEE J. Sel. Areas Commun.*, vol. 6, pp. 5–12, Jan. 1988.
81. W. H. Tranter and C. R. Ryan, "Simulation of communication systems using personal computers," *IEEE J. Sel. Areas Commun.*, vol. 6, pp. 13–23, Jan. 1988.
82. S. Yamamoto, H. Sakaguchi, M. Nunokawa, and Y. Iwamoto, "1.55-μm fiber optic transmission experiments for long-span submarine cable system design," *J. Lightwave Tech.*, vol. 6, pp. 380–391, Mar. 1988.
83. K. Kikushima and K. Hogari, "Statistical dispersion budgeting method for single-mode fiber transmission systems," *J. Lightwave Tech.*, vol. 8, pp. 11–15, Jan. 1990.

CHAPTER
9

ANALOG
SYSTEMS

In telecommunication networks the trend has been to link telephone exchanges with digital circuits. A major reason for this was the introduction of digital integrated-circuit technology which offered a reliable and economic method of transmitting both voice and data signals. Since the initial applications of fiber optics was to telecommunication networks, its first widespread usage has involved digital links. However, in many instances it is more advantageous to transmit information in analog form instead of first converting it to a digital format. Some examples of this are microwave-multiplexed signals,[1] subscriber-loop applications,[2] video distribution,[3-5] antenna remoting,[6] and radar signal processing.[7-9] For most analog applications one uses laser diode transmitters, so we shall concentrate on this optical source here.

When implementing an analog fiber optic system, the main parameters one needs to consider are the carrier-to-noise ratio, bandwidth, and signal distortion resulting from nonlinearities in the transmission system. Section 9.1 describes the general operational aspects and components of an analog fiber optic link. In an analog system, a carrier-to-noise ratio analysis is used instead of a signal-to-noise ratio analysis, since the information signal is normally superimposed on a radio-frequency (RF) carrier. Thus in Sec. 9.2 we examine carrier-to-noise ratio requirements taking into account transmitter, receiver, and quantum noise effects. This is first done for a single channel under the assumption that the information signal is directly modulated onto an optical carrier. Here we pay particular attention to intensity noise in the optical source.

For transmitting multiple signals over the same channel, one can use a subcarrier modulation technique. In this method, which is described in Sec. 9.3,

357

the information signals are first superimposed on ancillary RF subcarriers. These carriers are then combined and the resulting electrical signal is used to modulate the optical carrier. A limiting factor in these systems is the noises arising from harmonic and intermodulation distortions.

9.1 OVERVIEW OF ANALOG LINKS

Figure 9-1 shows the basic elements of an analog link. The transmitter contains either an LED or a laser diode optical source. As noted in Sec. 4.4 and shown in Fig. 4-35, in analog applications one first sets a bias point on the source approximately at the midpoint of the linear output region. The analog signal can then be sent using one of several modulation techniques. The simplest form for optical fiber links is direct intensity modulation, wherein the optical output from the source is modulated simply by varying the current around the bias point in proportion to the message signal level. Thus the information signal is transmitted directly in the baseband.

A somewhat more complex but often more efficient method is to translate the baseband signal onto an electrical subcarrier prior to intensity modulation of the source. This is done using standard amplitude modulation (AM), frequency modulation (FM), or phase modulation (PM) techniques.[10-12] No matter which method is implemented, one must pay careful attention to nonlinearities in the optical source. These include harmonic distortions, intermodulation products, and relative intensity noise (RIN) in the laser.

In relation to the fiber-optic element shown in Fig. 9-1, one must take into account the frequency dependence of the amplitude, phase, and group delay in the fiber. Thus the fiber should have a flat amplitude and group-delay response within the passband required to send the signal free of linear distortion. In addition, since modal-distortion-limited bandwidth is difficult to equalize, it is best to choose a single-mode fiber. The fiber attenuation is also important, since the carrier-to-noise performance of the system will change as a function of the received optical power.

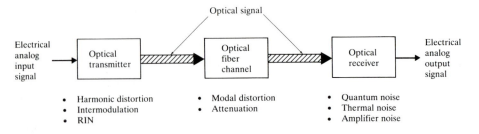

FIGURE 9-1
Basic elements of an analog link and the major noise contributors.

In the receiver the primary issue of concern is its noise performance. Photodetectors with excellent linearity and with bandwidths exceeding several gigahertz are readily available. The principal noise sources are quantum and thermal noise in the photodiode and amplifier noises in the receiver electronics.

9.2 CARRIER-TO-NOISE RATIO

In analyzing the performance of analog systems, one usually calculates the ratio of rms carrier power to rms noise power at the output of the optical receiver. This is known as the *carrier-to-noise ratio* (CNR) and is given by

$$\text{CNR} = \frac{\text{carrier power}}{\text{source + photodiode + amplifier + intermodulation noises}} \quad (9\text{-}1)$$

For links in which only a single information channel is transmitted, the first three noise terms are important. The intermodulation factor arises when multiple message channels operating at different carrier frequencies are sent simultaneously over the same fiber. In this section we shall first examine a simple single-channel amplitude-modulated signal sent at baseband frequencies. Section 9.3 addresses multichannel systems in which the intermodulation noise term becomes important.

9.2.1 Carrier Power

To find the carrier power, let us first look at the signal generated at the transmitter. As shown in Fig. 9-2, the drive current through the optical source is the sum of the fixed bias current and a time-varying sinusoid. The source acts as a square-law device, so that the envelope of the output optical power $P(t)$ has the same form as the input drive current. If the time-varying analog drive signal is $s(t)$, then

$$P(t) = P_t[1 + ms(t)] \quad (9\text{-}2)$$

where P_t is the optical output power at the bias current level and the modulation index m is defined by Eq. (4-54). In terms of optical power, the modulation index is given by

$$m = \frac{P_{\text{peak}}}{P_t} \quad (9\text{-}3)$$

where P_{peak} and P_t are defined in Fig. 9-2. Typical values of m for analog applications range from 0.25 to 0.50.

For a sinusoidal received signal, the carrier power C at the output of the receiver (in units of A^2) is

$$C = \tfrac{1}{2}(m\mathcal{R}_0 M\bar{P})^2 \quad (9\text{-}4)$$

where $\mathcal{R}_0$ is the unity gain responsivity of the photodetector, M is the

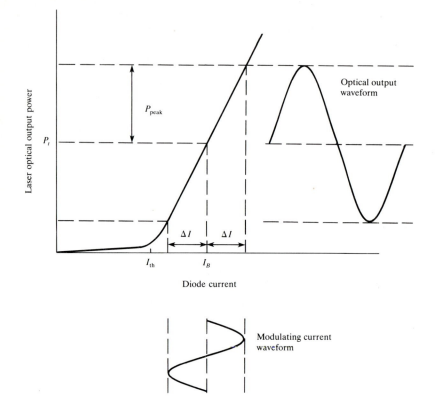

FIGURE 9-2
Biasing conditions of a laser diode and its response to analog signal modulation.

photodetector gain ($M = 1$ for *pin* photodiodes), and $\overline{P}$ is the average received optical power.

9.2.2   Photodetector and Preamplifier Noises

The expressions for the photodiode and preamplifier noises are given by Eqs. (6-16) and (6-17), respectively. That is, for the photodiode noise we have

$$\langle i_N^2 \rangle = 2q(I_p + I_D)M^2F(M)B + 2qI_L B \qquad (9\text{-}5)$$

Here, as defined in Chap. 6, $I_p = \mathscr{R}_0 \overline{P}$ is the primary photocurrent, I_D is the detector bulk dark current, M is the photodiode gain with $F(M)$ being its associated noise figure, I_L is the detector surface leakage current, and B is the receiver bandwidth. For well-designed photodiodes the leakage-current term is small compared to the first term and can be neglected.

Generalizing Eq. (6-17) for the preamplifier noise, we have

$$\langle i_T^2 \rangle = \frac{4k_B T}{R_{eq}} BF_t \tag{9-6}$$

Here R_{eq} is the equivalent resistance of the photodiode load and the preamplifier, and F_t is the noise factor of the preamplifier.

9.2.3 Relative Intensity Noise (RIN)

The source noise in Eq. (9-1) is given by

$$\langle i_{source}^2 \rangle = \text{RIN} \left(\mathcal{R}_0 \bar{P} \right)^2 B \tag{9-7}$$

where the laser *relative intensity noise* (RIN), which is measured in dB/Hz, is defined by

$$\text{RIN} = \frac{\langle (\Delta P_L)^2 \rangle}{\bar{P}_L^2} \tag{9-8}$$

where $\langle (\Delta P_L)^2 \rangle$ is the mean square intensity fluctuation of the laser output and $\bar{P}_L$ is the average laser light intensity. This noise decreases as the injection-current level increases according to the relationship

$$\text{RIN} \propto \left(\frac{I_B}{I_{th}} - 1 \right)^{-3} \tag{9-9}$$

Example 9-1. Figure 9-3 shows an example of Eq. (9-9) for two buried-heterostructure lasers.[13] The noise level was measured at 100 MHz. For injection currents sufficiently above threshold (that is, for $I_B/I_{th} > 1.2$), the RIN of these index-guided lasers lies between -140 and -150 dB/Hz.

Example 9-2. Figure 9-4 shows the RIN of an InGaAsP buried-heterostructure laser as a function of modulation frequency at several different bias levels.[1] The relative intensity noise is essentially independent of frequency below several hundred megahertz, and peaks at the resonant frequency. In this case, at a bias level of 60 mA, which gives a 5-mW output, the RIN is typically less than -135 dB/Hz for modulation frequencies up to 8 GHz. For received optical signal levels of -13 dBm (50 μW) or less, the RIN of buried-heterostructure InGaAsP lasers lies sufficiently below the noise level of a 50-Ω amplifier with a 3-dB noise figure.

Substituting Eqs. (9-4) through (9-7) into Eq. (9-1) yields the following carrier-to-noise ratio for a single-channel AM system:

$$\frac{C}{N} = \frac{\frac{1}{2} \left(m \mathcal{R}_0 M \bar{P} \right)^2}{\text{RIN} \left(\mathcal{R}_0 \bar{P} \right)^2 B + 2q(I_p + I_D) M^2 F(M) B + (4k_B T/R_{eq}) BF_t} \tag{9-10}$$

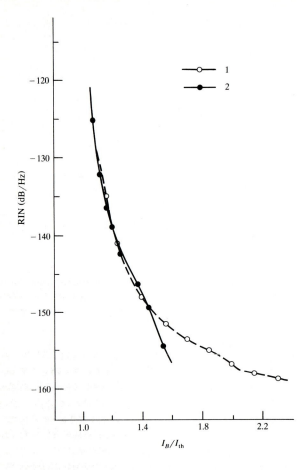

FIGURE 9-3
Example of the relative intensity noise (RIN) for two buried-hetero-structure laser diodes. The noise level was measured at 100 MHz. (Reproduced with permission from Sato,[13] © 1983, IEEE.)

9.2.4 Reflection Effects on RIN

In implementing a high-speed analog link, one must take special precautions to minimize optical reflections back into the laser.[2] Back-reflected signals can increase the RIN by 10 to 20 dB as shown in Fig. 9-5. These curves show the increase in relative intensity noise for bias points ranging from 1.24 to 1.62 times the threshold-current level. The feedback power ratio in Fig. 9-5 is the amount of optical power reflected back into the laser relative to the light output from the source. As an example, the dashed line shows that at $1.33I_{th}$ the feedback ratio must be less than -60 dB in order to maintain an RIN of less than -140 dB/Hz.

9.2.5 Limiting Conditions

Let us now look at some limiting conditions. When the optical power level at the receiver is low, the preamplifier circuit noise dominates the system noise.

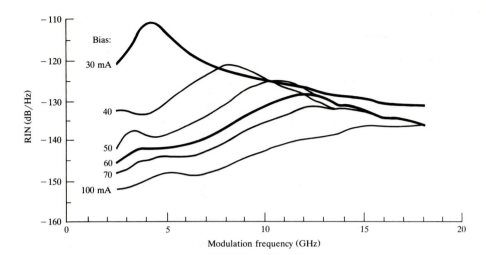

FIGURE 9-4
The RIN of an InGaAsP buried-heterostructure laser as a function of modulation frequency at several different bias levels. (Reproduced with permission from Olshansky, Lanzisera, and Hill,[1] © 1989, IEEE.)

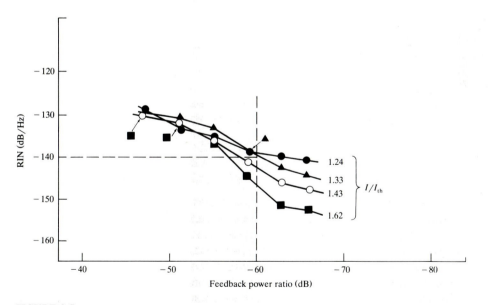

FIGURE 9-5
The increase in RIN due to the back-reflected optical signals. (Reproduced with permission from Sato,[13] © 1983, IEEE.)

For this we have

$$\left(\frac{C}{N}\right)_{\text{limit 1}} = \frac{\frac{1}{2}\left(m\mathscr{R}_0 M\overline{P}\right)^2}{\left(4k_B T/R_{\text{eq}}\right)BF_t} \qquad (9\text{-}11)$$

In this case the carrier-to-noise ratio is directly proportional to the square of the received optical power, so that for each 1-dB variation in received optical power, C/N will change by 2 dB.

For well-designed photodiodes, the bulk and surface dark currents are small compared to the shot (quantum) noise for intermediate optical signal levels at the receiver. Thus at intermediate power levels the quantum-noise term of the photodiode will dominate the system noise. In this case we have

$$\left(\frac{C}{N}\right)_{\text{limit 2}} = \frac{\frac{1}{2}m^2\mathscr{R}_0\overline{P}}{2qF(M)B} \qquad (9\text{-}12)$$

so that the carrier-to-noise ratio will vary by 1 dB for every 1-dB change in the received optical power.

If the laser has a high RIN value so that the reflection noise dominates over other noise terms, then the carrier-to-noise ratio becomes

$$\left(\frac{C}{N}\right)_{\text{limit 3}} = \frac{\frac{1}{2}(mM)^2}{\text{RIN } B} \qquad (9\text{-}13)$$

which is a constant. In this case the performance cannot be improved unless the modulation index is increased.

Example 9-3. As an example of the limiting conditions, consider a link with a laser transmitter and a *pin* photodiode receiver having the following characteristics:

Transmitter	Receiver
$m = 0.25$	$\mathscr{R}_0 = 0.6$ A/W
RIN $= -143$ dB/Hz	$B = 10$ MHz
$P_c = 0$ dBm	$I_D = 10$ nA
	$R_{\text{eq}} = 750\ \Omega$
	$F_t = 3$ dB

where P_c is the optical power coupled into the fiber. To see the effects of the different noise terms on the carrier-to-noise ratio, Fig. 9-6 shows a plot of C/N as a function of the optical power level at the receiver. In this case we see that at high received powers the source noise dominates to give a constant C/N. At intermediate levels the quantum noise is the main contributor, with a 1-dB drop in C/N for every 1-dB decrease in received optical power. For low light levels the thermal noise of the receiver is the limiting noise term, yielding a 2-dB rolloff in C/N for each 1-dB drop in received optical power. It is important to note that the limiting factors can vary significantly depending on the transmitter and receiver characteristics. For example, for low-impedance amplifiers the thermal noise of the receiver can be the dominating performance limiter for all practical link lengths (see Prob. 9-1).

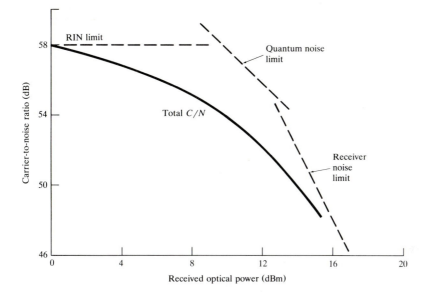

FIGURE 9-6
Carrier-to-noise ratio as a function of optical power level at the receiver. In this case, RIN dominates at high powers, quantum noise gives a 1-dB drop in C/N for each 1-dB power decrease at intermediate levels, and receiver thermal noise yields a 2-dB C/N rolloff per 1-dB drop in received power at low light levels.

9.3 MULTICHANNEL TRANSMISSION TECHNIQUES

So far we have examined only the case of a single signal being transmitted over a channel. In broadband analog applications, such as cable television (CATV) supertrunks, one needs to send multiple analog signals over the same fiber. To do this one can employ a multiplexing technique where a number of baseband signals are superimposed on a set of N subcarriers having different frequencies $f_1, f_2, \ldots, f_N$. These modulated subcarriers are then combined electrically through frequency-division multiplexing (FDM) to form a composite signal that directly modulates a single optical source. Methods for achieving this include vestigal-sideband amplitude modulation (VSB-AM), frequency modulation (FM), and subcarrier multiplexing (SCM).

Of these, AM is simple and cost-effective in that it is compatible with the equipment interfaces of a large number of CATV customers, but its signal is very sensitive to noise and nonlinear distortion. Although FM requires a larger bandwidth than AM, it provides a higher signal-to-noise ratio and is less sensitive to source nonlinearities. Thus FM systems are being considered as a candidate for cable television distribution of the late 1990s. Microwave subcarrier multiplexing (SCM) operates at higher frequencies than AM or FM and is an interesting approach for broadband distribution of both analog and digital signals.

9.3.1 Multichannel Amplitude Modulation

The initial widespread application of analog fiber optic links, which started in the late 1980s, was to CATV networks.[14–17] These coax-based television networks operate in a frequency range from 50 to 88 MHz and from 120 to 550 MHz. The band from 88 to 120 MHz is not used, since it is reserved for FM radio broadcast. These networks can deliver over 80 amplitude-modulated vestigal-sideband (AM-VSB) video channels, each having a noise bandwidth of 4 MHz within a channel bandwidth of 6 MHz, with signal-to-noise ratios exceeding 47 dB. To remain compatible with existing coax-based networks, a multichannel AM-VSB format was also chosen for the fiber optic system.

Figure 9-7 depicts the technique for combining N independent messages. An information-bearing signal on channel i amplitude-modulates a carrier wave having a frequency f_i, where $i = 1, 2, \ldots, N$. An RF power combiner then sums these N amplitude-modulated carriers to yield a composite frequency-division-multiplexed (FDM) signal which intensity modulates a laser diode. Following the optical receiver, a bank of parallel bandpass filters separates the combined carriers back into individual channels. The individual message signals are recovered from the carriers by standard RF techniques.

For a large number of FDM carriers with random phases, the carriers add on a power basis. Thus for N channels the optical modulation index m is related to the per-channel modulation index m_i by

$$m = \left(\sum_{i=1}^{N} m_i^2 \right)^{1/2} \qquad (9\text{-}14a)$$

If each channel modulation index m_i has the same value m_c, then

$$m = m_c N^{0.5} \qquad (9\text{-}14b)$$

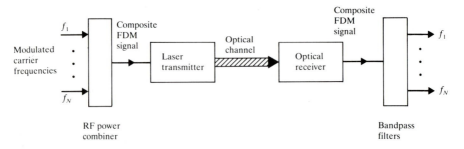

FIGURE 9-7
Standard technique for frequency-division multiplexing of N independent information-bearing signals.

As a result, when N signals are frequency-multiplexed and used to modulate a single optical source, the carrier-to-noise ratio of a single channel is degraded by $10 \log N$. If only a few channels are combined, the signals will add in voltage rather than power, so that the degradation will have a $20 \log N$ characteristic.

When multiple carrier frequencies pass through a nonlinear device such as a laser diode, signal products other than the original frequencies can be produced. As noted in Sec. 4.4, these undesirable signals are called *intermodulation products* and they can cause serious interference in both in-band and out-of-band channels. The result is a degradation of the transmitted signal. Among the intermodulation products, generally only the second-order and third-order terms are considered, since higher-order products tend to be significantly smaller.

If the operating frequency band of the channel is less than one octave, all harmonic distortions and even-order intermodulation (IM) distortion products will fall outside the passband. Thus third-order types at frequencies $f_i + f_j - f_k$ (which are known as *triple-beat IM products*) and $2f_i - f_j$ (which are known as *two-tone third-order IM products*) are the most dominant, since many of these fall within the bandwidth of the channel. The amplitudes of the triple-beat products are 3 dB higher than the two-tone third-order IM products. In addition, since there are $N(N-1)(N-2)/2$ triple-beat terms compared to $N(N-1)$ two-tone third-order terms, the triple-beat products tend to be the major source of IM noise.

If a signal passband contains a large number of equally spaced carriers, several IM terms will exist at or near the same frequency. This so-called *beat stacking* is additive on a power basis. For example, for N equally spaced equal-amplitude carriers, the number of third-order IM products that fall right on the rth carrier is given by[18, 19]

$$D_{1,2} = \tfrac{1}{2}\left\{ N - 2 - \tfrac{1}{2}\left[1 - (-1)^N\right](-1)^r \right\} \tag{9-15}$$

for two-tone terms of the type $2f_i - f_j$, and by

$$D_{1,1,1} = \frac{r}{2}(N - r + 1) + \tfrac{1}{4}\left\{(N-3)^2 - 5 - \tfrac{1}{2}\left[1 - (-1)^N\right](-1)^{N+r}\right\} \tag{9-16}$$

for triple-beat terms of the type $f_i + f_j - f_k$.

Whereas the two-tone third-order terms are fairly evenly spread through the operating passband, the triple-beat products tend to be concentrated in the middle of the channel passband, so that the center carriers receive the most intermodulation interference. Tables 9-1 and 9-2 show the distributions of the third-order triple-beat and two-tone IM products for the number of channels N ranging from 1 to 8.

The results of beat stacking are commonly referred to as *composite second order* (*CSO*) and *composite triple beat* (*CTB*), and are used to describe the

TABLE 9-1
Distribution of the number of third-order triple-beat intermodulation products for the number of channels N ranging from 1 to 8

N	1	2	3	4	5	6	7	8
1	0							
2	0	0						
3	0	1	0					
4	1	2	2	1				
5	2	4	4	4	2			
6	4	6	7	7	6	4		
7	6	9	10	11	10	9	6	
8	9	12	14	15	15	14	12	9

performance of multichannel AM links. These are defined as[20]

$$\text{CSO} = \frac{\text{peak carrier power}}{\text{peak power in composite 2nd-order IM tone}} \tag{9-17}$$

and

$$\text{CTB} = \frac{\text{peak carrier power}}{\text{peak power in composite 3rd-order IM tone}} \tag{9-18}$$

Example 9-4. Figures 9-8 and 9-9 show the predicted relative second-order and third-order intermodulation performance, respectively, for sixty CATV channels in the frequency range of 50 to 450 MHz. The effect of CSO is most significant at the passband edges, whereas CTB contributions are most critical at the center of the band.

TABLE 9-2
Distribution of the number of third-order two-tone intermodulation products for the number of channels N ranging from 1 to 8

N	1	2	3	4	5	6	7	8
1	0							
2	0	0						
3	1	0	1					
4	1	1	1	1				
5	2	1	2	1	2			
6	2	2	2	2	2	2		
7	3	2	3	2	3	2	3	
8	3	3	3	3	3	3	3	3

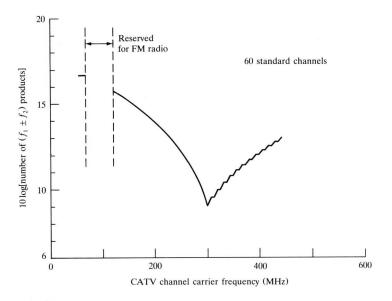

FIGURE 9-8
Predicted relative CSO performance for 60 amplitude-modulated CATV channels. The 88- to 120-MHz band is reserved for FM radio broadcast. (Reproduced with permission from Darcie, Lipson, Roxlo, and McGrath,[4] © 1990, IEEE.)

9.3.2 Multichannel Frequency Modulation

The use of AM-VSB signals for transmitting multiple analog channels is in principle straightforward and simple. However, it has a C/N requirement (or equivalently, for AM, an S/N requirement) of at least 40 dB for each AM channel, which places very stringent requirements on laser and receiver linearity. An alternative technique is frequency modulation (FM), wherein each subcarrier is frequency-modulated by a message signal.[2,21,22] This requires a wider bandwidth (30 MHz versus 4 MHz for AM), but yields a signal-to-noise ratio improvement over the carrier-to-noise ratio.

The S/N at the output of an FM detector is much larger than the C/N at the input of the detector. The improvement is given by[2]

$$\left(\frac{S}{N}\right)_{\text{out}} = \left(\frac{C}{N}\right)_{\text{in}} + 10\log\left[\frac{3}{2}\frac{B}{f_v}\left(\frac{\Delta f_{\text{pp}}}{f_v}\right)^2\right] + w \qquad (9\text{-}19)$$

where B is the required bandwidth, Δf_{pp} is the peak-to-peak frequency deviation of the modulator, f_v is the highest video frequency, and w is a weighting factor used to account for the nonuniform response of the eye pattern to white noise in the video bandwidth. The total S/N improvement depends on the system design, but is generally in the 36 to 44-dB range.[23,24] The reduced

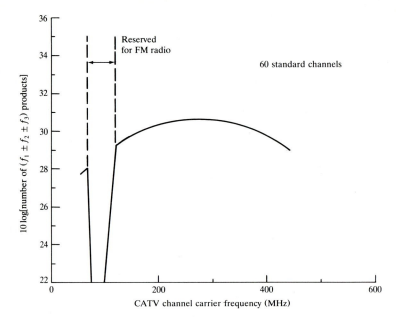

FIGURE 9-9
Predicted relative CTB performance for 60 amplitude-modulated CATV channels. The 88- to 120-MHz band is reserved for FM radio broadcast. (Reproduced with permission from Darcie, Lipson, Roxlo, and McGrath,[4] © 1990, IEEE.)

C/N requirements thus make an FM system much less susceptible to laser and receiver noises than an AM system.

> **Example 9-5.** Figure 9-10 shows a plot of RIN versus optical modulation index per channel, comparing AM and FM broadcast TV systems.[2] The following assumptions were made in this calculation:
>
> > The RIN noise dominates.
> > $S/N = C/N + 40$ dB for the FM system.
> > The AM bandwidth per channel is 4 MHz.
> > The FM bandwidth per channel is 30 MHz.
>
> If the per-channel optical modulation index is 5 percent, then a RIN of less than -120 dB/Hz is needed for each FM TV program to have studio-quality reception, requiring $S/N \geq 56$ dB. This is easily met with a typical packaged laser diode which has a nominal RIN value of -130 dB/Hz. On the other hand, for an AM system a laser with an RIN value of -140 dB/Hz can barely meet the CATV reception requirement of $S/N \geq 40$ dB.
>
> **Example 9-6.** Another disadvantage of AM transmission compared to FM is the limited power margin. Figure 9-11 depicts the calculated power budget versus the

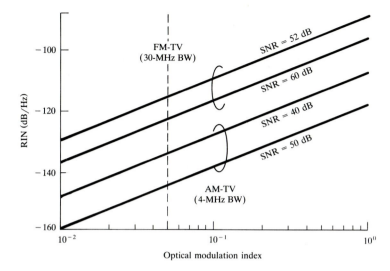

FIGURE 9-10
RIN versus the optical modulation index per channel for AM and FM video signals for several different signal-to-noise ratios (SNR). (Reproduced with permission from Way,[2] © 1989, IEEE.)

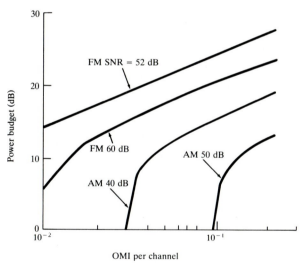

FIGURE 9-11
Power budget versus the optical modulation index (OMI) per channel for distribution of multichannel AM and FM video signals. (Reproduced with permission from Way,[2] © 1989, IEEE.)

optical modulation index per channel for distribution of multichannel AM and FM video signals. The curves are given for different signal-to-noise ratios. The following assumptions were made in this calculation:

Laser power coupled into single-mode fiber = 0 dBm
RIN = −140 dB/Hz

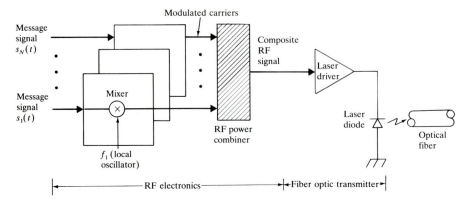

FIGURE 9-12
Basic concept of subcarrier multiplexing. One can simultaneously send analog and digital signals by frequency-division-multiplexing them on different subcarrier frequencies.

> *pin* photodiode receiver with a 50-Ω front end
> Preamplifier noise figure = 2 dB
> AM bandwidth per channel = 4 MHz
> FM bandwidth per channel = 30 MHz

Again assuming a per-channel optical modulation index of 5 percent, the AM system has a power margin of about 10 dB for a 40-dB signal-to-noise ratio, whereas the FM system has a power margin of 20 dB for $S/N = 52$ dB.

9.3.3 Subcarrier Multiplexing

In the mid 1980s great interest arose in using RF or *microwave subcarrier multiplexing* for high-capacity lightwave systems.[1, 2, 25–28] The term *subcarrier multiplexing* (SCM) is used to describe the capability of multiplexing both multichannel analog and digital signals within the same system.

Figure 9-12 shows the basic concept of an SCM system. The input to the transmitter consists of a mixture of N independent analog and digital baseband signals. These signals can carry either voice, data, video, digital audio, high-definition video, or any other analog or digital information. Each incoming signal $s_i(t)$ is mixed with a local oscillator (LO) having a frequency f_i. The local oscillator frequencies employed are in the 2- to 8-GHz range and are known as the *subcarriers*. Combining the modulated subcarriers gives a composite frequency-division-multiplexed signal which is used to drive a laser diode.

At the receiving end the optical signal is directly detected with a high-speed wideband InGaAs *pin* photodiode and reconverted to a microwave signal. For long-distance links one can also employ a wideband InGaAs avalanche photodiode with a 50- to 80-GHz gain-bandwidth product or by using an optical preamplifier. For amplifying the received microwave signal, one can use a commercially available wideband low-noise amplifier or a *pin*-FET receiver.[29-31] Commercial amplifiers have 2 to 8-GHz bandwidths and noise figures of about 3 dB. Similar performance can be achieved with a *pin*-FET receiver.

9.4 SUMMARY

Analog fiber optic links have found an important place in many telecommunication, video-distribution, and microwave transmission systems. Amplitude-modulated vestigal-sideband (AM-VSB) techniques are widely used for CATV applications, and frequency modulation (FM) is also under consideration for this. Microwave subcarrier multiplexing (SCM) operates at higher frequencies than AM or FM and is an interesting approach for broadband distribution of both analog and digital signals.

No matter which method is implemented, one must pay careful attention to nonlinearities in the optical source. These include harmonic distortions, intermodulation products, and relative intensity noise (RIN) in the laser. AM is a simple and cost-effective transmission scheme, but its signal is very sensitive to noise and nonlinear distortion. Although FM requires a larger bandwidth than AM, it provides a higher signal-to-noise ratio and is less sensitive to source nonlinearities.

PROBLEMS

9-1. Commercially available wideband receivers have equivalent resistances $R_{eq} = 75 \ \Omega$. With this value of R_{eq} and letting the remaining transmitter and receiver parameters be the same as in Example 9-3, plot the total carrier-to-noise ratio and its limiting expressions, as given by Eqs. (9-10) through (9-13), for received power levels ranging from 0 to -16 dBm. Show that the thermal noise of the receiver dominates over the quantum noise at all power levels when $R_{eq} = 75 \ \Omega$.

9-2. Consider a 5-channel frequency-division multiplexed (FDM) system having carriers at f_1, $f_2 = f_1 + \Delta$, $f_3 = f_1 + 2\Delta$, $f_4 = f_1 + 3\Delta$, and $f_5 = f_1 + 4\Delta$, where Δ is the spacing between carriers. On a frequency plot, show the number and location of the triple-beat and two-tone third-order intermodulation products.

9-3. Suppose we want to frequency-division multiplex 60 FM signals. If 30 of these signals have a per-channel modulation index $m_i = 3$ percent and the other 30 signals have $m_i = 4$ percent, find the optical modulation index of the laser.

9-4. Consider an SCM system having 120 channels, each modulated at 2.3 percent. The link consists of 12 km of single-mode fiber having a loss of 1 dB/km, plus a connector having an 0.5-dB loss on each end. The laser source couples 2 mW of optical power into the fiber and has RIN $= -135$ dB/Hz. The *pin* photodiode receiver has a responsivity of 0.6 A/W, $B = 5$ GHz, $I_D = 10$ nA, $R_{eq} = 50 \ \Omega$, and $F_t = 3$ dB. Find the carrier-to-noise ratio for this system.

9-5. What is the carrier-to-noise ratio for the system described in Prob. 9-4 if the *pin* photodiode is replaced with an InGaAs avalanche photodiode having $M = 10$ and $F(M) = M^{0.7}$?

9-6. Consider a 32-channel FDM system with a 4.4-percent modulation index per channel. Let RIN $= -135$ dB/Hz, and assume the *pin* photodiode receiver has a responsivity of 0.6 A/W, $B = 5$ GHz, $I_D = 10$ nA, $R_{eq} = 50\ \Omega$, and $F_t = 3$ dB.
(*a*) Find the carrier-to-noise ratio for this link if the received optical power is -10 dBm.
(*b*) Find the carrier-to-noise ratio if the modulation index is increased to 7 percent per channel and the received optical power is decreased to -13 dBm.

9-7. For a fiber optic link using a single-longitudinal-mode laser with a 3-dB linewidth of $\Delta\nu$ and having two fiber connectors with reflectivities R_1 and R_2, the worst-case RIN occurs when the direct and doubly reflected optical fields interfere in quadrature.[32] This is described by the equation

$$\text{RIN}(f) = \frac{4R_1R_2}{\pi}\frac{\Delta\nu}{f^2 + \Delta\nu^2}\left[1 + e^{-4\pi\Delta\nu\tau} - 2e^{-2\pi\Delta\nu\tau}\cos(2\pi f\tau)\right]$$

Show that this expression reduces to

$$\text{RIN}(f) = \frac{16R_1R_2}{\pi}\Delta\nu\,\tau^2 \qquad \text{for} \quad \Delta\nu\cdot\tau \ll 1,$$

and

$$\text{RIN}(f) = \frac{4R_1R_2}{\pi}\frac{\Delta\nu}{f^2 + \Delta\nu^2} \qquad \text{for} \quad \Delta\nu\cdot\tau \gg 1$$

9-8. A typical DFB laser has a linewidth 10 MHz $< \Delta\nu <$ 100 MHz, and with a 1- to 10-m optical jumper cable we have $0.05 < \Delta\nu\cdot\tau < 5$. Choosing the lower limits and letting $f = 10$ MHz, use the expression in Prob. 9-7 to show that to achieve an RIN of less than -140 dB/Hz, the average reflectivity per connector should be less than -30 dB.

REFERENCES

1. R. Olshansky, V. A. Lanzisera, and P. M. Hill, "Subcarrier multiplexed lightwave systems for broadband distribution," *J. Lightwave Tech.*, vol. 7, pp. 1329–1342, Sept. 1989.
2. W. I. Way, "Subcarrier multiplexed lightwave system design considerations for subscriber loop applications," *J. Lightwave Tech.*, vol. 7, pp. 1806–1818, Nov. 1989.
3. S. S. Wagner and R. C. Menendez, "Evolutionary architectures and techniques for video distribution on fiber," *IEEE Commun. Mag.*, vol. 27, pp. 17–25, Dec. 1989.
4. T. E. Darcie, J. Lipson, C. B. Roxlo, and C. J. McGrath, "Fiber optic device technology for broadband analog video systems," *IEEE Mag. Lightwave Commun. Sys.*, vol. 1, pp. 46–52, Feb. 1990.
5. D. E. Robinson and D. Grubb III, "A high-quality switched FM video system," *IEEE Mag. Lightwave Commun. Sys.*, vol. 1, pp. 53–59, Feb. 1990.
6. G. Grimes, "Remoting antennas with high-speed analog fiber optics," *Microwave Sys. News* (*MSN*), vol. 19, pp. 41–47, Aug. 1989.
7. J. J. Pan, "Fiber optics for microwave/millimeter-wave phased arrays," *Microwave Sys. News* (*MSN*), vol. 19, pp. 48–54, July 1989.
8. "Special Issue on Applications of Lightwave Technology to Microwave Devices, Circuits, and Systems," *IEEE Trans. Microwave Theory Tech.*, vol. 38, May 1990.
9. W. Ng, A. Walston, G. Tangonan, I. Newberg, and J. J. Lee, "Laser-switched wideband fiber optic delay network for phased array antenna steering," *OFC Conf. Proc.*, pp. 80–81, Jan. 1990.
10. E. A. Lee and D. G. Messerschmitt, *Digital Communication*, Kluwer Academic, Boston, 1988.

11. H. Taub and D. L. Schilling, *Principles of Communication Systems*, 2nd ed., McGraw-Hill, New York, 1986.
12. S. Haykin, *Digital Communication*, Wiley, New York, 1988.
13. K. Sato, "Intensity noise of semiconductor laser diodes in fiber optic analog video transmission," *IEEE J. Quantum Electron.*, vol. QE-19, pp. 1380–1391, Sept. 1983.
14. W. S. Ciciora, "An introduction to cable television in the United States," *IEEE Mag. Lightwave Commun. Sys.*, vol. 1, pp. 19–25, Feb. 1990.
15. L. Thompson, R. Pidgeon, and F. Little, "Supertrunking alternatives in CATV," *IEEE Mag. Lightwave Commun. Sys.*, vol. 1, pp. 26–31, Feb. 1990.
16. J. A. Chiddix, "Fiber backbone trunking in cable television networks," *IEEE Mag. Lightwave Commun. Sys.*, vol. 1, pp. 32–37, Feb. 1990.
17. G. M. Hart and N. F. Hamilton-Piercy, "A broadband urban hybrid coaxial/fiber telecommunications network, *IEEE Mag. Lightwave Commun. Sys.*, vol. 1, pp. 38–45, Feb. 1990.
18. Tri T. Ha, *Digital Satellite Communications*, McGraw-Hill, New York, 1990.
19. R. J. Westcott, "Investigation of multiple FM/FDM carriers through a satellite TWT operating near saturation," *Proc. IEE (London)*, vol. 114, pp. 726–740, June 1967.
20. *NCTA Recommended Practices for Measurements on Cable Television Systems*, National Cable Television Association, 1983.
21. W. I. Way, R. S. Wolff, and M. Krain, "A 1.3-μm 35-km fiber-optic microwave multicarrier transmission system for satellite earth stations," *J. Lightwave Tech.*, vol. LT-5, pp. 1325–1332, Sept. 1987.
22. R. Olshansky and V. A. Lanzisera, "60-channel FM video subcarrier multiplexed optical communication system," *Electron. Lett.*, vol. 23, pp. 1196–1198, 1987.
23. (*a*) F. V. C. Mendis and P. A. Rosher, "CNR requirements for subcarrier-multiplexed multichannel video FM transmission in optical fibre," *Electron. Lett.*, vol. 25, pp. 72–74, Jan. 1989.
    (*b*) F. V. C. Mendis, "Interpretation of signal/noise ratio expressions in FM video transmission," *Electron. Lett.*, vol. 25, pp. 67–69, Jan. 1989.
24. T. Pratt and C. W. Bostian, *Satellite Communications*, Wiley, New York, 1986.
25. J. E. Bowers, "Optical transmission using PSK-modulated subcarriers at frequencies to 16 GHz," *Electron. Lett.*, vol. 22, pp. 1119–1121, 1986.
26. T. E. Darcie, "Subcarrier multiplexing for multi-access lightwave networks," *J. Lightwave Tech.*, vol. LT-5, pp. 1103–1110, Aug. 1987.
27. W. I. Way and C. Castelli, "Simultaneous transmission of 2-Gb/s digital data and ten FM-TV analog signals over 16.5-km SM fiber," *Electron. Lett.*, vol. 24, pp. 611–612, 1988.
28. R. Olshansky, V. A. Lanzisera, and P. Hill, "Simultaneous transmission of 100 Mb/s at baseband and 60 FM video channels for a wideband optical communication network," *Electron. Lett.*, vol. 24, pp. 1234–1235, 1988.
29. J. Gimlett, "Low-noise 8-GHz *pin*-FET optical receiver," *Electron. Lett.*, vol. 23, pp. 281–283, 1987.
30. T. E. Darcie, B. L. Casper, J. R. Talman, and C. A. Burrus, Jr., "Resonant *pin*-FET receivers for lightwave subcarrier systems," *J. Lightwave Tech.*, vol. LT-6, pp. 582–589, Apr. 1988.
31. S. R. Cochran, "Low-noise receivers for fiber-optic microwave signal transmission," *J. Lightwave Tech.*, vol. LT-6, pp. 1328–1337, Aug. 1988.
32. R. W. Tkach and A. R. Chraplyvy, "Phase noise and linewidth in an InGaAsP DFB laser," *J. Lightwave Tech.*, vol. LT-4, pp. 1711–1716, Nov. 1986.

CHAPTER
10

COHERENT OPTICAL FIBER COMMUNICATIONS

So far in this text we have considered the most common and cost-effective transmission scheme being used in optical fiber transmission systems. In this scheme the light intensity of the optical source is modulated linearly with respect to the input electrical signal voltage. No attention is paid to the frequency or the phase of the optical carrier, since a photodetector only responds to changes in the power level (the intensity) of an optical signal, and not to its frequency or phase content. This is known as *intensity modulation* (IM). At the receiving end, one then uses *direct detection* (DD) to reconvert the optical signal into an electrical signal. In this respect a direct-detection optical system is analogous to the way a primitive crystal radio detects broadcast signals. These IM/DD methods offer system simplicity and relatively low cost, but they suffer from limited sensitivity and do not take full advantage of the tremendous bandwidth capabilities of optical fibers.

Around 1978 component researchers had improved the spectral purity and frequency stability of semiconductor lasers to the point where schemes using heterodyne or homodyne detection of the optical signal began to look feasible.[1-3] Optical communication systems which use heterodyne or homodyne detection are commonly referred to in the fiber optics literature as *coherent optical communication systems*. In this technique the light is treated as a carrier medium which can be amplitude-, frequency-, or phase-modulated similarly to the methods used in microwave radio systems. A number of review papers on

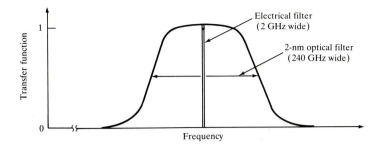

FIGURE 10-1
An optical filter with a 2-nm wavelength passband is equivalent to a 240-GHz frequency band.
Electrical filters with passbands of 2 GHz provide a much higher degree of frequency selectivity.

this subject are available[4-16] as well as a comprehensive text by Okoshi and
Kikuchi.[17]

The two main advantages of coherent optical communication schemes are
a nearly ideal receiver sensitivity (up to 20-dB improvement over direct detection) and a high degree of frequency selectivity. The selectivity of coherent
systems is due to the fact that narrowband electronic filters rather than broad
optical filters pick out the channels. This is illustrated in Fig. 10-1. First note
that a nominal 2-nm wavelength passband of an optical filter is equivalent to a
frequency passband of 240 GHz. Comparing this with a 2-GHz-wide electrical
filter, we see that electrical filtering provides several orders of magnitude more
frequency selectivity. Thus the channel density of a multichannel coherent
system (that is, the number of channels per unit bandwidth) can be 100 times
that for direct detection.

The two transmission regions of an optical fiber which can be used for
coherent systems are from 1270 to 1350 nm and from 1480 to 1600 nm. For
example, as shown in Fig. 10-2, with a 10-GHz channel spacing one could
theoretically have at least 1000 channels in the wavelength window from 1270 to
1350 nm, and 1500 channels in one from 1480 to 1600 nm. Coherent optical
detection is thus analogous to superheterodyne detection in modern radios.

The possibility of actually making such systems depends strongly on (*a*)
the ability to develop truly coherent semiconductor lasers that can be tuned
over a significant portion of the 30,000-GHz bandwidth of a single-mode optical
fiber, (*b*) the implementation of a proper coding method for imposing data on
the optical carrier, and (*c*) the development of a practical polarization control
method for the receiver. This chapter will thus address the following topics:

The definition and classification of coherent systems

Requirements on semiconductor lasers

Receiver sensitivities for different modulation techniques

The need for polarization control in coherent systems

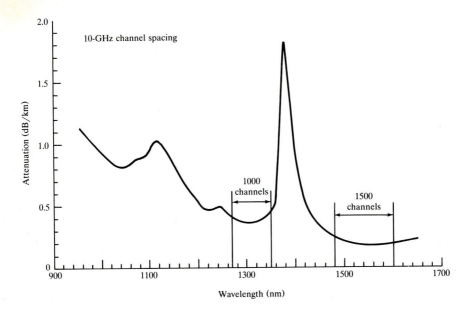

FIGURE 10-2
Assuming a 10-GHz channel spacing, one can theoretically transmit in at least 1000 channels each in the 1300-nm and 1550-nm windows.

10.1 DEFINITION AND CLASSIFICATION OF COHERENT SYSTEMS

In the optical fiber communications literature the term *coherent* refers to any technique employing nonlinear mixing between two optical waves. This is in contrast to the radio communications literature, where the term *coherent* refers to detection techniques in which the absolute phase of the incoming signal is tracked by the receiver. Consequently, many of the techniques which are called coherent in optical fiber communications are called incoherent in the radio communications literature.

10.1.1 Fundamental Concepts

Figure 10-3 illustrates the fundamental concept in coherent optical fiber systems. For simplicity, let us consider the electric field of the transmitted signal to be a plane wave having the form

$$E_s = A_s \cos[\omega_s t + \phi_s(t)] \tag{10-1}$$

where A_s is the amplitude of the optical signal field, ω_s is the optical signal carrier frequency, and $\phi_s(t)$ is the phase of the optical signal. To send information one can modulate either the amplitude, the frequency, or the phase of the optical carrier. Thus, one of the following three modulation techniques can be

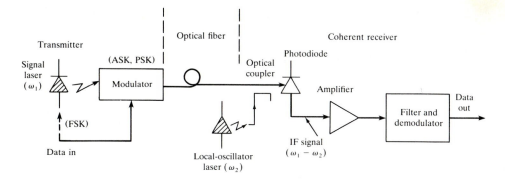

FIGURE 10-3
Fundamental concept of a coherent lightwave system. There are three possible modulation methods and four basic demodulation formats.

implemented:

1. *Amplitude shift keying (ASK) or on-off keying (OOK).* In this case ϕ_s is constant and the signal amplitude A_s takes one of two values during each bit period, depending on whether a 0 or a 1 is being transmitted.
2. *Frequency shift keying (FSK).* For FSK modulation the amplitude A_s is constant and $\phi_s(t)$ is either $\omega_1 t$ or $\omega_2 t$, where the frequencies ω_1 and ω_2 represent binary signal values.
3. *Phase shift keying (PSK).* In the PSK method, information is conveyed by varying the phase with a sine wave $\phi_s(t) = \beta \sin \omega_m t$, where β is the modulation index and ω_m is the modulation frequency.

In a *direct-detection* system the electrical signal coming into the transmitter *amplitude-modulates* the optical power level of the light source. Thus the optical power is proportional to the signal current level. At the receiver the incoming optical signal is converted directly into a demodulated electrical output. This directly detected current is proportional to the intensity I_{DD} (the square of the electric field) of the optical signal, yielding

$$I_{DD} = E_s E_s^* = \tfrac{1}{2}A_s^2[1 + \cos(2\omega_s t + 2\phi_s)] \qquad (10\text{-}2)$$

The term involving $\cos(2\omega_s t + 2\phi_s)$ gets eliminated from the receiver, since its frequency, which is twice the optical carrier frequency, is beyond the response capability of the detector. Thus for direct detection Eq. (10-2) becomes

$$I_{DD} = E_s E_s^* = \tfrac{1}{2}A_s^2 \qquad (10\text{-}3)$$

At the receiving end in coherent lightwave systems, the receiver first adds a locally generated optical wave to the incoming information-bearing signal and then detects the combination. There are four basic demodulation formats, depending on how the optical signal is mixed with the local oscillator (which gives heterodyne or homodyne detection) and how the electrical signal is detected (either synchronously or asynchronously). As we shall see in this section, for a given modulation format homodyne receivers are more sensitive than heterodyne receivers, and synchronous detection is more sensitive than asynchronous detection.

The mixing of the information-bearing and local-oscillator signals is done on the surface of the photodetector (before photodetection takes place). If the local-oscillator (LO) field has the form

$$E_{\text{LO}} = A_{\text{LO}} \cos[\omega_{\text{LO}} t + \phi_{\text{LO}}(t)] \tag{10-4}$$

where A_{LO} is the amplitude of the local oscillator field, and ω_{LO} and $\phi_{\text{LO}}(t)$ are the local-oscillator frequency and phase, respectively, then the detected current is proportional to the square of the total electric field of the signal falling on the photodetector. That is, the intensity $I_{\text{coh}}(t)$ is

$$
\begin{aligned}
I_{\text{coh}}(t) &= (\mathbf{E}_s + \mathbf{E}_{\text{LO}})^2 \\
&= \tfrac{1}{2}A_s^2 + \tfrac{1}{2}A_{\text{LO}}^2 + A_s A_{\text{LO}} \cos[(\omega_s - \omega_{\text{LO}})t + \phi(t)]\cos\theta(t) \tag{10-5}
\end{aligned}
$$

where $\phi(t) = \phi_s(t) - \phi_{\text{LO}}(t)$ is the relative phase difference between the incoming information-bearing signal and the local-oscillator signal, and

$$\cos\theta(t) = \frac{\mathbf{E}_s \cdot \mathbf{E}_{\text{LO}}}{|\mathbf{E}_s| |\mathbf{E}_{\text{LO}}|}$$

represents the polarization misalignment between the signal wave and the local-oscillator wave. Here again we have used the condition that the photodetector does not respond to terms oscillating near the frequency $2\omega_s$.

Since the optical power $P(t)$ is proportional to the intensity, at the photodetector we then have

$$P(t) = P_s + P_{\text{LO}} + 2\sqrt{P_s P_{\text{LO}}}\, \cos[(\omega_s - \omega_{\text{LO}})t + \phi(t)]\cos\theta(t) \tag{10-6}$$

where P_s and P_{LO} are the signal and local-oscillator optical powers, respectively, with $P_{\text{LO}} \gg P_s$. Thus we see that the angular-frequency difference $\omega_{\text{IF}} = \omega_s - \omega_{\text{LO}}$ is an intermediate frequency, and the phase angle $\phi(t)$ gives the time-varying phase difference between the signal and local-oscillator levels. The frequency ω_{IF} is normally in the radio-frequency range of a few tens or hundreds of megahertz.

10.1.2 Homodyne Detection

When the signal-carrier and local-oscillator frequencies are equal, that is, when $\omega_{\text{IF}} = 0$, we have the special case of *homodyne detection*. Equation (10-6) then becomes

$$P(t) = P_s + P_{\text{LO}} + 2\sqrt{P_s P_{\text{LO}}}\, \cos\phi(t)\cos\theta(t) \tag{10-7}$$

Thus one can use either OOK [varying the signal level P_s while keeping $\phi(t)$

constant] or PSK [varying the phase $\phi_s(t)$ of the signal and keeping P_s constant] modulation schemes to transmit information. Note that since $P_{LO} \gg P_s$ and P_{LO} is constant, the last term on the right-hand side of Eq. (10-7) contains the transmitted information. Since this term increases with increasing laser power, the local oscillator effectively acts as a signal amplifier, thereby giving greater receiver sensitivity than direct detection.

As can be seen from Eq. (10-7), homodyne detection brings the signal directly to the baseband frequency, so that no further electrical demodulation is required. Homodyne receivers yield the most sensitive coherent systems. However, they are also the most difficult to build, since the local oscillator must be controlled by an optical phase-locked loop. In addition, the need for the signal and the local-oscillator lasers to have the same frequencies puts very stringent requirements on these two optical sources. As Sec. 10.3 addresses, these criteria include an extremely narrow spectral width (linewidth) and a high degree of wavelength tunability.

10.1.3 Heterodyne Detection

In *heterodyne detection* the intermediate frequency ω_{IF} is nonzero and an optical phase-locked loop is not needed. Consequently heterodyne receivers are much easier to implement than homodyne receivers. However, the price for this simplification is a 3-dB degradation in sensitivity compared to homodyne detection.

Either OOK, FSK, or PSK modulation techniques can be used. Let us consider the output current at the receiver. Since $P_s \ll P_{LO}$, we can ignore the first term on the right-hand side of Eq. (10-6). The receiver output current then contains a dc term given by

$$i_{dc} = \frac{\eta q}{h\nu} P_{LO} \qquad (10\text{-}8)$$

and a time-varying IF term given by

$$i_{IF}(t) = \frac{2\eta q}{h\nu} \sqrt{P_s P_{LO}} \cos[\omega_{IF} + \phi(t)]\cos\theta(t) \qquad (10\text{-}9)$$

The dc current is normally filtered out in the receiver, and the IF current gets amplified. One then recovers the information from the amplified current using conventional RF demodulation techniques.

10.2 REQUIREMENTS ON SEMICONDUCTOR LASERS

For coherent optical fiber transmission systems one needs to use single-mode semiconductor lasers with narrow linewidths, stable frequencies, and wavelength-tuning capabilities. This applies to both the signal-generating laser and the local-oscillator laser.

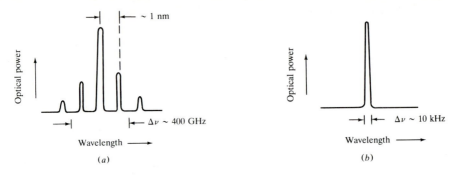

FIGURE 10-4
(*a*) Fabry-Perot laser with a linewidth of 3 nm or, equivalently, 400 GHz at 1550 nm; (*b*) external cavity laser with a linewidth of 10^{-7} nm or, equivalently, 10 kHz at 1550 nm.

10.2.1 Source Linewidths

As described in Chap. 4, the emission pattern of conventional Fabry-Perot semiconductor lasers consists of several spectral lines. These emission lines are spaced by about 0.1 to 1.0 nm within an optical wavelength range of about 1 to 5 nm, as shown in Fig. 10-4. From the fundamental relationship $c = \lambda \nu$, we have that the spread in optical frequency, $\Delta \nu$, which is known as the *linewidth*, is given by

$$\Delta \nu = \frac{c \, \Delta \lambda}{\lambda^2} \tag{10-10}$$

so that a wavelength spread $\Delta \lambda$ of 1 to 5 nm corresponds to optical linewidths in the range of 175 to 880 GHz.

In comparison, Fig. 10-4*b* shows the linewidths of an external cavity laser. With manually adjusted external gratings, these lasers have produced linewidths as narrow as 10 kHz.

Example 10-1

(*a*) A Fabry-Perot laser operating at 1300 nm has a 3-nm spectral width. Its linewidth (frequency spread) is

$$\Delta \nu = \frac{(3 \times 10^8 \text{ m/s})(3 \times 10^{-9} \text{ m})}{(1.3 \times 10^{-6} \text{ m})^2} = 533 \text{ GHz}$$

(*b*) A Fabry-Perot laser operating at 1550 nm has a 4-nm spectral width. Its linewidth is

$$\Delta \nu = \frac{(3 \times 10^8 \text{ m/s})(4 \times 10^{-9} \text{ m})}{(1.55 \times 10^{-6} \text{ m})^2} = 500 \text{ GHz}$$

For direct-detection systems which transmit at data rates of 2 Gb/s, for example, the linewidth of the laser diode is several orders of magnitude greater than the data bandwidth. Thus effectively a direct-detection system is based on intensity modulation of an optical noise source.

Coherent systems using optical phase detection have different needs, since the source linewidth must be narrow compared to the modulation bandwidth.[18-26] Let us see why this is so. As can be seen from Eq. (10-9), in a coherent system one should try to maintain a stable output phase $\phi(t)$ at the receiver. Unpredictable fluctuations in the phase create variations in the receiver IF output current, which lead to noise penalties in the receiver.

Errors owing to random variations in the phase are known as *phase errors* and are related to the coherence time of the laser. The coherence time of a source is approximately the time it takes for a significant phase error to occur, which is roughly one radian, or about 60°. Since the coherence time is inversely related to the spectral width of the laser output, a long coherence time (which is desired) corresponds to a narrow output wavelength spectrum. Normally one expresses coherence time in terms of spectral width (or, equivalently, frequency linewidth), since this is easier to measure than coherence time.

The exact linewidth requirements depend on the modulation format at the transmitter, the transmitted bit rate, and the demodulation technique at the receiver. Requirements for several different combinations are shown in Fig. 10-5. Here, given that the light sources have a lorentzian frequency distribution, the linewidth of the IF signal without modulation is equal to the sum of the source and local-oscillator linewidths. The most demanding requirements are

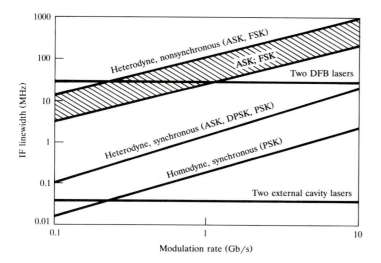

FIGURE 10-5

Laser linewidth requirements for various types of coherent lightwave systems. (Reproduced with permission from Wagner, Cheung, and Kaiser,[11] © 1987, IEEE.)

placed on systems using synchronous detection. For example, PSK heterodyne systems operating at 1 Gb/s require the linewidth of the IF signal to be 0.2 percent of the bit rate to achieve a receiver sensitivity that is within 1 dB of the theoretical limit (see Sec. 10.3). For PSK homodyne systems transmitting at 1 Gb/s, the linewidth needs to be 0.02 percent of the bit rate.

Less stringent linewidth requirements are placed on systems using asynchronous detection, such as envelope detection. For example, as shown in Fig. 10-5, for ASK and certain FSK formats which can use asynchronous detection, transmission rates of over 1 Gb/s can be achieved with lasers having linewidths of around 40 MHz. This is possible using a selected pair of distributed-feedback (DFB) lasers for the information source and the local oscillator as indicated by the top horizontal line. Much stricter requirements are imposed on synchronous systems. For example, 1-Gb/s heterodyne systems need lasers with linewidths of no more than 1 MHz. To achieve such narrow linewidths, one needs single-longitudinal-mode devices such as either a quarter-wavelength-shifted DFB laser, a distributed-Bragg-reflector (DBR) laser, or an external cavity laser.[28] The linewidth of a pair of 40-kHz external cavity lasers is indicated in the bottom horizontal line of Fig. 10-5.

10.2.2 Wavelength Tuning

As can be seen from Eq. (10-9), a constant intermediate frequency ω_{IF} is needed to prevent degradation of the bit error rate. Thus, in addition to narrow-linewidth requirements, some form of wavelength tuning (or, equivalently, frequency selectivity) is required to align the transmitter and local-oscillator lasers for heterodyne or homodyne detection.[29-34]

Example 10-2. From the fundamental relationship $c = \lambda \nu$, we see that operating at a wavelength of 1500 nm is equivalent to an optical carrier frequency of

$$\nu = \frac{c}{\lambda} = \frac{3 \times 10^8 \text{ m/s}}{1.50 \times 10^{-6} \text{ m}} = 2 \times 10^{14} \text{ Hz}$$

For data rates in the 200-Mb/s range, it is necessary for the laser to have a frequency stability of 1 MHz or less. Suppose we want a stability of $\Delta f = 1$ MHz. Then

$$\frac{\Delta f}{\nu} = \frac{1 \times 10^6 \text{ Hz}}{2 \times 10^{14} \text{ Hz}} = 5 \times 10^{-9}$$

which implies a fractional stability of 2 in 10^8.

To prevent bit error rate degradation, designers have developed various optical and IF frequency-control techniques. Figure 10-6 shows the general scheme for achieving frequency stability in a coherent system. At the transmitter a feedback loop is used to lock the frequency of the source to an optical frequency standard such as a Fabry-Perot interferometer, a fiber optic ring

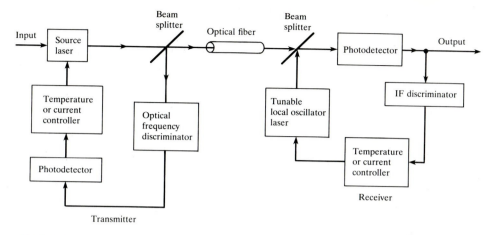

FIGURE 10-6
General scheme for achieving frequency stability in a coherent lightwave system.

resonator, or an atomic or molecular absorption line. Since the emission frequency of basic single-mode lasers depends on factors such as temperature, bias current, optical feedback, and modulation, one can use these properties to control the laser frequency.

At the receiver one can first examine the frequency of the heterodyne-detected IF signal, and then apply either temperature or current-injection control techniques to maintain a constant center frequency of the local-oscillator laser.

Typical values for the temperature and current dependence of a laser are 10 to 20 GHz/°C and 1 to 5 GHz/mA, respectively. Thus stabilization of the center frequency of a laser is achieved through either current injection or by changing its temperature. For current injection, this is equivalent to a wavelength tuning range of 0.75×10^{-2} to 3.8×10^{-2} nm/mA. An example for a three-electrode, wavelength-tunable DBR laser[34] is shown in Fig. 10-7. With the laser sections biased as shown in the figure inset, a continuous tuning range of 2 nm (240 GHz) having the characteristics shown in Fig. 10-7 can be achieved.

10.3 MODULATION TECHNIQUES

In Sec. 10.1 we saw that there are three fundamental ways by which information can be sent in a coherent optical transmission system. These are phase shift keying (PSK), frequency shift keying (FSK), or amplitude shift keying (ASK). In binary digital systems a common ASK technique is on-off keying (OOK). To recover the information at the receiver, one can use either homodyne or heterodyne optical detection techniques together with either synchronous or

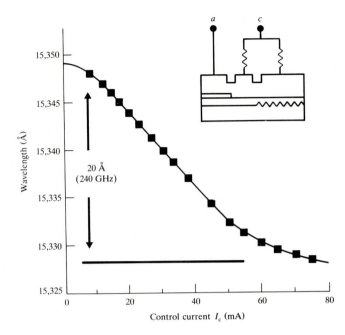

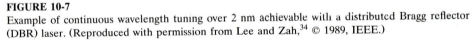

FIGURE 10-7
Example of continuous wavelength tuning over 2 nm achievable with a distributed Bragg reflector (DBR) laser. (Reproduced with permission from Lee and Zah,[34] © 1989, IEEE.)

asynchronous electrical detection. The particular choice of modulation and demodulation methods determines the fundamental receiver sensitivity.

Generally one characterizes the performance of a digital communication system in terms of the bit error rate (BER). The BER depends on the signal-to-noise ratio (S/N) and the probability density function (PDF) at the receiver output (at the input to the comparator). Since for high local-oscillator powers the PDF is gaussian for both heterodyne and homodyne techniques, the BER depends only on the signal-to-noise ratio. Thus one can describe receiver sensitivity in terms of the S/N available at the receiver output, which is directly proportional to the received optical signal power. Traditionally the receiver sensitivity for various coherent optical detection techniques has been described in terms of the average number of photons required to achieve a 10^{-9} BER.

10.3.1 Direct-Detection OOK

To make a comparison between various detection techniques, let us first look at a direct-detection OOK system. Suppose we send an OOK sequence of 1 and 0 pulses which occur with equal probability. Since the OOK data stream is on only half of the time on the average, the required number of photons per bit of information is half the number required per pulse. Thus if $\overline{N}$ and 0 electron-hole pairs are created during 1 and 0 pulses, respectively, then the average number of photons per bit $\overline{N}_p$ for unity quantum efficiency $(\eta = 1)$ is

$$\overline{N}_p = \tfrac{1}{2}\overline{N} + \tfrac{1}{2}(0) \qquad (10\text{-}11)$$

or $\overline{N} = 2\overline{N}_p$. From Eq. (7-26) we then have that the chance of making an error is

$$\tfrac{1}{2}P_r(0) = \tfrac{1}{2}e^{-2\overline{N}_p} \qquad (10\text{-}12)$$

Equation (10-12) implies that about 10 photons per bit are required to get a BER of 10^{-9} for a direct-detection OOK system.

   In practice this fundamental quantum limit is very difficult to achieve for direct-detection receivers. The amplification electronics that follow the photodetector add both thermal noise and shot noise, so that the required received power level lies between 13 and 20 dB above the quantum limit.

10.3.2 OOK Homodyne System

As noted in Sec. 10.1, either homodyne or heterodyne type receivers can be used with OOK modulation. Let us first look at the homodyne case. When a 0 pulse of duration T is received, the average number $\overline{N}_0$ of electron-hole pairs created is simply the number generated by the local oscillator; that is,

$$\overline{N}_0 = A_{\text{LO}}^2 T \qquad (10\text{-}13)$$

For a 1 pulse, the average number of electron-hole pairs, $\overline{N}_1$, is

$$\overline{N}_1 = (A_{\text{LO}} + A_s)^2 T \simeq (A_{\text{LO}}^2 + 2A_{\text{LO}}A_s)T \qquad (10\text{-}14)$$

where the approximation arises from the condition $A_{\text{LO}}^2 \gg A_s^2$. Since the local-oscillator output power is much higher than the received signal level, the voltage V seen by the decoder in the receiver during a 1 pulse is

$$V = \overline{N}_1 - \overline{N}_0 = 2A_{\text{LO}}A_s T \qquad (10\text{-}15)$$

and the associated rms noise σ is

$$\sigma \simeq \sqrt{\overline{N}_1} \simeq \sqrt{\overline{N}_0} \qquad (10\text{-}16)$$

Thus from Eq. (7-23) we have that the BER is

$$P_e = \text{BER} = \frac{1}{2}\left[1 - \text{erf}\left(\frac{V}{2\sqrt{2}\,\sigma}\right)\right] = \tfrac{1}{2}\,\text{erfc}\left(\frac{V}{2\sqrt{2}\,\sigma}\right) = \tfrac{1}{2}\,\text{erfc}\left(\frac{A_s T^{1/2}}{\sqrt{2}}\right) \qquad (10\text{-}17)$$

where $\text{erfc}(x) = 1 - \text{erf}(x)$ is the complementary error function.

   From Example 7-1 we recall that to achieve a BER of 10^{-9} we need $V/\sigma = 12$. Using Eqs. (10-15) and (10-16), this implies

$$A_s^2 T = 36 \qquad (10\text{-}18)$$

which is the expected number of signal photons created per pulse. Thus for OOK homodyne detection, the average energy of each pulse must produce 36 electron-hole pairs. In the ideal case when the quantum efficiency is unity, a 10^{-9} BER is achieved with an average received optical energy of 36 photons per pulse. Again, as in Sec. 10.3.1, if we assume an OOK sequence of 1 and 0 pulses which occur with equal probability, then the average number of received photons per bit of information, $\overline{N}_p$, is 18 (half the number required per pulse).

Thus for OOK homodyne detection the BER is given by

$$\text{BER} = \tfrac{1}{2}\, \text{erfc}\!\left(\sqrt{\eta \overline{N}_p}\,\right) \tag{10-19}$$

To simplify this we note that a useful approximation to $\text{erfc}(\sqrt{x}\,)$ for $x \geq 5$ is

$$\text{erfc}(\sqrt{x}\,) \simeq \frac{e^{-x}}{\sqrt{\pi x}} \tag{10-20}$$

so that

$$\text{BER} \approx \frac{e^{-\eta \overline{N}_p}}{\left(\pi \eta \overline{N}_p\right)^{1/2}} \tag{10-21}$$

for $\eta \overline{N}_p \geq 5$ in OOK homodyne detection.

10.3.3 PSK Homodyne System

Homodyne detection of PSK modulation gives the best theoretical receiver sensitivity, but it is also the most difficult method to implement.[35-37] Figure 10-8 shows the fundamental setup for a homodyne receiver. The incoming optical signal is first combined with a strong optical wave being emitted from the local oscillator. This is done using either a fiber directional coupler (see Chap. 11) or a partially reflecting plate called a *beam splitter*. When a beam splitter is used, it is made almost completely transparent, since the incoming signal is much weaker than the local-oscillator output.

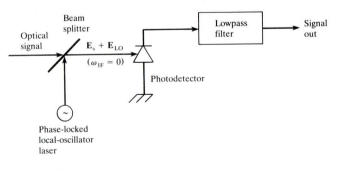

FIGURE 10-8
Fundamental setup for a homodyne receiver.

As we saw in Eq. (10-7), the information is sent by changing the phase of the transmitted wave. For a 0 pulse the signal and local oscillator are out of phase, so that the resultant number of electron-hole pairs generated is

$$\overline{N}_0 = (A_{\mathrm{LO}} - A_s)^2 T \tag{10-22}$$

Similarly, for a 1 pulse the signals are in phase, so that

$$\overline{N}_1 = (A_{\mathrm{LO}} + A_s)^2 T \tag{10-23}$$

Consequently we have that the voltage seen by the decoder in the receiver is

$$V = \overline{N}_1 - \overline{N}_0 = (A_{\mathrm{LO}} + A_s)^2 T - (A_{\mathrm{LO}} - A_s)^2 T = 4 A_{\mathrm{LO}} A_s T \tag{10-24}$$

and the associated rms noise is

$$\sigma = \sqrt{A_{\mathrm{LO}}^2 T} \tag{10-25}$$

Again, as in the case of homodyne OOK detection, the condition $V/\sigma = 12$ for a BER of 10^{-9} yields

$$A_{\mathrm{LO}}^2 T = 9 \tag{10-26}$$

This says that for ideal PSK homodyne detection ($\eta = 1$), an average of 9 photons per bit is required to achieve a 10^{-9} BER. Note that here we do not need to consider the difference between photons per pulse and photons per bit as in the OOK case, since a PSK optical signal is on all the time.

Again using Eq. (7-23), we have that

$$\mathrm{BER} = \tfrac{1}{2} \mathrm{erfc} \sqrt{2 \eta \overline{N}_p} \tag{10-27}$$

for PSK homodyne detection.

10.3.4 Heterodyne Detection Schemes

The analysis for heterodyne receivers is more complicated than in the homodyne case, because the photodetector output appears at an intermediate frequency ω_{IF}. The detailed derivations of the BER for various modulation schemes are given in numerous communication books[17,38-41] and in various journal papers,[6,7,26,42,43] so only the results are given here.

An attractive feature of heterodyne receivers is that they can employ either synchronous or asynchronous detection. Figure 10-9 shows the general receiver configuration. Let us look at this for PSK. In synchronous PSK detection (Fig. 10-9a) one uses a carrier-recovery circuit, which is usually a microwave phase-locked loop (PLL), to generate a local phase reference. The intermediate-frequency carrier is recovered by mixing the output of the PLL with the intermediate-frequency signal. One then uses a lowpass filter to recover the baseband signal. The BER for synchronous heterodyne PSK is given by

$$\mathrm{BER} = \tfrac{1}{2} \mathrm{erfc} \sqrt{\eta \overline{N}_p} \tag{10-28}$$

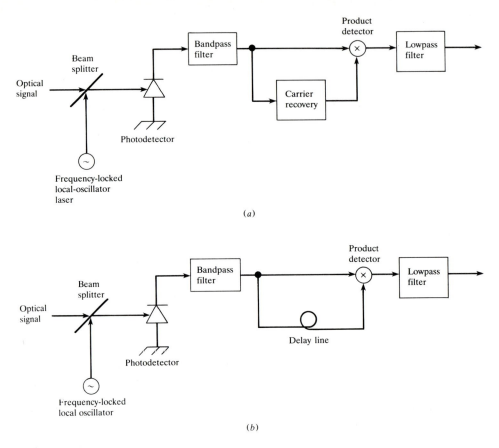

FIGURE 10-9
General heterodyne receiver configurations. (*a*) Synchronous detection uses a carrier-recovery circuit. (*b*) Asynchronous detection employs a simple one-bit delay line.

In this case the ideal PSK receiver requires 18 photons per bit for a 10^{-9} BER. Note that this is the same as for OOK homodyne detection.

A simpler but robust technique that does not use a PLL is *asynchronous detection*, as illustrated in Fig. 10-9*b*. This technique is called *differential PSK* or DPSK. Here the carrier-recovery circuit is replaced by a simple one-bit delay line. Since with a PSK method information is encoded by means of changes in the optical phase, the mixer will produce a positive or negative output depending on whether the phase of the received signal has changed from the previous bit. The transmitted information is thus recovered from this output. This DPSK technique has a sensitivity close to that of synchronous heterodyne detection of PSK, with a bit error rate of

$$\text{BER} = \tfrac{1}{2}\exp\left(-\eta \overline{N}_p\right) \tag{10-29}$$

TABLE 10-1
Summary of the probability of error as a function of the number of received photons per bit for coherent optical fiber systems

| | | Probability of error | | |
| | | Heterodyne | | |
Modulation	Homodyne	Synchronous detection	Asynchronous detection	Direct detection
On-off keying (OOK)	$\frac{1}{2}\,\mathrm{erfc}\,(\eta\overline{N}_p)^{1/2}$	$\frac{1}{2}\,\mathrm{erfc}\,(\frac{1}{2}\eta\overline{N}_p)^{1/2}$	$\frac{1}{2}\exp(-\frac{1}{2}\eta\overline{N}_p)$	$\frac{1}{2}\exp(-2\eta\overline{N}_p)$
Phase-shift keying (PSK)	$\frac{1}{2}\,\mathrm{erfc}\,(2\eta\overline{N}_p)^{1/2}$	$\frac{1}{2}\,\mathrm{erfc}\,(\eta\overline{N}_p)^{1/2}$	$\frac{1}{2}\exp(-\eta\overline{N}_p)$	—
Frequency-shift keying (FSK)	—	$\frac{1}{2}\,\mathrm{erfc}\,(\frac{1}{2}\eta\overline{N}_p)^{1/2}$	$\frac{1}{2}\exp(-\frac{1}{2}\eta\overline{N}_p)$	—

For a BER of 10^{-9} we thus require 20 photons per bit, which is a 0.5-dB penalty with respect to synchronous heterodyne detection of PSK.

Analogous to the PSK case, synchronous heterodyne OOK detection is 3 dB less sensitive than homodyne OOK. Thus the BER is given by

$$\mathrm{BER} = \tfrac{1}{2}\,\mathrm{erfc}\sqrt{\tfrac{1}{2}\eta\overline{N}_p} \tag{10-30}$$

Here one needs a minimum of 36 photons per bit for a 10^{-9} bit error rate. In the case of asynchronous heterodyne OOK detection, the bit error rate is given by

$$\mathrm{BER} = \tfrac{1}{2}\exp\!\left(-\tfrac{1}{2}\eta\overline{N}_p\right) \tag{10-31}$$

Thus asynchronous heterodyne OOK detection requires 40 photons per bit for a 10^{-9} BER, which is 3 dB less sensitive than DPSK.

The receiver sensitivities for the various modulation techniques are summarized in Tables 10-1 and 10-2. Table 10-1 gives the probability of error as a function of the number of received photons per bit, $\overline{N}_p$, and Table 10-2 shows the number of photons required for a 10^{-9} BER by an ideal receiver having a photodetector with a quantum efficiency of $\eta = 1$.

A summary of the requirements of linewidth versus photons per bit for heterodyne PSK, FSK, and OOK systems is given in Fig. 10-10 for a 10^{-9} BER. As shown in Fig. 10-5, PSK gives the best sensitivity for sources with very narrow linewidths. However, when the linewidth is greater than 0.2 percent of the bit rate, the sensitivity degrades quickly. In comparison, since FSK and OOK modulated signals can be detected using an optical power measurement which is not sensitive to phase noise, they maintain good performance with sensitivities below 60 photons per bit for linewidth-to-bit-rate ratios approaching unity.

TABLE 10-2
Summary of the number of photons required for a 10^{-9} BER by an ideal receiver having a photodetector with unity quantum efficiency

| | | Number of photons | | |
| | | Heterodyne | | |
Modulation	Homodyne	Synchronous detection	Asynchronous detection	Direct detection
On-off keying (OOK)	18	36	40	10
Phase-shift keying (PSK)	9	18	20	—
Frequency-shift keying (FSK)	—	36	40	—

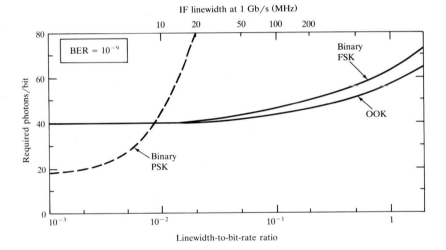

FIGURE 10-10
Calculated quantum-limited receiver sensitivity as a function of laser linewidth at 1 Gb/s or, equivalently, as a function of the linewidth-to-bit-rate ratio. (Reproduced with permission from Linke,[16] © 1989, IEEE.)

10.3.5 Performance Improvement with Coding

The above analysis has assumed a simple transmission code. In these schemes the laser linewidth requirement is one of the limiting factors in system performance, particularly when optical phase-locked loops are needed. However, by applying a suitable error-control technique, the laser linewidth can be greatly relaxed and the receiver sensitivity will be significantly improved. This was calculated by Wu, Wang, and Wu[44] using an (n, k) binary convolutional block code. Recall that in such codes a block of k information bits is transformed into a larger block of n bits by adding $n - k$ redundant bits which can be used for error control. In convolution coding, the information data are passed through a linear shift register with K stages which shift k bits at a time. For every K information bits stored in the shift register, there are n linear logic circuits which operate on the shift-register contents to produce n coded bits as the output of the encoder. The code rate R is therefore $R = k/n$. The parameter K is called the *constraint length* of the convolution code.

The results for several constraint lengths are depicted in Figs. 10-11 and 10-12. Here the horizontal axis gives the linewidth normalized to the data rate. Figure 10-11 shows that for a given BER, say 10^{-9}, the linewidth requirements can be relaxed by more than an order of magnitude for $K = 11$. In addition, Fig. 10-12 shows that for a given linewidth, the receiver sensitivity can increase by 1 to 10 dB, depending on the code constraint length.

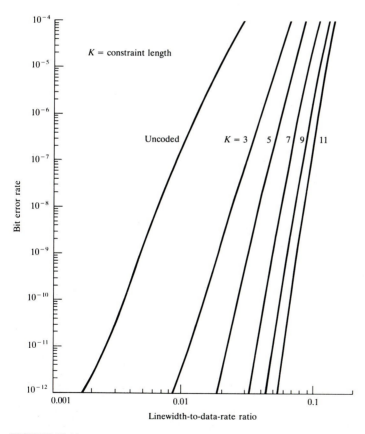

FIGURE 10-11

Bit error rate improvement for several $(2, 1)$ convolution codes at 30 photons per data bit. (Reproduced with permission from Wu, Wang, and Wu,[44] © 1990, IEEE.)

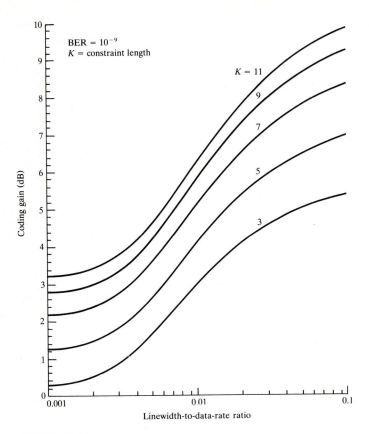

FIGURE 10-12
Coding gain versus linewidth for several $(2, 1)$ convolution codes at a 10^{-9} BER. (Reproduced with permission from Wu, Wang, and Wu,[44] © 1990, IEEE.)

10.4 POLARIZATION CONTROL REQUIREMENTS

As we see from Eq. (10-6), the optical power falling on the photodetector is a function of the polarization misalignment $\cos \theta(t)$ between the signal wave and the local-oscillator wave. To achieve a high degree of fidelity in the recovered signal, this term should remain constant and equal to one. The semiconductor light sources themselves are not a concern, since the optical output is generally linearly polarized. The problem that arises in a coherent optical communications link is related to the optical fiber. A significant degree of birefringence resulting from bends typically exists in the fiber path. Consequently the state of polarization at the receiver is usually not only elliptical but also changes with time as the fiber moves. The resulting polarization fluctuation can cause deep fades and loss of the signal.

Although there has been much activity in developing polarization-maintaining fibers and connectors,[45] a significant amount of symmetrical (non-polarization-maintaining) single-mode fiber is already installed. Thus researchers have expended a great deal of work in examining techniques for matching the time-varying state of polarization (SOP) of the incoming signal with that of the local oscillator,[46] and in developing a polarization-insensitive, or *polarization-diversity*, receiver.[47-49] Although a number of these methods have undergone field trials,[50, 51] further work still remains in the widespread implementation of coherent systems.

10.5 SUMMARY

Coherent optical fiber communication systems potentially offer improved receiver sensitivity and channel selectivity over direct-detection techniques. Although dramatic progress has been made in coherent techniques, their widespread application still awaits the development of a semiconductor laser which can be tuned at a 1-GHz rate over a 10 to 20-THz range. Potential applications of coherent systems may include high-sensitivity point-to-point links, dense multichannel broadcasting networks, and complex optical switching and routing architectures incorporating frequency-agile sources.

PROBLEMS

10-1. Verify the expression in Eq. (10-5) for the intensity due to the combined signal and local-oscillator fields.

10-2. A homodyne ASK receiver has a 100-MHz bandwidth and contains a 1300-nm *pin* photodiode with a responsivity of 0.6 A/W. It is shot-noise-limited and needs a signal-to-noise ratio of 12 to achieve a 10^{-9} BER. Find the photocurrent generated if the local-oscillator power is -3 dBm and the phase error is 10°. Assume the signal and local oscillator have the same polarization.

10-3. A distributed-feedback laser operating at 1550 nm has a 20-MHz linewidth. What is its spectral width?

10-4. For a bit error rate of 10^{-9} assume the combined spectral width of the signal carrier wave and the local oscillator should be 1 percent of the transmitted bit rate.
(*a*) What spectral width is needed at 1300 nm for a 100-Mb/s data rate?
(*b*) What is the maximum allowed width at 1.8 Gb/s?

10-5. A 1300-nm laser diode has a temperature-dependent wavelength change of 15 GHz/°C.
(*a*) What temperature fluctuation is permitted if the frequency must not change by more than 100 MHz?
(*b*) By how much does the wavelength change over this temperature excursion?

10-6. (*a*) Verify that 10 photons per bit are required to get a BER of 10^{-9} for a direct-detection OOK system.
(*b*) Show that for an ideal OOK homodyne system, one needs 36 photons per pulse to achieve a 10^{-9} BER.

10-7. Using the approximation given by Eq. (10-20) for erfc(x), find simplified expressions for PSK homodyne and PSK heterodyne detection. Letting $\eta = 1.0$, plot these expressions as a function of the number of received photons per bit for $5 < \overline{N}_p < 20$.

REFERENCES

1. T. Okoshi and K. Kikuchi, "Frequency stabilization of semiconductor lasers for heterodyne-type optical communication schemes," *Electron. Lett.*, vol. 16, pp. 179–181, Feb. 28, 1980.
2. F. Favre and D. LeGuen, "High frequency stability of laser diode for heterodyne communication systems," *Electron. Lett.*, vol. 16, pp. 709–710, Aug. 28, 1980.
3. Y. Yamamoto, "Receiver performance evaluation of various digital optical modulation-demodulation systems in the 0.5–10-μm-wavelength region," *IEEE J. Quantum Electron.*, vol. QE-16, pp. 1251–1259, Nov. 1980.
4. I. W. Stanley, "A tutorial review of techniques for coherent optical fiber transmission systems," *IEEE Commun. Mag.*, vol. 23, pp. 37–53, Aug. 1985.
5. P. S. Henry, "Lightwave primer," *IEEE J. Quantum Electron.*, vol. QE-21, pp. 1862–1879, Dec. 1985.
6. (*a*) J. Salz, "Coherent lightwave communications," *AT&T Tech. J.*, vol. 64, pp. 2153–2209, Dec. 1985.
   (*b*) J. Salz, "Modulation and detection for coherent lightwave communications," *IEEE Commun. Mag.*, vol. 24, pp. 38–49, June 1986.
7. L. G. Kazovsky, "Optical heterodyning versus optical homodyning: A comparison," *J. Optical Commun.*, vol. 6, pp. 18–24, Mar. 1985.
8. R. A. Linke and P. S. Henry, "Coherent optical detection: A thousand calls on one circuit," *IEEE Spectrum*, vol. 24, pp. 52–57, Feb. 1987.
9. T. Okoshi, "Recent advances in coherent optical fiber communication systems," *J. Lightwave Tech.*, vol. 5, pp. 44 52, Jan. 1987.
10. T. Kimura, "Coherent optical fiber transmission," *J. Lightwave Tech.*, vol. 5, pp. 414–428, Apr. 1987.
11. R. E. Wagner, N. K. Cheung, and P. Kaiser, "Coherent lightwave systems for interoffice and loop-feeder applications," *J. Lightwave Tech.*, vol. 5, pp. 429–439, Apr. 1987.
12. I. W. Stanley, G. R. Hill, and D. W. Smith, "The application of coherent optical techniques to wideband networks," *J. Lightwave Tech.*, vol. 5, pp. 440–451, Apr. 1987.
13. K. Nosu, "Advanced coherent lightwave technologies," *IEEE Commun. Mag.*, vol. 26, pp. 15–21, Feb. 1988.
14. Special Issues on Coherent Communications, *J. Lightwave Tech.*, (*a*) vol. 5, Apr. 1987; (*b*) vol. 8, Mar. 1990.
15. R. A. Linke and A. H. Gnauck, "High-capacity coherent lightwave systems," *J. Lightwave Tech.*, vol. 6, pp. 1750–1769, Nov. 1988.
16. R. A. Linke, "Optical heterodyne communications systems," *IEEE Commun. Mag.*, vol. 27, pp. 36–41, Oct. 1989.
17. T. Okoshi and K. Kikuchi, *Coherent Optical Fiber Communications*, Kluwer Academic, Boston, 1988.
18. C. H. Henry, "Phase noise in semiconductor lasers," *J. Lightwave Tech.*, vol. LT-4, pp. 298–311, Mar. 1986.
19. T. L. Koch and U. Koren, "Semiconductor lasers for coherent optical fiber communications," *J. Lightwave Tech.*, vol. 8, pp. 274–293, Mar. 1990.
20. C. H. Henry, "Theory of the linewidth of semiconductor lasers," *IEEE J. Quantum Electron.*, vol. QE-18, pp. 259–264, Feb. 1982.
21. K. Kojima, K. Kyuma, and T. Nakayama, "Analysis of the spectral linewidth of distributed feedback laser diodes," *J. Lightwave Tech.*, vol. LT-3, pp. 1048–1055, Oct. 1985.

22. G. Gray and R. Roy, "Noise in nearly single-mode semiconductor lasers," *Phys. Rev. A*, vol. 40, pp. 2452–2462, Sept. 1989.

23. K. Kikuchi, T. Okoshi, M. Nagamatsu, and N. Henmi, "Degradation of bit-error rate in coherent optical communications due to spectral spread of the transmitter and the local oscillator," *J. Lightwave Tech.*, vol. LT-2, pp. 1024–1033, Dec. 1984.

24. L. G. Kazovsky, "Impact of laser phase noise on optical heterodyne communication systems," *J. Optical Commun.*, vol. 7, pp. 66–78, 1986.

25. I. Garrett and G. Jacobsen, "The effect of laser linewidth on coherent optical receivers with nonsynchronous demodulation," *J. Lightwave Tech.*, vol. 5, pp. 551–560, Apr. 1987.

26. G. J. Foschini, L. J. Greenstein, and G. Vannucci, "Noncoherent detection of coherent lightwave signals corrupted by phase noise," *IEEE Trans. Commun.*, vol. 36, pp. 306–314, Mar. 1988.

27. E. E. Basch, R. F. Kearns, and T. G. Brown, *Microwave and Opt. Tech. Lett.*, vol. 1, pp. 184–186, July 1988.

28. R. Wyatt, "Spectral linewidth of external cavity semiconductor lasers with strong, frequency-selective feedback," *Electron. Lett.*, vol. 21, pp. 658–659, July 4, 1985.

29. A. Sollberger, A. Heinämäki, and H. Melchior, "Frequency stabilization of semiconductor lasers for applications in coherent communication systems," *J. Lightwave Tech.*, vol. 5, pp. 485–491, Apr. 1987.

30. (*a*) S. Sudo, Y. Sakai, H. Yasaka, T. Ikegami, "Frequency-stabilized DFB laser module using 1.53159-μm absorption line of C_2H_2," *IEEE Photon. Tech. Lett.*, vol. 1, pp. 281–284, Oct. 1989.

    (*b*) S. Sudo, Y. Sakai, H. Yasaka, T. Ikegami, "Frequency-stabilization of 1.55-μm DFB laser diode using vibrational-rotational absorption of C_2H_2 molecules," *Tech. Digest OFC-90*, p. 161, Jan. 1990.

31. S. Murata and I. Mito, "Tutorial review: Frequency-tunable semiconductor lasers," *Opt. Quantum Electron.*, vol. 22, pp. 1–15, Jan. 1990.

32. A. T. Schremer and C. L. Tang, "External-cavity semiconductor laser with 1000 GHz continuous piezoelectric tuning range," *IEEE Photon. Tech. Lett.*, vol. 2, pp. 3–5, Jan. 1990.

33. H. Kobrinski, M. P. Vecchi, T. E. Chapuran, J. B. Georges, C. E. Zah, C. Caneau, S. G. Menocal, P. S. D. Lin, A. S. Gozdz, and F. J. Favire, "Simultaneously fast wavelength switching and intensity modulation using a tunable DBR laser," *IEEE Photon. Tech. Lett.*, vol. 2, pp. 139–142, Feb. 1990.

34. T. P. Lee and C.-E. Zah, "Wavelength-tunable and single-frequency semiconductor lasers for photonic communications networks," *IEEE Commun. Mag.*, vol. 27, pp. 42–52, Oct. 1989.

35. L. G. Kazovsky, "Balance phase-locked loops for optical homodyne receivers: Performance analysis, design considerations, and laser linewidth requirements," *J. Lightwave Tech.*, vol. LT-4, pp. 182–195, Feb. 1986.

36. A. Schöpflin, S. Kugelmeier, E. Gottwald, D. Felicio, and G. Fischer, "PSK optical homodyne system with nonlinear phase-locked loop," *Electron. Lett.*, vol. 6, pp. 395–396, Mar. 15, 1990.

37. F. Derr, "Optical QPSK homodyne transmission of 280 Mb/s," *Electron. Lett.*, vol. 6, pp. 401–403, Mar. 15, 1990.

38. T. T. Ha, *Digital Satellite Communications*, McGraw-Hill, New York, 1990.

39. A. B. Carlson, *Communication Systems*, 3rd ed., McGraw-Hill, New York, 1986.

40. K. S. Shanmugan, *Digital and Analog Communication Systems*, Wiley, New York, 1979.

41. H. Taub and D. L. Schilling, *Principles of Communication Systems*, 2nd ed., McGraw-Hill, New York, 1986.

42. T. G. Hodgkinson, "Receiver analysis for synchronous coherent optical fiber transmission systems," *J. Lightwave Tech.*, vol. LT-5, pp. 573–586, Apr. 1987.

43. (*a*) L. G. Kazovsky and O. K. Tonguz, "ASK and FSK coherent lightwave systems: A simplified approximate analysis," *J. Lightwave Tech.*, vol. 8, pp. 338–352, Mar. 1990.

    (*b*) O. K. Tonguz and L. G. Kazovsky, "A novel analysis technique for coherent FSK lightwave systems," *IEEE Photon. Tech. Lett.*, vol. 2, pp. 72–74, Jan. 1990.

44. J. Wu, C.-K. Wang, and I.-F. Wu, "Coding to relax laser linewidth requirements and improve receiver sensitivity for coherent optical BPSK communications," *J. Lightwave Tech.*, vol. 8, pp. 545–553, Apr. 1990.
45. F. M. Sears, "Polarization-maintenance limits in polarization-maintaining fibers and measurements," *J. Lightwave Tech.*, vol. 8, pp. 684–690, May 1990.
46. N. G. Walker and G. R. Walker, "Polarization control for coherent communications," *J. Lightwave Tech.*, vol. 8, pp. 438–458, Mar. 1990.
47. B. Glance, "Polarization independent coherent optical receiver," *J. Lightwave Tech.*, vol. LT-5, pp. 274–276, Apr. 1987.
48. R. Noe, "Sensitivity comparison of coherent optical heterodyne, phase diversity, and polarization diversity receivers," *J. Optical Commun.*, vol. 10, pp. 11–18, Mar. 1989.
49. S. Betti, F. Curti, G. De Marchis, and E. Iannone, "Phase noise and polarization state insensitive optical coherent systems," *J. Lightwave Tech.*, vol. 8, pp. 756–767, May 1990.
50. S. Ryu, S. Yamamoto, Y. Namahira, K. Mochizuki, and H. Wakabayashi, "First sea trial of FSK heterodyne optical transmission system using polarization diversity," *Electron. Lett.*, vol. 24, pp. 399–400, Mar. 31, 1988.
51. M. C. Brain, M. J. Creaner, R. C. Steele, N. G. Walker, G. R. Walker, J. Mellis, S. Al-Chalabi, J. Davidson, M. Rutherford, and I. C. Sturgess, "Progress towards the field deployment of coherent optical fiber systems," *J. Lightwave Tech.*, vol. 8, pp. 423–437, Mar. 1990.

chapter 11

ADVANCED SYSTEMS AND TECHNIQUES

The explosive growth of lightwave technology since the first low-loss fiber was demonstrated in 1970 has resulted in the worldwide installation of more than 10 million kilometers of optical fiber transmission lines. These are being used not only for high-speed long-haul trunks with transmission speeds of several gigabits per second, but also for local communication networks. In the interest of making fuller use of their assets, engineers are continually seeing how to upgrade these links and networks as more sophisticated optoelectronic components and equipment are being devised. In addition to increasing the bandwidth and the distance between repeaters of long-haul, single-channel transmission links, the upgrading of short-distance systems is also underway.

Some areas of interest are the extension of local area networks from basic data transport into the realm of broadband integrated communication services (voice, data, and video on the same link) for local entities such as a large office building, an industrial complex, a research facility, or a university campus. Another application is the interconnection of supercomputers with high-resolution video workstations for displays of simulations of complex physics and engineering phenomena such as hypersonic fluid flow, for medical research, or for analyses of scientific experiments which can be shared in real time with several widely dispersed research groups.

To implement these ideas, technologists have investigated techniques such as wavelength division multiplexing, local area networks, optical amplifiers, photonic switching, and nonlinear optical effects such as the Raman effect,

399

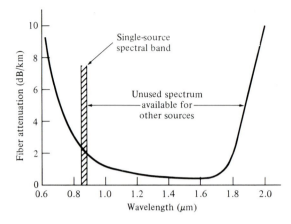

FIGURE 11-1

A single optical source uses a small part of the available spectral transmission band of a fiber. Wavelength division multiplexing makes simultaneous use of many spectral channels.

pulse compression, and soliton propagation. This chapter looks at these topics and points to areas of further research interest.

11.1 WAVELENGTH DIVISION MULTIPLEXING (WDM)

Having examined point-to-point links, we shall now consider more sophisticated systems which make a fuller utilization of the transmission capacity of an optical fiber. In standard point-to-point links a single-fiber line has one optical source at its transmitting end and one photodetector at the receiving end. Signals from different light sources require separate and uniquely assigned optical fibers. Since an optical source has a narrow spectral width, this type of transmission makes use of only a very narrow portion of the transmission bandwidth capability of a fiber, as is shown in Fig. 11-1 for a typical laser diode operating at 850 nm.

From Fig. 11-1 we see that many additional spectral operating regions are possible. Ideally, a dramatic increase in the information capacity of a fiber can thus be achieved by the simultaneous transmission of optical signals over the same fiber from many different light sources having properly spaced peak emission wavelengths. By operating each source at a different peak wavelength, the integrity of the independent messages from each source is maintained for subsequent conversion to electric signals at the receiving end. This is the basis of wavelength division multiplexing[1–5] (WDM). Conceptually the WDM scheme is the same as frequency division multiplexing (FDM) used in microwave radio and satellite systems. Various WDM techniques have been examined for optical fiber networks.[6–10]

Two different WDM setups are shown in Figs. 11-2 and 11-3. In Fig. 11-2 a unidirectional WDM device is used to combine different signal carrier wavelengths onto a single fiber at one end and to separate them into their corresponding detectors at the other end. A bidirectional WDM scheme is

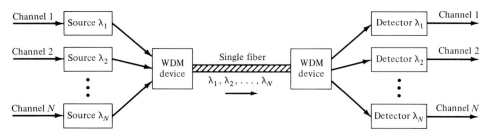

FIGURE 11-2
A unidirectional WDM system that combines N independent input signals for transmission over a single fiber.

shown in Fig. 11-3. This involves sending information in one direction at a wavelength λ_1 and simultaneously in the opposite direction at a wavelength λ_2.

Let us first examine the system requirements of the WDM technique. The three basic performance criteria are insertion loss, channel width, and cross talk. *Insertion loss* defines the amount of power loss that arises in the fiber optic line from the addition of a WDM coupling device. This includes losses occurring at the connection points of the WDM element to the fiber line and any intrinsic losses within the multiplexing element itself. In practice, designers can tolerate insertion losses of a few decibels at each end.

Channel width is the wavelength range that is allocated to a particular optical source. If laser diodes are used, channel widths of several tens of nanometers are required to ensure that no interchannel interference results from source instability (for example, drift of peak operating output wavelength with temperature changes). For LED sources, channel widths that are 10 to 20 times larger are required because of the wider spectral output width of these sources.

Cross talk refers to the amount of signal coupling from one channel to another. The tolerable interchannel cross-talk levels can vary widely depending on the application. However, in general, a -10-dB level is not sufficient whereas a -30-dB level is.

In implementing a unidirectional WDM system a multiplexer is needed at the transmitting end to combine optical signals from several light sources onto a

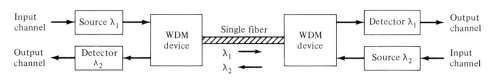

FIGURE 11-3
Schematic representation of a bidirectional WDM system in which two or more wavelengths are transmitted simultaneously in opposite directions over the same fiber.

single fiber. At the receiving end a demultiplexer is required to separate the signals into appropriate detection channels. Since the optical signals that are combined generally do not emit a significant amount of optical power outside of their designated channel spectral width, interchannel cross-talk factors are relatively unimportant at the transmitter end. The basic design problem here is that the multiplexer should provide a low-loss path from each optical source to the multiplexer output. A different requirement exists for the demultiplexer, since photodetectors are usually sensitive over a broad range of wavelengths which could include all the WDM channels. Thus, to prevent significant amounts of the wrong signal from entering each receiving channel, that is, to give good channel isolation of the different wavelengths being used, either the demultiplexer must be carefully designed or very stable optical filters with sharp wavelength cutoffs must be used.

In principle, any optical wavelength demultiplexer can also be used as a multiplexer. Thus, for simplicity, the word "multiplexer" is often used as a general term to refer to both multiplexers and demultiplexers, except when it is necessary to distinguish the two devices or functions.

The wavelength division multiplexers that are most widely used fall into two classes. These are angularly dispersive devices,[11-15] such as prisms or gratings, and filter-based devices,[16-18] such as multilayer thin-film interference filters or single-mode integrated-optical devices. The basis of an angularly dispersive multiplexer is shown schematically in Fig. 11-4 for a three-wavelength system, where $d\theta/d\lambda$ is the angular dispersion of the device. When the device is used as a demultiplexer, the light emerging from the left-hand fiber is collimated by lens L_1 (the collimating lens) and passed through the dispersive element which separates the various wavelength channels into different spatially oriented beams. Lens L_2 (the focusing lens) then focuses the output beams into the appropriate receiving fibers or detectors. The linear dispersion $dx/d\lambda$ at the output fibers shown on the right-hand side of Fig. 11-4 is[1]

$$\frac{dx}{d\lambda} = f\frac{d\theta}{d\lambda} \tag{11-1}$$

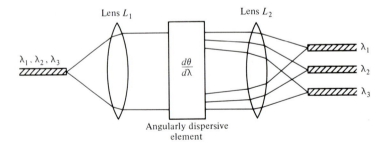

FIGURE 11-4
Schematic representation of an angularly dispersive WDM element shown for three wavelengths. Many wavelengths can be combined or separated with this type of device.

where f is the focal length of lens L_2. In the ideal case of aberrationless optics and zero source spectral width, the intrinsic insertion loss and cross talk will be zero if the output signals are separated by more than their diameter d, that is, if

$$\frac{dx}{d\lambda} \Delta\lambda \geq d \tag{11-2}$$

where $\Delta\lambda$ is the spectral separation between channels. We shall assume here that all fibers (input and output) have the same diameter d and numerical aperture NA.

To collect all the light from the input fiber, the collimating lens L_1 must have a diameter b satisfying the condition

$$b > 2f\frac{\text{NA}}{n'} \tag{11-3}$$

where n' is the refractive index of the medium between the dispersive device and the lens L_1. Combining Eqs. (11-1) through (11-3) then yields

$$b \geq \frac{2(\text{NA}/n')d}{\Delta\lambda\,(d\theta/d\lambda)} \tag{11-4}$$

In any real system the output beam spreads out because of the finite size of the light source and the angular dispersion resulting from the wavelength spread of the source spectral width. The fractional increase S in the beam diameter is approximately given by

$$S = \frac{b'-b}{b} \simeq (1+m)\frac{Wd(\text{NA})}{b^2 n'} \tag{11-5}$$

where m is the number of wavelength channels, b' is the diameter of lens L_2, and W is the total path length from the output of lens L_2 to the input of lens L_1. To avoid overfilling the numerical aperture of the output fiber, the total beam spread must be a small fraction of the collimating lens diameter, that is, $S \ll 1$.

A large number of channels can be combined and separated with angularly dispersive multiplexing elements. Most of these devices use a grating-plus-lens combination. Sometimes a prism is used for the angularly dispersive element. Insertion losses are typically in the 1- to 3-dB range, and cross-talk levels between -20 and -30 dB are routinely obtained.

The operation of a filter type multiplexing element is shown in Fig. 11-5 for a two-wavelength operation. The filters are designed to transmit light in a specific wavelength channel and either to absorb or to reflect all other wavelengths. Reflection type filters are normally used, since the losses of absorption filters tend to be high (greater than 1 dB). The reflection filter consists of a flat glass substrate upon which multiple layers of different dielectric films are deposited for wavelength selectivity. These filters can be used in series to separate additional wavelength channels. The complexity involved in stacking the filters in series and the increase in signal loss that occurs with the addition

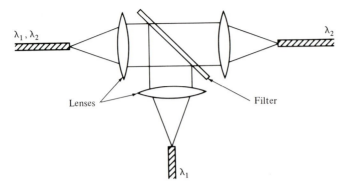

FIGURE 11-5
Multilayer thin-film-filter reflector used for WDM. This device is transparent at the wavelength λ_2 and reflects the wavelength λ_1.

of successive multiplexers generally limit operation to two or three filters (that is, three or four channels).

In designing a WDM system, care must be taken to minimize the factors causing link margin degradation. In addition to keeping a low insertion loss of the coupling devices, the designer must minimize reflections of optical power that occur at the coupling elements or at connectors and splices, since they can give rise to signal cross talk. These effects are of particular importance in bidirectional links.[19-22]

In addition to the passive WDM devices described above, in which the wavelength selectivity is fixed, active WDM elements have also been devised. These can be actively switched or tuned according to wavelength. Among these are multiwavelength source and detector integrated-optic arrays, wavelength-tunable lasers, and wavelength-tunable filters.[3, 23, 24] The concept for tunable filtering is illustrated in Fig. 11-6. In this procedure different information signals are sent in the individual frequency channels of bandwidth B. By using a filter with a passband of width B which is tunable over the frequency range of these channels, a user can select the desired channel.

Figure 11-7 shows an example of a wavelength-tunable-filter device.[25, 26] Here a birefringent multiple-order element, such as two quartz waveguides, is placed between two polarizing beam splitters. The optical output power I at the output ports A and B is related to the input power I_0 by

$$I = \frac{I_0}{2}\left(1 \pm \cos\frac{2\pi\,\Delta n\,L}{\lambda}\right) \qquad (11\text{-}6)$$

where Δn is the difference between the ordinary and the extraordinary refractive indices of the birefringent material, λ is the wavelength, and the plus and minus signs refer to ports A and B, respectively. The sinusoidal spectral output variation can be altered by changing the path length L through the crystal. This is achieved by moving one of the quartz plates up or down. Length changes on

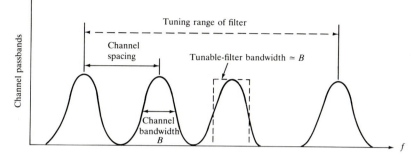

FIGURE 11-6

In a tunable filtering scheme, different signals are sent in individual frequency channels of bandwidth B. A tunable filter of passband B selects the desired channel.

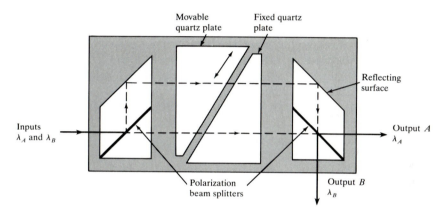

FIGURE 11-7

Example of a wavelength-tunable filter. A moving quartz plate changes the path length through the crystal to vary the sinusoidal spectral output.

the order of the polarization rotation period determine the channel position, whereas larger changes modify the channel spacing. Since the optical paths are reversible, this device can be used either as a multiplexer or as a demultiplexer.

11.2 LOCAL AREA NETWORKS

Although any type of transmission medium can be used, the increased demand for bandwidth in local environments has led to widespread implementation of fiber optics in local area networks (LANs).[27-30] The manner in which nodes are geometrically arranged and connected is known as the *topology* of the network.[31-33] Two major areas of concern for LANs in terms of the fiber optic transmission medium are the LAN topologies and fail-safe architectures for

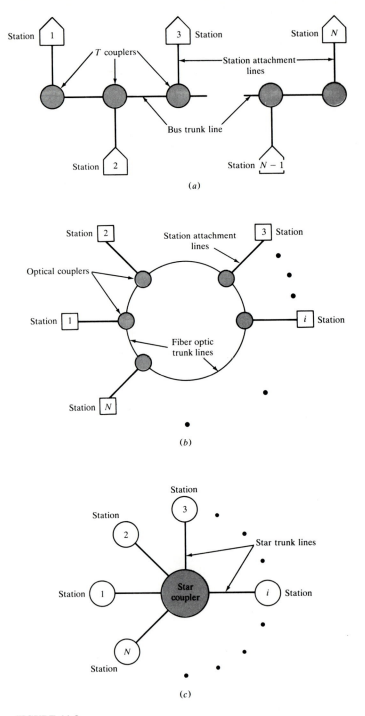

(a)

(b)

(c)

FIGURE 11-8
Three basic LAN topologies are the (a) bus, (b) ring, and (c) star configurations.

these topologies. Figure 11-8 depicts the three basic LAN topologies, which are the *bus*, *ring*, and *star* configurations. Each of these methodologies has its own particular advantages and limitations in terms of reliability, expandability, and performance characteristics. The following three subsections give details of these configurations, and Sec. 11.2.4 describes reliability enhancement techniques for ring networks.

11.2.1 Optical Fiber Bus

The bus network topology usually employs coaxial cable as the transmission medium. The primary advantages of such a network are the totally passive nature of the transmission medium and the ability to easily install low-perturbation (high-impedance) taps on the coaxial line without disrupting the operating network.

In contrast to a coaxial-cable bus, a fiber-optic-based bus network is more difficult to implement. The impediment is that low-perturbation bidirectional taps which efficiently couple optical signals into and out of the main optical fiber trunk are not as readily available as coax taps. Access to an optical data bus is achieved by means of a coupling element, which can be either active or passive.[34, 35] An *active coupler* converts the optical signal on the data bus to its electric baseband counterpart before any data processing (such as injecting additional data into the signal stream or merely passing on the received data) is carried out. A *passive coupler* employs no electronic elements. It is used passively to tap off a portion of the optical power from the bus.

ACTIVE OPTICAL FIBER BUS COUPLERS. Figure 11-9 gives an example of an active coupler. A photodiode receiver converts the optical signal from the bus into an electric signal. The processing element then can remove or copy part of

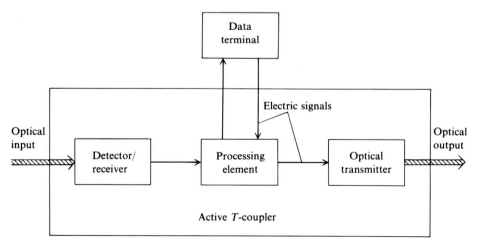

FIGURE 11-9
Example of an active T coupler for a fiber optic bus.

this signal for transmission to the terminal, while the remainder is forwarded to the optical transmitter. The processor can also insert additional information from the terminal into the data stream. The transmitter, in turn, reconverts the electric signal to an optical information stream which gets sent on to the next terminal via the optical fiber bus. Couplers of this type can be easily constructed by using any of a number of commercially available photodiodes and light sources.

The advantage of a linear fiber bus network of this type is that every accessing terminal acts as a repeater. This means that in principle the active bus can accommodate an unlimited number of terminals, since the signal is restored to its initial starting value at each node. However, the reliability of each repeater is critical to the operation of the overall network. For example, if one station on a single-fiber bus fails, all traffic will stop. A variety of optical bypassing schemes[36-38] or multifiber bus configurations have been proposed to alleviate this problem.

Example 11-1. An example of a passive fiber bypass scheme is shown in Fig. 11-10. Here every station is permanently bypassed with an optical fiber which is connected to opposite sides of a terminal via optical couplers. If one station fails, the bypass thus ensures optical continuity from the preceding transmitter to the next station. In order not to interfere with a signal being transmitted at the node during normal station operation, the bypassed signal must be sufficiently attenuated (to at least 7 dB below the optical signal at that node to get a 10^{-9} bit error rate) so that it only appears as interfering noise on a normally transmitted signal. This method is simple, reliable, and cost-effective, but it reduces the operating margin of the link somewhat.

The bypass function can also be performed with active devices such as electro-optic switches.[39] In this technique no signal is bypassed until the node

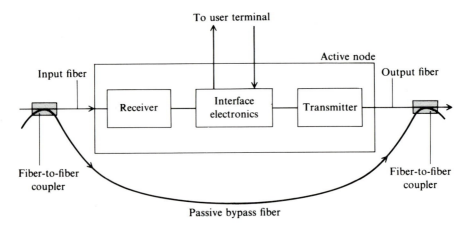

FIGURE 11-10
Example of a passive optical fiber bypass scheme for a linear data-bus configuration.

fails or is turned off. At that time the incoming fiber will be automatically disconnected from the equipment at that node and will be connected to the outgoing fiber by means of an electro-optic switch, thereby circumventing the node.

PASSIVE OPTICAL FIBER BUS COUPLERS. Passive couplers for an in-line bus are used at each terminal to remove a portion of the optical signal from the bus trunk line or to inject additional light signals onto the trunk. These couplers are known as either *in-line couplers* or T *couplers*. A major problem with the passive T coupler involves the optical power budget of the network. The difficulty arises from the fact that the optical signal is not regenerated at each terminal node. Insertion and output losses at each tap, plus the fiber losses between taps, limit the network size to a small number of terminals (generally less than 10) when symmetrical couplers are used. However, many more terminals can be attached to the bus by using asymmetric couplers.

A schematic of a generic T coupler is shown in Fig. 11-11. The coupler has four ports: two for connecting the device onto the bus, one for receiving tapped-off data, and one for inserting data onto the bus line. If optical power travels from left to right in Fig. 11-11 then, when a single T-coupler is used at a node, signals are injected onto the bus via port A and signals extracted from the bus are monitored at port B. The arrows in Fig. 11-11 show that some optical power is tapped off the primary fiber bus into the receiver at port B.

In many cases with this type of arrangement the transmitter would overload the receiver. To avoid this, two couplers can be put in tandem as shown in Fig. 11-12. Now the receiver port is upstream from the transmitter port and we avoid receiver overloading from the collocated transmitter. Note that there are two unused ports in this configuration.

The coupler shown in Fig. 11-11 can be simply made from two optical fibers by fusing them together over a short distance. This distance, which is designated as Γ_c in Fig. 11-11, is the *coupling length*. Its length and the spacing between the two fiber cores determine the degree of optical power coupling from one fiber to another. When the two fibers are geometrically similar, we

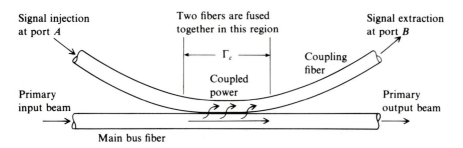

FIGURE 11-11
Schematic of a generic passive optical fiber T coupler.

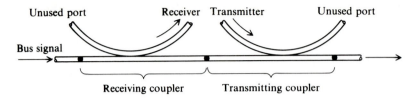
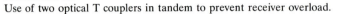

FIGURE 11-12
Use of two optical T couplers in tandem to prevent receiver overload.

have a *symmetric coupler*. An *asymmetric coupler* results when the two fibers are different; for example, the mainbus fiber could be multimode fiber and the coupling fiber could be a monomode fiber.[40–41]

To evaluate the performance of an in-line bus, let us examine the various sources of power loss along the transmission path. We consider this in terms of the fraction of power lost at a particular interface or within a particular component. First, over an optical fiber of length x (in kilometers), the ratio A of received to transmitted power levels is given by

$$A = 10^{-\alpha_f x/10}$$
$$= e^{-2.3\alpha_f x/10} \tag{11-7}$$

where α_f is the fiber attenuation in decibels per kilometer.

The losses encountered in a T coupler are shown schematically in Fig. 11-13. For simplicity we do not show the two unused ports here. The coupler thus has four functioning ports: two for connecting the device onto the fiber bus, one for receiving tapped-off data, and one for inserting data onto the line. We assume that a fraction F_c of optical power is lost at each port of the coupler. We take this fraction to be 20 percent, so that the connecting loss L_c is

$$L_c = -10 \log(1 - F_c) \simeq 1 \text{ dB} \tag{11-8}$$

That is, the optical power level gets reduced by 1 dB at any coupling junction.

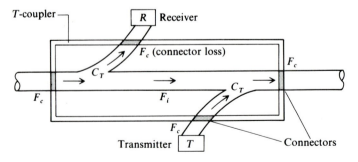

FIGURE 11-13
Optical power losses encountered in two tandem passive T couplers.

Let C_T represent the fraction of power removed from the bus and delivered to the detector port. Note that, in general, the fraction of power removed from the line in a T coupler is actually $2C_T$, since optical power is extracted at both the receiving and the transmitting taps of the device. The power removed at the transmitting tap is lost from the system. The coupling loss L_T in decibels is then given by

$$L_T = -10\log(1 - 2C_T) \qquad (11\text{-}9)$$

In addition to connection and tapping losses, there is an intrinsic transmission loss L_i associated with each T coupler. If the fraction of power lost in the coupler is F_i, then the intrinsic transmission loss L_i in decibels is

$$L_i = -10\log(1 - F_i) \qquad (11\text{-}10)$$

Generally, an in-line bus will consist of a number of stations separated by various lengths of bus line. However, here for analytical simplicity, we consider an in-line bus of N stations uniformly separated by a distance L. From Eq. (11-7) the fiber attenuation between adjacent stations is

$$A_0 = e^{-2.3\alpha_f L/10} \qquad (11\text{-}11)$$

We use the notation P_{jk} to denote the optical power received at the detector of the kth station from the transmitter of the jth station. For simplicity, we assume that a T coupler exists at every terminal of the bus including the two end stations.

Because of the serial nature of the in-line bus, the optical power available at a particular node decreases with increasing distance from the source. Thus a performance quantity of interest is the *system dynamic range*. This is the maximum optical power range to which any detector must be able to respond. The smallest difference in transmitted and received optical power occurs for adjacent stations, such as between station 1 and station 2 in Fig. 11-14. If P_0 is the optical power launched from a source flylead and E is the coupling

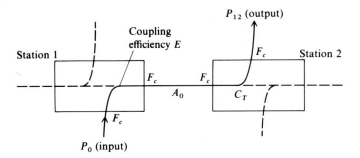

FIGURE 11-14
Optical path between adjacent stations for an in-line bus.

efficiency of optical power onto the bus line, then

$$P_{1,2} = A_0 C_T (1 - F_c)^4 E P_0 \tag{11-12}$$

The largest difference in transmitted and received optical power occurs between station 1 and station N. At the transmitting end the fractional power level coupled into the first length of cable is

$$F_1 = (1 - F_c)^2 E \tag{11-13}$$

At the receiving end the fraction of power from the T coupler input port that emerges from the detector port is

$$F_N = (1 - F_c)^2 C_T \tag{11-14}$$

For each of the $N - 2$ intermediate stations the fraction of power passed through each coupler is

$$F_{coup} = (1 - F_c)^2 (1 - 2C_T)(1 - F_i) \tag{11-15}$$

Combining these factors and the transmission losses of the $N - 1$ intervening fibers, we find that the power received at station N from station 1 is

$$P_{1N} = A_0^{N-1} F_1 F_{coup}^{N-2} F_N P_0$$

$$= A_0^{N-1}(1 - F_c)^{2N}(1 - 2C_T)^{N-2} C_T (1 - F_i)^{N-2} E P_0 \tag{11-16}$$

The worst-case dynamic range (DR) is then found from the ratio of Eq. (11-12) to Eq. (11-16):

$$\text{DR} = \frac{1}{\left[A_0 (1 - F_c)^2 (1 - 2C_T)(1 - F_i) \right]^{N-2}} \tag{11-17}$$

11.2.2   Ring Topology

In a ring topology,[42-44] consecutive nodes are connected by point-to-point links which are arranged to form a single closed path as is shown in Fig. 11-15. Information in the form of data packets (a group of information plus address bits) is transmitted from node to node around the ring, which can consist of either twisted-pair wire, baseband coaxial cable, or optical fibers. The interface at each node is an active device that has the ability to recognize its own address in a data packet in order to accept messages. The interface serves not only as a user attachment point but also as an active repeater for retransmitting messages that are addressed to other nodes.

From Fig. 11-15 we see that the interfaces in a ring network can operate in either a listen or a transmit mode, or the node can be bypassed if it is not operational. In the listen mode the bypassing bit stream is monitored for pertinent bit sequences such as a token or an address pattern. The incoming bit stream is thus simply copied to the output with a time delay of about one bit.

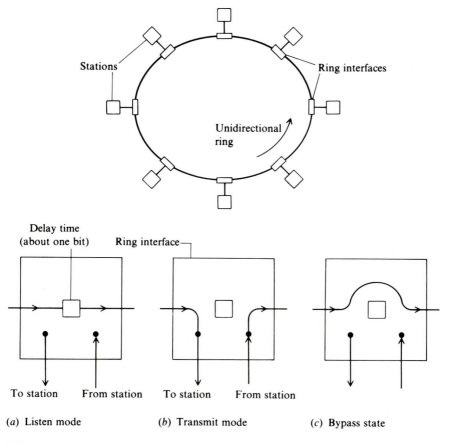

(a) Listen mode   (b) Transmit mode   (c) Bypass state

FIGURE 11-15
Ring topology and three possible operational modes.

When an interface in the listen mode recognizes an address field which specifies it as the information destination, each incoming bit in the packet is copied and sent to the attached station while simultaneously the bits are also retransmitted to the next station. This, of course, presents the question of when and where the data packets are removed from the ring. For a bus network the signals which travel along the line are absorbed by line terminators once they reach the endpoints of the bus. However, in a ring the packets would continue to circulate indefinitely if they were not removed by some station. This can take place either at the destination station, or at the sender once the packet has made one complete loop. Removal at the sending station is more advantageous, since it allows automatic acknowledgment that a packet has been received, and since it allows multicast addressing wherein one packet is simultaneously sent to many stations.

In the transmit mode, the interface breaks the connection between the input and output, and enters its own data onto the ring. A number of methods for switching from the listen to the transmit mode have been designed and implemented. Two popular ones are IEEE-802.5 token rings and the Fiber Distributed Data Interface (FDDI).[44]

A number of implementation considerations must be taken into account when configuring a ring network. First, rings must be physically arranged so that all nodes are fully connected. Whenever a node is added to support new devices, transmission lines have to be placed between this node and its two nearby, topologically adjacent nodes. Thus, it is often difficult to prewire a building for ring networks in anticipation of new nodes to be added in the future. In addition, a break in any line, the failure of a node, or adding a new node will disrupt network operation. A variety of steps can be taken to circumvent these problems, although this generally increases the complexity of the ring interface electronics as well as its cost. Among these methods are node bypass techniques and loopback schemes used in conjunction with multiple-line cables.

11.2.3   Star Architectures

In a star architecture[45-50] all nodes are joined at a single point called the *central node* or *hub*. The central node can be an active or a passive device, and the transmission medium can be either twisted-pair wire, coaxial cable, or optical fibers.

An active central node is generally used when control of the network is carried out in the central node. In this case the central node performs all routing of messages in the network, whether it is from the central node to outlying nodes, between outlying nodes, or from all nodes to remote points. Point-to-point links connect the central and outlying nodes, so that the outlying nodes are required to have complex network control functions. Star networks with an active central node are most useful when the bulk of the communication is between the central and the outlying nodes. If there is a great deal of message traffic between the outlying nodes, then a heavy switching burden is placed on the central node.

In a star network with a passive central node, a power splitter is used at the hub of the star to divide the incoming signals among all the stations. Power-splitting losses add up as $1/N$, where N represents the number of branches in the splitter. In addition there is always some coupling and throughput loss in each splitter. Thus if there are a large number of terminals, repeater amplifiers may be necessary to boost the power level to meet link power margin requirements. Since a star-coupled network is logically equivalent to a bus network, it offers an attractive implementation of a bus, because repeaters are eliminated and network reliability is improved.

> **Example 11-2.** An example of a single-ended coaxial network that employs a power splitter and a power combiner in a star configuration is shown in Fig. 11-16.

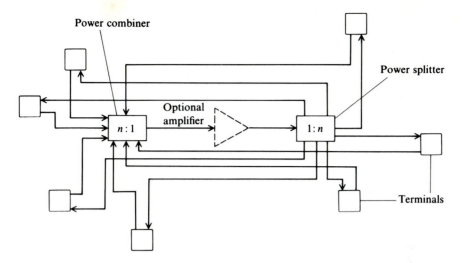

FIGURE 11-16
Single-ended coaxial network that uses power splitters and combiners in a star configuration.

An optional amplifier is shown between the combiner and the splitter. This would be used in those cases where the power losses in the splitter are too large to maintain an adequate power margin on all links. For impedance-matched combiners and splitters, any signal path can be treated as a point-to-point link with the additional losses from the couplers added.

A star network is also readily implemented using optical fibers. The couplers in this case can be either the transmission or reflection star couplers shown in Fig. 11-17. These couplers are passive mixing elements, that is, the optical powers from the input ports are mixed together and then divided equally among the output ports. They may be used to combine numerous signals together, to split a signal into a number of parts, or to tap optical power out of

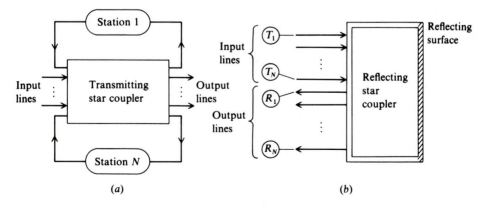

FIGURE 11-17
Transmitting (*a*) and reflecting (*b*) optical fiber star couplers.

or insert optical power into a fiber optic link. Either type of star is composed of a set of input fibers, a set of output fibers, and a mixing region.

In general, the reflection star is more versatile, because the relative number of input and output ports may be selected or varied after the device has been constructed (the total number of ports, in plus out, being fixed, however). By comparison, the number of transmission star input and output fibers is fixed by the initial design and fabrication. The reflection star is usually less efficient, since a portion of the light which has entered the coupler is injected back into the input fibers. Given the same number of input and output ports, the transmission star is twice as efficient as the reflection star. Therefore, reflection and transmission star couplers have their own particular advantages and disadvantages, and selection of a star coupler type for a particular application is largely determined by the network topology.

To see how star couplers can be applied to a given network, let us examine the various optical power losses associated with the coupler. The insertion loss L_S of a star coupler is defined as the fraction of optical power lost in the process of coupling light from the input port to all the output ports. It is given in decibels by

$$L_S = 10 \log \frac{\sum_{k=1}^{N} P_k}{P_i} \tag{11-18}$$

where the P_k are the output powers from all the ports, P_i is the input power at one port, and for a transmission star coupler N is the number of outputs, whereas for a reflection star coupler N is the number of inputs plus outputs.

The optical power that enters a star coupler gets divided equally among the N output ports. That is, the optical power at any one output port is $1/N$ times the total optical power emerging from all the coupler outputs. This ratio is known as the *power splitting factor* L_{sp}, and is given in decibels by

$$L_{sp} = 10 \log N \tag{11-19}$$

If P_S is the fiber-coupled output power from a source in dBm, P_R is the minimum optical power in dBm required at the receiver for a specific data rate, α_f is the fiber attenuation, and L_c is the connector loss in decibels, then the balance equation for a particular link having all stations located at the same distance L from the star coupler is

$$P_S - P_R = L_S + 2\alpha_f L + 4L_c + 10 \log N + \text{system margin} \tag{11-20}$$

where we have assumed connector losses at the transmitter, the receiver, and the input and output ports of the star coupler. For a transmission star N is the number of output ports, whereas for a reflection star N is the total number of input plus output ports.

11.2.4 Fail-Safe Fiber Optic Nodes

As an example of a bypass technique, let us look at some fail-safe nodes that have been used in fiber optic local area networks. The characteristics of these nodes are such that in a network containing a string of active repeaters, the

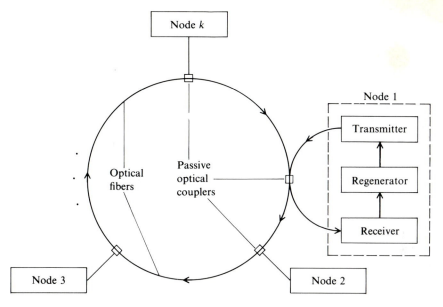

FIGURE 11-18
General schematic of a fiber optic ring network.

network keeps functioning when power is lost at one or more nodes. Figure 11-18 shows a ring network consisting of a transmitter and receiver pair, a regenerator connecting this pair, and a passive lightwave coupler. The nodes are connected by optical fibers. A similar scheme can be envisioned for a linear bus network.

In normal operation each node actively monitors the transmission line and regenerates (amplifies and retransmits) information flowing in the network. When a node wishes to transmit, it turns off its regenerator and inserts its information into the network at the transmitter. Thus, in case the node fails electronically or if components are removed for maintenance, the lightwave coupler provides optical-line continuity for uninterrupted network operation.

To operate properly, the fail-safe network must meet sensitivity and interference constraints. The *sensitivity constraint* is that any receiver must be sensitive enough to detect the signal from a preceding transmitter when several nodes between the transmitter and the receiver are off. The exact allowed number of adjacent failed nodes is derived from the system requirements. The *interference constraint* states that when two nodes are transmitting simultaneously, a downstream node should receive the signal from the closer node. This means that the less intense signal from the more distant transmitter must be below the threshold level of the receiver.

Each fail-safe node can have one or two optical couplers, depending on the particular operation desired. Figure 11-19 shows the concept of a one-coupler node. Here P_{in} is the optical power entering the coupler from the ring, and

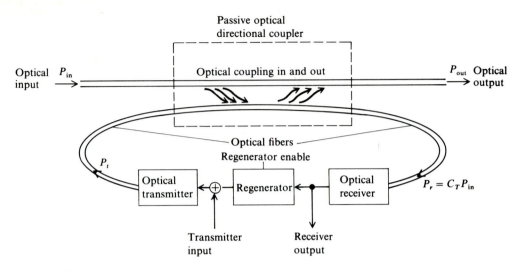

FIGURE 11-19
Concept for connecting to a fiber optic ring or bus using one passive coupler.

P_t is the optical power entering the coupler from the transmitter. In this configuration the receiver must be off when the transmitter is on to avoid saturation of the receiver. This is accomplished with a signal format having a duty cycle less than 50 percent, and having the transmitter operate out of phase with the receiver.

As we noted in Fig. 11-13, the coupler is characterized by a four-port device. Its basic parameters are C_T, which represents the fraction of power removed from the ring; F_i, which is the fraction of power lost in the coupler; and E, which is the coupling efficiency of optical power onto the ring. For simplicity of analysis, here we will ignore the factor F_c, which is the fraction of power lost at each port of the coupler (this is essentially the connector loss).

Under the condition that the transmitter is off, the optical power P_r entering the receiver is

$$P_r = C_T P_{in} \qquad (11\text{-}21)$$

When the transmitter is on, the optical power leaving the coupler and entering the ring is

$$P_{out} = (1 - F_i)(1 - C_T)P_{in} + EP_t \qquad (11\text{-}22)$$

where P_{in} and P_t fall in different time slots.

Now suppose Q adjacent upstream nodes have failed and are off. Given that the fibers between nodes have a length L and an attenuation α_f, the power received after these Q nodes is $P_S + P_I$. Here

$$P_S = C_T \alpha_f^{Q+1}(1 - F_i)^Q (1 - C_T)^Q EP_t \qquad (11\text{-}23)$$

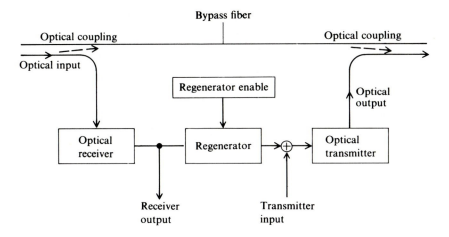

FIGURE 11-20
Concept for connecting to a fiber optic ring or bus using two passive couplers.

is the power received from the closest active transmitter, and

$$P_I = \sum_{j=Q}^{\infty} C_T \alpha_f^{j+2} (1 - F_i)^{j+1} (1 + C_T)^{j+1} EP_t$$

$$== \frac{\alpha_f^{Q+2} [(1 - F_i)(1 + C_T)]^{Q+1} C_T EP_t}{1 - (1 - F_i)(1 - C_T)\alpha_f} \tag{11-24}$$

is the sum of the powers from all the other upstream active transmitters that could cause interference.

Equation (11-24) takes into account the worst case of interference. The sensitivity and interference constraints are fulfilled when the effective received power, which is the difference of Eqs. (11-23) and (11-24), is greater than the sensitivity S of the receiver:

$$P_S - P_I \geq S \tag{11-25}$$

Figure 11-20 shows a two-coupler arrangement. In this case the receiver can always be on, because it does not receive optical power from the collocated transmitter.

A similar analysis to that in Eqs. (11-21) to (11-25) can be carried out for the two-coupler node. The only difference is that we replace $\beta = (1 - F_i)(1 - C_T)$ in Eqs. (11-23) and (11-24) by β^2.

11.3 OPTICAL AMPLIFIERS

Traditionally, when setting up an optical link, one formulates a power budget and adds repeaters when the path loss exceeds the available power margin. To amplify an optical signal with a conventional repeater one performs photon-to-electron conversion, electrical amplification, retiming, pulse shaping, and then electron-to-photon conversion. In many applications it would be advantageous to amplify the optical signal directly without having to go to the electrical domain. Thus there has been a great deal of interest in developing an all-optical amplifier.[51-60] The four main parameters used for characterizing the performance of optical amplifiers in a communication system are the noise figure, signal gain, frequency bandwidth, and saturation output power.

11.3.1 Basic Applications

There are three basic applications of optical amplifiers, as Fig. 11-21 shows. These are as follows:

1. In a single-mode link the effects of fiber dispersion may be small, so that the main limitation on repeater spacing is fiber attenuation. Since such a link

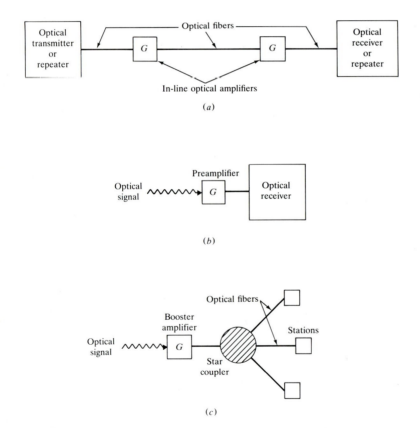

FIGURE 11-21
Three basic applications of optical amplifiers: (*a*) amplifier to increase repeater spacing, (*b*) preamplifier to improve receiver sensitivity, and (*c*) LAN booster amplifier.

does not necessarily require a complete regeneration of the signal; simple amplification of the optical signal is sufficient. Thus an optical amplifier can be used to reduce transmission loss and increase the distance between repeaters.

2. An optical amplifier can be used as a front-end preamplifier for an optical receiver. Thereby a weak optical signal is amplified before photodetection so that the signal-to-noise ratio degradation caused by thermal noise in the receiver electronics can be suppressed. Compared with other front-end devices such as avalanche photodiodes or optical heterodyne detectors, an optical preamplifier provides a larger gain factor and a broader bandwidth.

3. In a local area network one can employ an optical amplifier as a booster amplifier or a linear gain block to compensate for coupler insertion loss and power-splitting loss. Figure 11-21c shows an example for boosting the optical signal in front of a star coupler.

11.3.2 Optical Amplifier Types

A number of different types of optical amplifiers have been studied. These include semiconductor laser amplifiers (SLAs), rare-earth-ion-doped fiber amplifiers, fiber Raman amplifiers, and fiber Brillouin amplifiers. Of these the most widely investigated devices are the SLAs, which are discussed in this section. Research on fiber-based amplifiers is discussed in Sec. 11.5.

Semiconductor laser amplifiers are usually classified according to their facet reflectivity. Figure 11-22 shows a schematic diagram of a typical semiconductor laser amplifier. Here I is the bias current, and R_1 and R_2 are the facet reflectivities. In general the gain of the optical input signal depends on the bias current, facet reflectivities, material gain, and spontaneous recombination. The two major types of SLAs are the nonresonant, traveling-wave amplifier (TWA) and the resonant, Fabry-Perot amplifier (FPA). The FPA is basically a conventional laser diode which is biased just below the threshold level. Although FPAs are easy to fabricate, the optical signal gain is very sensitive to variations in amplifier temperature and input optical frequency. Thus they require very careful stabilization of temperature and injection current.

Traveling-wave amplifiers have been used more widely because they have a large optical bandwidth, high saturation power, and low polarization sensitiv-

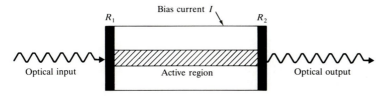

FIGURE 11-22
Schematic diagram of a typical semiconductor laser amplifier.

ity. However, they have a lower signal gain and higher spontaneous emission than FPAs because of their low facet reflectivity (around 0.1 to 0.2 percent). In addition they are difficult to fabricate because high-quality antireflection coatings are needed on the facets. Nevertheless, TWAs are superior to FPAs in terms of noise and saturation of signal gain.

11.3.3   Optical Amplifier Gain

Let us examine the optical gain of a traveling-wave semiconductor optical amplifier. If we assume the carrier density to be constant along the active region of the amplifier, then the signal gain G, which is defined as the ratio between the output and the input optical signal power, can be expressed as[61]

$$G = \frac{(1 - R_1)(1 - R_2)G_s}{\left(1 - G_s\sqrt{R_1R_2}\right)^2 + 4G_s\sqrt{R_1R_2}\,\sin^2\phi} \tag{11-26}$$

Here G_s is the single-pass gain, which from Eq. (4-35) is

$$G_s = \exp\left[(\Gamma g_m - \bar{\alpha})L\right]$$

$$= \exp\left[\left(\frac{\Gamma g_0}{1 + I/I_s} - \bar{\alpha}\right)L\right] \tag{11-27}$$

where Γ is the optical confinement factor, g_m is the material gain coefficient, g_0 is the unsaturated material gain coefficient in the absence of signal input, $\bar{\alpha}$ is the internal loss, I is the optical signal intensity, I_s is the saturation intensity (the intensity in the active layer at which the material gain coefficient g_m is reduced to half of g_0), and L is the amplifier length. The single-pass phase shift ϕ is defined by

$$\phi = \frac{2\pi n(\nu - \nu_0)L}{c}$$

$$= \phi_0 + \frac{g_0 Lb}{2}\left(\frac{I}{I + I_s}\right) \tag{11-28}$$

where ν is the incident signal frequency, ν_0 is the frequency of a resonant mode of the amplifier, n is the refractive index of the active-layer material, c is the speed of light, ϕ_0 is the nominal phase shift, and b is the line-width broadening factor ($b = 5$). Thus since G_s and ϕ are functions of the input intensity, the amplifier cavity will exhibit bistable and nonlinear characteristics at high optical input powers.

Example 11-3. Consider the following parameter values for a 1300-nm InGaAsP laser amplifier:

Symbol	Parameter	Value
g	Material gain coefficient	1000 cm^{-1}
L	Amplifier cavity length	$200 \ \mu\text{m}$
n	Refractive index	3.5
$R_1 = R_2$	Facet reflectivity	0.2%
$\bar{\alpha}$	Internal absorption	23 cm^{-1}
Γ	Confinement factor	0.3

To find the minimum and maximum optical amplifications we consider the single-pass phase shift ϕ to have the values 90° and 0°, respectively. First, the single-pass gain is

$$G_s = \exp\left\{\left[(0.3)(1000 \text{ cm}^{-1}) - 23 \text{ cm}^{-1}\right]200 \ \mu\text{m}\right\} = 255$$

The minimum optical amplification is thus

$$G_{min} = \frac{(1 - 0.002)(1 - 0.002)255}{\left[1 - 255(0.002)\right]^2 + 4(255)(0.002)\sin^2 90°} = 111.4$$

or $G_{min} = 20.5$ dB. With $\phi = 0$, the maximum optical amplification is

$$G_{max} = \frac{(1 - 0.002)(1 - 0.002)255}{\left[1 - 255(0.002)\right]^2} = 1058$$

or $G_{max} = 30.2$ dB.

11.3.4 Amplifier Noise Figure

Noise is the most important property in optical amplifiers, since it determines the signal-to-noise ratio in system applications. In particular, the noise power can limit the number of linear repeaters that can be cascaded. Since the mean noise power due to spontaneous emission in the optical amplifiers accumulates in proportion to the number of repeaters, gain saturation occurs when the accumulated noise power equals the signal power.

The noise figure F for optical amplifiers is defined as the signal-to-noise degradation due to amplification, just as for electrical amplifiers. That is,

$$F = \frac{(S/N)_{in}}{(S/N)_{out}} \tag{11-29}$$

This is governed by factors such as the population inversion, the excess noise factor, the number of transverse modes in the cavity, the optical bandwidth for the amplified spontaneous emissions, and the number of photons incident on the amplifier. A typical noise figure is 4.5 dB.

Example 11-4. Consider Fig. 11-23, which shows a system with k cascaded optical amplifiers. Let α_i denote the link attenuation in front of the ith amplifier, which has a signal gain G_i and a noise figure F_i. Here the input and output signal-to-noise

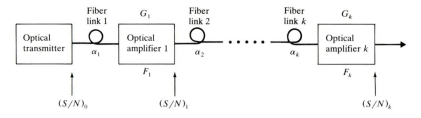

FIGURE 11-23
Signal-to-noise ratios in a cascaded amplifier chain.

ratios for the system are defined at the transmitter output and the kth amplifier output, respectively. Analogous to the case for electrical amplifiers, the total noise figure for this cascaded link is given by

$$F_{\text{total}} = \frac{(S/N)_0}{(S/N)_k}$$

$$= \frac{F_1}{\alpha_1} + \frac{F_2}{\alpha_1 G_1 \alpha_2} + \frac{F_3}{\alpha_1 G_1 \alpha_2 G_2 \alpha_3} + \cdots + \frac{F_k}{\prod_{i=1}^{k-1}(\alpha_i G_i)\alpha_k} \quad (11\text{-}30)$$

11.3.5   Optical Bandwidth

The ± 3-dB optical bandwidth $2\,\Delta f$ of a single longitudinal mode is given as a function of the amplifier gain and the facet reflectivities by

$$2\,\Delta f = \frac{c}{\pi nL} \arcsin\left\{\frac{1}{2}\left[\frac{(1-R_1)(1-R_2)}{\sqrt{R_1 R_2}\,G}\right]^{1/2}\right\} \quad (11\text{-}31)$$

where n is the active layer refractive index and L is the cavity length.

> **Example 11-5.** Given a semiconductor amplifier with a refractive index of 3.5, facet reflectivities of 0.3, a gain of 25 dB, and a cavity length of 200 μm, we have a bandwidth on the order of 5 GHz.

11.4   PHOTONIC SWITCHING

With the widespread implementation of optical fiber transmission lines has come the question of how best to switch or route high-speed optical signals. One driving force in this area is the rapid emergence of broadband integrated services digital networks (BISDN). Of interest here are services such as high-resolution image transfer, switched multimegabit data service, remote computer graphics systems, remote medical diagnosis, and TV-program on-demand distribution.

To keep pace with these communication demands, researchers have examined fast packet-switching technologies wherein information is formatted into

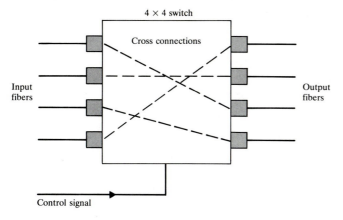

FIGURE 11-24
Simple schematic of a 4 × 4 crosspoint switch.

specific-length message units called *packets*. These packets are sent through a network as individual units and are reassembled into complete messages at the destination station. A necessary component for fast packet switching is a high-capacity switch for interconnecting a large number of subscribers. A simple schematic of a basic 4 × 4 *crosspoint switch* is shown in Fig. 11-24. This device sets up specific transmission paths between two groups of optical fibers, and allows these paths to be reconfigured as needed. Traditionally the network architecture consists of optical fiber links connected to electronic switches. Since the transmission bandwidth of an individual channel is limited to a few gigahertz, electronic switches can achieve a high throughput by using high dimensionality. However, the relatively low transmission bandwidth of electronic switches, the need to have intermediate optoelectronic interfaces, and the signal-processing electronics needed for routing control do not allow full utilization of the large bandwidth-distance product of an optical fiber.

These obstacles could be overcome if the signals remained in an optical form during the switching and signal-processing operations. Thus a great deal of activity has been generated in developing photonic switches.[39, 61-70] Not only do photonic switches potentially have bandwidths in excess of 1 THz and subpicosecond switching speeds, but their use would decrease hardware complexity by eliminating the need for optoelectronic signal conversion.

11.4.1 Mechanical Switches

Before examining high-speed switches, let us first look at some lower-speed applications. A number of mechanical switches in which a fiber is physically moved from one position to another have been developed. Figure 11-25 shows an example of a 2 × 2 mechanical switch. Here the center unit may consist of a metallized element which can be moved from one position to another by means

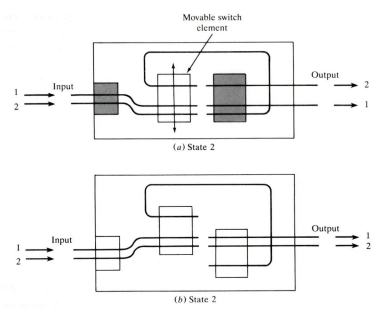

Movable switch element

Input

Output

(*a*) State 2

Input

Output

(*b*) State 2

FIGURE 11-25
Two states of a two-fiber mechanical optical switch.

of a magnet. Alternatively, a stepper motor can move the fibers to the desired position. These switches can have applications such as *protection switching* for bypassing broken fibers or failed equipment, *network reconfiguration switching* to accommodate changing customer needs or services, or *maintenance switching* for testing fiber transmission lines. Larger arrays, for example 16×16 or 32×32, have also been made. Other types of mechanical switches include moving prisms, plates, and connectors. The switching speeds of all these devices are on order of several milliseconds.

11.4.2 Integrated-Optical Switches

For large-scale implementation of high-speed photonic switches one needs to use integrated-optics technology. The most advanced electro-optic waveguide technology for photonic switches utilizes lithium niobate ($LiNbO_3$). This material offers the necessary low insertion loss, it is capable of high switching speed and large bandwidth with special electrode designs, it allows for large-scale integration with electronic circuits, and it can be made economically in large batches.

The various switch configurations which are possible with $LiNbO_3$ can be broadly classified into directional couplers, Mach-Zender type interferometers, and intersecting waveguides (also known as *x-type* waveguides). The most common devices are shown schematically in Fig. 11-26. Of these the directional

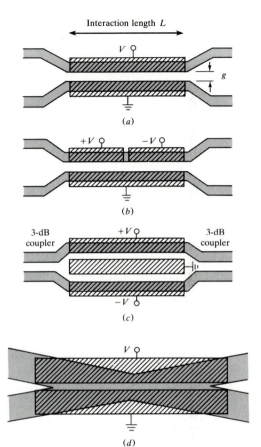

FIGURE 11-26
Different types of 2×2 integrated-optical crosspoint elements for an optical switch: (a) basic directional coupler; (b) reversed $\Delta\beta$ directional coupler; (c) Mach-Zender interferometer;(d) intersecting waveguide or x type switch.

coupler has been most widely studied and implemented. The intersecting waveguides are a viable alternative. For large arrays the Mach-Zender switches do not perform as well as the directional couplers and x types.

The directional coupler (Fig. 11-26a and b) consists of a pair of optical channel waveguides that are parallel and in close proximity over an interaction or coupling length L. Light entering one of these waveguides gets coupled to the second waveguide via evanescent coupling. By placing electrodes over the waveguides to induce a change in their refractive indices, the propagation constants $\Delta\beta$ in the two waveguides can be made sufficiently large so that no light couples between them.

The basic design parameters for directional couplers are the voltage required to change the state of a switch, the switching efficiency, and the switching speed. The change in the refractive index is given by

$$\Delta n = -\frac{n^3 r}{2}\frac{V}{g}\Gamma \tag{11-32}$$

where n is the refractive index, r is the electro-optic coefficient for the crystal orientation being used, V is the applied voltage, g is the electrode gap, and Γ is an overlap integral between the electric field and the optical mode, which is maximum when the two are matched.

The induced phase shift is then

$$\Delta \beta L = \frac{2\pi}{\lambda} \Delta n L \tag{11-33}$$

where λ is the wavelength of the light in free space and L is the electrode length.

For a uniform-$\Delta\beta$ coupler the length L is chosen so that when $V = 0$ the switch is in the *cross state* (the input and output lines are crossed). When a voltage of

$$V = \frac{\lambda g \sqrt{3}}{n^3 r L \Gamma} \tag{11-34}$$

is applied, the switch changes to the *bar state* (the input and output ports are on the same line). Thus the required switching voltage is directly proportional to the electrode gap g and inversely proportional to the electrode length L.

One problem with the device shown in Fig. 11-26a is that to achieve a perfect cross state with no applied voltage, fabrication tolerances are very critical in order to have low cross talk. The use of the so-called *reversed delta beta* ($\Delta\beta$) *coupler* shown in Fig. 11-26b overcomes this fabrication difficulty. Splitting the electrode into at least two sections, one can tune both the cross and the bar states over a relatively large range of coupling values. The only complication is that now two drive voltages are required per switch.

The Mach-Zender interferometer switch shown in Fig. 11-26c consists of a pair of 3-dB couplers connected by two waveguides which are sufficiently separated so that they do not couple. Light entering one port is divided equally by the first coupler. With no voltage applied to the electrodes, the optical path lengths of the two center arms are equal, so that the light in them enters the second coupler in phase. In the second coupler all the light is crossed over to the second waveguide to provide the cross state. The bar state is enabled by applying a voltage to produce a π phase difference between the two arms. These switches are typically 15 to 20 mm long and require a phase-shift voltage of 3 to 5 V. However, they have high cross-talk levels of -8 to -20 dB and thus are not suitable for integration.

The intersecting-waveguide switch shown in Fig. 11-26d can be understood either as a modal interferometer or as a zero-gap directional coupler. Some of its features are topological flexibility, moderate voltage requirements, and simple electrode configurations.

In contrast to using an electric field to induce a refractive-index change in an optical waveguide, research is also underway into methods for changing the refractive index through a variation in the intensity of the optical field itself.[70, 71] With such a technique the optical signal produces its own phase shift in a waveguide and hence selects its own output channel. The advantage is that switching can be done very rapidly without the need for electronics.

11.5 NONLINEAR OPTICAL EFFECTS

The widespread usage of single-mode fibers has also created great interest in the field of nonlinear effects in optical fibers.[72-73] In a nonlinear process, optical power launched into an optical fiber at a given wavelength is transferred to a set of longer wavelengths. The exact nature of the shift is determined by certain characteristic vibrations of the fiber materials which induce changes in its refractive index. The refractive index is due to the applied optical field perturbing the atoms or molecules of the material to induce various scattering processes. At low optical intensities the scattering is a linear function of the applied fields, and is known as Rayleigh scattering. The net perturbation is simply described by a constant refractive index. At higher intensities the perturbations are no longer linear functions of the applied field, so that inelastic scattering results. This leads to nonlinear phenomena such as the Kerr effect, Raman scattering, and Brillouin scattering. Applications of these nonlinear effects includes Raman amplifiers, Brillouin amplifiers, soliton propagation, pulse compression, and all-optical switching.

The Kerr effect refers to phenomena that can be simply described by an intensity-dependent refractive index. This results in an intensity-dependent phase shift in a fiber. Thus Kerr nonlinearities can alter the frequency spectrum of a pulse traveling through a fiber. The main application of the Kerr effect has been to pulse-compression techniques.[75] Although the pulse widths of 6 fs (6×10^{-15} s) which have been achieved are not applicable to communication systems, they are very useful in research on components and materials for high-bit-rate fiber transmission systems.

The use of optical amplifiers has created a great interest in the Raman and the Brillouin effects. In the *Raman effect* a molecule absorbs a photon at the original frequency and emits it again at a shifted frequency, while simultaneously making a transition between vibrational states. In fused silica the primary vibration has a frequency of about 1.3×10^{13} Hz. For 1-μm pump light the Raman gain gets peaked at around 1.045 μm. The important feature of the Raman effect is that it not only results in a frequency shift of the incident light, but for a sufficiently high optical intensity it gives rise to an *optical gain* at the shifted frequency. For a typical fiber, a pump power of about 1 W in 100 m of fiber yields a Raman amplification of about 2. Consequently much research is underway on fiber Raman amplifiers.[76, 77] The advantages of Raman amplification are self-phase-matching between the pump and the signal, and a high-speed response compared with other nonlinear processes. However, they require substantial amounts of pump power. As an example, a fiber Raman amplifier acting as a power booster for a 1-Gb/s system operating at 1550 nm might yield a 3-dB increase in receiver sensitivity for 100-mW pump power.

High-intensity fields can also interact with the acoustic vibrations of the glass, thereby shifting the frequency of the light by the acoustic frequency, which is around 10^{10} Hz. This is known as the *Brillouin effect*.[78, 79] Fiber Brillouin amplifiers have a very narrow bandwidth (typically 20 MHz) and require small amounts of pump power. Threshold powers for Brillouin scattering can be as low as 5 mW at 1320 nm. To make these types of amplifiers useful for fiber optic communication systems, a variety of schemes are being considered for extending the intrinsic narrow bandwidth. Bandwidths as high as 150 MHz have been achieved.[79]

An interesting phenomenon occurs when one examines the Kerr effect in the negative-dispersion region of a fiber. For a high-intensity pulse the Kerr nonlinearity causes the leading edge of the pulse to develop a red shift (a shift toward lower optical frequencies) and the trailing edge to develop a blue shift (toward higher optical frequencies). In addition, because of the negative group-velocity dispersion, the red-shifted light will travel slower than the unshifted center of the pulse and the blue-shifted light will travel faster. By carefully choosing a proper combination of pulse shape, pulse intensity, and pulse width, the effects of the Kerr nonlinearity and of the negative dispersion will cancel exactly. Such a pulse, which will travel without any changes, is referred to as a *soliton*.[80-83] Since the transmission of solitons would be extremely useful in communication systems, extensive research is underway in this area.

11.6 SUMMARY

Along with the maturing of optical fiber communications has come the upgrading of older links and networks together with the search for more advanced techniques to take full advantage of the transmission potential of an optical fiber. Among some of these techniques are wavelength division multiplexing, local area networking on fibers, optical amplifiers, photonic switching, and the exploitation of nonlinear effects in optical fibers.

The simplest optical fiber system is a point-to-point link having a transmitter on one end and a receiver on the other. Since the optical source has a relatively narrow spectral width, the full transmission capability of the fiber is not used in this case. A fuller utilization of the transmission capacity of a fiber can be achieved with wavelength division multiplexing (WDM). In this scheme one simultaneously transmits optical signals over the same fiber from a number of different light sources having properly spaced peak emission wavelengths. By operating each source at a different peak wavelength, the integrity of the independent messages from each source is maintained for subsequent conversion to electrical signals at the receiving end.

Alternatives to point-to-point links are local area network (LAN) architectures. By using bus, ring, or star network configurations, a number of physically separated terminals can communicate with one another by multiplexed signals. Access to an optical fiber LAN is achieved with a coupling element, which can be either active or passive. An active coupler converts the optical signal coming off the network to its electrical baseband counterpart before any data processing is carried out. A passive coupler contains no electronics. It passively taps off a portion of the optical power from the network line, or it can insert an optical signal onto the line. Each of the three basic LAN topologies has its own advantages and limitations in terms of reliability, expandability, and performance characteristics.

Traditionally, when setting up an optical link, one formulates a power budget and adds repeaters when the path loss exceeds the available power margin. To amplify an optical signal with a conventional repeater one performs photon-to-electron conversion, electrical amplification, retiming, pulse shaping, and then electron-to-photon conversion. In many applications it would be advantageous to amplify the optical signal directly without having to go to the electrical domain. Thus there has been a great deal of interest in developing an all-optical amplifier. The optical amplifiers that have been examined include semiconductor laser amplifiers, rare-earth-ion-doped fiber amplifiers, fiber Raman amplifiers, and fiber Brillouin amplifiers. Of these the most widely used device is the semiconductor laser amplifier. Applications include in-line signal amplification for increasing repeater spacing, front-end preamplifiers for increasing receiver sensitivity, and booster amplifiers for local area networks.

To keep pace with increasing communication demands, researchers have turned to optical switching techniques for routing high-speed optical signals. For large-scale implementation of high-speed photonic switches one needs to use integrated-optics technology. The most advanced electro-optic waveguide technology for photonic switches is lithium niobate ($LiNbO_3$). This material offers the necessary low insertion loss, it is capable of high switching speed and large bandwidth with special electrode designs, it allows for large-scale integration with electronic circuits, and it can be made economically in large batches.

The widespread usage of single-mode fibers has also created great interest in the field of nonlinear effects in optical fibers. In a nonlinear process, optical power launched into an optical fiber at a given wavelength is transferred to a set of longer wavelengths. The exact nature of the shift is determined by certain characteristic vibrations of the fiber materials which induce changes in its refractive index. The refractive index is due to the applied optical field perturbing the atoms or molecules of the material to induce various scattering processes. At low optical intensities the scattering is a linear function of the applied fields, and is known as Rayleigh scattering. The net perturbation is simply described by a constant refractive index. At high intensities the perturbations are no longer linear functions of the applied field, so that inelastic scattering results. This leads to nonlinear phenomena such as the Kerr effect, Raman scattering, and Brillouin scattering. Applications of these nonlinear effects includes Raman amplifiers, Brillouin amplifiers, soliton propagation, pulse compression, and all-optical switching.

PROBLEMS

11-1. A convenient wavelength division multiplexer is a plane reflection grating which is mounted as shown in Fig. P11-1. The angular properties of this grating are given

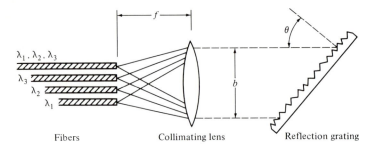

Fibers   Collimating lens   Reflection grating

FIGURE P11-1

by the grating equation

$$\sin \phi + \sin \theta = \frac{k\lambda}{n'\Lambda}$$

where Λ is the grating period, k is the interference order, n' is the refractive index of the medium between the lens and grating, and ϕ and θ are the angles of the incident and reflected beams, respectively, measured normal to the grating.

(a) Using the grating equation, show that the angular dispersion is given by

$$\frac{d\theta}{d\lambda} = \frac{k}{n'\Lambda \cos \theta} = \frac{2 \tan \theta}{\lambda}$$

where the last equality was derived from the condition $\phi \approx \theta$ (this condition minimizes distortion).

(b) Show that the minimum collimating lens diameter is

$$b \geq \frac{\lambda(NA/n')d}{\Delta\lambda \tan \theta}$$

(c) Letting W in Eq. (11-5) be the maximum path length of the collimated beam, show that the fractional beam spread is given by

$$S = 2(1 + m)\tan^2 \theta \, \frac{\Delta\lambda}{\lambda}$$

(d) Given that $\Delta\lambda = 26$ nm, $\lambda = 850$ nm, and $m = 3$, what is the upper limit on θ for beam spreading of less than 1 percent?

11-2. A four-channel WDM system using laser diodes has the optical level parameters shown in Table P11-2. The system operates over 10 km at a 20-Mb/s rate with a

TABLE P11-2

Source wavelength (nm)	810	835	860	890
Source output (dBm)	0.0	−1.0	−0.5	−1.8
Multiplexer insertion loss (dB)	2.9	3.5	4.1	3.1
Fiber attenuation (dB/km)	3.9	3.6	3.4	3.5
Demultiplexer insertion loss (dB)	3.6	3.1	3.6	3.8
Receiver sensitivity (dBm)	−56.0	−55.5	−55.9	−55.0

10^{-9} BER. What is the system operating margin for each channel? Assume connector losses at the multiplexer, source, and detector are 1.0 dB.

11-3. Consider a star-coupled optical fiber network with 16 inputs and 16 outputs operating at 10 Mb/s. Assume this system has the following parameters: $L_c = 1.0$ dB, star coupler insertion loss $L_s = 5$ dB, and a fiber loss of 2 dB/km. Let the sources be InGaAsP LEDs having an output of -16 dBm from a fiber flylead, and assume InGaAsP *pin* photodiodes with a -49-dBm sensitivity are used. Assume that a 6-dB system margin is required.

(*a*) What is the maximum transmission distance if a transmission star coupler is used?

(*b*) What is the maximum distance if a reflection star coupler is used?

11-4. Consider an optical fiber transmission star coupler having seven inputs and seven outputs. Suppose the coupler is constructed by arranging the seven fibers in a circular pattern (a ring of six with one in the center) and butting them against the end of a glass rod which serves as the mixing element.

(*a*) If the fibers have 50-μm core diameters and 125-μm outer cladding diameters, what is the coupling loss resulting from light escaping between the fiber cores? Let the rod diameter be 300 μm. Assume the fiber cladding is not removed.

(*b*) What is the coupling loss if the fiber ends are arranged in a row and a 50-μm by 800-μm glass plate is used as the star coupler?

11-5. Repeat Prob. 11-4 for seven fibers having 200-μm core diameters and 400-μm outer cladding diameters. What should the sizes of the glass rod and the glass plate be in this case?

11-6. An engineer wishes to construct an in-line optical fiber data bus operating at 10 Mb/s. The stations are to be separated by 100 m, for which optical fibers with a 3-dB/km attenuation are used. The optical sources are laser diodes having an output of 1 mW from a fiber flylead, and the detectors are avalanche photodiodes with a -58-dBm (1.6-nW) sensitivity. The couplers have a power-coupling efficiency $E = 10$ percent, a power tapoff factor of $C_T = 5$ percent, and a 10-percent fractional intrinsic loss F_i. The power loss at the connectors is 20 percent (1 dB).

(*a*) Make a plot of P_{1N} in dBm as a function of the number of stations N from 0 to -58 dBm.

(*b*) What is the system operating margin for eight stations?

(*c*) What is the worst-case dynamic range for the maximum allowable number of stations if a 6-dB power margin is required?

11-7. A two-story office building has two 10-foot-wide hallways per floor which connect four rows of offices with eight offices per row as is shown in Fig. P11-7. Each office is 15 feet square. The office ceiling height is 9 feet with a false ceiling hung 1 foot below the actual ceiling. Also, as shown in Fig. P11-7, there is a wiring room for LAN interconnection and control equipment in one corner of each floor. Every office has a local-area-network socket on each of the two walls that are perpendicular to the hallway wall. If we assume that cables can be run only in the walls and in the ceilings, estimate the length of cable (in feet) that is required for a fiber optic star that connects each outlet to the wiring room on the corresponding floor and a vertical fiber optic riser that connects the stars in each wiring room.

11-8. Consider the *M*-by-*N* grid of stations shown in Fig. P11-8 which are to be connected by a local area network. Let the stations be spaced a distance d apart,

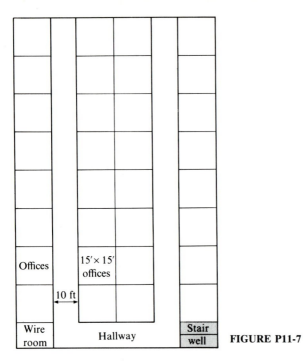

FIGURE P11-7

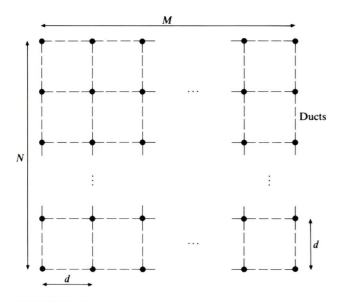

FIGURE P11-8

and assume that interconnection cables will be run in ducts that connect nearest-neighbor stations (that is, ducts are not run diagonally in Fig. P11-8). Show that for the following configurations, the cable length for interconnecting the stations is as stated:

(*a*) $(MN - 1)d$ for a bus configuration

(*b*) MNd for a ring topology

(*c*) $MN(M + N - 2)d/2$ for a star topology where each subscriber is connected *individually* to the network hub, which is located *in one corner* of the grid.

11-9. Consider the *M*-by-*N* rectangular grid of computer stations shown in Fig. P11.8, where the spacing between stations is d. Assume these stations are to be connected by a star-configured LAN using the duct network shown in the figure. Further assume that each station is connected to the central star by means of its own dedicated cable.

(*a*) If m and n denote the relative position of the star, show that the total cable length L needed to connect the stations is given by

$$L = [MN(M + N + 2)/2 - Nm(M - m + 1) - Mn(N - n + 1)]d$$

(*b*) Show that if the star is located in one corner of the grid, then this expression becomes

$$L = MN(M + N - 2)d/2$$

(*c*) Show that the shortest cable length is obtained when the star is at the center of the grid.

11-10. Suppose we have a series of k optical amplifiers in tandem. Assume $\alpha_i G_i = 1$, so that the transmission loss α_i is compensated for by the amplifier gain G_i.

(*a*) Find an expression in terms of G_i and F_i for the total noise figure.

(*b*) If all amplifiers are identical, that is, $G_i = G$ and $F_i = F$ for all amplifiers, make a general plot of the degradation in S/N as a function of the number of repeaters k.

REFERENCES

1. W. J. Tomlinson, "Wavelength multiplexing in multimode optical fibers," *Appl. Opt.*, vol. 16, pp. 2180–2194, Aug. 1977.
2. T. Kanada, Y. Okano, K.-I. Aoyama, and T. Kitami, "Design and performance of WDM transmission systems at 6.3 Mb/s," *IEEE Trans. Commun.*, vol. COM-31, pp. 1095–1102, Sept. 1983.
3. (*a*) G. Winzer, "What has WDM to offer for long-haul single-mode fiber-optic links?" *Siemens Forsch. u. Entwickl. Ber.*, vol. 12, no. 5, pp. 332–339, 1983.
   (*b*) G. Winzer, "Wavelength-multiplexing components—A review of single-mode devices and their applications," *J. Lightwave Tech.*, vol. LT-2, pp. 369–378, Aug. 1984.
4. H. Ishio, J. Minowa, and K. Nosu, "Review and status of wavelength-division-multiplexing technology and its application," *J. Lightwave Tech.*, vol. LT-2, pp. 448–463, Aug. 1984.
5. J. M. Senior and S. D. Cusworth, "Devices for wavelength multiplexing and demultiplexing," *IEE Proc.*, vol. 136, Pt. J, pp. 183–202, June 1989.
6. D. B. Payne and J. R. Stern, "Transparent single-mode fiber optical networks," *J. Lightwave Tech.*, vol. LT-4, pp. 864–869, July 1986.
7. B. S. Glance, K. Pollack, C. A. Burrus, B. L. Casper, G. Eisenstein, and L. W. Stulz, "WDM coherent optical star network," *J. Lightwave Tech.*, vol. 6, pp. 67–72, Jan. 1988.

8. S. S. Wagner and H. L. Lemberg, "Technology and system issues for a WDM-based fiber loop architecture," *J. Lightwave Tech.*, vol. 7, pp. 1759–1768, Nov. 1989.

9. S. S. Wagner and H. Kobrinski, "WDM applications in broadband telecommunication networks," *IEEE Commun. Mag.*, vol. 29, pp. 22–30, Mar. 1989.

10. M. L. Loeb and G. R. Stilwell, Jr., "An algorithm for bit-skew correction in byte-wide WDM optical fiber systems," *J. Lightwave Tech.*, vol. 8, pp. 239–242, Feb. 1990.

11. R. Watanabe, K. Nosu, and Y. Fujii, "Optical grating multiplexer in the 1.1–1.5 μm wavelength region," *Electron. Lett.*, vol. 16, pp. 108–109, Jan. 1980.

12. K. Kobayashi and M. Seki, "Microoptic grating multiplexers and optical isolators for fiber optic communications," *Electron Lett.*, vol. 16, pp. 11–22, Jan. 1980.

13. (*a*) L. Jou and B. Metcalf, "Wavelength division multiplexing," *Proc. SPIE*, vol. 340, pp. 69–74, 1982.

   (*b*) B. Metcalf and L. Jou, "Dual GRIN lens wavelength multiplexer," *Appl. Opt.*, vol. 22, pp. 455–459, Feb. 1983.

14. I. Nishi, T. Oguchi, and K. Kato, "Broad passband multi/demultiplexer for multimode fibers using a diffraction grating and retroflectors," *J. Lightwave Tech.*, vol. LT-5, pp. 1695–1700, Dec. 1987.

15. C. Cremer, G. Heise, R. März, M. Schienle, G. Schulte-Roth, and H. Unzeitig, "Bragg gratings on InGaAsP/InP waveguides; use as polarization-independent filters for wavelength-division multiplexing," *J. Lightwave Tech.*, vol. 7, pp. 1641–1645, Nov. 1989.

16. A. Frenkel and C. Lin, "Angle-tuned etalon filters for optical channel selection in high density wavelength-division multiplexed systems," *J. Lightwave Tech.*, vol. 7, pp. 615–624, Apr. 1989.

17. V. A. Bhagavatula, G. T. Holmes, D. A. Nolan, R. Jansen, and M. McCourt, "Planar technology enhances coupler performance," *Laser Focus World*, vol. 25, pp. 155–160, Oct. 1989.

18. C. H. Henry, R. F. Kazarinov, Y. Shani, R. C. Kistler, V. Pol, and K. J. Orlowsky, "Four-channel wavelength-division multiplexers and bandpass filters based on elliptical Bragg reflectors," *J. Lightwave Tech.*, vol. 8, pp. 748–755, May 1990.

19. W. H. Wells, "Crosstalk in a bidirectional optical fiber," *Fiber Integ. Optics*, vol. 1, pp. 243–287, 1978.

20. J. Conradi and R. Maciejko, "Digital optical receiver sensitivity degradation caused by crosstalk in a bidirectional fiber optic systems," *IEEE Trans. Commun.*, vol. COM-29, pp. 1012–1016, July 1981.

21. P. P. Bohn and S. K. Das, "Return loss requirements for optical duplex transmission," *J. Lightwave Tech.*, vol. LT-5, pp. 243–254, Feb. 1987.

22. A. A. Al-Orainy and J. J. O'Reilly, "New error probability bounds for the influence of crosstalk on wavelength-division multiplexed systems," *J. Optic. Commun.*, vol. 11, pp. 7–10, Mar. 1990.

23. S. R. Mallinson, "Wavelength-selective filters for single-mode fiber WDM systems using Fabry-Perot interferometers," *Appl. Opt.*, vol. 26, pp. 430–436, Feb. 1987.

24. H. Kobrinski and K. W. Cheung, "Wavelength-tunable optical filters: Applications and technologies," *IEEE Commun. Mag.*, vol. 27, pp. 53–63, Oct. 1989.

25. (*a*) P. Melman, W. J. Carlsen, and B. Foley, "Tunable birefringent wavelength-division multiplexer/demultiplexer," *Electron. Lett.*, vol. 21, pp. 634–635, July 1985.

   (*b*) W. J. Carlsen and C. F. Buhrer, "Flat passband birefringent wavelength-division multiplexers," *Electron. Lett.*, vol. 23, pp. 106–107, 1987.

26. B. Foley, S. Cousins, and P. Melman, "Analysis of bit error rate and power penalty for a birefringent wavelength-division multiplexer," *Fiber & Integ. Optics*, vol. 7, pp. 109–114, 1988.

27. G. E. Keiser, *Local Area Networks*, McGraw-Hill, New York, 1989.

28. P. S. Henry, "High-capacity lightwave local area networks," *IEEE Commun. Mag.*, vol. 27, pp. 20–26, Oct. 1989.

29. Y.-K. M. Lin, D. R. Spears, and M. Yin, "Fiber-based local access network architectures," *IEEE Commun. Mag.*, vol. 27, pp. 64–73, Oct. 1989.

30. G. R. Hill, "Wavelength domain optical network techniques," *Proc. IEEE*, vol. 78, pp. 121–132, Jan. 1990.

31. D. D. Clark, K. T. Pogran, and D. P. Reed, "An introduction to local area networks," *Proc. IEEE*, vol. 66, pp. 1497–1517, Nov. 1978.

32. T. V. Feng, "A survey of interconnection networks," *Computer*, vol. 14, pp. 12–27, Dec. 1981.

33. M. M. Nassehi, F. A. Tobagi, and M. E. Marhic, "Fiber-optic configurations for local area networks," *IEEE J. Sel. Areas Commun.*, vol. SAC-3, pp. 941–949, Nov. 1985.

34. H. H. Witte and V. Kulich, "Branching elements for optical data buses," *Appl. Optics*, vol. 20, pp. 715–718, Feb. 1981.

35. W. J. Tomlinson, "Passive and low-speed active optical components for fiber systems," in S. E. Miller and I. P. Kaminow, eds., *Optical Fiber Telecommunications—II*, Academic, New York, 1988.

36. A. Albanese, "Fail-safe nodes for lightguide digital networks," *Bell Sys. Tech. J.*, vol. 16, pp. 247–256, Feb. 1982.

37. R. R. Talbott, "Feed forward rings," *Fiber Integ. Optics*, vol. 7, no. 4, pp. 275–298, 1988.

38. P. J. W. Severin, "A fail-safe self-routing fiber-optic ring network using multitailed receiver-transmitter units as nodes," *J. Lightwave Tech.*, vol. 7, pp. 358–363, May 1989.

39. S. F. Su, L. Jou, and J. Lenart, "A review on classification of optical switching systems," *IEEE Commun. Mag.*, vol. 24, pp. 50–55, May 1986.

40. T. H. Wood, "Increased power injection in multimode optical-fiber buses through mode-selective coupling," *IEEE J. Lightwave Tech.*, vol. LT-3, pp. 537–543, June 1985.

41. H. S. Huang, H. C. Chang, and J. S. Wu, "Power transfer between single-mode and multimode optical fibers," *1986 IEEE MTT-S Microwave Symp. Digest*, pp. 519–522, Baltimore, MD, 2–4 June 1986.

42. J. H. Saltzer, K. T. Pogran, and D. D. Clark, "Why a ring?" *Computer Networks*, vol. 7, pp. 223–231, 1983.

43. W. Bux, F. H. Closs, K. Kuemmerle, H. J. Keller, and H. R. Mueller, "Architecture and design of a reliable token-ring network," *IEEE J. Select. Areas Commun.*, vol. SAC-1, pp. 756–765, Apr. 1983.

44. F. E. Ross, "An overview of FDDI," *IEEE J. Select. Areas Commun.*, vol. 7, pp. 1043–1051, Sept. 1989.

45. S. Moustakas, H. H. Witte, V. Bodlaj, and V. Kulich, "Passive optical star bus with collision detection for CSMA/CD-based local area networks," *J. Lightwave Tech.*, vol. LT-3, pp. 93–100, Feb. 1985.

46. D. R. Porter, P. R. Couch, and J. W. Schelin, "A high-speed fiber optic data bus for local data communication," *IEEE J. Select. Areas Commun.*, vol. SAC-1, pp. 479–488, Apr. 1983.

47. F. W. Scholl and M. H. Coden, "Passive optical star systems for fiber optic local area networks," *IEEE J. Select. Areas Commun.*, vol. 6, pp. 913–923, July 1988.

48. S. Y. Suh, S. W. Granlund, A. J. Lumsdaine, D. A. Snyder, S. J. Wetzel, C. J. Daniels, and K. W. Haag, "Active star coupler based fiber optic local area network," *J. Lightwave Tech.*, vol. LT-5, pp. 1050–1061, Aug. 1987.

49. I. P. Kaminow, P. P. Iannone, J. Stone, and L. W. Stulz, "FDMA-FSK star network with a tunable optical filter demultiplexer," *J. Lightwave Tech.*, vol. 6, pp. 1406–1414, Sept. 1988.

50. N. Mehravari, "Performance and protocol improvements for very high speed optical fiber local area networks using a passive star topology," *J. Lightwave Tech.*, vol. 8, pp. 520–530, Apr. 1990.

51. J. Wang, H. Olesen, and K. E. Stubkjaer, "Recombination, gain, and bandwidth characteristics of 1.3-μm semiconductor laser amplifiers," *J. Lightwave Tech.*, vol. LT-5, pp. 184–189, Jan. 1987.

52. S. Yamamoto, K. Mochizuki, H. Wakabayashi, and Y. Iwamoto, "Long-haul high-speed optical communication systems using a semiconductor laser amplifier," *J. Lightwave Tech.*, vol. 6, pp. 1554–1558, Oct. 1988.

53. H. E. Lassen, P. B. Hansen, and K. E. Stubkjaer, "Crosstalk in 1.5-μm InGaAsP optical amplifiers," *J. Lightwave Tech.*, vol. 6, pp. 1559–1565, Oct. 1988.

54. M. J. O'Mahony, "Semiconductor laser optical amplifiers for use in future fiber systems," *J. Lightwave Tech.*, vol. 6, pp. 531–544, Apr. 1988.

55. N. A. Olsson, "Lightwave systems with optical amplifiers," *J. Lightwave Tech.*, vol. 7, pp. 1071–1082, July 1989.

56. (*a*) T. Saitoh and T. Mukai, "Recent progress in semiconductor laser amplifiers," *J. Lightwave Tech.*, vol. 6, pp. 1656–1664, Nov. 1988.

   (*b*) Y. Yamamoto and T. Mukai, "Fundamentals of optical amplifiers," *Fiber Integ. Optics*, vol. 21, Special Issue on Optical Amplifiers, pp. S1–S14, 1989.

   (*c*) Eight additional papers in "Special Issue on Optical Amplifiers," *Fiber Integ. Optics*, vol. 21, 1989.

57. H. Taga, S. Yamamoto, K. Mochizuki, and H. Wakabayashi, "Power penalty due to optical back reflection in semiconductor optical amplifier repeater systems," *IEEE Photonics Tech. Lett.*, vol. 2, pp. 279–281, Apr. 1990.

58. R. S. Fyath, A. J. McDonald, and J. J. O'Reilly, "Sensitivity and power penalty considerations for laser preamplified direct detection optical receivers," *IEE Proc.*, vol. 136, pt. J, pp. 238–248, Aug. 1989.

59. E. Dietrich, B. Enning, G. Grosskopf, L. Küller, R. Ludwig, R. Molt, E. Patzak, and H. G. Weber, "Semiconductor laser optical amplifiers for multichannel coherent optical transmission," *J. Lightwave Tech.*, vol. 7, pp. 1941–1956, Dec. 1989.

60. A. G. Failla, G. P. Bava, and I. Montrosset, "Structural design criteria for polarization insensitive semiconductor optical amplifiers," *J. Lightwave Tech.*, vol. 8, pp. 302–308, Mar. 1990.

61. *IEEE Commun. Mag.*, Special Issue on Photonic Switching, vol. 25, May 1987.

62. *IEEE J. Sel. Areas Commun.*, Special Issue on Photonic Switching, vol. 6, Aug. 1988.

63. J. E. Midwinter, "Photonic switching technology: Component characteristics versus network requirements," *J. Lightwave Tech.*, vol. 6, pp. 1512–1519, Oct. 1988.

64. A. Selvarajan and J. E. Midwinter, "Tutorial Review: Photonic switches and switch arrays on $LiNbO_3$," *Opt. Quantum Electron.*, vol. 21, no. 1, pp. 1–15, Jan. 1989.

65. T.-H. Wu and S. F. Habiby, "Strategies and technologies for planning a cost-effective survivable fiber network architecture using optical switches," *J. Lightwave Tech.*, vol. 8, pp. 152–159, Feb. 1990.

66. P. A. Kirby, "Multichannel wavelength-switched transmitters and receivers—new component concepts for broadband networks and distributed switching systems," *J. Lightwave Tech.*, vol. 8, pp. 202–211, Feb. 1990.

67. *Optical Engineering*, Special Issue on Photonic Switching and Interconnects, vol. 29, Mar. 1990.

68. H. S. Hinton, "Photonic switching fabrics," *IEEE Commun. Mag.*, vol. 28, pp. 71–89, Apr. 1990.

69. S. Suzuki, M. Nishio, T. Numai, M. Fujiwara, M. Itoh, S. Murata, and M. Shimosaka, "A photonic wavelength-division switching system using tunable laser diode filters," *J. Lightwave Tech.*, vol. 8, pp. 660–666, May 1990.

70. P. R. Prucnal and P. A. Perrier, "Optically processed routing for fast packet switching," *IEEE Mag. Lightwave Commun. Sys.*, vol. 1, no. 2, pp. 54–67, May 1990.

71. G. I. Stegeman and E. M. Wright, "Tutorial review: All-optical waveguide switching," *Opt. Quantum Electron.*, vol. 22, no. 2, pp. 95–122, Mar. 1990.

72. H. G. Winful, "Nonlinear optical phenomena in single-mode fibers," in E. E. Basch, ed., *Optical Fiber Transmission*, H. W. Sams, Indianapolis, IN, 1987.

73. W. J. Tomlinson and R. H. Stolen, "Nonlinear phenomena in optical fibers," *IEEE Commun. Mag.*, vol. 26, pp. 36–44, Apr. 1988.

74. G. P. Agrawal, *Nonlinear Fiber Optics*, Academic, New York, 1989.

75. W. J. Tomlinson, R. H. Stolen, and C. V. Shank, "Compression of optical pulses by self-phase modulation in fibers," *J. Opt. Soc. Amer. B*, vol. 1, pp. 139–149, 1984.

76. (*a*) Y. Aoki, "Properties of fiber Raman amplifiers and their applicability to digital optical communication systems," *J. Lightwave Tech.*, vol. 6, pp. 1225–1239, July 1988.

(*b*) Y. Aoki, "Fiber Raman amplifier properties for application to long-distance optical communications," *Opt. Quantum Electron.*, vol. 21, Special Issue on Optical Amplifiers, pp. S89–S104, 1989.

77. B. Foley, M. L. Dakss, R. W. Davies, and P. Melman, "Gain saturation in fiber Raman amplifiers due to stimulated Brillouin scattering," *J. Lightwave Tech.*, vol. 7, pp. 2024–2032, Dec. 1989.

78. R. W. Tkach and A. R. Chraplyvy, "Fibre Brillouin amplifiers," *Opt. Quantum Electron.*, vol. 21, Special Issue on Optical Amplifiers, pp. S105–S112, 1989.

79. N. A. Olsson and J. P. van der Ziel, "Characteristics of a semiconductor laser pumped Brillouin amplifier with electronically controlled bandwidth," *J. Lightwave Tech.*, vol. 5, pp. 147–153, Jan. 1987.

80. A. Hasagawa and F. Tappert, "Transmission of stationary nonlinear optical pulses in dispersive dielectric fibers. I. Anomalous dispersion," *Appl. Phys. Lett.*, vol. 23, pp. 142–144, 1973.

81. R. K. Dodd, J. C. Eilbeck, J. D. Gibbons, and H. C. Morris, *Solitons and Nonlinear Wave Equations*, Academic, New York, 1982.

82. (*a*) Y. Lai and H. A. Haus, "Quantum theory of solitons in optical fibers. I. Time-dependent Hartree approximation," *Phys. Rev. A*, vol. 40, pp. 844–853, July 15, 1989.

    (*b*) Y. Lai and H. A. Haus, "Quantum theory of solitons in optical fibers. II. Exact solution," *Phys. Rev. A*, vol. 40, pp. 854–866, July 15, 1989.

83. B.-J. Hong, L. Wang, and C. C. Yang, "Using nonsoliton pulses for soliton-based fiber communications," *J. Lightwave Tech.*, vol. 8, pp. 568–575, Apr. 1990.

84. R. G. Waarts, A. A. Friesem, E. Lichtman, H. H. Jaffe, and R. P. Braun, "Nonlinear effects in coherent multichannel transmission through optical fibers," *Proc. IEEE*, vol. 78, pp. 1344–1368, Aug. 1990.

85. J. R. Barry and E. A. Lee, "Perfomance of coherent optical receivers," *Proc. IEEE*, vol. 78, pp. 1369–1394, Aug. 1990.

86. Thirty papers in Special Issue on Dense Wavelength Division Multiplexing Techniques for High Capacity and Multiple Access Communications Systems, *IEEE J. Sel. Areas Commun.*, vol. 8, Aug. 1990.

87. Ten papers in Special Section on Optical Amplifiers for Communication, *IEE Proc. J.*, vol. 137, no. 4, Aug. 1990.

APPENDIX
A

INTERNATIONAL SYSTEM OF UNITS

Quantity	Unit	Symbol	Dimensions
Length	meter	m	
Mass	kilogram	kg	
Time	second	s	
Temperature	kelvin	K	
Current	ampere	A	
Frequency	hertz	Hz	$1/s$
Force	newton	N	$(kg \cdot m)/s^2$
Pressure	pascal	Pa	N/m^2
Energy	joule	J	$N \cdot m$
Power	watt	W	J/s
Electric charge	coulomb	C	$A \cdot s$
Potential	volt	V	J/C
Conductance	siemens	S	A/V
Resistance	ohm	Ω	V/A
Capacitance	farad	F	C/V
Magnetic flux	weber	Wb	$V \cdot s$
Magnetic induction	tesla	T	Wb/m^2
Inductance	henry	H	Wb/A

APPENDIX
B

USEFUL
MATHEMATICAL
RELATIONS

Some of the mathematical relations encountered in this text are listed for convenient reference. More comprehensive listings are available in various handbooks.[1-4]

B.1 TRIGONOMETRIC IDENTITIES

$$e^{\pm j\theta} = \cos\theta \pm j\sin\theta$$
$$\sin^2\theta + \cos^2\theta = 1$$
$$\cos^2\theta - \sin^2\theta = \cos 2\theta$$
$$4\sin^3\theta = 3\sin\theta - \sin 3\theta$$
$$4\cos^3\theta = 3\cos\theta + \cos 3\theta$$
$$8\sin^4\theta = 3 - 4\cos 2\theta + \cos 4\theta$$
$$8\cos^4\theta = 3 + 4\cos 2\theta + \cos 4\theta$$
$$\sin(\alpha \pm \beta) = \sin\alpha\cos\beta \pm \cos\alpha\sin\beta$$
$$\cos(\alpha \pm \beta) = \cos\alpha\cos\beta \mp \sin\alpha\sin\beta$$
$$\tan(\alpha \pm \beta) = \frac{\tan\alpha \pm \tan\beta}{1 \mp \tan\alpha\tan\beta}$$

B.2 VECTOR ANALYSIS

The symbols $\mathbf{e}_x$, $\mathbf{e}_y$, and $\mathbf{e}_z$ denote unit vectors lying parallel to the x, y, and z axes, respectively, of the rectangular coordinate system. Similarly, $\mathbf{e}_r$, $\mathbf{e}_\phi$, and $\mathbf{e}_z$ are unit vectors for cylindrical coordinates. The unit vectors $\mathbf{e}_r$ and $\mathbf{e}_\phi$ vary in

441

direction as the angle ϕ changes. The conversion from cylindrical to rectangular coordinates is made through the relationships

$$x = r \cos \phi \qquad y = r \sin \phi \qquad z = z$$

B.2.1 Rectangular Coordinates

$$\text{Gradient } \nabla f = \frac{\partial f}{\partial x}\mathbf{e}_x + \frac{\partial f}{\partial y}\mathbf{e}_y + \frac{\partial f}{\partial z}\mathbf{e}_z$$

$$\text{Divergence } \nabla \cdot \mathbf{A} = \frac{\partial A_x}{\partial x} + \frac{\partial A_y}{\partial y} + \frac{\partial A_z}{\partial z}$$

$$\text{Curl } \nabla \times \mathbf{A} = \begin{vmatrix} \mathbf{e}_x & \mathbf{e}_y & \mathbf{e}_z \\ \dfrac{\partial}{\partial x} & \dfrac{\partial}{\partial y} & \dfrac{\partial}{\partial z} \\ A_x & A_y & A_z \end{vmatrix}$$

$$\text{Laplacian } \nabla^2 f = \frac{\partial^2 f}{\partial x^2} + \frac{\partial^2 f}{\partial y^2} + \frac{\partial^2 f}{\partial z^2}$$

B.2.2 Cylindrical Coordinates

$$\text{Gradient } \nabla f = \frac{\partial f}{\partial r}\mathbf{e}_r + \frac{1}{r}\frac{\partial f}{\partial \phi}\mathbf{e}_\phi + \frac{\partial f}{\partial z}\mathbf{e}_z$$

$$\text{Divergence } \nabla \cdot \mathbf{A} = \frac{1}{r}\frac{\partial(rA_r)}{\partial r} + \frac{1}{r}\frac{\partial A_\phi}{\partial \phi} + \frac{\partial A_z}{\partial z}$$

$$\text{Curl } \nabla \times \mathbf{A} = \begin{vmatrix} \dfrac{1}{r}\mathbf{e}_r & \mathbf{e}_\phi & \dfrac{1}{r}\mathbf{e}_z \\ \dfrac{\partial}{\partial r} & \dfrac{\partial}{\partial \phi} & \dfrac{\partial}{\partial z} \\ A_r & rA_\phi & A_z \end{vmatrix}$$

$$\text{Laplacian } \nabla^2 f = \frac{1}{r}\frac{\partial}{\partial r}\left(r\frac{\partial f}{\partial r}\right) + \frac{1}{r^2}\frac{\partial^2 f}{\partial \phi^2} + \frac{\partial^2 f}{\partial z^2}$$

B.2.3 Vector Identities

$$\nabla \times (\nabla \times \mathbf{A}) = \nabla(\nabla \cdot \mathbf{A}) - \nabla^2 \mathbf{A}$$

$$\nabla^2 \mathbf{A} = \nabla^2 A_x \mathbf{e}_x + \nabla^2 A_y \mathbf{e}_y + \nabla^2 A_z \mathbf{e}_z$$

B.3 INTEGRALS

$$\int \sin x \, dx = -\cos x$$

$$\int \cos x \, dx = \sin x$$

$$\int \sqrt{a^2 - x^2} \, dx = \frac{1}{2}\left(x\sqrt{a^2 - x^2} + a^2 \arcsin \frac{x}{a}\right)$$

$$\int x\sqrt{a^2 - x^2} \, dx = -\tfrac{1}{3}(a^2 - x^2)^{3/2}$$

$$\int x^2 \sin^2 x \, dx = \frac{x^3}{6} - \left(\frac{x^2}{6} - \frac{1}{8}\right)\sin 2x - \frac{x \cos 2x}{4}$$

$$\int \frac{dx}{\cos^n x} = \frac{1}{n-1}\frac{\sin x}{\cos^{n-1} x} + \frac{n-2}{n-1}\int \frac{dx}{\cos^{n-2} x}$$

$$\int u \, dv = uv - \int v \, du$$

$$\int e^{ax} \, dx = \frac{1}{a}e^{ax}$$

$$\int \sin^2 x \, dx = \frac{x}{2} - \tfrac{1}{4}\sin 2x$$

$$\int \sin^n x \, dx = -\frac{\sin^{n-1} x \cos x}{n} + \frac{n-1}{n}\int \sin^{n-2} x \, dx$$

$$\int \cos^2 x \, dx = \frac{x}{2} + \tfrac{1}{4}\sin 2x$$

$$\int \cos^n x \, dx = \frac{1}{n}\cos^{n-1} x \sin x + \frac{n-1}{n}\cos^{n-2} x \, dx$$

$$\int_{-\infty}^{\infty} \frac{e^{jpx}}{(\beta + jx)^n} \, dx = \begin{cases} 0 & \text{if } p < 0 \\ \dfrac{2\pi(p)^{n-1}e^{-\beta p}}{\Gamma(n)} & \text{if } p \geq 0 \quad \text{where } \Gamma(n) = (n-1)! \end{cases}$$

$$\int_{-\infty}^{\infty} e^{-p^2 x^2 + qx} \, dx = e^{q^2/4p^2}\frac{\sqrt{\pi}}{p}$$

$$\int_0^{\infty} \frac{1}{1 + (x/a)^2} \, dx = \frac{\pi a}{2}$$

$$\frac{2}{\sqrt{\pi}}\int_0^t e^{-x^2} \, dx = \text{erf}(t)$$

B.4 SERIES EXPANSIONS

$$(1 + x)^n = 1 + nx + \frac{n(n - 1)}{2!}x^2 + \frac{n(n - 1)(n - 2)}{3!}x^3 + \cdots \qquad \text{for} \quad |nx| < 1$$

$$e^x = 1 + x + \frac{x^2}{2!} + \frac{x^3}{3!} + \cdots$$

$$\sin x = x - \frac{x^3}{3!} + \frac{x^5}{5!} - \cdots$$

$$\cos x = 1 - \frac{x^2}{2!} + \frac{x^4}{4!} - \cdots$$

REFERENCES

1. W. H. Beyer, *Standard Mathematical Tables*, 28th ed., CRC Press, Boca Raton, FL, 1987.
2. J. J. Tuma, *Engineering Mathematics Handbook*, 2nd ed., McGraw-Hill, New York, 1982.
3. M. Abramowitz and I. A. Stegun, *Handbook of Mathematical Functions*, Dover, New York, 1965.
4. I. S. Gradshteyn and I. M. Ryzhik, *Table of Integrals, Series, and Products*, Academic, New York, 1980.

BESSEL
FUNCTIONS

This appendix lists the definitions and some recurrence relations for integer-order Bessel functions of the first kind $J_\nu(z)$ and modified Bessel functions $K_\nu(z)$. Detailed mathematical properties of these and other Bessel functions can be found in Refs. 24 through 26 of Chap. 2. Here the parameter ν is any integer and n is a positive integer or zero. The parameter $z = x + jy$.

C.1 BESSEL FUNCTIONS OF THE FIRST KIND

C.1.1 Various Definitions

A Bessel function of the first kind of order n and argument z, commonly denoted by $J_n(z)$, is defined by

$$J_n(z) = \frac{1}{2\pi} \int_{-\pi}^{\pi} e^{jz \sin\theta - jn\theta} \, d\theta$$

or, equivalently,

$$J_n(z) = \frac{1}{\pi} \int_0^{\pi} \cos(z \sin\theta - n\theta) \, d\theta$$

Just as the trigonometric functions can be expanded in power series, so can the

Bessel function $J_\nu(z)$:

$$J_\nu(z) = \sum_{k=0}^{\infty} \frac{(-1)^k \left(\frac{1}{2}z\right)^{\nu+2k}}{k!(\nu+k)!}$$

In particular, for $\nu = 0$,

$$J_0(z) = 1 - \frac{\frac{1}{4}z^2}{(1!)^2} + \frac{\left(\frac{1}{4}z^2\right)^2}{(2!)^2} - \frac{\left(\frac{1}{4}z^2\right)^3}{(3!)^2} + \cdots$$

For $\nu = 1$,

$$J_1(z) = \frac{1}{2}z - \frac{\left(\frac{1}{2}z\right)^3}{2!} + \frac{\left(\frac{1}{2}z\right)^5}{2!3!} - \cdots$$

and so on for higher values of ν.

C.1.2 Recurrence Relations

$$J_{\nu-1}(z) + J_{\nu+1}(z) = \frac{2\nu}{z}J_\nu(z)$$

$$J_{\nu-1}(z) - J_{\nu+1}(z) = 2J_\nu'(z)$$

$$J_\nu'(z) = J_{\nu-1}(z) - \frac{\nu}{z}J_\nu(z)$$

$$J_\nu'(z) = -J_{\nu+1}(z) + \frac{\nu}{z}J_\nu(z)$$

$$J_0'(z) = -J_1(z)$$

C.2 MODIFIED BESSEL FUNCTIONS

C.2.1 Integral Representations

$$K_0(z) = \frac{-1}{\pi}\int_0^\pi e^{\pm z\cos\theta}\left[\gamma + \ln(2z\sin^2\theta)\right]d\theta$$

where Euler's constant $\gamma = 0.57722$.

$$K_\nu(z) = \frac{\pi^{1/2}\left(\frac{1}{2}z\right)^\nu}{\Gamma(\nu + \frac{1}{2})}\int_0^\infty e^{-z\cosh t}\sinh^{2\nu} t\, dt$$

$$K_0(x) = \int_0^\infty \cos(x\sinh t)\, dt = \int_0^\infty \frac{\cos(xt)}{\sqrt{t^2+1}}\, dt \qquad (x>0)$$

$$K_\nu(x) = \sec(\tfrac{1}{2}\nu\pi)\int_0^\infty \cos(x\sinh t)\cosh(\nu t)\, dt \qquad (x>0)$$

C.2.2 Recurrence Relations

If $L_\nu = e^{j\pi\nu}K_\nu$, then

$$L_{\nu-1}(z) - L_{\nu+1}(z) = \frac{2\nu}{z}L_\nu(z)$$

$$L'_\nu(z) = L_{\nu-1}(z) - \frac{\nu}{z}L_\nu(z)$$

$$L_{\nu-1}(z) + L_{\nu+1}(z) = 2L'_\nu(z)$$

$$L'_\nu(z) = L_{\nu+1}(z) + \frac{\nu}{z}L_\nu(z)$$

C.3 ASYMPTOTIC EXPANSIONS

For fixed ν ($\neq -1, -2, -3, \ldots$) and $z \to 0$,

$$J_\nu(z) \simeq \frac{\left(\frac{1}{2}z\right)^\nu}{\Gamma(\nu+1)}$$

For fixed ν and $|z| \to \infty$,

$$J_\nu(z) \simeq \left(\frac{2}{\pi z}\right)^{1/2}\cos\left(z - \frac{\nu\pi}{2} - \frac{\pi}{4}\right)$$

For fixed ν and large $|z|$,

$$K_\nu(z) \simeq \left(\frac{\pi}{2z}\right)^{1/2}e^{-z}\left[1 - \frac{\mu-1}{8z} + \frac{(\mu-1)(\mu-9)}{2!(8z)^2} + \cdots\right]$$

where $\mu = 4\nu^2$.

C.4 GAMMA FUNCTION

$$\Gamma(z) = \int_0^\infty t^{z-1}e^{-t}\,dt$$

For integer n,

$$\Gamma(n+1) = n!$$

For fractional values,

$$\Gamma\left(\tfrac{1}{2}\right) = \pi^{1/2} = \left(-\tfrac{1}{2}\right)! \simeq 1.77245$$

$$\Gamma\left(\tfrac{3}{2}\right) = \tfrac{1}{2}\pi^{1/2} = \left(\tfrac{1}{2}\right)! \simeq 0.88623$$

APPENDIX
D

DECIBELS

D.1 DEFINITION

In designing and implementing an optical fiber link, it is of interest to establish, measure, and/or interrelate the signal levels at the transmitter, at the receiver, at the cable connection and splice points, and in the cable. A convenient method for this is to reference the signal level either to some absolute value or to a noise level. This is normally done in terms of a power ratio measured in *decibels* (dB) defined as

$$\text{Power} = 10 \log \frac{P_2}{P_1} \text{ dB} \qquad \text{(D-1)}$$

where P_1 and P_2 are electric or optical powers.

The logarithmic nature of the decibel allows a large ratio to be expressed in a fairly simple manner. Power levels differing by many orders of magnitude can be simply compared when they are in decibel form. Some very helpful figures to remember are given in Table D-1. For example, doubling the power means a 3-dB gain (the power level increases by 3 dB), halving the power means a 3-dB loss (the power level decreases by 3 dB), and power levels differing by

TABLE D-1
Examples of decibel measures of power ratios

Power ratio	10^N	10	2	1	0.5	0.1	10^{-N}
dB	$+10N$	$+10$	$+3$	0	-3	-10	$-10N$

factors of 10^N or 10^{-N} have decibel differences of $+10N$ dB and $-10N$ dB, respectively.

D.2   THE dBm

The decibel is used to refer to ratios or relative units. For example, we can say that a certain optical fiber has a 6-dB loss (the power level gets reduced by 75 percent in going through the fiber) or that a particular connector has a 1-dB loss (the power level gets reduced by 20 percent at the connector). However, the decibel gives no indication of the absolute power level. One of the most common derived units for doing this in optical fiber communications is the *dBm*. This is the decibel power level referred to 1 mW. In this case the power in dBm is an absolute value defined by

$$\text{Power level} = 10 \log \frac{P}{1 \text{ mW}} \tag{D-2}$$

An important relationship to remember is that 0 dBm = 1 mW. Other examples are shown in Table D-2.

TABLE D-2
Examples of dBm units (decibel measure of power relative to 1 mW)

Power (mW)	100	10	2	1	0.5	0.1	0.01	0.001
Value (dBm)	$+20$	$+10$	$+3$	0	-3	-10	-20	-30

APPENDIX
E

TOPICS
FROM
COMMUNICATION
THEORY

E.1 CORRELATION FUNCTIONS

A spectral density function is often used in communication theory to describe signals in the frequency domain. To define this, we first introduce the *autocorrelation function* $R_v(\tau)$, defined as

$$R_v(\tau) = \frac{1}{T_0} \int_{-T_0/2}^{T_0/2} v(t)v(t - \tau)\, dt \qquad \text{(E-1)}$$

where $v(t)$ is a periodic signal of period T_0. The autocorrelation function measures the dependence on the time τ of $v(t)$ and $v(t - \tau)$. An important property of $R_v(\tau)$ is that, if $v(t)$ is an energy or power signal, then

$$R_v(0) = \|v\|^2 = \frac{1}{T_0} \int_{-T_0/2}^{T_0/2} v^2(t)\, dt \qquad \text{(E-2)}$$

where $\|v\|^2$ represents the signal energy or power, respectively.

E.2 SPECTRAL DENSITY

Since $R_v(\tau)$ gives information about the time domain behavior of $v(t)$, the spectral behavior of $v(t)$ in the frequency domain can be found from the

Fourier transform of $R_v(\tau)$. We thus define the *spectral density* $G_v(f)$ by

$$G_v(f) = F[R_v(\tau)] = \int_{-\infty}^{\infty} R_v(\tau) e^{-j2\pi f\tau} \, d\tau \qquad \text{(E-3)}$$

The fundamental property of $G_v(f)$ is that by integrating over all frequencies we obtain $\|v\|^2$:

$$\int_{-\infty}^{\infty} G_v(f) \, df = R_v(0) = \|v\|^2 \qquad \text{(E-4)}$$

The spectral density thus tells how energy or power is distributed in the frequency domain. When $v(t)$ is a current or voltage waveform feeding a 1-Ω load resistor, then $G_v(f)$ is measured in units of watts/hertz. For loads other than 1 Ω, $G_v(f)$ is expressed either in volts²/hertz or amperes²/hertz. The symbol S is often used in electronics books to denote the spectral density.

Consider a signal $x(t)$ that is passed through a linear system having a transfer function $h(t)$. The output signal is given by

$$y(t) = h(t)x(t)$$

If $x(t)$ is modeled in the frequency domain by a spectral density $G_x(f)$ and the linear system has a transfer function $H(f)$, then the spectral density $G_y(f)$ of the output signal $y(t)$ is

$$G_y(f) = |H(f)|^2 G_x(f) \qquad \text{(E-5)}$$

E.3 NOISE-EQUIVALENT BANDWIDTH

A *white-noise* signal $n(t)$ is characterized by a spectral density that is flat or constant over all frequencies, that is,

$$G_n(f) = \frac{\eta}{2} = \text{constant} \qquad \text{(E-6)}$$

The factor $\frac{1}{2}$ indicates that half the power is associated with positive frequencies and half with negative frequencies.

If white noise is applied at the input of a linear system having a transfer function $H(f)$, the spectral density $G_0(f)$ of the output noise is

$$G_0(f) = |H(f)|^2 G_n(f) = \frac{\eta}{2} |H(f)|^2 \qquad \text{(E-7)}$$

Thus the output noise power N_0 is given by

$$N_0 = \int_{-\infty}^{\infty} G_0(f) \, df = \eta \int_{0}^{\infty} |H(f)|^2 \, df \qquad \text{(E-8)}$$

If the same noise comes from an ideal low-pass filter of bandwidth B and amplitude $H(0)$, that is, the magnitude of an arbitrary filter transfer function at zero frequency, then

$$N_0 = \eta |H(0)|^2 B \tag{E-9}$$

By combining Eqs. (E-8) and (E-9), we define a *noise-equivalent bandwidth B* as

$$B = \frac{1}{|H(0)|^2} \int_0^\infty |H(f)|^2 \, df \tag{E-10}$$

E.4   CONVOLUTION

Convolution is an important mathematical operation used by communication engineers. The convolution of two real-valued functions $p(t)$ and $q(t)$ of the same variable is defined as

$$
\begin{aligned}
p(t) * q(t) &= \int_{-\infty}^{\infty} p(x) q(t-x) \, dx \\
&= \int_{-\infty}^{\infty} q(x) p(t-x) \, dx \\
&= q(t) * p(t)
\end{aligned}
\tag{E-11}
$$

where the symbol $*$ denotes convolution. Note that convolution is commutative. Two important properties of convolutions are

$$F[p(t) * q(t)] = P(f)Q(f) \tag{E-12}$$

that is, the convolution of two signals in the time domain corresponds to the multiplication of their Fourier transforms in the frequency domain, and

$$F[p(t)q(t)] = P(f) * Q(f) \tag{E-13}$$

that is, the multiplication of two functions in the time domain corresponds to their convolution in the frequency domain.

INDEX